I0055635

Innate Immunity and Inflammation

A subject collection from *Cold Spring Harbor Perspectives in Biology*

OTHER SUBJECT COLLECTIONS FROM *COLD SPRING HARBOR PERSPECTIVES IN BIOLOGY*

The Genetics and Biology of Sexual Conflict
The Origin and Evolution of Eukaryotes
Endocytosis
Mitochondria
Signaling by Receptor Tyrosine Kinases
DNA Repair, Mutagenesis, and Other Responses to DNA Damage
Cell Survival and Cell Death
Immune Tolerance
DNA Replication
Endoplasmic Reticulum
Wnt Signaling
Protein Synthesis and Translational Control
The Synapse
Extracellular Matrix Biology
Protein Homeostasis
Calcium Signaling
The Golgi
Germ Cells
The Mammary Gland as an Experimental Model
The Biology of Lipids: Trafficking, Regulation, and Function
Auxin Signaling: From Synthesis to Systems Biology

SUBJECT COLLECTIONS FROM *COLD SPRING HARBOR PERSPECTIVES IN MEDICINE*

Human Fungal Pathogens
Tuberculosis
The Biology of Heart Disease
The Skin and Its Diseases
MYC *and the Pathway to Cancer*
Bacterial Pathogenesis
Transplantation
Cystic Fibrosis: A Trilogy of Biochemistry, Physiology, and Therapy
Hemoglobin and Its Diseases
Addiction
Parkinson's Disease
Type 1 Diabetes
Angiogenesis: Biology and Pathology
HIV: From Biology to Prevention and Treatment
The Biology of Alzheimer Disease

Innate Immunity and Inflammation

A subject collection from *Cold Spring Harbor Perspectives in Biology*

EDITED BY

Ruslan Medzhitov

Yale School of Medicine

COLD SPRING HARBOR LABORATORY PRESS
Cold Spring Harbor, New York • www.cshlpress.org

Innate Immunity and Inflammation

A Subject Collection from *Cold Spring Harbor Perspectives in Biology*
Articles online at www.cshperspectives.org

Executive Editor	Richard Sever
Managing Editor	Maria Smit
Senior Project Manager	Barbara Acosta
Permissions Administrator	Carol Brown
Production Editor	Diane Schubach
Production Manager/Cover Designer	Denise Weiss
Publisher	John Inglis

Front cover artwork: The image is a colorized scanning electron micrograph showing a human neutrophil 30 minutes after infection with *Bacillus anthracis.* (Cover image kindly provided by Dr. Volker Brinkmann, Mikroskopie, Max Planck-Institut für Infektionsbiologie.)

Library of Congress Cataloging-in-Publication Data

Innate immunity and inflammation/edited by Ruslan M. Medzhitov.
p. ; cm.
"A subject collection from Cold Spring Harbor perspectives in biology."
Includes bibliographical references and index.
ISBN 978-1-62182-029-1 (hardcover : alk. paper)
I. Medzhitov, Ruslan, editor. II. Cold Spring Harbor perspectives in biology.
[DNLM: 1. Immunity, Innate--Collected Works. 2. Inflammation--immunology--Collected Works. QW 700]

616'.0473--dc23

2014027280

All World Wide Web addresses are accurate to the best of our knowledge at the time of printing.

For a complete catalog of all Cold Spring Harbor Laboratory Press publications, visit our website at www.cshlpress.org.

Contents

Preface, vii

Microbial Sensing by Toll-Like Receptors and Intracellular Nucleic Acid Sensors, 1
Surya Pandey, Taro Kawai, and Shizuo Akira

Emerging Principles Governing Signal Transduction by Pattern-Recognition Receptors, 19
Jonathan C. Kagan and Gregory M. Barton

Transcriptional Control of Inflammatory Responses, 35
Stephen T. Smale and Gioacchino Natoli

Inflammasomes, 45
Marcel R. de Zoete, Noah W. Palm, Shu Zhu, and Richard A. Flavell

Tumor Necrosis Family Superfamily in Innate Immunity and Inflammation, 67
John Šedý, Vasileios Bekiaris, and Carl F. Ware

IL-6 in Inflammation, Immunity, and Disease, 85
Toshio Tanaka, Masashi Narazaki, and Tadamitsu Kishimoto

The Chemokine System in Innate Immunity, 101
Caroline L. Sokol and Andrew D. Luster

Lipid Mediators in the Resolution of Inflammation, 121
Charles N. Serhan, Nan Chiang, Jesmond Dalli, and Bruce D. Levy

DNA Degradation and Its Defects, 141
Kohki Kawane, Kou Motani, and Shigekazu Nagata

Group 2 Innate Lymphoid Cells in Health and Disease, 155
Brian S. Kim and David Artis

Allergic Inflammation—Innately Homeostatic, 169
Laurence E. Cheng and Richard M. Locksley

Inflammation and the Blood Microvascular System, 183
Jordan S. Pober and William C. Sessa

Sinusoidal Immunity: Macrophages at the Lymphohematopoietic Interface, 195
Siamon Gordon, Annette Plüddemann, and Subhankar Mukhopadhyay

Approaching the Next Revolution? Evolutionary Integration of Neural and Immune Pathogen Sensing and Response, 215
Kevin J. Tracey

Index, 227

Preface

The fields of innate immunity and inflammation have undergone dramatic transformation over the past 15 years. From relatively minor research areas, they have developed into large, vibrant, and exciting fields of investigation that span multiple traditional disciplines ranging from cell biology and gene regulation to neurophysiology and chronic diseases. The discoveries of innate immune recognition mechanisms and their role in controlling adaptive immunity have placed the innate immune system at the foundation of modern immunology. We now have a reasonably good idea of the complexity of the innate immune sensing pathways. However, we also appreciate that much remains to be learned, especially regarding the initiation of type 2 immunity and allergic responses. Similarly, the mechanisms of sensing of pathogen virulence activities remain enigmatic, which complicates our understanding of host–microbiota interactions. It is likely that new concepts that go beyond pattern recognition theory will be required to develop a more general paradigm of the innate immune system.

The discovery of the innate immune signaling pathways also had a major impact on our understanding of inflammation. Indeed, the best-characterized mechanisms of initiating inflammatory responses are the ones involved in pathogen sensing during an infection. It has been well appreciated for almost a century that inflammation plays a fundamental role in host defense from pathogens. However, over the past decade or so, it has become increasingly clear that inflammation plays a more fundamental role in mammalian biology and human diseases. Indeed, almost every human disease is accompanied by inflammation even in the absence of infection or tissue injury. The origins of these inflammatory responses remain enigmatic, but advances in multiple areas of inflammation research suggest that we may be on the verge of a reconceptualization of the entire phenomenon, perhaps along the lines envisioned by Élie Metchnikoff when he introduced the concept of physiological inflammation about a century ago.

Given these developments in both empirical and conceptual knowledge, the goal of this collection of articles is to summarize what is now firmly established in key areas of innate immunity and inflammation. Given the space and time limitations, a collection like this is necessarily incomplete, and dozens of additional topics could have been covered. However, without being comprehensive, this book presents some of the most mature areas of innate immunity and inflammation written by the authorities who have made essential contributions to their respective research areas.

I would like to thank all the authors who contributed their time and knowledge to this volume. I would also like to thank Richard Sever, Barbara Acosta, and their colleagues at Cold Spring Harbor Laboratory Press for their excellent editorial and managerial efforts to envision, supervise, and complete this project.

Ruslan Medzhitov

Microbial Sensing by Toll-Like Receptors and Intracellular Nucleic Acid Sensors

Surya Pandey[1,2,3], Taro Kawai[1,2], and Shizuo Akira[2,3]

[1]Laboratory of Molecular Immunobiology, Graduate School of Biological Sciences, Nara Institute of Science and Technology (NAIST), Nara 630-0192, Japan

[2]Laboratory of Host Defense, WPI Immunology Frontier Research Center, Osaka University, Osaka 565-0871, Japan

[3]Department of Host Defense, Research Institute for Microbial Diseases, Osaka University, Osaka 565-0871, Japan

Correspondence: tarokawai@bs.naist.jp; sakira@biken.osaka-u.ac.jp

Recognition of an invading pathogen is critical to elicit protective responses. Certain microbial structures and molecules, which are crucial for their survival and virulence, are recognized by different families of evolutionarily conserved pattern recognition receptors (PRRs). This recognition initiates a signaling cascade that leads to the transcription of inflammatory cytokines and chemokines to eliminate pathogens and attract immune cells, thereby perpetuating further adaptive immune responses. Considerable research on the molecular mechanisms underlying host–pathogen interactions has resulted in the discovery of multifarious PRRs. In this review, we discuss the recent developments in microbial recognition by Toll-like receptors (TLRs) and intracellular nucleic acid sensors and the signaling pathways initiated by them.

Microbes are disease-causing entities that have exerted powerful selective pressure throughout the evolution of eukaryotes (Beutler 2009). As a consequence, eukaryotes have evolved a complex immune system to counteract microbial attacks. This immune system has two arms, termed innate immunity and adaptive immunity. Initially, innate immunity was considered to be nonspecific and less complex. However, this notion was changed by the revolutionary discovery of the first Toll-like receptor (TLR) in the mid-1990s. Since then, there has been tremendous growth in our understanding of the previously understated roles of the innate immune system in the recognition of microbial pathogens, their elimination, and inflammation. It has been established that innate immune recognition engages germline-encoded pattern recognition receptors (PRRs) that recognize conserved microbial structures of pathogens known as pathogen-associated molecular patterns (PAMPs) (Akira et al. 2001; Janeway and Medzhitov 2002) and activate the expression of major histocompatibility (MHC) proteins, costimulatory molecules, and inflammatory mediators in the form of cytokines and chemokines by macrophages, dendritic cells (DCs), neutrophils, and other nonprofessional immune cells (Table 1). These processes not only trigger immediate and early mechanisms of host defense,

Cite this article as *Cold Spring Harb Perspect Biol* doi: 10.1101/cshperspect.a016246

Table 1. Pattern recognition receptors (PRRs) that recognize conserved microbial structures of pathogens

PRRs	Localization	PAMP recognized	Key adaptors	Effector response
TLRs				
TLR1	Cell surface	Triacylated lipopeptides	MyD88	IL-6, TNF-α
TLR2	Cell surface	Di/triacylated lipopeptides	MyD88, TIRAP	IL-6, TNF-α, IL-8, MCP-1, RANTES
TLR3	Endosomes	dsRNA	TRIF	IFN-β
TLR4	Cell surface	LPS	MyD88, TRIF, TIRAP, TRAM	IL-6, TNF-α, IFNβ, IP-10
TLR5	Cell surface	Flagellin	MyD88	TNF-α
TLR6	Cell surface	Diacylated lipopeptides	MyD88, TIRAP	TNF-α, IL-6, IL-8, MCP-1, RANTES
TLR7	Endosomes	ssRNA	MyD88	IFN-α
TLR8	Endosomes	ssRNA	MyD88	IFN-α
TLR9	Endosomes	CpG DNA	MyD88	IFN-α
TLR11	Endosomes	Profilin, flagellin	MyD88	IL-12, TNF-α
TLR12	Endosomes	Profilin	MyD88	IL-12p40, IFN-α
TLR13	Endosomes	23s rRNA	MyD88	IL-6, IL-12p40
RLRs				
RIG-I	Cytoplasm	Short dsRNA, ssRNA	IPS-1, STING	IFN-β, IL-6
MDA5	Cytoplasm	Long dsRNA	IPS-1	IFN-β
LGP2	Cytoplasm	dsRNA	IPS-1	IFN-β
DDX3	Cytoplasm	Viral RNA	IPS-1	IFN-β
Cytosolic DNA sensors				
DAI	Cytoplasm	dsDNA	STING	IFN-β
RNA Pol III		AT rich dsDNA	IPS1	IFN-β
IFI16	Nucleus and cytoplasm	dsDNA	STING	IFN-β, IP-10, IL-6, IL-1β
AIM2	Cytoplasm	dsDNA	ASC	IL-1β, IL-18
Ku70	Cytoplasm	dsDNA	?	IFN-γ
MRE11		dsDNA, ISD	STING	IFN-β, IL-6, IP-10
cGAS	Cytoplasm	dsDNA	STING	IFN-β
LRRFIP1	Cytoplasm	dsDNA, dsRNA	β-catenin	IFN-β
DHX36	Cytoplasm	dsDNA	MyD88	TNF-α
DHX9	Cytoplasm	dsDNA	MyD88	TNF-α
DDX41	Cytoplasm	c-di-GMP, c-di-AMP	STING	IFN-α, IFN-β
STING	Cytoplasm	c-di-GMP		IFN-β
HMGB	Cytoplasm	dsDNA, ssDNA	?	IFN-β, IL-6, RANTES
Histone H2B	Nucleus and cytoplasm	Poly (dA:dT), genomic DNA	IPS1	IFN-β

Innate immune recognition engages germline-encoded pattern recognition receptors (PRRs) that recognize conserved microbial structures of pathogens known as pathogen-associated molecular patterns (PAMPs) and activate the expression of major histocompatibility (MHC) proteins, costimulatory molecules, and inflammatory mediators in the form of cytokines and chemokines by macrophages, dendritic cells (DCs), neutrophils, and other nonprofessional immune cells.

IL, interleukin; IFN, interferon; TNF, tumor necrosis factor.

but also prime and orchestrate antigen-specific adaptive immune responses.

Microbial species, such as bacteria, viruses, fungi, and parasites, are uniquely accessorized with diverse kinds of PAMPs, which are fundamentally important for their survival. PAMPs can vary in their molecular nature from lipids, lipoproteins, and proteins to nucleic acids

(Akira et al. 2006) and are redundantly or nonredundantly detected by specific PRRs. PRRs belong to several classes depending on their structural similarities, including TLRs, RIG-I-like receptors (RLRs), NOD-like receptors (NLRs), C-type lectin receptors (CLRs), and cytosolic DNA sensors. Recently, PRRs have also been reported to recognize host-derived danger signals, which are known as damage-associated molecular patterns (DAMPs) (Tang et al. 2012).

This article reviews the current information on the detection of microbial pathogens by PRRs, especially TLRs, RLRs, and intracellular DNA sensors and their signaling mechanisms that culminate in inflammation.

MICROBIAL SENSING BY TLRs

To date, 10 and 12 members of the TLR family have been identified in humans and mice, respectively. TLRs are type I transmembrane proteins composed of extracellular leucine-rich repeats (LRRs) that mediate recognition of PAMPs, transmembrane domains, and cytoplasmic Toll interleukin (IL)-1 receptor (TIR) domains that interact with downstream adaptor proteins required for signaling. TLR1 to TLR9 are conserved in both mice and humans, whereas mouse TLR10 is nonfunctional and TLR11, TLR12, and TLR13 have been deleted from the human genome (Kawai and Akira 2010). TLRs exist as either heterodimers or homodimers, and ligand binding to TLRs induces conformational changes for their activation. TLRs are broadly classified into two categories depending on their cellular localizations and the PAMPs they recognize. TLR1, TLR2, TLR4, TLR5, TLR6, and TLR10 are located on the plasma membrane and recognize lipids, lipoproteins, and proteins, whereas TLR3, TLR7, TLR8, TLR9, TLR11, TLR12, and TLR13 are localized in endosomal compartments where they recognize microbial nucleic acids and, under some special conditions (autoimmunity), self-nucleic acids (Kawai and Akira 2010; Celhar et al. 2012; Oldenburg et al. 2012; Koblansky et al. 2013). On specific ligand recognition, TLRs activate multiple signaling pathways by recruiting adaptor proteins, which initiate signal transduction pathways that culminate in the activation of transcription factors, such as nuclear factor-kappa B (NF-κB), mitogen-activated protein kinases (MAPKs), and members of the interferon (IFN) regulatory factor family, to regulate the expressions of cytokines, chemokines, and IFNs that eventuate in the host defense to microbial infection. TLRs are localized in different subcellular locations to provide optimum access to their ligands, and this specific localization is very important for the precise signaling of TLRs.

Cell Surface TLRs—Expression, Structure, and Ligands

TLR4 was identified as a receptor for bacterial lipopolysaccharide (LPS), a cell wall component of Gram-negative bacteria known to cause septic shock (Kawai and Akira 2010). TLR4 associates with myeloid differentiation factor 2 (MD2) on the cell surface to recognize LPS. Insights from a crystal structure study of the TLR4–MD2–LPS complex revealed that two copies of TLR4–MD2–LPS interact symmetrically to form a TLR4 homodimer (Park et al. 2009). After ligand binding, TLR4 translocates to the endosome through a dynein-dependent mechanism to induce a TRIF-dependent pathway (Kagan et al. 2008). TLR4 also recognizes the F protein of RSV (Kurt-Jones et al. 2000), mouse mammary tumor virus envelope proteins, *Streptococcus pneumoniae* pneumolysin, paclitaxel, and glycoinositol phosphate from *Trypanosoma* spp. (Kawai and Akira 2010; Broz and Monack 2013).

TLR2 has specificity for multiple microbial components derived from bacteria, fungi, viruses, and mycoplasma. It senses lipoproteins, peptidoglycans, lipotechoic acids from Gram-positive bacteria, zymosan, mannan, tGPI-mucin from *Trypanosoma cruzi*, and hemagglutinin of measles virus (Schwandner et al. 1999; Underhill et al. 1999; Kawai and Akira 2010). TLR2-mediated recognition of PAMPs and subsequent signaling occur via heterodimerization of TLR2 with either TLR1 or TLR6 on the plasma membrane. TLR2/TLR1 and TLR2/TLR6 recognize triacylated lipoproteins and diacylated lipoproteins, respectively, and induce the production of various proinflammatory cytokines, but not type I IFNs. The crystal structures of the two heterodimers revealed that each heterodimer forms an "m"-shaped complex with its ligand, thereby stabilizing the two receptors (Jin et al. 2007; Oliveira-Nascimento et al. 2012). A recent study showed the existence of TLR2/TLR10 preformed dimers, although their function is unknown (Guan et al. 2010). Many accessory molecules and coreceptors concentrate microbial products on the cell surface or inside phagosomes to facilitate TLR2 responses. One such coreceptor is CD36, which binds to ligands and transfers them to an accessory molecule, CD14, which finally loads the ligands onto TLR2/TLR6 heterodimers (FSL-1, MALP-2, and LTA) or TLR2/TLR1 heterodimers (lipomannan) (Hoebe et al. 2005; Jimenez-Dalmaroni et al. 2009). In a cell type–specific manner, TLR2 can trigger type I IFN production in response to Vaccinia viruses specifically by inflammatory monocytes (Barbalat et al. 2009).

TLR5 is a receptor for flagellin, a protein component of bacterial flagella (Akira et al. 2006). TLR5 is highly expressed in $CD11c^+$ $CD11b^+$ lamina propria DCs (LPDCs) in the small intestine. Following flagellin recognition, LPDCs induce the differentiation of naïve B cells into IgA-producing plasma cells and promote the differentiation of naïve T cells into IL-17-producing helper T cells (T_H17) and T helper type 1 (T_H1) cells (Uematsu et al. 2008). TLR11 shares a close homology with TLR5 and recognizes flagellin independently of TLR5 (Mathur et al. 2012). It also recognizes an unknown proteinaceous component of uropathogenic *Escherichia coli* (UPEC) (Zhang et al. 2004) and a profilin-like molecule derived from *Toxoplasma gondii* (Yarovinsky et al. 2005).

Endosomal TLRs—Expression, Structure, and Ligands

The nucleic acid–recognizing TLRs are TLR3, TLR7, TLR8, TLR9, and TLR13, which recognize DNA or RNA derived from bacteria and viruses as well as self-nucleic acids in autoimmune conditions. TLR3 is highly expressed in innate immune cells, except for neutrophils and plasmacytoid DCs (pDCs), and recognizes viral double-stranded RNA (dsRNA) and a synthetic analogue of dsRNA polyinosinic–polycytidylic acid (poly I:C) in endolysosomes (Takeuchi and Akira 2010; Thompson et al. 2011). Cocrystallization studies of a TLR3–dsRNA complex revealed that TLR3 has a horseshoe-shaped solenoid structure and that dsRNA binds to the amino-terminal and carboxy-terminal portions on the lateral convex surface of the TLR3 ectodomain (Choe et al. 2005; Liu et al. 2008). Recently, Bruton's tyrosine kinase (BTK) was shown to phosphorylate the cytoplasmic domain of TLR3, particularly the Tyr759 residue, following ligand binding to initiate downstream signaling (Lee et al. 2012). TLR3 is one of the major RNA sensors, and recognizes a number of microbial RNAs including genomic RNA of reoviruses, dsRNA produced during replication of single-stranded RNA (ssRNA) of RSV, encephalomyocarditis virus (EMCV), and West Nile virus (WNV), some small-interfering RNAs, murine cytomegalovirus, and herpes simplex virus type I (HSV-1) (Zhang et al. 2007; Kawai and Akira 2010).

TLR7 and TLR8 are closely related and confined to endosomal compartments where they recognize ssRNA, mainly viral RNAs. TLR7 is predominantly expressed in pDCs and can recognize small purine analog imidazoquinoline derivatives imiquimod and resiquimod (R848), guanine analogues, and uridine or uridine/guanosine-rich ssRNA. The viral ssRNAs recognized by TLR7 are from vesicular stomatitis virus (VSV), influenza A virus, human immunodeficiency virus, and coxsackievirus B (Wang et al. 2007). RSV, Sendai virus (SV) partially, and human metapneumovirus (HMPV) are also recognized by TLR7 in a cell-specific manner (Melchjorsen et al. 2005; Lee et al. 2007; Phipps et al. 2007; Goutagny et al. 2010). Additionally, RNA from streptococcus B bacteria is sensed by TLR7 in conventional DCs (cDCs) (Mancuso et al. 2009). Although little is known about the ligands recognized by TLR8, human TLR8 was shown to recognize R848 and viral ssRNA, whereas TLR8-deficient mice responded normally to these ligands. Other ligands, like Vaccinia viral RNA, were reported to be TLR8 agonists, but remain controversial (Cervantes et al. 2012). Moreover, several studies have shown that human TLR8 responds to total bacterial RNA (Broz and Monack 2013). Crystal structure studies of unliganded and ligand-induced activated human TLR8 revealed that, the unliganded TLR8 showed a preformed dimeric form and the Z-loop between LRR14 and LRR15 was cleaved, whereas amino- and carboxy-terminal halves remained associated and confer ligand recognition and dimerization. On ligand binding, the preformed TLR8 dimer undergoes reorganization to bring the two carboxyl termini to close proximity to enable subsequent dimerization with TIR domain and downstream signaling (Tanji et al. 2013).

Notably, likewise TLR5, TLR11 recognizes flagellin, but is localized in different subcellular compartment than TLR5 (i.e., TLR5 is on the cell membrane whereas TLR11 in endolysosomes) (Mathur et al. 2012). TLR12 is predominantly expressed in myeloid cells and is highly homologous to TLR11. It can recognize profilin from *T. gondii* (Koblansky et al. 2013). TLR12 can function either as a heterodimer with TLR11 or alone (Broz and Monack 2013).

Recent studies have shed some light on TLR13 ligands. It was reported that TLR13 recognizes a conserved CGGAAAGACC motif in *Staphylococcus aureus* 23S rRNA and that *E. coli* 23S rRNA can induce a TLR13-dependent transcriptional response resulting in pro-IL-1β induction (Li and Chen 2012; Oldenburg et al. 2012). Similarly, both heat-killed and live *Streptococcus pyogenes* are recognized by a TLR13-dependent pathway (Hidmark et al. 2012). Because both TLR8 and TLR13 can recognize bacterial RNAs, the possibility of redundancy between TLR8 and TLR13 is a question for future studies.

Another member of the TLR family recognizing nucleic acids is TLR9, which, unlike other TLRs, recognizes bacterial and viral DNA that is rich in unmethylated CpG DNA motifs. TLR9 is highly expressed in pDCs, macrophages, and B cells, and can be activated by synthetic CpG oligonucleotides. Compared with the sequence-independent recognition of the 2′-deoxyribose sugar backbone of natural phosphodiester oligodeoxynucleotides by TLR9, the CpG DNA motif is required for recognition of synthetic phosphorothioate oligodeoxynucleotides (Haas et al. 2008; Wagner 2008). Viral DNAs recognized by TLR9 arise from murine cytomegalovirus (MCMV), HSV-1, HSV-2, and adenoviruses (Gurtler and Bowie 2013). Apart from its specificity for DNA, TLR9 has been shown to directly recognize hemozoin, an insoluble crystalline byproduct generated by *Plasmodium falciparum* during the process of detoxification after host hemoglobin is digested (Coban et al. 2010). Additionally, TLR3, TLR7, TLR9, and UNC93B1 render host resistance to *Leishmania major* infection (Schamber-Reis et al. 2013).

Proper cellular localization is critical for efficient signaling by nucleic acid-recognizing TLRs. It is also important for avoiding recognition of host self-DNA, which, if recognized by TLRs, can lead to autoimmunity. TLR9 and TLR7 are localized in the endoplasmic reticulum (ER) in unstimulated cells, but are recruited to endolysosomes after ligand stimulation (Kim et al. 2008). UNC93B1 is an important protein that controls the trafficking of TLR3, TLR7, and TLR9 from the ER to endolysosomes (Tabeta et al. 2006). Further, UNC93B1 actively and continuously regulates excessive TLR7 activation of immune cells by using TLR9 to counteract TLR7 as shown by mice harboring an amino acid substitution (D34 to A) in UNC93B1, that shows TLR7-hyperreactive and TLR9-hyporeactive phenotype and subsequent TLR7-dependent systemic lethal inflammation, thereby supporting the understanding that an opposing relationship between TLR7 and TLR9 is a potential mechanism regulating autoimmunity (Fukui et al. 2011).

In endolysosomes, TLR9 undergoes proteolytic cleavage by cathepsins B, S, L, H, and K and asparginyl endopeptidase to acquire a functional form that mediates ligand recognition and initiates subsequent signaling cascades (Asagiri et al. 2008; Ewald et al. 2008; Matsumoto et al. 2008; Park et al. 2008; Sepulveda et al. 2009). However, this functional cleavage of TLR9 remains controversial based on the importance of the amino-terminal region of TLR9 for CpG DNA recognition and binding (Peter et al.

2009). Interestingly, a study reported TLR9 to be expressed in the cell surface of murine splenic DCs using TLR9N- and TLR9C-specific antibodies and further explained that, succeeding the cleavage of TLR9 in the endolysosome, the amino-terminal cleaved fragment (TLR9N) remains associated with truncated TLR9 forming the complex TLR9+C, which acts as a bona fide DNA sensor; however, TLR9C alone could not recognize DNA (Onji et al. 2013).

TLR SIGNALING

TLR signaling starts with the organization and recruitment of four TIR domain–containing adaptor proteins differentially. They are myeloid differentiation factor 88 (MyD88), MyD88 adaptor-like protein (MAL/TIRAP), TIR domain–containing adaptor–inducing IFN-β (TRIF; also known as TICAM-1), and TRIF-related adaptor molecule (TRAM) (Takeuchi and Akira 2010). TLR signaling is majorly divided into two pathways: MyD88 dependent and TRIF dependent, depending on the primary adaptor usage. With the exception of TLR3, all TLRs use MyD88 for activation of NF-κB and MAPKs to induce proinflammatory genes. TLR2 and TLR4 signaling depends on TIRAP, which mediate interaction between MyD88 and activated TLRs. The promiscuity of lipid binding by phosphoinositide-binding domain of TIRAP allows it to sample multiple compartments for the presence of activated TLRs and, hence, diversifies the subcellular sites of TLR signal transduction (Bonham et al. 2014). MyD88 recruitment activates a series of IL-1R1-associated protein kinases (IRAKs), IRAK4, IRAK2, and IRAK1, to form "Myddsome," which further interacts with tumor necrosis factor (TNF)-α receptor–associated factor 6 (TRAF6) (Lin et al. 2010). TRAF6 activates TAK1 complex, which further phosphorylates IκB kinase (IKK)-β and MAP kinase (MAPKs). IKK complex (IKK-α/IKK-β/NEMO) catalyzes the phosphorylation of NF-κB inhibitory protein IκBα, which undergoes proteasome degradation to render NF-κB free to translocate into nucleus and induce proinflammatory gene expression (Fig. 1) (Kawai and Akira 2010).

TLR3 as well as TLR4 use a TRIF-dependent pathway that results in activation of IRF3 and NF-κB for subsequent induction of type I interferons and inflammatory cytokines, respectively (Akira et al. 2006). TLR4 requires TRAM to interact with TRIF; however, to directly interact with TRIF, the two tyrosine residues in the cytoplasmic domain of TLR3, Tyr858, and Tyr759 are phosphorylated by the epidermal growth factor ErbB1 and Btk, respectively (Lee et al. 2012; Yamashita et al. 2012). TRIF interacts with TRAF6 and activates TAK1 to activate NF-κB in a manner similar to MyD88. TRIF also recruits TRAF3, which activates TANK-binding kinase 1 (TBK1) and IKKi to phosphorylate IRF3 and IRF7 for subsequent induction of type I IFN gene expression (Hacker et al. 2006; Kawai and Akira 2010)

Continuing studies have uncovered plenty of regulators critical for TLR signaling that differ in their cellular location and mode of action (Qian and Cao 2013). Membrane-resident molecules, such as CD14 and CD36, modulate TLR signaling. CD14, a glycophosphotidylinositol-anchored protein, which acts as a coreceptor for LPS along with TLR4 and MD2, induces ITAM-mediated Syk and PLC-γ2-dependent endocytosis, thus promoting TLR4 internalization to endosomes to facilitate TRIF-dependent signaling (Zanoni et al. 2011). CD14 is also necessary for TLR7- and TLR9-dependent induction of proinflammatory cytokines (Baumann et al. 2010). CD36, a protein of class B scavenger receptor family, acts as a coreceptor for oxidized low-density lipoprotein (LDL) and amyloid β peptide and enables formation of TLR4/6 heterodimer through Src kinases ensuing to sterile proinflammatory events (Stewart et al. 2010). Ubiquitin-modifying enzymes are also emerging as important regulators of TLR signaling. Nrdp-1, an E3 ligase, directly binds and polyubiquitinates MyD88 and TBK1, inducing degradation of former and augmenting activation of later to attenuate production of inflammatory cytokines and promote preferential type I IFN production, respectively (Wang et al. 2009). Furthermore, heat shock cognate 70 (HSC70)-interacting protein (CHIP) can enhance TLR signaling by recruiting, polyubiqui-

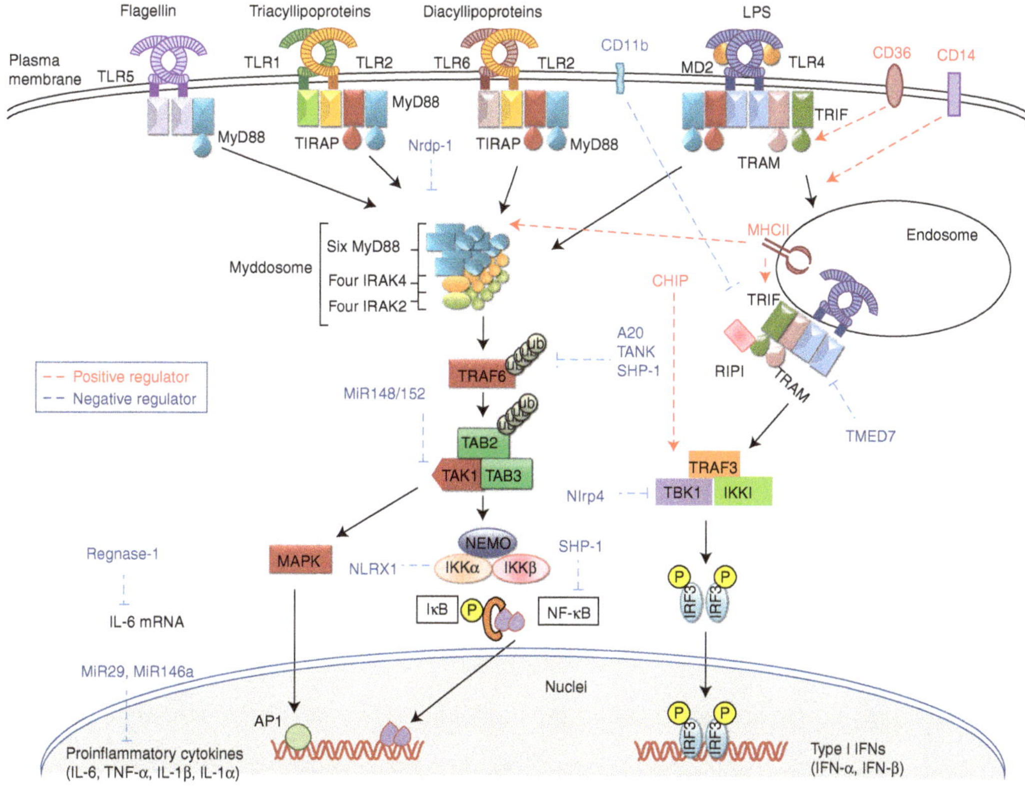

Figure 1. Signaling pathway of cell surface TLRs. TLR4 recognizes LPS in complex with MD2 to start a signaling cascade by recruiting adaptors MyD88 and TIRAP and forming a complex of IRAK4, 1, and 2, and TRAF6. TRAF6 catalyzes formation of K-63 linked polyubiquitin chains on TRAF6 itself and generates an unconjugated polyubiquitin chain. TRAF6 activates TAK1 complex, which further activates IKK complex and MAP kinases. IKK complex phosphorylates IκB, which undergoes proteosomal degradation to release NF-κB for further translocation to nucleus and subsequent induction of proinflammatory cytokines. MAP kinases phosphorylate Jun kinases (JNK), p38 kinase, extracellular signal regulated kinase 1(ERK1), and ERK2. TLR4 also signals through TRIF-dependent pathways with the help of adaptor TRAM after translocation to endosome and activates IRF3 for type I IFN production. TLR1/2 and TLR6/2 heterodimers recognize triacylated and diacylated lipoproteins, respectively, whereas TLR5 recognizes flagellin and all of them initiate a MyD88-dependent signaling pathway that culminates in the induction of proinflammatory cytokines. TLR signaling is modulated by CD14, CD36, Nrdp1, CHIP, MHCII, Regnase-1, A20, SHP-1, TMED7, CD11b, NLRP4, NLRX1, and miRNAs (miR-146a, miR-29, miR-148/152).

nating and activating the tyrosine kinase Src and atypical protein kinase C ζ (PKCζ) to the TLR complex, henceforth, leading to activation of IRAK-1, TBK1, IRF3, and IRF7 (Fig. 1) (Yang et al. 2011). Mitochondrial ubiquitin ligase MARCH5 positively regulates TLR7 signaling and catalyzes the K63-linked ubiquitination of TANK, making it incapable of inhibiting TRAF3 (Fig. 2) (Shi et al. 2011).

A novel function of MHC class II molecules is reported, where intracellular MHC class II molecules interacted with Btk via the costimulatory molecule CD40 and maintained Btk activation. Activated Btk interacts with MyD88 and TRIF to promote enhanced production of inflammatory cytokines and type I IFNs (Liu et al. 2011). Antiviral protein viperin mediates antiviral function through TLR7/9- IRAK1 signaling axis, wherein, viperin that is induced after TLR7 or TLR9 stimulation interacts with IRAK1 and TRAF6 to recruit them to lipid bodies and to facilitate K63-linked ubiquitination of IRAK1, which induces nuclear translocation of IRF7 and subsequent type I IFN production

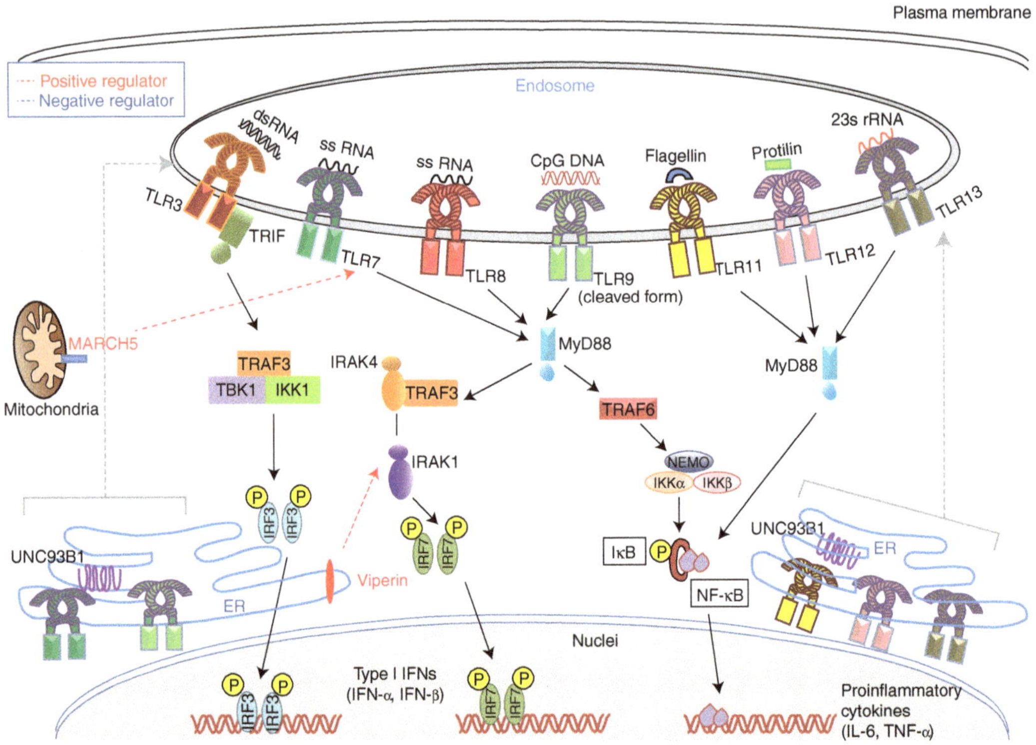

Figure 2. Signaling pathway of endosomal TLRs. Endosomal residents of TLR family are TLR3, TLR7, TLR8, TLR9, TLR11, TLR12, and TLR13. dsRNA recognition by TLR3 recruits TRIF to initiate signaling cascade that further activates TBK1 to induce IRF3-mediated type I IFN production. Concurrently, TRIF can also interact with TRAF6 to activate NF-κB for transcription of proinflammatory cytokines. TLR7 and TLR8 recognize viral ssRNA, whereas TLR9 recognizes CpG DNA from both bacteria and viruses. Ligand stimulation facilitates UNC93B1-dependent trafficking of TLR7 and TLR9 from ER to endosomes, where TLR9 undergoes proteolytic cleavage. Following this TLR7 and TLR9 recruit MyD88 to activate NF-κB and IRF7 to induce transcription of proinflammatory cytokines and type I IFN genes, respectively. TLR11, TLR12, and TLR13 recognize flagellin, profilin, and bacterial 23sRNA, respectively, and induce NF-κB signaling through MyD88 and TIRAP. Endosomal TLR signaling can be modulated by MARCH5 and viperin.

by pDCs (Saitoh et al. 2011). Utilizing a combinatorial approach including transcriptomics, genetic/chemical perturbations, and phosphoproteomics, signaling components involved in the TLR response in dendritic cells were discovered systematically, particularly, Polo-like kinases (Plks) 2 and 4 were found to be important for antiviral responses both in vitro and in vivo (Chevrier et al. 2011).

TLR signaling is negatively regulated by various molecules at numerous levels to keep check on the excessive immune responses that can lead to detrimental consequences. TANK, IRAK-M, Atg16L, Regnase-1, tristeraprolin, A20, SHP-1, SENP6, GOLD domain–containing TMED7, Trim30α, CD11b, NLRP4, NLRX1, and miRNAs (miR-146a, miR-199a, miR-155, miR-126, miR-21, miR-29, miR-148/152, miR-466l) negatively regulate TLR signaling by different mechanisms (Fig. 1) (Kawai and Akira 2010; Doyle et al. 2012; Kondo et al. 2012; Liu et al. 2013; Olivieri et al. 2013; Qian and Cao 2013).

INTRACELLULAR VIRAL RECOGNITION AND SIGNALING BY RLRs

Viral RNAs are also recognized by another family of cytoplasmic receptors known as RLRs. The three members that constitute the RLR family are retinoic acid-inducible gene I (RIG-I),

 Cite this article as *Cold Spring Harb Perspect Biol* doi: 10.1101/cshperspect.a016246

melanoma differentiated gene 5 (MDA5), and laboratory of genetics and physiology 2 (LGP2). RLRs are expressed in a variety of cell types, including myeloid cells, epithelial cells, fibroblasts, and cells of the central nervous system, although RLR function is not necessary for IFN production by pDCs despite their expression in this cell type (Loo and Gale 2011). RLRs are structurally similar and harbor three distinct domains: an amino-terminal region composed of tandem caspase activation and recruitment domains (CARDs), a central DEAD box helicase/ATPase domain, and a carboxy-terminal regulatory domain. LGP2 is homologous to RIG-I and MDA5, but lacks the amino-terminal CARD domain, and functions as a regulator of RIG-I and MDA5 signaling (Fig. 3) (Saito et al. 2007; Yoneyama and Fujita 2008).

A number of studies have shown that RIG-I recognizes members of the Paramyxoviridae, Rhabdoviridae, Orthomyxoviridae, and flavivirus genera, and, consequently, RIG-I-deficient cells have impaired type I and inflammatory cytokine production in response to NDV, SV, VSV, and JEV (Kato et al. 2006). RIG-I also recognizes influenza A virus, measles virus, and Ebola virus (Thompson et al. 2011). *Listeria monocytogenes* actively secretes small RNAs via a SecA2 secre-

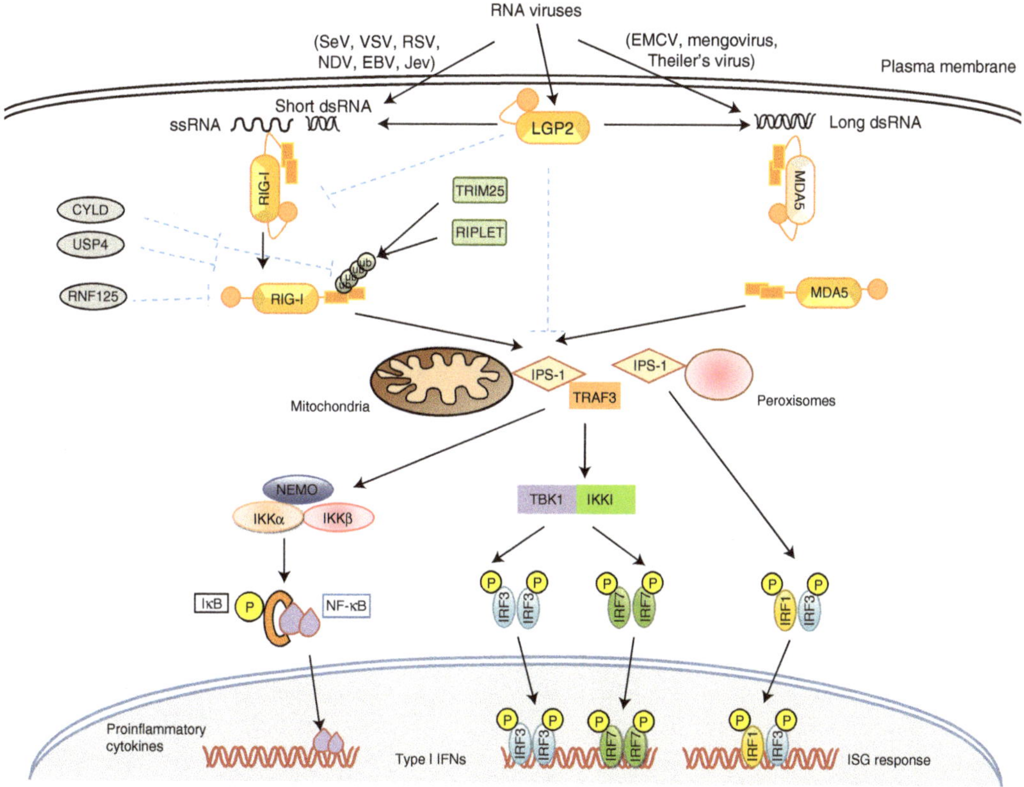

Figure 3. RLR Signaling pathway. RLR family contains of RIG-I, MDA5, and LGP2. RIG-I and MDA5 recognize viruses differentially. RIG-I recognizes SeV, VSV, RSV, NDV, EBV, and JEV, whereas MDA5 recognizes viruses of picornavirus family members. After virus infection, long viral dsRNA or short dsRNA activates MDA5 or RIG-I, respectively, and they undergo conformational change to obtain an open confirmation and their CARD domain interacts with CARD domain of IPS-1 localized in mitochondria and peroxisomes to start signaling. Mitochondrial IPS-1 signaling leads to activation of NF-κB and RF3/IRF7 resulting in the production of type I interferons and proinflammatory cytokines, whereas peroxisomal IPS-1 activates IRF1- and IRF3-dependent signaling leading to interferon stimulatory genes (ISGs) expression. LGP2 is a positive regulator of RIG-I and MDA5. RIG-I signaling is also positively regulated by Trim25 and RIPLET, whereas it is negatively regulated by CYLD, USP4, and RNF125.

tion system, thereby triggering strong RIG-I activation that leads to type I IFN production (Abdullah et al. 2012; Hagmann et al. 2013). Compared with RIG-I, MDA5 recognizes a different class of viruses, such as Picornaviridae, and MDA5-deficient mice show abrogated type I IFN production in response to EMCV, Theiler's virus, hepatitis C virus, mengovirus, and murine norovirus (Kato et al. 2006; McCartney et al. 2008). Moreover, dengue virus, WNV, and reovirus are recognized by both RIG-I and MDA5 in concert (Fredericksen et al. 2008; Loo et al. 2008). The preferences for target RNA recognition differ between RIG-I and MDA5. RIG-I preferentially recognizes short dsRNA (from 19- or 21-mers to 1 kb) with 5′-triphosphorylated ssRNA, a common characteristic of viral RNA (Hornung et al. 2006; Pichlmair et al. 2006; Takeuchi and Akira 2010; Thompson et al. 2011). However, MDA5 recognizes high molecular weight dsRNA- (long dsRNA of more than 2 kb) like poly I:C (Kato et al. 2006). LGP2 was initially suggested to function as a negative regulator of RIG-I and MDA5 signaling (Yoneyama et al. 2005; Saito et al. 2007). In contrast, LGP2-deficient mice and cells expressing an LGP2 mutant with a point mutation, D30A, that abrogates the ATPase activity, had impaired type I IFN production in response to RIG-I and MDA5 ligands, implying positive regulation of RIG-I and MDA5 by LGP2 (Satoh et al. 2010).

Structural studies have shown that RIG-I in unstimulated cells is sequestered in the cytoplasm in an inactive ("closed" conformation) form that is autoinhibited by its regulatory domain (Saito et al. 2007). However, after virus infection, RIG-I undergoes conformational changes to an "open" conformation to become active and multimerizes in an ATP-dependent manner (Scott 2010). The CARD domains of activated RIG-I and MDA5 then homotypically interact with the CARD of interferon promoter stimulator 1 (IPS-1; also known as MAVS, VISA, or Cardif) (Kawai and Akira 2006), which is localized in the outer membranes of mitochondria and the peroxisomal membranes (Dixit et al. 2010; Scott 2010). This interaction relocates RLRs to IPS-1-associated membranes to form an IPS-1 signalosome with other downstream molecules (Loo and Gale 2011). To elucidate the activation mechanism of MAVS, it was shown that viral infection induces a prion-like conformational switch forming very large functional aggregates of MAVS-like prions on the mitochondrial membrane catalyzed by RIG-I in the presence of unanchored K63 ubiquitin chains, thereby activating and propagating antiviral signaling cascade (Hou et al. 2011).

RLR signaling is tightly regulated both positively and negatively by differential ubiquitination. TRIM25 and RNF135 (Riplet or REUL) are E3 ubiquitin ligases that enhance RIG-I activation by K63 polyubiquitination (Gack et al. 2007; Pichlmair et al. 2009). In contrast, RNF125 inhibits RIG-I signaling by K48 polyubiquitination and proteasome-mediated degradation of RIG-I (Arimoto et al. 2007). CYLD (cylindromatosis) and ubiquitin-specific protease 4 (USP4) are deubiquitinases that remove K63-linked or K48-linked polyubiquitin chains, respectively, from RIG-I to inhibit RIG-I-dependent IFN production (Friedman et al. 2008; Wang et al. 2013). Additionally, several microRNAs, such as miR-146a, miR-4661, miR-122, and miR-24, have been shown to regulate the RIG-I pathway. Intriguingly, recently reported that *MAVS* mRNA is bicistronic and codes for both full-length MAVS and a truncated variant, miniMAVS. miniMAVS inhibits MAVS-induced antiviral responses, however, both proteins positively regulate cell death (Brubaker et al. 2014).

Recently, many studies have shown that cytosolic synthetic dsRNA, poly I:C, virus-derived RNA, and bacteria-derived RNA can activate the NLRP3 inflammasome independently of RLRs (Kanneganti et al. 2006; Rajan et al. 2010; Eigenbrod et al. 2012). DExD/H box helicase family member DHX33 was identified as an RNA sensor that activates the NLRP3 inflammasome in response to cytosolic poly I:C, reoviral RNA, and bacterial RNA (Mitoma et al. 2013).

INTRACELLULAR DNA SENSING

DNA was initially shown to be recognized by TLR9. However, several studies using TLR9 antagonists, TLR9, and DNase II–deficient mice soon suggested the presence of additional DNA sensors and pathways that function independently of TLR9 and are present in the cytoplasm. New advances are revealing a whole repertoire of much sought-after cytosolic DNA sensors, and the list is continuing to expand. These DNA sensors recognize microbial or self-DNA present in the cytoplasm as a sign of infection or cell damage and induce the production of type I IFNs, type III IFNs, or IL-1β. Most of the DNA sensors recognize foreign DNA in the cytoplasm and generally utilize stimulator of interferon genes (STING; also known as TMEM173, MPYS, MITA, and ERIS) and TBK1 to induce type I IFN production (Fig. 4) (Ishii et al. 2006; Ishikawa and Barber 2008).

DNA-dependent activator of IRFs (DAI) (Takaoka et al. 2007; Ishii et al. 2008), RNA polymerase III (Pol III) (Ablasser et al. 2009; Chiu et al. 2009), IFN-γ-inducible protein 16 (IFI16) (Unterholzner et al. 2010), leucine-rich repeat flightless-interacting protein 1 (LRRFP1) (Yang et al. 2010), extrachromosomal histone H2B (Kobiyama et al. 2010), DNA-PK (Ferguson et al. 2012), and MRE11 (Kondo et al. 2013) recognize dsDNA to induce type I IFN production. Members of the DExD/H-box helicase family, DHX9, DHX36, and DDX41, were also identified as candidate DNA sensors, although they have different specificities for microbial DNA recognition (Kim et al. 2010). Unlike other cytosolic DNA sensors, AIM2 and nuclear IFI16

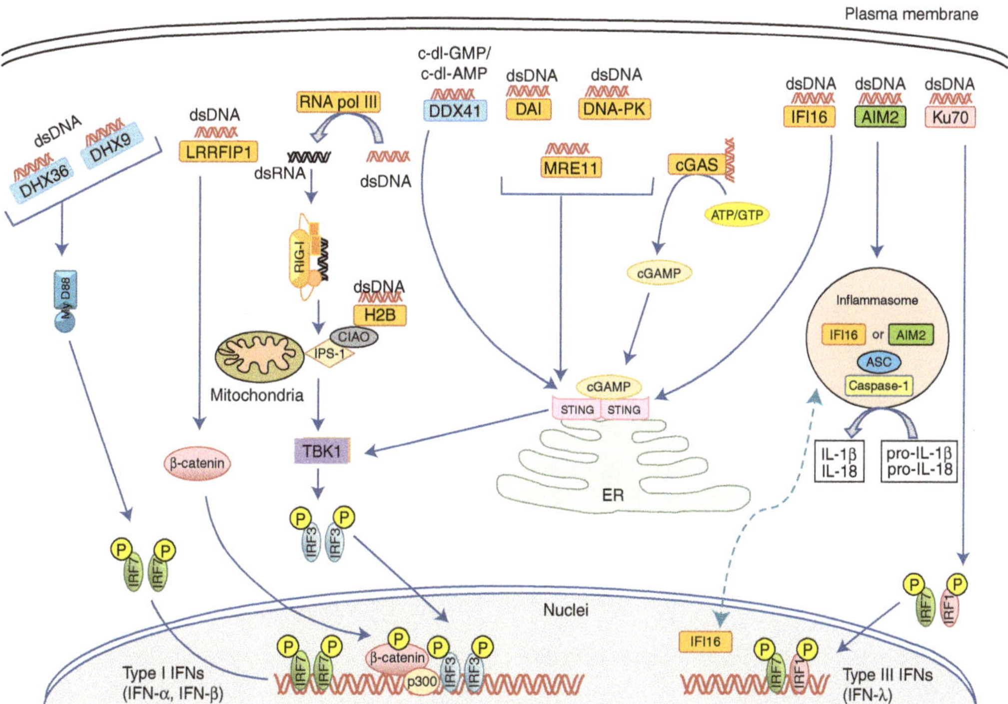

Figure 4. Cytosolic DNA sensors and their signaling pathways. DNA present in the cytoplasm derived from either hosts or microbes are recognized by various kinds of DNA sensors. DDX41, DAI, DNA-PK, IFI16, MRE11, and cGAS recognize dsDNA and induce type I IFN production through STING via TBK1- and IRF3-dependent pathways. AT-rich dsDNA recognized by RNA polymerase III is transcribed to 5′-triphosphate RNA to be further sensed by RIG-I to induce IPS-1-dependent type I production; similarly, H2B also recognizes dsDNA and signals through IPS1. LRRFIP1 binds both dsDNA and dsRNA and activates β-catenin, which enhances a TBK1-independent and IRF3-dependent IFN-β transcription. DHX9 and DHX36 are MyD88-dependent DNA sensors. AIM2 and nuclear IFI16 can form inflammasomes together with ASC and caspase-1 on recognition of dsDNA to mediate a caspase-1-dependent cleavage of IL-1β and IL-18 from pro-IL-1β and pro-1L-18. Finally, Ku-70 recognizes dsDNA to induce IRF1/IRF7-dependent type III IFN.

recognize cytosolic DNA and form an active inflammasome by recruiting ASC and pro-caspase-1, which cleaves pro-IL-1β and pro-IL-18 to their mature forms for secretion (Fernandes-Alnemri et al. 2010; Kerur et al. 2011). Ku70 senses cytosolic DNA and induces type III IFN production (Zhang et al. 2011).

Recently, by using biochemical strategies, a crucial study by Chen and colleagues showed that cyclic GMP-AMP (cGAMP) synthase (cGAS), a member of the nucleotidyltransferase family acts as a novel sensor for cytosolic DNA and provides the endogenous second messenger cGAMP for STING activation. On encountering DNA in the cytoplasm, cGAS synthesizes cyclic-di-GMP-AMP (c-di-GAMP) from ATP and GTP, which then binds to STING to induce IRF3-mediated IFN-β production. cGAS have been shown to directly interact with dsDNA independently of its sequence through the amino-terminal domain and activates the synthesis of cGAMP. cGAS knockdown leads to impaired IFN-β induction and IRF3 activation in response to DNA transfection or DNA virus infection (Sun et al. 2013; Wu et al. 2013). Furthermore, by generating *cGAS* knockout mice, the Chen group showed that multiple cell types including fibroblasts, macrophages, and DCs from cGAS-deficient mice had impaired type I IFN and other proinflammatory cytokine production in response to transfected immunostimulatory DNA or DNA virus (HSV-1, VACV) infection, whereas the responses to poly I:C, poly dAdT, and RNA virus infection remained intact. Moreover, cGAS-deficient mice were more susceptible to lethal HSV-1 infection than wild-type mice. cGAMP also acts as an immune adjuvant to stimulate antigen-specific T cells and antibody responses in a STING-dependent manner (Li et al. 2013). Previous studies have shown that a mutant allele of murine STING (R231A) was unresponsive to cyclic dinucleotides (CDNs) but normally responsive to dsDNA, thus presenting a paradox (Burdette et al. 2011). Later, following the discovery of endogenous cGAMP, the R231A mutant was found to respond normally to endogenous cGAMP, but did not respond to the microbial counterpart cGAMP synthesized by cGAS from *Vibrio cholera* or chemically synthesized cGAMP (Davies et al. 2012; Ablasser et al. 2013; Diner et al. 2013). Taken together, these observations suggested that a clear mechanism exists for classification of endogenous cGAMP and other microbial CDNs. In a series of elegant biochemical, biophysical, and structural studies, four independent groups showed that endogenous cGAMP synthesized by cGAS is a noncanonical CDN with unique phosphodiester linkages between 2′OH of GMP and 5′-phosphate of AMP and between 3′OH of AMP and 5′-phosphate of GMP, thus referred to as 2′3′-cGAMP, and that 2′3′-cGAMP is a far more potent stimulator of STING than canonical CDNs from microbial origin that have 3′–5′ linkages (Ablasser et al. 2013; Diner et al. 2013; Gao et al. 2013; Zhang et al. 2013). These structural studies and in vitro enzymatic reactions further showed that cGAS first catalyzes the synthesis of a linear 2′–5′-linked dinucleotide, followed by its cGAS-dependent cyclization in a second step through a 3′–5′ phosphodiester linkage to form 2′3′-cGAMP (Ablasser et al. 2013; Gao et al. 2013; Xiao and Fitzgerald 2013). Further investigation of the structural insights revealed that cGAS contains a less conserved amino-terminal stretch, followed by a highly conserved Mab21 domain of the nucleotidyl transferase (NTase) superfamily having a bilobal scaffold, and that a unique zinc-binding motif called a "zinc thumb" is present across both lobes of cGAS, providing an essential B-DNA-binding platform (Civril et al. 2013).

STING Is Cardinal for Intracellular DNA Sensing

With continuing research, STING has emerged as the key adaptor in the cytosolic DNA-sensing pathway, as several studies have shown its indispensable roles in innate immune responses to DNA derived from viral pathogens, bacterial pathogens, and eukaryotic pathogens, thus ingraining STING as a prime signaling adaptor in the intracellular DNA-sensing pathway (Burdette and Vance 2013). STING is also implicated in certain autoimmune diseases and adjuvanticity of DNA vaccines (Ishikawa et al.

2009; Gall et al. 2012). Structurally, STING is composed of an amino-terminal domain consisting of four transmembrane domains that anchor STING in the ER and a carboxy-terminal domain (CTD) that is speculated to be cytosolic (Burdette and Vance 2013; Paludan and Bowie 2013). In resting cells, STING is localized in the mitochondrial-associated membrane (MAM)/ER, but after HSV-1 infection, it rapidly traffics via unclear mechanisms to an uncharacterized perinuclear region, although this endosomal compartment containing STING and TBK1 does involve autophagy-related gene 9a (Atg9a) (Ishikawa et al. 2009; Saitoh et al. 2010). Furthermore, on recognition of DNA by upstream DNA sensors, STING binds to both IRF3 and TBK1 via its carboxy-terminal region and thus provides a scaffold to promote TBK1-mediated phosphorylation of IRF3 (Tanaka and Chen 2012). Ubiquitination is also suggested to be important for STING signaling, as the ubiquitin ligases TRIM56 and RNF5 have been shown to regulate STING signaling positively and negatively, respectively (Zhong et al. 2009; Tsuchida et al. 2010). Furthermore, STING was reported to directly recognize CDNs, such as c-di-GMP from bacteria (Burdette et al. 2011). Structural studies of human STING by different research groups have shown that the STING CTD exists in a symmetrical dimeric state when unbound to CDNs and that CDN binding does not bring about any conformational change of the CTD, meaning that a definite mechanism for how STING is activated to move from the ER to engage with TBK1 remains elusive (Paludan and Bowie 2013). Besides providing a scaffold for TBK1 and IRF3 on virus infection or nucleic acid stimulation, STING regulates a unique antiviral pathway by recruiting STAT6 to the ER for subsequent phosphorylation by TBK1, independently of Janus kinases (JAKs), thereby inducing STAT6-specific target genes for immune cell homing (Chen et al. 2011). Recently, negative feedback control of STING activity by CDNs was reported. After initial activation of the STING–TBK1–IRF3 axis, STING is phosphorylated by UNC-51-like kinase (ULK1/ATG1), which suppresses IRF3 activity to avoid sustained production of inflammatory cytokines. ULK1 activation is triggered by CDNs generated by cGAS (Konno et al. 2013).

CONCLUDING REMARKS

Studies on pathogen recognition by the host innate immune system have come a long way since the first description of PRRs. Diverse kinds of PRR families, such as TLRs, RLRs, NLRs, CLRs, AIM2-like receptors (ALRs), peptidoglycan recognition proteins (PGRPs), and an expanding number of cytosolic DNA sensors have been identified, each having specificity for distinct PAMPs. This review has refurbished the present knowledge on pathogen sensing by PRRs, as well as their ligands, structures, expression patterns, and signaling mechanisms that are crucial for host defense. Nevertheless, a thrust area for research on pathogen recognition requires investigation of multiple PRRs activated in response to pathogens and the crosstalk among these PRRs that coordinates host immune responses. Furthermore, in the case of cytosolic DNA sensors, with the identification of a number of DNA sensor candidates, the scope of the questions is increasing with regard to the extent of redundancy between these receptors, their expression patterns in different cell types, and the mechanisms by which they recognize different kinds of dsDNA (i.e., their ligand specificity). Hence, detailed research encompassing the molecular and cellular mechanisms of PRR signaling will provide a better understanding, thereby helping to combat microbial infections effectively.

REFERENCES

Abdullah Z, Schlee M, Roth S, Mraheil MA, Barchet W, Bottcher J, Hain T, Geiger S, Hayakawa Y, Fritz JH, et al. 2012. RIG-I detects infection with live *Listeria* by sensing secreted bacterial nucleic acids. *EMBO J* **31:** 4153–4164.

Ablasser A, Bauernfeind F, Hartmann G, Latz E, Fitzgerald KA, Hornung V. 2009. RIG-I-dependent sensing of poly(dA:dT) through the induction of an RNA polymerase III–transcribed RNA intermediate. *Nat Immunol* **10:** 1065–1072.

Ablasser A, Goldeck M, Cavlar T, Deimling T, Witte G, Rohl I, Hopfner KP, Ludwig J, Hornung V. 2013. cGAS produces a 2′-5′-linked cyclic dinucleotide second messenger that activates STING. *Nature* **498:** 380–384.

Akira S, Takeda K, Kaisho T. 2001. Toll-like receptors: Critical proteins linking innate and acquired immunity. *Nat Immunol* **2:** 675–680.

Akira S, Uematsu S, Takeuchi O. 2006. Pathogen recognition and innate immunity. *Cell* **124:** 783–801.

Arimoto K, Takahashi H, Hishiki T, Konishi H, Fujita T, Shimotohno K. 2007. Negative regulation of the RIG-I signaling by the ubiquitin ligase RNF125. *Proc Natl Acad Sci* **104:** 7500–7505.

Asagiri M, Hirai T, Kunigami T, Kamano S, Gober HJ, Okamoto K, Nishikawa K, Latz E, Golenbock DT, Aoki K, et al. 2008. Cathepsin K-dependent toll-like receptor 9 signaling revealed in experimental arthritis. *Science* **319:** 624–627.

Barbalat R, Lau L, Locksley RM, Barton GM. 2009. Toll-like receptor 2 on inflammatory monocytes induces type I interferon in response to viral but not bacterial ligands. *Nat Immunol* **10:** 1200–1207.

Baumann CL, Aspalter IM, Sharif O, Pichlmair A, Bluml S, Grebien F, Bruckner M, Pasierbek P, Aumayr K, Planyavsky M, et al. 2010. CD14 is a coreceptor of Toll-like receptors 7 and 9. *J Exp Med* **207:** 2689–2701.

Beutler B. 2009. Microbe sensing, positive feedback loops, and the pathogenesis of inflammatory diseases. *Immunol Rev* **227:** 248–263.

Bonham KS, Orzalli MH, Hayashi K, Wolf AI, Glanemann C, Weninger W, Iwasaki A, Knipe DM, Kagan JC. 2014. A promiscuous lipid-binding protein diversifies the subcellular sites of Toll-like receptor signal transduction. *Cell* **156:** 705–716.

Broz P, Monack DM. 2013. Newly described pattern recognition receptors team up against intracellular pathogens. *Nat Rev Immunol* **13:** 551–565.

Brubaker SW, Gauthier AE, Mills EW, Ingolia NT, Kagan JC. 2014. A bicistronic *MAVS* transcript highlights a class of truncated variants in antiviral immunity. *Cell* **156:** 800–811.

Burdette DL, Vance RE. 2013. STING and the innate immune response to nucleic acids in the cytosol. *Nat Immunol* **14:** 19–26.

Burdette DL, Monroe KM, Sotelo-Troha K, Iwig JS, Eckert B, Hyodo M, Hayakawa Y, Vance RE. 2011. STING is a direct innate immune sensor of cyclic di-GMP. *Nature* **478:** 515–518.

Celhar T, Magalhaes R, Fairhurst AM. 2012. TLR7 and TLR9 in SLE: When sensing self goes wrong. *Immunol Res* **53:** 58–77.

Cervantes JL, Weinerman B, Basole C, Salazar JC. 2012. TLR8: The forgotten relative revindicated. *Cell Mol Immunol* **9:** 434–438.

Chen H, Sun H, You F, Sun W, Zhou X, Chen L, Yang J, Wang Y, Tang H, Guan Y, et al. 2011. Activation of STAT6 by STING is critical for antiviral innate immunity. *Cell* **147:** 436–446.

Chevrier N, Mertins P, Artyomov MN, Shalek AK, Iannacone M, Ciaccio MF, Gat-Viks I, Tonti E, DeGrace MM, Clauser KR, et al. 2011. Systematic discovery of TLR signaling components delineates viral-sensing circuits. *Cell* **147:** 853–867.

Chiu YH, Macmillan JB, Chen ZJ. 2009. RNA polymerase III detects cytosolic DNA and induces type I interferons through the RIG-I pathway. *Cell* **138:** 576–591.

Choe J, Kelker MS, Wilson IA. 2005. Crystal structure of human toll-like receptor 3 (TLR3) ectodomain. *Science* **309:** 581–585.

Civril F, Deimling T, de Oliveira Mann CC, Ablasser A, Moldt M, Witte G, Hornung V, Hopfner KP. 2013. Structural mechanism of cytosolic DNA sensing by cGAS. *Nature* **498:** 332–337.

Coban C, Igari Y, Yagi M, Reimer T, Koyama S, Aoshi T, Ohata K, Tsukui T, Takeshita F, Sakurai K, et al. 2010. Immunogenicity of whole-parasite vaccines against Plasmodium falciparum involves malarial hemozoin and host TLR9. *Cell Hos Microbe* **7:** 50–61.

Davies BW, Bogard RW, Young TS, Mekalanos JJ. 2012. Coordinated regulation of accessory genetic elements produces cyclic di-nucleotides for *V. cholerae* virulence. *Cell* **149:** 358–370.

Diner EJ, Burdette DL, Wilson SC, Monroe KM, Kellenberger CA, Hyodo M, Hayakawa Y, Hammond MC, Vance RE. 2013. The innate immune DNA sensor cGAS produces a noncanonical cyclic dinucleotide that activates human STING. *Cell Rep* **3:** 1355–1361.

Dixit E, Boulant S, Zhang Y, Lee AS, Odendall C, Shum B, Hacohen N, Chen ZJ, Whelan SP, Fransen M, et al. 2010. Peroxisomes are signaling platforms for antiviral innate immunity. *Cell* **141:** 668–681.

Doyle SL, Husebye H, Connolly DJ, Espevik T, O'Neill LA, McGettrick AF. 2012. The GOLD domain-containing protein TMED7 inhibits TLR4 signalling from the endosome upon LPS stimulation. *Nat Commun* **3:** 707.

Eigenbrod T, Franchi L, Munoz-Planillo R, Kirschning CJ, Freudenberg MA, Nunez G, Dalpke A. 2012. Bacterial RNA mediates activation of caspase-1 and IL-1β release independently of TLRs 3, 7, 9 and TRIF but is dependent on UNC93B. *J Immunol* **189:** 328–336.

Ewald SE, Lee BL, Lau L, Wickliffe KE, Shi GP, Chapman HA, Barton GM. 2008. The ectodomain of Toll-like receptor 9 is cleaved to generate a functional receptor. *Nature* **456:** 658–662.

Ferguson BJ, Mansur DS, Peters NE, Ren H, Smith GL. 2012. DNA-PK is a DNA sensor for IRF-3-dependent innate immunity. *eLife* **1:** e00047.

Fernandes-Alnemri T, Yu JW, Juliana C, Solorzano L, Kang S, Wu J, Datta P, McCormick M, Huang L, McDermott E, et al. 2010. The AIM2 inflammasome is critical for innate immunity to *Francisella tularensis. Nat Immunol* **11:** 385–393.

Fredericksen BL, Keller BC, Fornek J, Katze MG, Gale M Jr. 2008. Establishment and maintenance of the innate antiviral response to West Nile Virus involves both RIG-I and MDA5 signaling through IPS-1. *J Virol* **82:** 609–616.

Friedman CS, O'Donnell MA, Legarda-Addison D, Ng A, Cardenas WB, Yount JS, Moran TM, Basler CF, Komuro A, Horvath CM, et al. 2008. The tumour suppressor CYLD is a negative regulator of RIG-I-mediated antiviral response. *EMBO Rep* **9:** 930–936.

Fukui R, Saitoh S, Kanno A, Onji M, Shibata T, Ito A, Matsumoto M, Akira S, Yoshida N, Miyake K. 2011. Unc93B1 restricts systemic lethal inflammation by orchestrating Toll-like receptor 7 and 9 trafficking. *Immunity* **35:** 69–81.

Gack MU, Shin YC, Joo CH, Urano T, Liang C, Sun L, Takeuchi O, Akira S, Chen Z, Inoue S, et al. 2007. TRIM25 RING-finger E3 ubiquitin ligase is essential for RIG-I-mediated antiviral activity. *Nature* **446:** 916–920.

Gall A, Treuting P, Elkon KB, Loo YM, Gale M Jr, Barber GN, Stetson DB. 2012. Autoimmunity initiates in nonhematopoietic cells and progresses via lymphocytes in an interferon-dependent autoimmune disease. *Immunity* **36:** 120–131.

Gao P, Ascano M, Wu Y, Barchet W, Gaffney BL, Zillinger T, Serganov AA, Liu Y, Jones RA, Hartmann G, et al. 2013. Cyclic [G(2′,5′)pA(3′,5′)p] is the metazoan second messenger produced by DNA-activated cyclic GMP-AMP synthase. *Cell* **153:** 1094–1107.

Goutagny N, Jiang Z, Tian J, Parroche P, Schickli J, Monks BG, Ulbrandt N, Ji H, Kiener PA, Coyle AJ, et al. 2010. Cell type-specific recognition of human metapneumoviruses (HMPVs) by retinoic acid-inducible gene I (RIG-I) and TLR7 and viral interference of RIG-I ligand recognition by HMPV-B1 phosphoprotein. *J Immunol* **184:** 1168–1179.

Guan Y, Ranoa DR, Jiang S, Mutha SK, Li X, Baudry J, Tapping RI. 2010. Human TLRs 10 and 1 share common mechanisms of innate immune sensing but not signaling. *J Immunol* **184:** 5094–5103.

Gurtler C, Bowie AG. 2013. Innate immune detection of microbial nucleic acids. *Trends Microbiol* **21:** 413–420.

Haas T, Metzger J, Schmitz F, Heit A, Muller T, Latz E, Wagner H. 2008. The DNA sugar backbone 2′ deoxyribose determines toll-like receptor 9 activation. *Immunity* **28:** 315–323.

Hacker H, Redecke V, Blagoev B, Kratchmarova I, Hsu LC, Wang GG, Kamps MP, Raz E, Wagner H, Hacker G, et al. 2006. Specificity in Toll-like receptor signalling through distinct effector functions of TRAF3 and TRAF6. *Nature* **439:** 204–207.

Hagmann CA, Herzner AM, Abdullah Z, Zillinger T, Jakobs C, Schuberth C, Coch C, Higgins PG, Wisplinghoff H, Barchet W, et al. 2013. RIG-I detects triphosphorylated RNA of Listeria monocytogenes during infection in non-immune cells. *PLoS ONE* **8:** e62872.

Hidmark A, von Saint Paul A, Dalpke AH. 2012. Cutting edge: TLR13 is a receptor for bacterial RNA. *J Immunol* **189:** 2717–2721.

Hoebe K, Georgel P, Rutschmann S, Du X, Mudd S, Crozat K, Sovath S, Shamel L, Hartung T, Zahringer U, et al. 2005. CD36 is a sensor of diacylglycerides. *Nature* **433:** 523–527.

Hornung V, Ellegast J, Kim S, Brzozka K, Jung A, Kato H, Poeck H, Akira S, Conzelmann KK, Schlee M, et al. 2006. 5′-Triphosphate RNA is the ligand for RIG-I. *Science* **314:** 994–997.

Hou F, Sun L, Zheng H, Skaug B, Jiang QX, Chen ZJ. 2011. MAVS forms functional prion-like aggregates to activate and propagate antiviral innate immune response. *Cell* **146:** 448–461.

Ishii KJ, Coban C, Kato H, Takahashi K, Torii Y, Takeshita F, Ludwig H, Sutter G, Suzuki K, Hemmi H, et al. 2006. A Toll-like receptor-independent antiviral response induced by double-stranded B-form DNA. *Nat Immunol* **7:** 40–48.

Ishii KJ, Kawagoe T, Koyama S, Matsui K, Kumar H, Kawai T, Uematsu S, Takeuchi O, Takeshita F, Coban C, et al. 2008. TANK-binding kinase-1 delineates innate and adaptive immune responses to DNA vaccines. *Nature* **451:** 725–729.

Ishikawa H, Barber GN. 2008. STING is an endoplasmic reticulum adaptor that facilitates innate immune signalling. *Nature* **455:** 674–678.

Ishikawa H, Ma Z, Barber GN. 2009. STING regulates intracellular DNA-mediated, type I interferon-dependent innate immunity. *Nature* **461:** 788–792.

Janeway CA Jr, Medzhitov R. 2002. Innate immune recognition. *Ann Rev Immunol* **20:** 197–216.

Jimenez-Dalmaroni MJ, Xiao N, Corper AL, Verdino P, Ainge GD, Larsen DS, Painter GF, Rudd PM, Dwek RA, Hoebe K, et al. 2009. Soluble CD36 ectodomain binds negatively charged diacylglycerol ligands and acts as a co-receptor for TLR2. *PLoS ONE* **4:** e7411.

Jin MS, Kim SE, Heo JY, Lee ME, Kim HM, Paik SG, Lee H, Lee JO. 2007. Crystal structure of the TLR1-TLR2 heterodimer induced by binding of a tri-acylated lipopeptide. *Cell* **130:** 1071–1082.

Kagan JC, Su T, Horng T, Chow A, Akira S, Medzhitov R. 2008. TRAM couples endocytosis of Toll-like receptor 4 to the induction of interferon-β. *Nat Immunol* **9:** 361–368.

Kanneganti TD, Body-Malapel M, Amer A, Park JH, Whitfield J, Franchi L, Taraporewala ZF, Miller D, Patton JT, Inohara N, et al. 2006. Critical role for Cryopyrin/Nalp3 in activation of caspase-1 in response to viral infection and double-stranded RNA. *J Biol Chem* **281:** 36560–36568.

Kato H, Takeuchi O, Sato S, Yoneyama M, Yamamoto M, Matsui K, Uematsu S, Jung A, Kawai T, Ishii KJ, et al. 2006. Differential roles of MDA5 and RIG-I helicases in the recognition of RNA viruses. *Nature* **441:** 101–105.

Kawai T, Akira S. 2006. Innate immune recognition of viral infection. *Nat Immunol* **7:** 131–137.

Kawai T, Akira S. 2010. The role of pattern-recognition receptors in innate immunity: Update on Toll-like receptors. *Nat Immunol* **11:** 373–384.

Kerur N, Veettil MV, Sharma-Walia N, Bottero V, Sadagopan S, Otageri P, Chandran B. 2011. IFI16 acts as a nuclear pathogen sensor to induce the inflammasome in response to Kaposi Sarcoma-associated herpesvirus infection. *Cell Host Microbe* **9:** 363–375.

Kim YM, Brinkmann MM, Paquet ME, Ploegh HL. 2008. UNC93B1 delivers nucleotide-sensing Toll-like receptors to endolysosomes. *Nature* **452:** 234–238.

Kim T, Pazhoor S, Bao M, Zhang Z, Hanabuchi S, Facchinetti V, Bover L, Plumas J, Chaperot L, Qin J, et al. 2010. Aspartate-glutamate-alanine-histidine box motif (DEAH)/RNA helicase A helicases sense microbial DNA in human plasmacytoid dendritic cells. *Proc Natl Acad Sci* **107:** 15181–15186.

Kobiyama K, Takeshita F, Jounai N, Sakaue-Sawano A, Miyawaki A, Ishii KJ, Kawai T, Sasaki S, Hirano H, Ishii N, et al. 2010. Extrachromosomal histone H2B mediates innate antiviral immune responses induced by intracellular double-stranded DNA. *J Virol* **84:** 822–832.

Koblansky AA, Jankovic D, Oh H, Hieny S, Sungnak W, Mathur R, Hayden MS, Akira S, Sher A, Ghosh S. 2013. Recognition of profilin by Toll-like receptor 12 is critical for host resistance to *Toxoplasma gondii*. *Immunity* **38:** 119–130.

Kondo T, Kawai T, Akira S. 2012. Dissecting negative regulation of Toll-like receptor signaling. *Trends Immunol* **33:** 449–458.

Kondo T, Kobayashi J, Saitoh T, Maruyama K, Ishii KJ, Barber GN, Komatsu K, Akira S, Kawai T. 2013. DNA damage sensor MRE11 recognizes cytosolic double-stranded DNA and induces type I interferon by regulating STING trafficking. *Proc Natl Acad Sci* **110:** 2969–2974.

Konno H, Konno K, Barber GN. 2013. Cyclic dinucleotides trigger ULK1 (ATG1) phosphorylation of STING to prevent sustained innate immune signaling. *Cell* **155:** 688–698.

Kurt-Jones EA, Popova L, Kwinn L, Haynes LM, Jones LP, Tripp RA, Walsh EE, Freeman MW, Golenbock DT, Anderson LJ, et al. 2000. Pattern recognition receptors TLR4 and CD14 mediate response to respiratory syncytial virus. *Nat Immunol* **1:** 398–401.

Lee HK, Lund JM, Ramanathan B, Mizushima N, Iwasaki A. 2007. Autophagy-dependent viral recognition by plasmacytoid dendritic cells. *Science* **315:** 1398–1401.

Lee KG, Xu S, Kang ZH, Huo J, Huang M, Liu D, Takeuchi O, Akira S, Lam KP. 2012. Bruton's tyrosine kinase phosphorylates Toll-like receptor 3 to initiate antiviral response. *Proc Natl Acad Sci* **109:** 5791–5796.

Li XD, Chen ZJ. 2012. Sequence specific detection of bacterial 23S ribosomal RNA by TLR13. *eLife* **1:** e00102.

Li XD, Wu J, Gao D, Wang H, Sun L, Chen ZJ. 2013. Pivotal roles of cGAS-cGAMP signaling in antiviral defense and immune adjuvant effects. *Science* **341:** 1390–1394.

Lin SC, Lo YC, Wu H. 2010. Helical assembly in the MyD88-IRAK4-IRAK2 complex in TLR/IL-1R signalling. *Nature* **465:** 885–890.

Liu L, Botos I, Wang Y, Leonard JN, Shiloach J, Segal DM, Davies DR. 2008. Structural basis of toll-like receptor 3 signaling with double-stranded RNA. *Science* **320:** 379–381.

Liu X, Zhan Z, Li D, Xu L, Ma F, Zhang P, Yao H, Cao X. 2011. Intracellular MHC class II molecules promote TLR-triggered innate immune responses by maintaining activation of the kinase Btk. *Nat Immunol* **12:** 416–424.

Liu X, Chen W, Wang Q, Li L, Wang C. 2013. Negative regulation of TLR inflammatory signaling by the SUMO-deconjugating enzyme SENP6. *PLoS Pathog* **9:** e1003480.

Loo YM, Gale M Jr. 2011. Immune signaling by RIG-I-like receptors. *Immunity* **34:** 680–692.

Loo YM, Fornek J, Crochet N, Bajwa G, Perwitasari O, Martinez-Sobrido L, Akira S, Gill MA, Garcia-Sastre A, Katze MG, et al. 2008. Distinct RIG-I and MDA5 signaling by RNA viruses in innate immunity. *J Virol* **82:** 335–345.

Mancuso G, Gambuzza M, Midiri A, Biondo C, Papasergi S, Akira S, Teti G, Beninati C. 2009. Bacterial recognition by TLR7 in the lysosomes of conventional dendritic cells. *Nat Immunol* **10:** 587–594.

Mathur R, Oh H, Zhang D, Park SG, Seo J, Koblansky A, Hayden MS, Ghosh S. 2012. A mouse model of *Salmonella typhi* infection. *Cell* **151:** 590–602.

Matsumoto F, Saitoh S, Fukui R, Kobayashi T, Tanimura N, Konno K, Kusumoto Y, Akashi-Takamura S, Miyake K. 2008. Cathepsins are required for Toll-like receptor 9 responses. *Biochem Biophys Res Commun* **367:** 693–699.

McCartney SA, Thackray LB, Gitlin L, Gilfillan S, Virgin HW, Colonna M. 2008. MDA-5 recognition of a murine norovirus. *PLoS Pathog* **4:** e1000108.

Melchjorsen J, Jensen SB, Malmgaard L, Rasmussen SB, Weber F, Bowie AG, Matikainen S, Paludan SR. 2005. Activation of innate defense against a paramyxovirus is mediated by RIG-I and TLR7 and TLR8 in a cell-type-specific manner. *J Virol* **79:** 12944–12951.

Mitoma H, Hanabuchi S, Kim T, Bao M, Zhang Z, Sugimoto N, Liu YJ. 2013. The DHX33 RNA helicase senses cytosolic RNA and activates the NLRP3 inflammasome. *Immunity* **39:** 123–135.

Oldenburg M, Kruger A, Ferstl R, Kaufmann A, Nees G, Sigmund A, Bathke B, Lauterbach H, Suter M, Dreher S, et al. 2012. TLR13 recognizes bacterial 23S rRNA devoid of erythromycin resistance-forming modification. *Science* **337:** 1111–1115.

Oliveira-Nascimento L, Massari P, Wetzler LM. 2012. The role of TLR2 in infection and immunity. *Frontiers Immunol* **3:** 79.

Olivieri F, Rippo MR, Prattichizzo F, Babini L, Graciotti L, Recchioni R, Procopio AD. 2013. Toll like receptor signaling in "inflammaging": MicroRNA as new players. *Immun Ageing* **10:** 11.

Onji M, Kanno A, Saitoh S, Fukui R, Motoi Y, Shibata T, Matsumoto F, Lamichhane A, Sato S, Kiyono H, et al. 2013. An essential role for the N-terminal fragment of Toll-like receptor 9 in DNA sensing. *Nat Commun* **4:** 1949.

Paludan SR, Bowie AG. 2013. Immune sensing of DNA. *Immunity* **38:** 870–880.

Park B, Brinkmann MM, Spooner E, Lee CC, Kim YM, Ploegh HL. 2008. Proteolytic cleavage in an endolysosomal compartment is required for activation of Toll-like receptor 9. *Nat Immunol* **9:** 1407–1414.

Park BS, Song DH, Kim HM, Choi BS, Lee H, Lee JO. 2009. The structural basis of lipopolysaccharide recognition by the TLR4-MD-2 complex. *Nature* **458:** 1191–1195.

Peter ME, Kubarenko AV, Weber AN, Dalpke AH. 2009. Identification of an N-terminal recognition site in TLR9 that contributes to CpG-DNA-mediated receptor activation. *J Immunol* **182:** 7690–7697.

Phipps S, Lam CE, Mahalingam S, Newhouse M, Ramirez R, Rosenberg HF, Foster PS, Matthaei KI. 2007. Eosinophils contribute to innate antiviral immunity and promote clearance of respiratory syncytial virus. *Blood* **110:** 1578–1586.

Pichlmair A, Schulz O, Tan CP, Naslund TI, Liljestrom P, Weber F, Reis e Sousa C. 2006. RIG-I-mediated antiviral responses to single-stranded RNA bearing 5′-phosphates. *Science* **314:** 997–1001.

Pichlmair A, Schulz O, Tan CP, Rehwinkel J, Kato H, Takeuchi O, Akira S, Way M, Schiavo G, Reis e Sousa C. 2009. Activation of MDA5 requires higher-order RNA structures generated during virus infection. *J Virol* **83:** 10761–10769.

Qian C, Cao X. 2013. Regulation of Toll-like receptor signaling pathways in innate immune responses. *Ann NY Acad Sci* **1283:** 67–74.

Rajan JV, Warren SE, Miao EA, Aderem A. 2010. Activation of the NLRP3 inflammasome by intracellular poly I:C. *FEBS Lett* **584:** 4627–4632.

Saito T, Hirai R, Loo YM, Owen D, Johnson CL, Sinha SC, Akira S, Fujita T, Gale M Jr. 2007. Regulation of innate antiviral defenses through a shared repressor domain in RIG-I and LGP2. *Proc Natl Acad Sci* **104:** 582–587.

Saitoh T, Fujita N, Yoshimori T, Akira S. 2010. Regulation of dsDNA-induced innate immune responses by membrane trafficking. *Autophagy* **6:** 430–432.

Satoh T, Kato H, Kumagai Y, Yoneyama M, Sato S, Matsushita K, Tsujimura T, Fujita T, Akira S, Takeuchi O. 2010. LGP2 is a positive regulator of RIG-I- and MDA5-mediated antiviral responses. *Proc Natl Acad Sci* **107:** 1512–1517.

Saitoh T, Satoh T, Yamamoto N, Uematsu S, Takeuchi O, Kawai T, Akira S. 2011. Antiviral protein viperin promotes Toll-like receptor 7- and Toll-like receptor 9-mediated type I interferon production in plasmacytoid dendritic cells. *Immunity* **34:** 352–363.

Schamber-Reis BL, Petritus PM, Caetano BC, Martinez ER, Okuda K, Golenbock D, Scott P, Gazzinelli RT. 2013. UNC93B1 and nucleic acid-sensing Toll-like receptors mediate host resistance to infection with *Leishmania major*. *J Biol Chem* **288:** 7127–7136.

Schwandner R, Dziarski R, Wesche H, Rothe M, Kirschning CJ. 1999. Peptidoglycan- and lipoteichoic acid-induced cell activation is mediated by toll-like receptor 2. *J Biol Chem* **274:** 17406–17409.

Scott I. 2010. The role of mitochondria in the mammalian antiviral defense system. *Mitochondrion* **10:** 316–320.

Sepulveda FE, Maschalidi S, Colisson R, Heslop L, Ghirelli C, Sakka E, Lennon-Dumenil AM, Amigorena S, Cabanie L, Manoury B. 2009. Critical role for asparagine endopeptidase in endocytic Toll-like receptor signaling in dendritic cells. *Immunity* **31:** 737–748.

Shi HX, Liu X, Wang Q, Tang PP, Liu XY, Shan YF, Wang C. 2011. Mitochondrial ubiquitin ligase MARCH5 promotes TLR7 signaling by attenuating TANK action. *PLoS Pathog* **7:** e1002057.

Stewart CR, Stuart LM, Wilkinson K, van Gils JM, Deng J, Halle A, Rayner KJ, Boyer L, Zhong R, Frazier WA, et al. 2010. CD36 ligands promote sterile inflammation through assembly of a Toll-like receptor 4 and 6 heterodimer. *Nat Immunol* **11:** 155–161.

Sun L, Wu J, Du F, Chen X, Chen ZJ. 2013. Cyclic GMP-AMP synthase is a cytosolic DNA sensor that activates the type I interferon pathway. *Science* **339:** 786–791.

Tabeta K, Hoebe K, Janssen EM, Du X, Georgel P, Crozat K, Mudd S, Mann N, Sovath S, Goode J, et al. 2006. The *Unc93b1* mutation 3d disrupts exogenous antigen presentation and signaling via Toll-like receptors 3, 7 and 9. *Nat Immunol* **7:** 156–164.

Takaoka A, Wang Z, Choi MK, Yanai H, Negishi H, Ban T, Lu Y, Miyagishi M, Kodama T, Honda K, et al. 2007. DAI (DLM-1/ZBP1) is a cytosolic DNA sensor and an activator of innate immune response. *Nature* **448:** 501–505.

Takeuchi O, Akira S. 2010. Pattern recognition receptors and inflammation. *Cell* **140:** 805–820.

Tanaka Y, Chen ZJ. 2012. STING specifies IRF3 phosphorylation by TBK1 in the cytosolic DNA signaling pathway. *Sci Signal* **5:** ra20.

Tang D, Kang R, Coyne CB, Zeh HJ, Lotze MT. 2012. PAMPs and DAMPs: Signal 0s that spur autophagy and immunity. *Immunol Rev* **249:** 158–175.

Tanji H, Ohto U, Shibata T, Miyake K, Shimizu T. 2013. Structural reorganization of the Toll-like receptor 8 dimer induced by agonistic ligands. *Science* **339:** 1426–1429.

Thompson MR, Kaminski JJ, Kurt-Jones EA, Fitzgerald KA. 2011. Pattern recognition receptors and the innate immune response to viral infection. *Viruses* **3:** 920–940.

Tsuchida T, Zou J, Saitoh T, Kumar H, Abe T, Matsuura Y, Kawai T, Akira S. 2010. The ubiquitin ligase TRIM56 regulates innate immune responses to intracellular double-stranded DNA. *Immunity* **33:** 765–776.

Uematsu S, Fujimoto K, Jang MH, Yang BG, Jung YJ, Nishiyama M, Sato S, Tsujimura T, Yamamoto M, Yokota Y, et al. 2008. Regulation of humoral and cellular gut immunity by lamina propria dendritic cells expressing Toll-like receptor 5. *Nat Immunol* **9:** 769–776.

Underhill DM, Ozinsky A, Smith KD, Aderem A. 1999. Toll-like receptor-2 mediates mycobacteria-induced proinflammatory signaling in macrophages. *Proc Natl Acad Sci* **96:** 14459–14463.

Unterholzner L, Keating SE, Baran M, Horan KA, Jensen SB, Sharma S, Sirois CM, Jin T, Latz E, Xiao TS, et al. 2010. IFI16 is an innate immune sensor for intracellular DNA. *Nat Immunol* **11:** 997–1004.

Wagner H. 2008. The sweetness of the DNA backbone drives Toll-like receptor 9. *Curr Opin Immunol* **20:** 396–400.

Wang JP, Asher DR, Chan M, Kurt-Jones EA, Finberg RW. 2007. Cutting Edge: Antibody-mediated TLR7-dependent recognition of viral RNA. *J Immunol* **178:** 3363–3367.

Wang C, Chen T, Zhang J, Yang M, Li N, Xu X, Cao X. 2009. The E3 ubiquitin ligase Nrdp1 "preferentially" promotes TLR-mediated production of type I interferon. *Nat Immunol* **10:** 744–752.

Wang L, Zhao W, Zhang M, Wang P, Zhao K, Zhao X, Yang S, Gao C. 2013. USP4 positively regulates RIG-I-mediated antiviral response through deubiquitination and stabilization of RIG-I. *J Virol* **87:** 4507–4515.

Wu J, Sun L, Chen X, Du F, Shi H, Chen C, Chen ZJ. 2013. Cyclic GMP-AMP is an endogenous second messenger in innate immune signaling by cytosolic DNA. *Science* **339:** 826–830.

Xiao TS, Fitzgerald KA. 2013. The cGAS-STING pathway for DNA sensing. *Mol Cell* **51:** 135–139.

Yamashita M, Chattopadhyay S, Fensterl V, Saikia P, Wetzel JL, Sen GC. 2012. Epidermal growth factor receptor is essential for Toll-like receptor 3 signaling. *Sci Signal* **5:** ra50.

Yang P, An H, Liu X, Wen M, Zheng Y, Rui Y, Cao X. 2010. The cytosolic nucleic acid sensor LRRFIP1 mediates the production of type I interferon via a β-catenin-dependent pathway. *Nat Immunol* **11:** 487–494.

Yang M, Wang C, Zhu X, Tang S, Shi L, Cao X, Chen T. 2011. E3 ubiquitin ligase CHIP facilitates Toll-like receptor signaling by recruiting and polyubiquitinating Src and atypical PKCζ. *J Exp Med* **208:** 2099–2112.

Yarovinsky F, Zhang D, Andersen JF, Bannenberg GL, Serhan CN, Hayden MS, Hieny S, Sutterwala FS, Flavell RA, Ghosh S, et al. 2005. TLR11 activation of dendritic cells by a protozoan profilin-like protein. *Science* **308:** 1626–1629.

Yoneyama M, Fujita T. 2008. Structural mechanism of RNA recognition by the RIG-I-like receptors. *Immunity* **29:** 178–181.

Yoneyama M, Kikuchi M, Matsumoto K, Imaizumi T, Miyagishi M, Taira K, Foy E, Loo YM, Gale M Jr, Akira S, et al. 2005. Shared and unique functions of the DExD/H-box helicases RIG-I, MDA5, and LGP2 in antiviral innate immunity. *J Immunol* **175:** 2851–2858.

Zanoni I, Ostuni R, Marek LR, Barresi S, Barbalat R, Barton GM, Granucci F, Kagan JC. 2011. CD14 controls the LPS-induced endocytosis of Toll-like receptor 4. *Cell* **147:** 868–880.

Zhang D, Zhang G, Hayden MS, Greenblatt MB, Bussey C, Flavell RA, Ghosh S. 2004. A toll-like receptor that prevents infection by uropathogenic bacteria. *Science* **303:** 1522–1526.

Zhang SY, Jouanguy E, Ugolini S, Smahi A, Elain G, Romero P, Segal D, Sancho-Shimizu V, Lorenzo L, Puel A, et al. 2007. TLR3 deficiency in patients with herpes simplex encephalitis. *Science* **317:** 1522–1527.

Zhang X, Brann TW, Zhou M, Yang J, Oguariri RM, Lidie KB, Imamichi H, Huang DW, Lempicki RA, Baseler MW, et al. 2011. Cutting edge: Ku70 is a novel cytosolic DNA sensor that induces type III rather than type I IFN. *J Immunol* **186:** 4541–4545.

Zhang X, Shi H, Wu J, Sun L, Chen C, Chen ZJ. 2013. Cyclic GMP-AMP containing mixed phosphodiester linkages is an endogenous high-affinity ligand for STING. *Mol Cell* **51:** 226–235.

Zhong B, Zhang L, Lei C, Li Y, Mao AP, Yang Y, Wang YY, Zhang XL, Shu HB. 2009. The ubiquitin ligase RNF5 regulates antiviral responses by mediating degradation of the adaptor protein MITA. *Immunity* **30:** 397–407.

Emerging Principles Governing Signal Transduction by Pattern-Recognition Receptors

Jonathan C. Kagan[1] and Gregory M. Barton[2]

[1]Harvard Medical School and Division of Gastroenterology, Boston Children's Hospital, Boston, Massachusetts 02115

[2]Department of Molecular & Cell Biology, University of California, Berkeley, Berkeley, California 94720-3200

Correspondence: Jonathan.Kagan@childrens.harvard.edu; barton@berkeley.edu

The problem of recognizing and disposing of non-self-organisms, whether for nutrients or defense, predates the evolution of multicellularity. Accordingly, the function of the innate immune system is often intimately associated with fundamental aspects of cell biology. Here, we review our current understanding of the links between cell biology and pattern-recognition receptors of the innate immune system. We highlight the importance of receptor localization for the detection of microbes and for the initiation of antimicrobial signaling pathways. We discuss examples that illustrate how pattern-recognition receptors influence, and are influenced by, the general membrane trafficking machinery of mammalian cells. In the future, cell biological analysis likely will rival pure genetic analysis as a tool to uncover fundamental principles that govern host–microbe interactions.

The innate immune system uses families of pattern-recognition receptors (PRRs) to recognize diverse microbial ligands (Janeway 1989; Janeway and Medzhitov 2002). During infection, these receptors provide signals that up-regulate general antimicrobial features of the innate immune system as well as instruct and initiate adaptive immunity (Iwasaki and Medzhitov 2010). A significant challenge faced by innate immune recognition is the reliable detection of highly diverse, rapidly evolving microbial organisms, many of which possess virulence mechanisms that enable survival within distinct host niches. Moreover, recognition must be linked to induction of contextual signals appropriate for the type of infection. The specificity, signal transduction, and cell biology of PRRs have evolved under these selective pressures to enable broad recognition of microbes within each host niche.

Although the collection of PRRs is decidedly less diverse than antigen receptors of the adaptive immune system, the list of players has grown considerably over the past decade (Kawai and Akira 2010). If one classifies these receptors based on common structure and functional domains, then six families emerge: Toll-like receptors (TLRs), C-type lectin receptors (CLRs), RIG-I-like receptors (RLRs), AIM-like receptors (ALRs), Nod-like receptors (NLRs), and OAS-like receptors (OLRs) (Geijtenbeek and Gringhuis 2009; Kawai and Akira 2010; Rathi-

Cite this article as *Cold Spring Harb Perspect Biol* doi: 10.1101/cshperspect.a016253

nam and Fitzgerald 2011; Lamkanfi and Dixit 2012; Kranzusch et al. 2013). Collectively, these receptors bind a diverse array of targets, including lipoproteins, polysaccharides, nucleic acids, carbohydrate structures, and a few highly conserved microbial proteins. These ligands are typically shared across large microbial classes, which facilitate broad recognition with such a limited number of PRRs. Moreover, alteration or masking of these ligands to avoid PRR activation often results in reduced microbial fitness.

The molecular recognition challenge faced by PRRs is all the more complex when one considers the need to detect microbes within distinct subcellular niches. Microbes can be extracellular or intracellular within membrane-bound organelles, within the cytosol, or in the nucleus. In addition, both the innate and adaptive immune mechanisms appropriate for eliminating microbes within these distinct environments are quite distinct, so it is vital that PRR signaling communicate the location of a microbe as well its nature. We now understand that members of the PRR families highlighted above localize to distinct subcellular compartments, and, in some cases, localization can change in a dynamic fashion that regulates or influences recognition and signaling. Moreover, in some cases, signal transduction and resulting gene induction can be dramatically influenced by the organelle from which signaling initiates. Thus, the innate immune system has harnessed the organization inherent to cells as a means of achieving regulation and signaling specificity. Activation of PRRs can also feed back on basic cell biological processes, such as phagocytosis and autophagy, to enhance or accelerate the response to microbial infection.

In the following sections, we discuss these links between cell biology and PRRs of mammalian innate immunity. Our discussions of PRR function and signal transduction will be limited to this theme, as a result, in part, of space constraints but also because in-depth reviews of each PRR family have appeared elsewhere. For discussion purposes, we have grouped the transmembrane PRRs together and the cytosolic PRRs together.

TRANSMEMBRANE PATTERN-RECOGNITION RECEPTORS

Basic Features of TLRs and CLRs

The 13 mammalian TLRs (10 in humans, 12 in mice) consist of leucine-rich repeat containing ectodomains and cytosolic Toll-IL-1 receptor (TIR) domains (Kawai and Akira 2010). CLRs are a much larger family, whose members share a carbohydrate recognition domain (CRD) (Sancho and Reis e Sousa 2012). In many proteins, this domain mediates carbohydrate binding but it is now clear that not all CRD-containing proteins bind sugars. Here, we will consider only the CLRs that bind carbohydrates and function clearly as PRRs, such as the dectins. As such, the emphasis in this section will mostly be on TLRs, but we will highlight newer findings especially as they relate to links between CLR function and cell biology.

Both TLRs and CLRs can link the recognition of diverse microbes to signaling pathways that promote both innate and adaptive immunity. TLRs induce a core signaling pathway via recruitment of adaptors with TIR domains: MyD88, Trif, Tram, and Tirap (Kawai and Akira 2010). Distinct signal transduction and gene induction can arise from the differential use of adaptors or selective recruitment of specific signaling components, and examples of such specialization will be discussed later in this article. The ultimate outcome of this signal transduction is the activation of NF-κB and interferon regulatory factor (IRF) transcription factors and induction of genes that promote the subsequent immune response (Kawai and Akira 2010). CLRs are more varied in their signal transduction, but we will focus on the dectin family. Dectin-1, -2, and -3 all activate the tyrosine kinase Syk and, subsequently, the transcription factors NF-κB and nuclear factor of activated T cells (Sancho and Reis e Sousa 2012). Although there are certainly differences in gene induction by CLRs and TLRs, conceptually the outcome of their activation is similar.

TLRs and CLRs are expressed by many types of immune cells (Kawai and Akira 2010; Sancho and Reis e Sousa 2012). The CLRs discussed in this review primarily regulate phagocytic cells, such as macrophages, dendritic cells (DCs), and neutrophils, whereas TLRs are expressed on these cell types as well as lymphocytes and some nonhematopoietic cell types. Expression of individual TLRs can vary between these cell types, and certain cell types only express a few family members. This restricted expression is one mechanism by which distinct responses can be generated to specific microbes. For example, plasmacytoid DCs selectively express TLR7 and TLR9, which, because of the features of these specialized cells, enables these TLRs to induce substantial production of type I interferons (IFNs) (Kadowaki et al. 2001). For the purposes of this review, it is also important to note that most studies of TLR and CLR cell biology have used macrophages. It is quite possible that other specialized cell types possess distinct cell biological features that may impact the function of these receptors.

Individual TLR family members traffic to distinct subcellular locations and, in some cases, receptor localization changes on activation. In general, the TLRs specific for ligands associated with the exterior surfaces of microbes (Table 1), such as bacterial lipopolysaccharides and lipoproteins, are found in the plasma membrane of unstimulated cells (Barton and Kagan 2009). This subset of TLRs includes TLR4 and TLR5 and the heterodimers TLR1/TLR2, TLR2/TLR6, and TLR2/TLR10 (TLR10 is a pseudogene in mice). A second subset of TLRs resides within endosomes. All of these TLRs (TLR3, TLR7, TLR8, TLR9, and TLR13) share specificity for various forms of nucleic acids (Table 1). As discussed in the next section, endosomal localization is thought to facilitate ligand recognition, as microbes must be degraded before their genetic material can be recognized by a TLR. Endosomal localization has also been reported for TLR11 and TLR12, yet these TLRs recognize the protein ligands flagellin and profilin (Yarovinsky et al. 2005; Pifer et al. 2011; Koblansky et al. 2012; Mathur et al. 2012). Although such localization may facilitate ligand recognition by these TLRs, the evidence supporting their exclusive endosomal localization is fairly limited, and it remains possible that one or both of these receptors also localize to the cell surface. In contrast to the diverse subcellular

Table 1. The TLRs specific for ligands associated with the exterior surfaces of microbes

Receptor	Best defined ligand(s)	Key features
TLR1	Triacylated bacterial lipoproteins	Forms heterodimers with TLR2
TLR2	Di- and triacylated bacterial lipoproteins	Forms heterodimers with TLR1 or TLR6 (TLR10 in humans)
TLR3	Double-stranded RNA	Ectodomain must be cleaved to form an active receptor
TLR4	Bacterial lipopolysaccharide	Recognizes ligands indirectly via MD2
TLR5	Bacterial flagellin	
TLR6	Diacylated bacterial lipoproteins	Forms heterodimers with TLR2
TLR7	Single-stranded RNA	Ectodomain must be cleaved to form an active receptor
TLR8	Single-stranded RNA	Ectodomain must be cleaved to form an active receptor
TLR9	Unmethylated CpG-containing DNA	Ectodomain must be cleaved to form an active receptor
TLR10	Unknown	Pseudogene in mice; present in humans; forms heterodimers with TLR2
TLR11	Toxoplasma profilin/bacterial flagellin	Forms heterodimers with TLR12
TLR12	Toxoplasma profilin	Forms heterodimers with TLR11
TLR13	Bacterial 23s ribosomal RNA	
Dectin-1	Fungal β-glucan	
Dectin-2	Fungal α-mannans	Forms heterodimers with Dectin-3
Dectin-3	Fungal α-mannans	Forms heterodimers with Dectin-2
RIG-I	Short dsRNA with 5′ triphosphate	Detects ligands in the cytosol
MDA5	Long dsRNA	Detects ligands in the cytosol
cGAS	DNA	Detects ligands in the cytosol

sites of TLR residence, it is generally thought that the dectin family of CLRs is located at the cell surface. This localization is likely linked to their function as phagocytic receptors. TLRs also can promote phagocytosis, but this activity is not as robust as that observed for the dectins (see discussion below).

Identification of the trafficking machinery responsible for delivery of TLRs to the correct subcellular location is an area of active research. Correct folding of TLR ectodomains in the endoplasmic reticulum (ER) requires the function of at least two chaperones, gp96 and PRAT4A (Takahashi et al. 2007; Yang et al. 2007b). All TLRs exit the ER and enter the secretory pathway but certain TLRs require a dedicated trafficking chaperone UNC93b1 for this step (Kim et al. 2008). UNC93b1 associates with TLRs in the ER and facilitates their loading into COPII vesicles (Brinkmann et al. 2007; Lee et al. 2013). In cells lacking UNC93b1 function, all of the normally endosomal TLRs fail to exit the ER. UNC93b1 binding to TLRs is at least partially determined by interactions between transmembrane domains and also requires certain TLR juxtamembrane residues (Brinkmann et al. 2007; Kim et al. 2013), but common molecular features that mediate UNC93b1 binding or necessitate UNC93b1 chaperone function remain largely undefined. Why only this subset of TLRs requires UNC93b1 for ER exit is not understood, and it is quite possible that other, as yet unidentified, accessory molecules play a similar role in ER exit for the UNC93b1-independent TLRs. Indeed, the efficiency of ER export can dramatically impact the threshold of receptor activation, so it is likely that this step is subject to complex regulation (Fukui et al. 2009, 2011; Hart and Tapping 2012).

We are only beginning to understand which trafficking factors are necessary for proper post-Golgi sorting of TLRs. Over the past several years, several Rab GTPase family members have been implicated in sorting of TLR4 to subcellular compartments, either at steady state or in response to stimulation (Wang et al. 2007, 2010; Husebye et al. 2010). Similarly, proper trafficking of TLR7 and TLR9 to endosomes requires adaptor protein (AP) sorting complexes (Blasius et al. 2010; Sasai et al. 2010; Lee et al. 2013). These examples will be discussed in greater detail in the following sections.

Recognition of TLR Ligands from/within Distinct Subcellular Compartments

The finding that different TLRs are found in different subcellular locations has prompted studies to identify the significance of differential localization for receptor function. For example, mutant alleles of endosomal TLR9 have been generated that direct this protein to the plasma membrane (Barton et al. 2006; Mouchess et al. 2011). TLR9 at the cell surface can respond to the synthetic ligand CpG DNA (Barton et al. 2006; Mouchess et al. 2011). This result indicates that endosomal localization of TLR9 is not formally required for its signaling functions, although ligand binding may be enhanced at acidic pH (Rutz et al. 2004). Interestingly, cell surface–localized TLR9 was unable to recognize the DNA virus HSV-2, whereas WT TLR9 efficiently detected this virus (Barton et al. 2006). These findings established that the localization of TLR9 to endosomes was not important for ligand binding per se, rather, localization to endosomes was important to ensure that TLR9 detects pathogen-associated DNA. It is therefore likely that two selective pressures resulted in the localization of nucleic acid–sensing TLRs in endosomes. First was the need for rapid detection of viral nucleic acids, which are not likely to be displayed on the surface of any virulent pathogen. Thus, if nucleic acid–sensing TLRs evolved to be located at the cell surface, those cells could not detect their pathogen-associated ligands that are hidden within the virion. The second selective pressure to direct nucleic acid–sensing TLRs into endosomes probably arose from the need to "ignore" self-nucleic acids to prevent autoimmunity. As stated above, nucleic acid–sensing TLRs do not need to be in endosomes to activate signal transduction; they can be localized to the cell surface (at least in the case of TLR9). The ability of these receptors to function from the cell surface may be useful in some situations, as there are reports of cell types that display nucleic acid–sensing TLRs at their plasma membrane (Lee et al. 2006b; Lindau et al. 2013). However, the risk of potentially detecting self-nucleic acids may be so great that the vast majority of cells that express these receptors restrict their localization to endosomes. Evidence in direct support of this potential risk comes from in vivo studies of mice whose hematopoietic cells displayed TLR9 at the cell surface (Mouchess et al. 2011). These mice display systemic autoinflammatory symptoms.

Additional means of preventing access of nucleic acid–sensing TLRs to self-DNA or RNA exist. For example, the ectodomain of most nucleic acid–sensing TLRs must be cleaved to generate a signaling-competent receptor (Ewald et al. 2008, 2011; Park et al. 2008; Garcia-Cattaneo et al. 2012). The requirement for TLR cleavage before they become signaling competent is similar to the regulatory processes that control other potentially harmful endosomal proteins, such as the cathepsins. Both cathepsins and endosomal TLRs are created as pro-proteins that must be cleaved by endosomal proteases to become active. This cleavage event is mediated by several endosomal proteases including cathepsin L, cathepsin S, and asparagine endopeptidase, but the relative importance of each of these proteins may vary between cell types (Ewald et al. 2008, 2011; Park et al. 2008; Sepulveda et al. 2009). For example, cathepsin L (not S) is required for TLR9 cleavage in B cells, whereas both proteases are necessary in macrophages (Avalos et al. 2013). Recent work has implicated furin proteases in processing of the human TLR7 ectodomain, indicating that the proteolytic regulation of individual nucleic acid–sensing TLRs may be different and perhaps occur in distinct compartments (Hipp et al. 2013). Interestingly, the amino-terminal fragment of TLR9 that is separated by cleavage appears to be required for TLR9 signaling, although the mechanism underlying this requirement is unclear (Peter et al. 2009; Onji et al. 2013).

Recent studies that examine the transport route taken by newly synthesized nucleic acid–sensing TLRs suggest the importance of these cleavage events. Newly synthesized TLR9 is first delivered to the plasma membrane via interactions with the aforementioned chaperone Unc93b1 (Lee et al. 2013). The strong endocytosis motif within Unc93b1 then delivers TLR9 to endosomes, where it is cleaved and poised for activation. That TLR9 transits to endosomes via a plasma membrane intermediate highlights the risk of detecting self-nucleic acids. TLR3 also associates with UNC93b1 and traffics to the plasma membrane before internalization into endosomes (Matsumoto et al. 2003; Pohar et al. 2013). Not all nucleic acid–sensing TLRs follow this transport pathway. For example, TLR7 is delivered directly to endosomes, yet it is also cleaved to generate a functional receptor (Lee et al. 2013). Additional work must be performed to explain the importance of the transport routes taken by TLRs, and the cleavage events that control their activation.

The section above describes the efforts taken by the innate immune system to prevent nucleic acid–sensing TLRs from responding to all possible ligands, especially extracellular (i.e., self) DNA. In this section, we will highlight how the opposite approach is taken for TLRs located at the plasma membrane, in which several regulatory events are in place to ensure rapid and highly sensitive detection of bacterial cell-surface components. The best-studied example of a bacterial cell-surface component that activates innate immunity is lipopolysaccharide (LPS), which is found in the outer membrane of Gram-negative bacteria. LPS activates TLR4 to induce the expression of numerous immunomodulatory genes. However, TLR4 forms few direct contacts with LPS (Kim et al. 2007; Park et al. 2009). Thus, several proteins facilitate LPS-induced signal transduction by TLR4. These include LPS-binding protein (LBP), albumin, CD14, and MD2 (Gioannini et al. 2004; Prohinar et al. 2007; Teghanemt et al. 2007, 2008; Resman et al. 2009; Esparza et al. 2012). LBP is capable of binding LPS from the outer membrane of Gram-negative bacteria by a process facilitated by albumin (Gioannini et al. 2002). LPB then transfers LPS to CD14, which, in turn, transfers LPS to MD2. The relative affinities of LBP, CD14, and MD2 for LPS differ, with each successive protein in this cascade displaying a higher binding constant for this bacterial product. This sequential increase in LPS affinity probably facilitates the unidirectional flow of ligand between these LPS-binding proteins. LPS-bound MD2 helps to cross-link TLR4, which is the first step in the initiation of signal transduction. CD14 has also been reported to facilitate ligand binding by TLR2, TLR3, and TLR9 (Henneke et al. 2001; Lee et al. 2006a; Baumann et al. 2010). Interestingly, CD36 and the mannose-binding lectin (MBL) are also important for ligand binding by TLR2 (Hoebe et al. 2005; Ip et al. 2008). The relative roles of CD14, CD36, and MBL in the control of TLR2 signaling remain to be defined. Overall, these observations highlight the differing means by which plasma membrane localized TLRs, and endosomal TLRs are regulated at the level of ligand binding. It appears that numerous regulatory factors are in place to increase the sensitivity of ligand binding by cell-surface TLRs that detect microbial products. In contrast, it appears that the unique regulatory mechanisms of endosomal TLR activation (e.g., the need for ectodomain cleavage) are designed to restrict access to their respective ligands. The benefit of this latter approach is likely to ensure that only microbial nucleic acids are detected, thus limiting the possibility of autoimmunity. In this regard, it is worth noting that, relative to plasma membrane localized TLRs, endosomal TLRs are most commonly implicated in autoimmune syndromes.

Induction of Distinct TLR-Dependent Signaling Pathways from Distinct Subcellular Locations

Although the TLRs appear to be positioned within mammalian cells as a means of regulating ligand binding, recent studies have indicated that they do not, by themselves, define the subcellular sites of innate immune signal transduction. For example, TLR4 can bind to LPS at the plasma membrane, but needs to be transported into detergent-resistant microdomains of the cell surface called lipid rafts to promote MyD88-dependent signal transduction. The transport of TLR4 into rafts is poorly defined, but is dependent on CD14. After signaling via MyD88 at the plasma membrane, TLR4 is then transported into the cell via an endocytosis pathway that is notable for two reasons. First, it is not activated by TLR4, thus establishing that mammalian cells can respond to LPS via TLR4-independent means. TLR4 endocytosis is instead mediated by CD14. On exposure to LPS, CD14 induces the endocytosis of TLR4 and itself by a process dependent on several factors, including the transmembrane adaptors DAP12 and FcεRγ, the tyrosine kinase Syk, and its downstream effector PLC-γ2 (Zanoni et al. 2011; Lin et al. 2013). In addition, reactive oxygen species have been implicated in the regulation of CD14-dependent endocytosis (Chiang et al. 2012). This CD14-dependent pathway represents the first example of a TLR4-independent response to LPS that operates in numerous mammalian cell types, including macrophages, DCs, and fibroblasts.

The second notable feature of this new LPS response pathway is that TLR4 delivery into endosomes is required for its ability to both inactivate MyD88-dependent signaling and promote a second signaling pathway mediated by the TRIF adaptor (Kagan et al. 2008). This TRIF-dependent pathway results in the expression of type I IFNs and helps stabilize the strong transcriptional activity of NF-κB. Thus, in the case of MyD88- and TRIF-dependent pathways, TLR4 must be transported to a region of the cell after ligand binding to promote signal transduction. Evidence exists that TLR4 is also located on Rab11-positive recycling endosomes in resting cells (Husebye et al. 2010). TLR4 can be delivered from recycling endosomes to phagosomes as a means of further enhancing TRIF-mediated signal transduction. The separation of the sites of ligand binding from signal transduction is not unique to TLR4. In fact, TLR2 signaling on inflammatory monocytes also displays this activity. In these cells, TLR2 must be internalized into endosomes before it is able to activate type I IFN expression (Barbalat et al. 2009). Interestingly, within these cells, TLR2 can only induce type I IFNs in response to viral, not bacterial, ligands.

A final example of the dissociation between the sites of ligand binding and signal transduction can be provided by endosomal TLR9. In plasmacytoid DCs, TLR9 recognizes unmethylated CpG from viruses and induces the expression of inflammatory cytokines and type I IFNs. Both of these responses are dependent on MyD88. It appears that TLR9 activates inflammatory cytokine expression and IFN expression from distinct populations of endosomes (Honda et al. 2005). TLR9-induced IFN expression occurs from lysosome-related organelles (LROs) rich for phosphatidylinositol 3,5-bisphosphate PI(3,5)P2 (Sasai et al. 2010). Delivery of TLR9 to this IFN-inducing LRO population depends on the trafficking adaptor complex AP3 (Sasai et al. 2010). Consequently, AP3 mutant cells do not permit TLR9 to induce IFN expression in response to CpG DNA or DNA viruses, and at least one group has reported that MyD88-dependent inflammatory cytokine expression (e.g., IL-12) proceeds normally in these cells (Blasius et al. 2010; Sasai et al. 2010).

Sorting Adaptor Proteins: Factors that Define the Subcellular Sites of TLR Signal Transduction

The section above highlights several examples in the TLR network in which the subcellular site of ligand binding is distinct from the site of signal transduction. These observations raise the question of what defines the subcellular sites of signaling, if not the receptors. This task is accomplished by the actions of a structurally unrelated family of proteins called sorting adaptors (Kagan 2012). Sorting adaptors are defined by several functional criteria. First, these are proteins that act at the receptor proximal level, having the ability to bind directly to ligand-bound PRRs. In the case of the TLR network, the known sorting adaptors are TIRAP and TRAM, both of which act at the receptor proximal level to control signaling (Kagan and Medzhitov 2006; Kagan et al. 2008). Second, sorting adaptors are the only signaling proteins in a given pathway that are present at the site of signaling before any microbial encounter. These proteins are therefore poised to rapidly detect the delivery of a ligand-bound TLR into the region of the cell that is permissive for signal transduction. These proteins can therefore be considered the cytosolic "sensors" of activated TLRs.

TIRAP was the first sorting adaptor defined (Kagan and Medzhitov 2006). This protein contains an amino-terminal lipid-binding domain that interacts with several phosphoinositides (Kagan and Medzhitov 2006). Its ability to bind to PI(4,5)P2 permits localization to the inner leaflet of the plasma membrane (probably in lipid rafts), the site to which CD14 delivers TLR4 (Triantafilou et al. 2002). Once TLR4 enters this region of the plasma membrane, TIRAP can detect the activated receptor and engage MyD88 to promote inflammatory cytokine expression (Kagan and Medzhitov 2006). Concomitant with the initiation of MyD88-dependent signaling from the plasma membrane, CD14 initiates the endocytosis of TLR4 (Zanoni et al. 2011). This process results in the delivery of TLR4 to the second sorting adaptor in the TLR pathway, TRAM. TRAM is localized to both the plasma membrane and endosomes via an amino-terminal bipartite domain that contains a myristate group adjacent to a phosphoinositide-binding domain (Rowe et al. 2006; Kagan et al. 2008). Delivery of TLR4 to endosomal TRAM triggers the TRIF-dependent signaling pathway leading the expression of type I IFNs (Kagan et al. 2008; Tanimura et al. 2008; Tseng et al. 2010). Mislocalization of TIRAP or TRAM to the cytosol results in a defective cellular response to LPS, yet TLR4 endocytosis proceeds normally, thus underscoring the importance of sorting adaptor localization in defining the subcellular sites of signal transduction (Kagan and Medzhitov 2006; Kagan et al. 2008). The fact that TIRAP and TRAM localization is controlled by interactions with phosphoinositides is notable for two reasons. First, a similar mechanism of sorting adaptor localization in *Drosophila melanogaster* exists, suggesting that the use of sorting adaptors in innate immunity is conserved throughout evolution (Marek and Kagan 2012). Second, several posttranslational modifications have been reported within the critical residues of TIRAP's lipid-binding domain (Mansell et al. 2006). These modifications are induced on TLR signaling, and occur on residues known to influence the interactions between TIRAP and phosphoinositides (Kagan and Medzhitov 2006), perhaps serving as a novel means of regulation via altering protein localization. Consistent with this idea, PI-3 kinase activity, which converts PI(4,5)P2 to PI(3,4,5)P2, can displace TIRAP from membranes in vitro and within mammalian cells (Aksoy et al. 2012). Of note, all of the sorting adaptors in the Toll pathways (TIRAP, TRAM, and *Drosophila* MyD88) are promiscuous lipid-binding proteins in vitro, yet functional studies within cells indicate the importance of specific lipid targets in the regulation of signal transduction (Kagan and Medzhitov 2006; Kagan et al. 2008; Marek and Kagan 2012). It remains unclear why sorting adaptors show this promiscuity of interactions with multiple lipids. Finally, as will be described later in this article, sorting adaptors have been identified in other innate immune signaling pathways, most notably the RLR pathway that detects cytosolic viruses.

Links between TLRs, Phagocytosis, and Autophagy

Phagocytosis is one of the most ancient mechanisms of host defense, so it is not surprising that PRR function is intimately linked to this process. Phagocytosis involves a complex series of events, including recognition of cargo, membrane, and cytoskeleton remodeling to facilitate engulfment, and finally phagosome maturation (Flannagan et al. 2012). PRRs would appear ideally suited for the task of cargo recognition and initiation of engulfment, and certain CLRs can promote phagocytosis. In particular, heterologous expression of Dectin-1 is sufficient to facilitate uptake of yeast particles (Herre et al. 2004). This activity is likely because of Syk activation, which is downstream from most CLRs (Osorio and Reis e Sousa 2011); although, at least one report suggests that Syk-independent signals may also be involved (Herre et al. 2004). When compared with CLRs, TLRs do not promote phagocytosis robustly, and macrophages or DCs lacking TLR function engulf microbes with comparable efficiency to wild-type cells (Blander and Medzhitov 2004). However, TLR activation has been linked to an up-regulation in macropinocytosis, and some reports have reported increased bacterial uptake associated with TLR signaling (West et al. 2004; Jain et al. 2008). In addition, multiple groups have reported delayed phagosome maturation in TLR-deficient cells based on a variety of readouts (Blander and Medzhitov 2004; Arpaia et al. 2011). TLR signaling accelerates phagosomal acidification and enhances processing of phagosomal proteins for antigen presentation. Remarkably, the TLR effect is phagosome autonomous, meaning phagosomes that do not contain TLR ligands do not undergo accelerated maturation even if TLR activation occurs in other phagosomes in the same cell (Blander and Medzhitov 2006b). The mechanism by which individual phagosomes are "marked" has not been further elucidated. Notably, a role for TLRs in phagosome maturation has not been observed by every group, and the reason for these disparate results remains unexplained (Blander and Medzhitov 2006a; Russell and Yates 2007). Nevertheless, most of the published reports support a role for TLRs in cargo recognition and enhanced phagosome maturation.

The function of PRRs is also linked to autophagy. Autophagy is the process by which cytosolic contents or damaged organelles are surrounded by a nascent double membrane structure (Levine et al. 2011). These autophagosomes undergo a maturation process similar to phagosome maturation. Although originally described as a response to starvation to reclaim nutrients, it is now clear that autophagy also contributes to host defense (Levine et al. 2011). In some instances, autophagy can directly enhance PRR function. Specifically, during certain viral infections, autophagy is necessary to deliver viral nucleic acids from the cytosol to TLR7 in endosomes (Lee et al. 2007). TLR activation can also regulate autophagic processes. Early work suggested that autophagy can be induced directly by TLR signaling. Indeed, increased association of LC3 with endosomal and phagosomal membranes has been observed in macrophages activated by TLR2 and TLR4 ligands (Sanjuan et al. 2007; Xu et al. 2007). Although originally interpreted as evidence of classic autophagy, it now appears that this LC3 recruitment may be a noncanonical form of autophagy that uses a subset of the autophagic machinery. This process does not require the initiating components of classic autophagy (e.g., Ulk-1), and the LC3-positive compartments lack the classic double membrane autophagosome structure (Martinez et al. 2011). How the autophagic machinery alters the composition of endosomal or phagosomal compartments remains unclear, nor is it clear that all instances of noncanonical autophagy are equivalent. TLR-dependent induction of noncanonical autophagy can enhance phagosome maturation and promote microbial killing (Sanjuan et al. 2007). In plasmacytoid DCs, this machinery is required for trafficking of DNA-immune complexes to a TLR9-containing LRO with type I IFN-signaling capability (Henault et al. 2012). This specialized compartment does not require AP-3, indicating that noncanonical autophagy contributes to further heterogeneity and functional specialization of endosomes. The signaling pathways that induce noncanonical autophagy, downstream from TLRs as well as other receptors, remain poorly characterized.

CYTOSOLIC RECEPTORS THAT DETECT NUCLEIC ACIDS

In addition to surveying the extracellular and endosomal environments for microbial products, PRRs that survey the cytosol for bacteria and viruses exist. Cytosolic PRRs include the RLRs, which detect viral RNA, the NLRs, which activate various inflammasomes (described below), and the newly defined sensor of viral DNA, cyclic GAMP-AMP synthase (cGAS) (Rathinam and Fitzgerald 2011; Lamkanfi and Dixit 2012; Burdette and Vance 2013). Generally, these cytosolic sensors of microbes can be divided into two groups based on their effector functions. One group will be discussed in this section (RLRs and cGAS), which commonly induces transcriptional responses on microbial detection. The second group consists of the NLRs, most of which do not activate potent transcriptional responses but rather induces immediate responses in the cell. The best characterization of these responses is the activation of inflammasomes, which are cytosolic protein complexes that function to promote the processing and secretion of IL-1 family cytokines and induce cell death (Lamkanfi and Dixit 2012). There is very little known about the cell biology processes that regulate NLR functions, and, as such, we will focus our attention on the cytosolic RNA and DNA sensors described above.

The RLRs differ from the TLRs and dectins in two fundamental ways. First, these receptors are expressed in most (perhaps all) mammalian cells, whereas the aforementioned receptors are expressed on subsets of cells in the mouse and human. As such, RLRs provide a comprehensive means of detecting cytosolic microbes. Second, the fact that RLRs survey the cytosol places them in a position in which they can gauge the virulence of the microbe they encounter. The reason for this is that all pathogens (even extracellular pathogens) must have the ability to manipulate the host cells they encounter. Several examples of host cell manipulation exist, ranging from the use of bacterial toxins, specialized secretion systems, or the expression of immune evasion genes by viral pathogens. Although these strategies differ mechanistically, they share the common attribute of delivering some molecule to the cytosol of the host to disrupt host defenses. Consequently, every pathogen has the need to interact with the cytosolic environment in some way. Because RLRs are present in the cytosol, these receptors are ideally positioned only to be activated during pathogenic encounters. In contrast, TLRs and dectins survey the extracellular environments, a location that both pathogens and nonpathogens may occupy. Thus, TLRs and dectin receptors can be classified as microbe-detection receptors, whereas RLRs (and other cytosolic sensors) can be classified as pathogen-detection receptors.

The RLR family includes RIG-I, MDA5, and LGP2. Of these, RIG-I and MDA5 are the best characterized. Although RIG-I and MDA5 are both RNA-binding proteins, they bind different types of RNA. RIG-I binds to small double-stranded RNA species that contain a 5′ triphosphate group and a 3′ polyuridine-rich region (Pichlmair et al. 2006; Saito et al. 2008; Uzri and Gehrke 2009; Rehwinkel et al.

2010). MDA5, in contrast, binds to long double-stranded RNA species that may contain branched high-order structures (Pichlmair et al. 2009). Structural analysis has indicated that, whereas RIG-I binds the terminus of double-stranded RNA, MDA5 uses protein–protein contacts to oligomerize along the length of double-stranded RNA (Berke and Modis 2012; Peisley et al. 2013; Wu et al. 2013a). In both instances, RLR interactions with RNA result in interactions between the CARD domains present in the receptors with the CARD domain present in the protein mitochondrial antiviral signaling (MAVS). These CARD–CARD interactions result in the formation of a large prion-like aggregate of MAVS, which is thought to promote a signaling pathway important for expression of type I IFNs, interferon-stimulated genes (ISGs), cytokines, and chemokines (Hou et al. 2011). MAVS accomplishes this task by promoting interactions between the kinase TBK1 and its substrate, the transcription factor IRF3 (Belgnaoui et al. 2011). Phosphorylated IRF3 then acts together with other inflammatory transcription factors to induce inflammatory gene expression.

Cell biological analysis of the MAVS protein revealed its function as a sorting adaptor, in that it is localized to the eventual sites of RLR signaling before any microbial encounter (Seth et al. 2005). MAVS contains a transmembrane domain at its carboxyl terminus that directs localization to mitochondria, peroxisomes, and mitochondria-associated membranes (MAM) of the endoplasmic reticulum (Seth et al. 2005; Dixit et al. 2010; Horner et al. 2011). Localization of MAVS to each of these compartments is important for the antiviral activities of the RLRs, as mutant alleles of MAVS that are mislocalized to the cytosol are unable to signal (Seth et al. 2005; Dixit et al. 2010). In addition, several viral proteases have been identified that cleave MAVS from membranes, resulting in a cytosolic species that is signaling deficient (Li et al. 2005; Chen et al. 2007; Yang et al. 2007a).

Studies on the significance of differential MAVS localization found that its function differs depending on the organelle on which it resides (Dixit et al. 2010). For example, mitochondria-localized MAVS is capable of inducing the RLR-dependent expression of type I IFNs, ISGs, cytokines, and chemokines. In contrast, peroxisome-localized MAVS primarily induces the expression of ISGs and chemokines, but cannot induce the expression of type I IFNs. The actions of MAVS on both of these compartments are coordinated by MAM-localized MAVS (Horner et al. 2011), which creates an innate immune synapse that helps synergize compartment-specific signaling events and promote robust antiviral innate immunity. Thus, a common feature of the RLR and TLR networks is the ability of these receptors to induce different and complementary responses in a location-specific manner.

Although the RLR pathway is best known for its ability to detect viral RNA, it can also detect DNA viruses, albeit indirectly. In some instances, the DNA of viruses can be transcribed by the host-encoded RNA polymerase III to create RNA ligands that activate RIG-I dependent antiviral responses (Ablasser et al. 2009; Chiu et al. 2009). In this regard, RLRs may have a role in the detection of both RNA and DNA viruses; however, their role in the detection of DNA viruses is likely limited. As such, numerous studies have attempted to identify a general sensor of cytosolic DNA viruses, and many candidate sensors have been reported, including cGAS, IFI16, DAI, DDX41, MRE11, and DNA-PK (Burdette and Vance 2013). Although all of these proteins remain candidates, the cGAS protein has received the most attention as genetic studies in humans and mice support its role as the major sensor of cytosolic DNA (Sun et al. 2013). cGAS is a broadly expressed cytosolic protein that binds directly to double-stranded DNA, irrespective of sequence (Sun et al. 2013). DNA binding by cGAS activates its intrinsic enzymatic activity that catalyzes the creation of cyclic GMP-AMP (cGAMP) (Wu et al. 2013b). cGAMP contains phosphodiester linkages between the 2′-hydroxyl group of GMP and the 5′-phosphate of AMP, as well as a phosphodiester link between the 3′-hydroxyl of AMP and 5′-phosphate of GMP (Ablasser et al. 2013; Diner et al. 2013; Gao et al. 2013b; Zhang et al. 2013). cGAMP produced by cGAS binds to an endoplasmic reticulum–localized protein called STING, which is an adaptor that promotes interactions between TBK1 and IRF3 to induce the expression of type I IFNs (Tanaka and Chen 2012). Consequently, cGAS- or STING-deficient mice or cells are highly sensitive to DNA virus infection (Ishikawa and Barber 2008; Gao et al. 2013a; Li et al. 2013). Recent work has extended the role of the STING-dependent pathway to include the detection of *Mycobacterium tuberculosis* (Manzanillo et al. 2012; Watson et al. 2012). Interestingly, detection of Mycobacteria by the STING pathway activates an autophagic response that promotes the maturation of bacteria-containing phagosomes into lysosomes, a process that facilitates killing of the pathogen (Watson et al. 2012).

Cell biological analysis of the cGAS-STING pathway has indicated a complex coordination between signal transduction and protein localization. For example, transfection of mammalian cells (as a mimic of viral infection) triggers the relocalization of STING from the ER to small vesicles that stain positive for TBK1, the exocyst component Sec5, and the autophagy regulators ATG9a and LC-3 (Ishikawa et al. 2009; Saitoh et al. 2009). The identity of these vesicles remains unclear but the movement of STING to this unusual compartment is thought to be associated with the onset of signal transduction (Ishikawa et al. 2009; Saitoh et al. 2009).

CONCLUDING REMARKS HIGHLIGHTING THEMES

This article was designed to not only provide an overview of the PRR pathways activated in mammalian cells, but to highlight themes that appear to operate in these pathways, regardless of their specific genetic components. The simplest theme that has emerged is that the pathways of the innate immune system are most different at the receptor proximal level. Each pathway is activated by a distinct receptor in response to a distinct ligand, yet most converge on common responses such as gene transcription or autophagy. It is unknown how many other cellular responses are activated by PRRs, and a major challenge faced by the community is to develop novel assays to study more immediate (nontranscriptional) responses that occur on microbial detection. A second theme that has emerged is that regulatory processes are in place to govern the transport and localization of PRRs and their associated sorting adaptors, the latter of which defines the subcellular sites of innate immune signal transduction. A second major challenge faced by the community lies in this area, in which tools are still lacking to monitor the movement of endogenous proteins in primary cells. The use of such tools will be needed to better understand the cell type–specific actions of PRRs, some of which are highlighted in this article. A third theme that emerges is that of the comprehensiveness of immune surveillance by the PRR families. Most subcellular compartments are surveyed by one or more PRRs, including the endolysosomal network, the plasma membrane, and the cytosol. This ubiquity of surveillance poses a formidable challenge to the microbial world, but may have also resulted in the evolution of virulence strategies designed to interfere with PRR functions. Additional means of interrogating the PRR-induced signaling pathways will likely reveal novel means of microbial immunoevasion, and perhaps provide clues for therapeutic intervention.

ACKNOWLEDGMENTS

G.M.B. is supported by the National Institutes of Health (NIH Grants AI072429, AI104914, AI095587, and AI063302) and the Lupus Research Institute. The NIH Grants AI093589, AI072955, P30DK34854, and an unrestricted gift from Mead Johnson & Company support the work performed in the laboratory of J.C.K. J.C.K. is also supported by the Bill and Melinda Gates Foundation (OPP1066203). G.M.B. and J.C.K. hold Investigators in the Pathogenesis of Infectious Disease Awards from the Burroughs Wellcome Fund.

REFERENCES

Ablasser A, Bauernfeind F, Hartmann G, Latz E, Fitzgerald KA, Hornung V. 2009. RIG-I-dependent sensing of poly(dA:dT) through the induction of an RNA polymerase III-transcribed RNA intermediate. *Nat Immunol* **10:** 1065–1072.

Ablasser A, Goldeck M, Cavlar T, Deimling T, Witte G, Rohl I, Hopfner KP, Ludwig J, Hornung V. 2013. cGAS produces a 2′-5′-linked cyclic dinucleotide second messenger that activates STING. *Nature* **498:** 380–384.

Aksoy E, Taboubi S, Torres D, Delbauve S, Hachani A, Whitehead MA, Pearce WP, Berenjeno IM, Nock G, Filloux A, et al. 2012. The p110δ isoform of the kinase PI(3)K controls the subcellular compartmentalization of TLR4 signaling and protects from endotoxic shock. *Nat Immunol* **13:** 1045–1054.

Arpaia N, Godec J, Lau L, Sivick KE, McLaughlin LM, Jones MB, Dracheva T, Peterson SN, Monack DM, Barton GM. 2011. TLR signaling is required for *Salmonella typhimurium* virulence. *Cell* **144:** 675–688.

Avalos AM, Kirak O, Oelkers JM, Pils MC, Kim YM, Ottinger M, Jaenisch R, Ploegh HL, Brinkmann MM. 2013. Cell-specific TLR9 trafficking in primary APCs of transgenic TLR9-GFP mice. *J Immunol* **190:** 695–702.

Barbalat R, Lau L, Locksley RM, Barton GM. 2009. Toll-like receptor 2 on inflammatory monocytes induces type I interferon in response to viral but not bacterial ligands. *Nat Immunol* **10:** 1200–1207.

Barton GM, Kagan JC. 2009. A cell biological view of Toll-like receptor function: Regulation through compartmentalization. *Nat Rev Immunol* **9:** 535–542.

Barton GM, Kagan JC, Medzhitov R. 2006. Intracellular localization of Toll-like receptor 9 prevents recognition of self-DNA but facilitates access to viral DNA. *Nat Immunol* **7:** 49–56.

Baumann CL, Aspalter IM, Sharif O, Pichlmair A, Bluml S, Grebien F, Bruckner M, Pasierbek P, Aumayr K, Planyavsky M, et al. 2010. CD14 is a coreceptor of Toll-like receptors 7 and 9. *J Exp Med* **207:** 2689–2701.

Belgnaoui SM, Paz S, Hiscott J. 2011. Orchestrating the interferon antiviral response through the mitochondrial antiviral signaling (MAVS) adapter. *Curr Opin Immunol* **23:** 564–572.

Berke IC, Modis Y. 2012. MDA5 cooperatively forms dimers and ATP-sensitive filaments upon binding double-stranded RNA. *EMBO J* **31:** 1714–1726.

Blander JM, Medzhitov R. 2004. Regulation of phagosome maturation by signals from Toll-like receptors. *Science* **304:** 1014–1018.

Blander JM, Medzhitov R. 2006a. On regulation of phagosome maturation and antigen presentation. *Nat Immunol* **7:** 1029–1035.

Blander JM, Medzhitov R. 2006b. Toll-dependent selection of microbial antigens for presentation by dendritic cells. *Nature* **440:** 808–812.

Blasius AL, Arnold CN, Georgel P, Rutschmann S, Xia Y, Lin P, Ross C, Li X, Smart NG, Beutler B. 2010. Slc15a4, AP-3, and Hermansky-Pudlak syndrome proteins are required for Toll-like receptor signaling in plasmacytoid dendritic cells. *Proc Natl Acad Sci* **107:** 19973–19978.

Brinkmann MM, Spooner E, Hoebe K, Beutler B, Ploegh HL, Kim Y-M. 2007. The interaction between the ER membrane protein UNC93B and TLR3, 7, and 9 is crucial for TLR signaling. *J Cell Biol* **177:** 265–275.

Burdette DL, Vance RE. 2013. STING and the innate immune response to nucleic acids in the cytosol. *Nat Immunol* **14:** 19–26.

Chen Z, Benureau Y, Rijnbrand R, Yi J, Wang T, Warter L, Lanford RE, Weinman SA, Lemon SM, Martin A, et al. 2007. GB virus B disrupts RIG-I signaling by NS3/4A-mediated cleavage of the adaptor protein MAVS. *J Virol* **81:** 964–976.

Chiang CY, Veckman V, Limmer K, David M. 2012. Phospholipase Cγ-2 and intracellular calcium are required for lipopolysaccharide-induced Toll-like receptor 4 (TLR4) endocytosis and interferon regulatory factor 3 (IRF3) activation. *J Biol Chem* **287:** 3704–3709.

Chiu Y-H, MacMillan JB, Chen ZJ. 2009. RNA polymerase III detects cytosolic DNA and induces type I interferons through the RIG-I pathway. *Cell* **138:** 576–591.

Diner EJ, Burdette DL, Wilson SC, Monroe KM, Kellenberger CA, Hyodo M, Hayakawa Y, Hammond MC, Vance RE. 2013. The innate immune DNA sensor cGAS produces a noncanonical cyclic dinucleotide that activates human STING. *Cell Rep* **3:** 1355–1361.

Dixit E, Boulant S, Zhang Y, Lee AS, Odendall C, Shum B, Hacohen N, Chen ZJ, Whelan SP, Fransen M, et al. 2010. Peroxisomes are signaling platforms for antiviral innate immunity. *Cell* **141:** 668–681.

Esparza GA, Teghanemt A, Zhang D, Gioannini TL, Weiss JP. 2012. Endotoxin•albumin complexes transfer endotoxin monomers to MD-2 resulting in activation of TLR4. *Innate Immun* **18:** 478–491.

Ewald SE, Lee BL, Lau L, Wickliffe KE, Shi G-P, Chapman HA, Barton GM. 2008. The ectodomain of Toll-like receptor 9 is cleaved to generate a functional receptor. *Nature* **456:** 658–662.

Ewald SE, Engel A, Lee J, Wang M, Bogyo M, Barton GM. 2011. Nucleic acid recognition by Toll-like receptors is coupled to stepwise processing by cathepsins and asparagine endopeptidase. *J Exp Med* **208:** 643–651.

Flannagan RS, Jaumouille V, Grinstein S. 2012. The cell biology of phagocytosis. *Annu Rev Pathol* **7:** 61–98.

Fukui R, Saitoh S-I, Matsumoto F, Kozuka-Hata H, Oyama M, Tabeta K, Beutler B, Miyake K. 2009. Unc93B1 biases Toll-like receptor responses to nucleic acid in dendritic cells toward DNA- but against RNA-sensing. *J Exp Med* **206:** 1339–1350.

Fukui R, Saitoh S, Kanno A, Onji M, Shibata T, Ito A, Matsumoto M, Akira S, Yoshida N, Miyake K. 2011. Unc93B1 restricts systemic lethal inflammation by orchestrating Toll-like receptor 7 and 9 trafficking. *Immunity* **35:** 69–81.

Gao D, Wu J, Wu YT, Du F, Aroh C, Yan N, Sun L, Chen ZJ. 2013a. Cyclic GMP-AMP synthase is an innate immune sensor of HIV and other retroviruses. *Science* **341:** 903–906.

Gao P, Ascano M, Wu Y, Barchet W, Gaffney BL, Zillinger T, Serganov AA, Liu Y, Jones RA, Hartmann G, et al. 2013b. Cyclic [G(2′,5′)pA(3′,5′)p] is the metazoan second messenger produced by DNA-activated cyclic GMP-AMP synthase. *Cell* **153:** 1094–1107.

Garcia-Cattaneo A, Gobert FX, Muller M, Toscano F, Flores M, Lescure A, Del Nery E, Benaroch P. 2012. Cleavage of Toll-like receptor 3 by cathepsins B and H is essential for signaling. *Proc Natl Acad Sci* **109:** 9053–9058.

Geijtenbeek TB, Gringhuis SI. 2009. Signalling through C-type lectin receptors: Shaping immune responses. *Nat Rev Immunol* **9:** 465–479.

Gioannini TL, Zhang D, Teghanemt A, Weiss JP. 2002. An essential role for albumin in the interaction of endotoxin with lipopolysaccharide-binding protein and sCD14 and resultant cell activation. *J Biol Chem* **277:** 47818–47825.

Gioannini TL, Teghanemt A, Zhang D, Coussens NP, Dockstader W, Ramaswamy S, Weiss JP. 2004. Isolation of an endotoxin-MD-2 complex that produces Toll-like receptor 4-dependent cell activation at picomolar concentrations. *Proc Natl Acad Sci* **101:** 4186–4191.

Hart BE, Tapping RI. 2012. Cell surface trafficking of TLR1 is differentially regulated by the chaperones PRAT4A and PRAT4B. *J Biol Chem* **287:** 16550–16562.

Henault J, Martinez J, Riggs JM, Tian J, Mehta P, Clarke L, Sasai M, Latz E, Brinkmann MM, Iwasaki A, et al. 2012. Noncanonical autophagy is required for type I interferon secretion in response to DNA-immune complexes. *Immunity* **37:** 986–997.

Henneke P, Takeuchi O, van Strijp JA, Guttormsen HK, Smith JA, Schromm AB, Espevik TA, Akira S, Nizet V, Kasper DL, et al. 2001. Novel engagement of CD14 and multiple Toll-like receptors by group B streptococci. *J Immunol* **167:** 7069–7076.

Herre J, Marshall AS, Caron E, Edwards AD, Williams DL, Schweighoffer E, Tybulewicz V, Reis e Sousa C, Gordon S, Brown GD. 2004. Dectin-1 uses novel mechanisms for yeast phagocytosis in macrophages. *Blood* **104:** 4038–4045.

Hipp MM, Shepherd D, Gileadi U, Aichinger MC, Kessler BM, Edelmann MJ, Essalmani R, Seidah NG, Reis ESC, Cerundolo V. 2013. Processing of human Toll-like receptor 7 by furin-like proprotein convertases is required for its accumulation and activity in endosomes. *Immunity* **39:** 711–721.

Hoebe K, Georgel P, Rutschmann S, Du X, Mudd S, Crozat K, Sovath S, Shamel L, Hartung T, Zahringer U, et al. 2005. CD36 is a sensor of diacylglycerides. *Nature* **433:** 523–527.

Honda K, Ohba Y, Yanai H, Negishi H, Mizutani T, Takaoka A, Taya C, Taniguchi T. 2005. Spatiotemporal regulation of MyD88-IRF-7 signalling for robust type-I interferon induction. *Nature* **434:** 1035–1040.

Horner SM, Liu HM, Park HS, Briley J, Gale M Jr. 2011. Mitochondrial-associated endoplasmic reticulum membranes (MAM) form innate immune synapses and are targeted by hepatitis C virus. *Proc Natl Acad Sci* **108:** 14590–14595.

Hou F, Sun L, Zheng H, Skaug B, Jiang QX, Chen ZJ. 2011. MAVS forms functional prion-like aggregates to activate and propagate antiviral innate immune response. *Cell* **146:** 448–461.

Husebye H, Aune MH, Stenvik J, Samstad E, Skjeldal F, Halaas O, Nilsen NJ, Stenmark H, Latz E, Lien E, et al. 2010. The Rab11a GTPase controls Toll-like receptor 4-induced activation of interferon regulatory factor-3 on phagosomes. *Immunity* **33:** 583–596.

Ip WK, Takahashi K, Moore KJ, Stuart LM, Ezekowitz RA. 2008. Mannose-binding lectin enhances Toll-like receptors 2 and 6 signaling from the phagosome. *J Exp Med* **205:** 169–181.

Ishikawa H, Barber GN. 2008. STING is an endoplasmic reticulum adaptor that facilitates innate immune signalling. *Nature* **455:** 674–678.

Ishikawa H, Ma Z, Barber GN. 2009. STING regulates intracellular DNA-mediated, type I interferon-dependent innate immunity. *Nature* **461:** 788–792.

Iwasaki A, Medzhitov R. 2010. Regulation of adaptive immunity by the innate immune system. *Science* **327:** 291–295.

Jain V, Halle A, Halmen KA, Lien E, Charrel-Dennis M, Ram S, Golenbock DT, Visintin A. 2008. Phagocytosis and intracellular killing of MD-2 opsonized Gram-negative bacteria depend on TLR4 signaling. *Blood* **111:** 4637–4645.

Janeway CA Jr. 1989. Approaching the asymptote? Evolution and revolution in immunology. *Cold Spring Harb Symp Quant Biol* **54:** 1–13.

Janeway CA Jr, Medzhitov R. 2002. Innate immune recognition. *Annu Rev Immunol* **20:** 197–216.

Kadowaki N, Ho S, Antonenko S, Malefyt RW, Kastelein RA, Bazan F, Liu YJ. 2001. Subsets of human dendritic cell precursors express different Toll-like receptors and respond to different microbial antigens. *J Exp Med* **194:** 863–869.

Kagan JC. 2012. Signaling organelles of the innate immune system. *Cell* **151:** 1168–1178.

Kagan J, Medzhitov R. 2006. Phosphoinositide-mediated adaptor recruitment controls Toll-like receptor signaling. *Cell* **125:** 943–955.

Kagan JC, Su T, Horng T, Chow A, Akira S, Medzhitov R. 2008. TRAM couples endocytosis of Toll-like receptor 4 to the induction of interferon-β. *Nat Immunol* **9:** 361–368.

Kawai T, Akira S. 2010. The role of pattern-recognition receptors in innate immunity: Update on Toll-like receptors. *Nat Immunol* **11:** 373–384.

Kim HM, Park BS, Kim J-I, Kim SE, Lee J, Oh SC, Enkhbayar P, Matsushima N, Lee H, Yoo OJ, et al. 2007. Crystal structure of the TLR4-MD-2 complex with bound endotoxin antagonist Eritoran. *Cell* **130:** 906–917.

Kim YM, Brinkmann MM, Paquet ME, Ploegh HL. 2008. UNC93B1 delivers nucleotide-sensing Toll-like receptors to endolysosomes. *Nature* **452:** 234–238.

Kim J, Huh J, Hwang M, Kwon EH, Jung DJ, Brinkmann MM, Jang MH, Ploegh HL, Kim YM. 2013. Acidic amino acid residues in the juxtamembrane region of the nucleotide-sensing TLRs are important for UNC93B1 binding and signaling. *J Immunol* **190:** 5287–5295.

Koblansky AA, Jankovic D, Oh H, Hieny S, Sungnak W, Mathur R, Hayden MS, Akira S, Sher A, Ghosh S. 2012. Recognition of profilin by Toll-like receptor 12 is critical for host resistance to *Toxoplasma gondii*. *Immunity* **38:** 119–130.

Kranzusch PJ, Lee AS, Berger JM, Doudna JA. 2013. Structure of human cGAS reveals a conserved family of second-messenger enzymes in innate immunity. *Cell Rep* **3:** 1362–1368.

Lamkanfi M, Dixit VM. 2012. Inflammasomes and their roles in health and disease. *Annu Rev Cell Dev Biol* **28:** 137–161.

Lee HK, Dunzendorfer S, Soldau K, Tobias PS. 2006a. Double-stranded RNA-mediated TLR3 activation is enhanced by CD14. *Immunity* **24:** 153–163.

Lee J, Mo J-H, Katakura K, Alkalay I, Rucker AN, Liu Y-T, Lee H-K, Shen C, Cojocaru G, Shenouda S, et al. 2006b. Maintenance of colonic homeostasis by distinctive apical TLR9 signalling in intestinal epithelial cells. *Nat Cell Biol* **8:** 1327–1336.

Lee HK, Lund JM, Ramanathan B, Mizushima N, Iwasaki A. 2007. Autophagy-dependent viral recognition by plasmacytoid dendritic cells. *Science* **315:** 1398–1401.

Lee BL, Moon JE, Shu JH, Yuan L, Newman ZR, Schekman R, Barton GM. 2013. UNC93B1 mediates differential trafficking of endosomal TLRs. *eLife* **2:** e00291.

Levine B, Mizushima N, Virgin HW. 2011. Autophagy in immunity and inflammation. *Nature* **469:** 323–335.

Li XD, Sun L, Seth RB, Pineda G, Chen ZJ. 2005. Hepatitis C virus protease NS3/4A cleaves mitochondrial antiviral signaling protein off the mitochondria to evade innate immunity. *Proc Natl Acad Sci* **102:** 17717–17722.

Li XD, Wu J, Gao D, Wang H, Sun L, Chen ZJ. 2013. Pivotal roles of cGAS–cGAMP signaling in antiviral defense and immune adjuvant effects. *Science* **341:** 1390–1394.

Lin YC, Huang DY, Chu CL, Lin YL, Lin WW. 2013. The tyrosine kinase Syk differentially regulates Toll-like receptor signaling downstream of the adaptor molecules TRAF6 and TRAF3. *Sci Signal* **6:** ra71.

Lindau D, Mussard J, Wagner BJ, Ribon M, Ronnefarth VM, Quettier M, Jelcic I, Boissier MC, Rammensee HG, Decker P. 2013. Primary blood neutrophils express a functional cell surface Toll-like receptor 9. *Eur J Immunol* **43:** 2101–2113.

Mansell A, Smith R, Doyle SL, Gray P, Fenner JE, Crack PJ, Nicholson SE, Hilton DJ, O'Neill LA, Hertzog PJ. 2006. Suppressor of cytokine signaling 1 negatively regulates Toll-like receptor signaling by mediating Mal degradation. *Nat Immunol* **7:** 148–155.

Manzanillo PS, Shiloh MU, Portnoy DA, Cox JS. 2012. *Mycobacterium tuberculosis* activates the DNA-dependent cytosolic surveillance pathway within macrophages. *Cell Host Microbe* **11:** 469–480.

Marek LR, Kagan JC. 2012. Phosphoinositide binding by the Toll adaptor dMyD88 controls antibacterial responses in *Drosophila*. *Immunity* **36:** 612–622.

Martinez J, Almendinger J, Oberst A, Ness R, Dillon CP, Fitzgerald P, Hengartner MO, Green DR. 2011. Microtubule-associated protein 1 light chain 3 α(LC3)-associated phagocytosis is required for the efficient clearance of dead cells. *Proc Natl Acad Sci* **108:** 17396–17401.

Mathur R, Oh H, Zhang D, Park SG, Seo J, Koblansky A, Hayden MS, Ghosh S. 2012. A mouse model of *Salmonella typhi* infection. *Cell* **151:** 590–602.

Matsumoto M, Funami K, Tanabe M, Oshiumi H, Shingai M, Seto Y, Yamamoto A, Seya T. 2003. Subcellular localization of Toll-like receptor 3 in human dendritic cells. *J Immunol* **171:** 3154–3162.

Mouchess ML, Arpaia N, Souza G, Barbalat R, Ewald SE, Lau L, Barton GM. 2011. Transmembrane mutations in Toll-like receptor 9 bypass the requirement for ectodomain proteolysis and induce fatal inflammation. *Immunity* **35:** 721–732.

Onji M, Kanno A, Saitoh S, Fukui R, Motoi Y, Shibata T, Matsumoto F, Lamichhane A, Sato S, Kiyono H, et al. 2013. An essential role for the N-terminal fragment of Toll-like receptor 9 in DNA sensing. *Nat Commun* **4:** 1949.

Osorio F, Reis e Sousa C. 2011. Myeloid C-type lectin receptors in pathogen recognition and host defense. *Immunity* **34:** 651–664.

Park B, Brinkmann MM, Spooner E, Lee CC, Kim Y-M, Ploegh HL. 2008. Proteolytic cleavage in an endolysosomal compartment is required for activation of Toll-like receptor 9. *Nat Immunol* **9:** 1407–1414.

Park BS, Song DH, Kim HM, Choi BS, Lee H, Lee JO. 2009. The structural basis of lipopolysaccharide recognition by the TLR4-MD-2 complex. *Nature* **458:** 1191–1195.

Peisley A, Wu B, Yao H, Walz T, Hur S. 2013. RIG-I forms signaling-competent filaments in an ATP-dependent, ubiquitin-independent manner. *Mol Cell* **51:** 573–583.

Peter ME, Kubarenko AV, Weber ANR, Dalpke AH. 2009. Identification of an N-terminal recognition site in TLR9 that contributes to CpG-DNA-mediated receptor activation. *J Immunol* **182:** 7690–7697.

Pichlmair A, Schulz O, Tan CP, Näslund TI, Liljeström P, Weber F, Reis e Sousa C. 2006. RIG-I-mediated antiviral responses to single-stranded RNA bearing 5′-phosphates. *Science* **314:** 997–1001.

Pichlmair A, Schulz O, Tan CP, Rehwinkel J, Kato H, Takeuchi O, Akira S, Way M, Schiavo G, Reis e Sousa C. 2009. Activation of MDA5 requires higher-order RNA structures generated during virus infection. *J Virol* **83:** 10761–10769.

Pifer R, Benson A, Sturge CR, Yarovinsky F. 2011. UNC93B1 is essential for TLR11 activation and IL-12-dependent host resistance to *Toxoplasma gondii*. *J Biol Chem* **286:** 3307–3314.

Pohar J, Pirher N, Bencina M, Mancek-Keber M, Jerala R. 2013. The role of UNC93B1 in surface localization of TLR3 and in cell priming to nucleic acid agonists. *J Biol Chem* **288:** 442–454.

Prohinar P, Re F, Widstrom R, Zhang D, Teghanemt A, Weiss JP, Gioannini TL. 2007. Specific high affinity interactions of monomeric endotoxin•protein complexes with Toll-like receptor 4 ectodomain. *J Biol Chem* **282:** 1010–1017.

Rathinam VA, Fitzgerald KA. 2011. Cytosolic surveillance and antiviral immunity. *Curr Opin Virol* **1:** 455–462.

Rehwinkel J, Tan CP, Goubau D, Schulz O, Pichlmair A, Bier K, Robb N, Vreede F, Barclay W, Fodor E, et al. 2010. RIG-I detects viral genomic RNA during negative-strand RNA virus infection. *Cell* **140:** 397–408.

Resman N, Vasl J, Oblak A, Pristovsek P, Gioannini TL, Weiss JP, Jerala R. 2009. Essential roles of hydrophobic residues in both MD-2 and Toll-like receptor 4 in activation by endotoxin. *J Biol Chem* **284:** 15052–15060.

Rowe DC, McGettrick AF, Latz E, Monks BG, Gay NJ, Yamamoto M, Akira S, O'Neill LA, Fitzgerald KA, Golenbock DT. 2006. The myristoylation of TRIF-related adaptor molecule is essential for Toll-like receptor 4 signal transduction. *Proc Natl Acad Sci* **103:** 6299–6304.

Russell DG, Yates RM. 2007. Toll-like receptors and phagosome maturation. *Nat Immunol* **8:** 217; author reply 217–218.

Rutz M, Metzger J, Gellert T, Luppa P, Lipford GB, Wagner H, Bauer S. 2004. Toll-like receptor 9 binds single-stranded CpG-DNA in a sequence- and pH-dependent manner. *Eur J Immunol* **34:** 2541–2550.

Saito T, Owen DM, Jiang F, Marcotrigiano J, Gale M Jr. 2008. Innate immunity induced by composition-dependent RIG-I recognition of hepatitis C virus RNA. *Nature* **454:** 523–527.

Saitoh T, Fujita N, Hayashi T, Takahara K, Satoh T, Lee H, Matsunaga K, Kageyama S, Omori H, Noda T, et al. 2009. Atg9a controls dsDNA-driven dynamic translocation of STING and the innate immune response. *Proc Natl Acad Sci* **106:** 20842–20846.

Sancho D, Reis e Sousa C. 2012. Signaling by myeloid C-type lectin receptors in immunity and homeostasis. *Annu Rev Immunol* **30:** 491–529.

Sanjuan MA, Dillon CP, Tait SW, Moshiach S, Dorsey F, Connell S, Komatsu M, Tanaka K, Cleveland JL, Withoff S, et al. 2007. Toll-like receptor signalling in macrophages links the autophagy pathway to phagocytosis. *Nature* **450:** 1253–1257.

Sasai M, Linehan MM, Iwasaki A. 2010. Bifurcation of Toll-like receptor 9 signaling by adaptor protein 3. *Science* **329:** 1530–1534.

Sepulveda FE, Maschalidi S, Colisson R, Heslop L, Ghirelli C, Sakka E, Lennon-Duménil A-M, Amigorena S, Cabanie L, Manoury B. 2009. Critical role for asparagine endopeptidase in endocytic Toll-like receptor signaling in dendritic cells. *Immunity* **31:** 737–748.

Seth RB, Sun L, Ea CK, Chen ZJ. 2005. Identification and characterization of MAVS, a mitochondrial antiviral signaling protein that activates NF-κB and IRF3. *Cell* **122:** 669–682.

Sun L, Wu J, Du F, Chen X, Chen ZJ. 2013. Cyclic GMP-AMP synthase is a cytosolic DNA sensor that activates the type I interferon pathway. *Science* **339:** 786–791.

Takahashi K, Shibata T, Akashi-Takamura S, Kiyokawa T, Wakabayashi Y, Tanimura N, Kobayashi T, Matsumoto F, Fukui R, Kouro T, et al. 2007. A protein associated with Toll-like receptor (TLR) 4 (PRAT4A) is required for TLR-dependent immune responses. *J Exp Med* **204:** 2963–2976.

Tanaka Y, Chen ZJ. 2012. STING specifies IRF3 phosphorylation by TBK1 in the cytosolic DNA signaling pathway. *Sci Signal* **5:** ra20.

Tanimura N, Saitoh S, Matsumoto F, Akashi-Takamura S, Miyake K. 2008. Roles for LPS-dependent interaction and relocation of TLR4 and TRAM in TRIF-signaling. *Biochem Biophys Res Commun* **368:** 94–99.

Teghanemt A, Prohinar P, Gioannini TL, Weiss JP. 2007. Transfer of monomeric endotoxin from MD-2 to CD14: Characterization and functional consequences. *J Biol Chem* **282:** 36250–36256.

Teghanemt A, Widstrom RL, Gioannini TL, Weiss JP. 2008. Isolation of monomeric and dimeric secreted MD-2. Endotoxin.sCD14 and Toll-like receptor 4 ectodomain selectively react with the monomeric form of secreted MD-2. *J Biol Chem* **283:** 21881–21889.

Triantafilou M, Miyake K, Golenbock DT, Triantafilou K. 2002. Mediators of innate immune recognition of bacteria concentrate in lipid rafts and facilitate lipopolysaccharide-induced cell activation. *J Cell Sci* **115:** 2603–2611.

Tseng P-H, Matsuzawa A, Zhang W, Mino T, Vignali DAA, Karin M. 2010. Different modes of ubiquitination of the adaptor TRAF3 selectively activate the expression of type I interferons and proinflammatory cytokines. *Nat Immunol* **11:** 70–75.

Uzri D, Gehrke L. 2009. Nucleotide sequences and modifications that determine RIG-I/RNA binding and signaling activities. *J Virol* **83:** 4174–4184.

Wang Y, Chen T, Han C, He D, Liu H, An H, Cai Z, Cao X. 2007. Lysosome-associated small Rab GTPase Rab7b negatively regulates TLR4 signaling in macrophages by promoting lysosomal degradation of TLR4. *Blood* **110:** 962–971.

Wang D, Lou J, Ouyang C, Chen W, Liu Y, Liu X, Cao X, Wang J, Lu L. 2010. Ras-related protein Rab10 facilitates TLR4 signaling by promoting replenishment of TLR4 onto the plasma membrane. *Proc Natl Acad Sci* **107:** 13806–13811.

Watson RO, Manzanillo PS, Cox JS. 2012. Extracellular *M. tuberculosis* DNA targets bacteria for autophagy by activating the host DNA-sensing pathway. *Cell* **150:** 803–815.

West MA, Wallin RP, Matthews SP, Svensson HG, Zaru R, Ljunggren HG, Prescott AR, Watts C. 2004. Enhanced dendritic cell antigen capture via Toll-like receptor-induced actin remodeling. *Science* **305:** 1153–1157.

Wu B, Peisley A, Richards C, Yao H, Zeng X, Lin C, Chu F, Walz T, Hur S. 2013a. Structural basis for dsRNA recognition, filament formation, and antiviral signal activation by MDA5. *Cell* **152:** 276–289.

Wu J, Sun L, Chen X, Du F, Shi H, Chen C, Chen ZJ. 2013b. Cyclic GMP-AMP is an endogenous second messenger in innate immune signaling by cytosolic DNA. *Science* **339:** 826–830.

Xu Y, Jagannath C, Liu XD, Sharafkhaneh A, Kolodziejska KE, Eissa NT. 2007. Toll-like receptor 4 is a sensor for autophagy associated with innate immunity. *Immunity* **27:** 135–144.

Yang Y, Liang Y, Qu L, Chen Z, Yi M, Li K, Lemon SM. 2007a. Disruption of innate immunity due to mitochondrial targeting of a picornaviral protease precursor. *Proc Natl Acad Sci* **104:** 7253–7258.

Yang Y, Liu B, Dai J, Srivastava PK, Zammit DJ, Lefrançois L, Li Z. 2007b. Heat shock protein gp96 is a master chaperone for Toll-like receptors and is important in the innate function of macrophages. *Immunity* **26:** 215–226.

Yarovinsky F, Zhang D, Andersen J, Bannenberg G, Serhan C, Hayden M, Hieny S, Sutterwala F, Flavell R, Ghosh S, et al. 2005. TLR11 activation of dendritic cells by a protozoan profilin-like protein. *Science* **308:** 1626–1629.

Zanoni I, Ostuni R, Marek LR, Barresi S, Barbalat R, Barton GM, Granucci F, Kagan JC. 2011. CD14 controls the LPS-induced endocytosis of Toll-like receptor 4. *Cell* **147:** 868–880.

Zhang X, Shi H, Wu J, Sun L, Chen C, Chen ZJ. 2013. Cyclic GMP-AMP containing mixed phosphodiester linkages is an endogenous high-affinity ligand for STING. *Mol Cell* **51:** 226–235.

Transcriptional Control of Inflammatory Responses

Stephen T. Smale[1] and Gioacchino Natoli[2]

[1]Department of Microbiology, Immunology, and Molecular Genetics, University of California, Los Angeles, California 90095

[2]Department of Experimental Oncology, European Institute of Oncology (IEO), I-20139 Milan, Italy

Correspondence: gioacchino.natoli@ieo.eu

The inflammatory response requires the activation of a complex transcriptional program that is both cell-type- and stimulus-specific and involves the dynamic regulation of hundreds of genes. In the context of an inflamed tissue, extensive changes in gene expression occur in both parenchymal cells and infiltrating cells of the immune system. Recently, basic transcriptional mechanisms that control inflammation have been clarified at a genome scale, particularly in macrophages and conventional dendritic cells. The regulatory logic of distinct groups of inflammatory genes can be explained to some extent by identifiable sequence-encoded features of their chromatin organization, which impact on transcription factor (TF) accessibility and impose different requirements for gene activation. Moreover, it has become apparent that the interplay between TFs activated by inflammatory stimuli and master regulators exerts a crucial role in controlling cell-type-specific transcriptional outputs.

Survival of any living organism, from simple prokaryotes to complex multicellular eukaryotes, critically relies on its ability to sense virtually every environmental change and to mount highly specialized responses that are tailored on both the precise nature of the stimulus and its intensity. Among environmental responses, inflammation is especially complex as a consequence of several factors: (1) its essential role as a first line of defense against microbes, as a source of instructions to the adaptive immune system, and as the starting point for resolution and repair; (2) the enormous variety of potentially dangerous microbes and their extreme ability to evolve under the pressure of the host immune system; and (3) finally, the extensive exploitation of molecules and mechanisms invented in evolution for antimicrobial defense and to solve nonmicrobial tissue damage.

At the cellular level, the very basis of inflammation is the deployment of complex gene expression programs that include hundreds of genes and are activated within minutes after the primary stimulus (Medzhitov and Horng 2009). Although these programs include a core set of genes that are almost invariably activated in most cell types and in response to most inflammatory stimuli, they extensively differ from each other depending on the cell type and tissue in which they are elicited, the nature and the intensity of the trigger, as well as the preexisting cellular or organismal conditions (Smale 2010).

Cite this article as *Cold Spring Harb Perspect Biol* doi: 10.1101/cshperspect.a016261

For instance, in distinct cell types, a specific inflammatory stimulus can activate the same set of transcription factors (TFs) and yet produce different transcriptional responses. As explained below, this occurs because each cell type has a unique repertoire of available genomic transcriptional regulatory elements that have been specified during its differentiation, and, thus, differ from those of any other differentiated cell type (Natoli 2010). Moreover, the same cell type (e.g., a macrophage) can be differentially conditioned by the tissue milieu to the point that its response to the same stimulus (e.g., a microbial component) will be even dramatically different depending on its anatomical location.

An additional aspect of the inflammatory gene expression program that has been extensively analyzed in the last years is its kinetic complexity (Saccani et al. 2001). In fact, products of different inflammatory genes may have completely different cellular targets and may have to be released by migratory innate immune cells (such as dendritic cells [DCs]) in different tissues and at different times. Consider for instance a DC recruited at a peripheral site on microbial invasion and activated by microbial products. Among the first genes to be transcribed are those encoding chemokines (e.g., IL-8) that attract neutrophils to the inflammatory site to amplify the response and contain the invaders; then, genes encoding chemokine receptors enabling DC migration to lymph nodes (CCR7); and, eventually, genes relevant for T- and B-lymphocyte stimulation in the lymph node, such as CD80, CD86, and IL-6. In this regard, it is remarkable that in nonmigratory cells, such as fibroblasts, IL-6 is a rapidly induced gene, which indirectly implies the evolution of active mechanisms to delay its activation in DCs and macrophages.

At the level of individual genes, specific rules determining selectivity of induction in response to different inflammatory agonists as well as distinct kinetics of activation are increasingly being clarified. The recent availability of genomic approaches to analyze transcription and chromatin properties (Metzker 2010) has represented an enabling condition to move from single-gene analyses to global models describing the complexity of the response in an integrated manner. Moreover, with the refinement of technologies for screening based on RNA interference, it is now possible to evaluate the requirement for a large panel of transcriptional regulators in the inflammatory response. It is useful to classify inflammatory genes into primary and secondary response genes (PRGs and SRGs, respectively) based on their different requirements for new protein synthesis. PRG induction depends on preexisting signaling molecules and/or TFs, whereas SRG induction requires the de novo synthesis of signaling molecules and/or transcriptional regulators (such as interferon [IFN]-β, C/EBPδ, and IκBζ) during the early phases of the response (Doyle et al. 2002; Yamamoto et al. 2004; Thomas et al. 2006; Litvak et al. 2009). Therefore, the expression of SRGs will be selectively impaired by protein synthesis inhibitors. For obvious reasons, PRGs tend to be induced faster than SRGs as a class, but individual exceptions have been described.

THE INFLAMMATORY GENE EXPRESSION PROGRAM

The gene expression cascade induced by inflammatory stimuli in mouse macrophages and DCs has been described with increasing resolution. The highest resolution studies have involved RNA-sequencing (RNA-seq) to examine either mature, polyadenylated mRNA or nascent transcripts (Rabani et al. 2011; Bhatt et al. 2012). RNA-seq of mRNA allows an examination of transcripts that may be available for translation into protein and considers the influences of regulation at the levels of transcription, RNA stability, and pre-mRNA processing. In contrast, nascent transcripts, which can be monitored by metabolic labeling, genome-wide nuclear runon (GRO-seq), or an examination of chromatin-associated transcripts, provides greater insight into transcriptional control independent of the influences of pre-mRNA processing and stability. An examination of nascent transcripts is especially valuable for defining the precise kinetics of transcriptional activation and inactivation of inducible genes. Comparisons of nascent transcript kinetics and mRNA kinetics in macrophages or DCs following lipopolysaccharide (LPS) stimulation have revealed that transcription is the primary mode of regulating inducible gene expression (Rabani et al. 2011; Bhatt et al. 2012). However, stability and pre-mRNA processing play important roles in refining the expression pattern of many genes through the activities of miRNAs and other regulators of stability and processing (Hao and Baltimore 2009). Although transcription is the dominant regulator of inflammatory gene expression, transcription itself can be regulated at multiple levels, including the levels of transcription initiation, promoter escape, elongation, and termination. This topic is complicated and beyond the scope of this article. However, it is noteworthy that transcriptional regulation of a specific gene at any of these levels is ultimately dependent on the recognition of specific DNA sequences by sequence-specific TFs. Furthermore, even when gene induction is associated with release of a transcriptional pause or an increase in elongation efficiency, the frequency of transcription initiation also usually increases.

Despite the high-resolution, quantitative nature of recent RNA-seq studies of inflammatory gene induction, it is difficult to provide precise numbers of inducible genes, because such numbers are influenced by arbitrary decisions about transcript abundances and induction magnitudes required for the inclusion of genes in the set. One recent study reported 560 genes induced by more than fivefold at the nascent transcript level within 2 h of stimulation with lipid A (the active component of LPS) (Bhatt et al. 2012). However, more than 1000 genes were induced to a statistically significant extent during this same time period, with some genes induced by less than twofold and others induced by more than 1000-fold from their basal levels. These quantitative analyses further revealed that both the basal and maximal transcript levels for the set of induced genes spanned three orders of magnitude, with readily detected basal transcription of a large percentage of inducible genes. This extensive heterogeneity in basal and maximum transcription levels, as well as in fold-induction values, will ultimately need to be incorporated into both biological and mechanistic models of inflammatory gene expression. In addition, future studies will need to address the extensive heterogeneity in the gene expression cascade that has been documented in recent single-cell RNA-seq studies (RNA-seq) (Shalek et al. 2013). Despite the clear documentation of differential responses at a single-cell level within a population of phenotypically homogeneous cells, the biological and mechanistic significance of this heterogeneity remains unknown.

Notably, the hundreds of genes induced by LPS in macrophages and DCs can be divided into several defined clusters in which all genes within each cluster show similar temporal kinetics of transcription (Rabani et al. 2011; Bhatt et al. 2012). Given their similar kinetics, the genes within each cluster may be regulated by common sets of TFs. Consistent with this possibility, different transcription-factor-binding motifs were found to be overrepresented in the promoters of genes within different temporal clusters. Although these observations provide a path toward further dissection of the transcriptional program, each temporal cluster is undoubtedly heterogeneous, with different subsets of genes within each cluster regulated by different mechanisms.

INTEGRATION OF SIGNALS AT THE CHROMATIN LEVEL

Transcription Factors Controlling the Inflammatory Response

TFs activated in response to inflammatory stimuli belong to a few main families with distinct binding specificities (Fig. 1), including the nuclear factor of the κ light chain enhancer of B cells (NF-κBs) (Hayden and Ghosh 2012), interferon regulatory factors (IRFs) (Tamura et al. 2008), signal transducers and activators of transcription (STAT) (Stark and Darnell 2012), and activator protein 1 (AP-1) (Wagner and Eferl 2005) families. Wiring of different receptors for microbial and endogenous danger signals

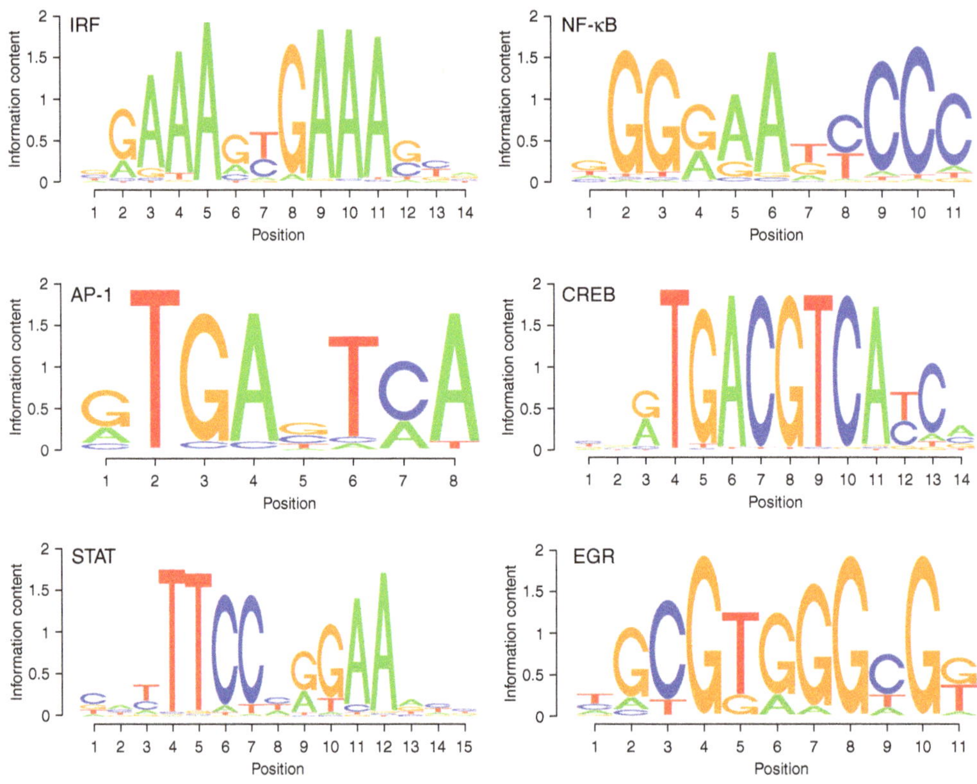

Figure 1. DNA recognition specificities of TFs that control inflammatory gene expression. Representative examples are shown.

to distinct signal transducers lays the groundwork for selective or preferential activation of subsets of these TFs. For instance, activation of IRF3 is selectively coupled to Toll-like receptor (TLR)3 and TLR4 (Doyle et al. 2002; Toshchakov et al. 2002) and explains their ability to induce the IFN-β gene and the downstream autocrine/paracrine IFN response, a property that is unique among TLRs. Conversely, cytokines acting primarily through the activation of STAT TFs such as IFN-γ, are in general unable to activate NF-κB and AP-1, which are broadly responsive to a large panel of inflammatory agonists (ranging from LPS and most other microbial products to tumor necrosis factor [TNF]-α) (O'Shea and Plenge 2012).

The activity of these inflammatory TFs, however, is restrained and controlled by two main groups of factors: (1) nucleosomes, and (2) constitutively expressed TFs collectively defined here as master regulators that control the usage of both regulatory and coding genomic information in differentiated cells.

The Interplay between Nucleosomes and TFs in the Inflammatory Response

The great structural differences among the various classes of DNA-binding domains (Garvie and Wolberger 2001) explain why TFs have inherent differences in their ability to bind nucleosomal DNA. A broad class of TFs defined as pioneer factors (Zaret and Carroll 2011) and exemplified by the FoxA TFs, are operationally characterized by their ability to bind sites in a nucleosomal context and make them accessible. At the opposite side of the spectrum is TATA-binding protein, which is unable to bind the TATA sequence in a nucleosomal context (Imbalzano et al. 1994). TFs activated by inflammatory agonists also appear to show a different ability to bind nucleosomal sites. STAT and IRF family TFs bind chromatin remodelers and, in this way, they may be able to promote access to otherwise occluded sequences (Huang et al. 2002; Liu et al. 2002; Cui et al. 2004). The scenario with NF-κB is more controversial. From a structural point of view (Huxford et al. 1999), NF-κB is predicted to be unable to bind nucleosomal recognition sites (Natoli et al. 2005). However, an initial report suggested the surprising notion that the nucleosome does not even minimally impair NF-κB binding (Angelov et al. 2004). This apparent inconsistency was probably because of the experimental conditions used in these early experiments, which in fact relaxed octamer-DNA association and made underlying binding sites accessible to NF-κB (Lone et al. 2013). Taken together, data are consistent with the notion that the DNA-binding mode of NF-κB makes it unable to contact sites associated with, and occluded by, nucleosomes.

Determinants of Nucleosome Positions and the Role of CpG Islands

Nucleosome positions are dictated by a combination of factors, including the underlying DNA sequence and the presence of *trans*-acting factors (mainly TFs) that compete with histones for binding the DNA and/or recruit chromatin remodeling activities that keep surrounding DNA accessible (Struhl and Segal 2013). In general, a moderate G/C content favors nucleosome assembly (Segal et al. 2006; Tillo and Hughes 2009) but the extreme guanine-cytosine content of some CpG islands is not compatible with efficient bending around the histone octamer and thus favors the formation of nucleosome-depleted areas (Ramirez-Carrozzi et al. 2009; Fenouil et al. 2012) that are rapidly accessible to stimulus-activated TFs as well as to the basal transcriptional machinery. The binding of ubiquitous TFs, such as Sp1, to many CpG islands is also likely to play a major role in the depletion of nucleosomes in these regions. Indeed, inflammatory genes containing a CpG island promoter are constitutively associated with RNA polymerase (Pol II) (Hargreaves et al. 2009) have other features of active chromatin before stimulation, and often have higher basal transcriptional activity and are poised for further activation on stimulation (Ramirez-Carrozzi et al. 2009). Importantly, relaxed nucleosomal constraints to TF access to CpG island–containing genes favor promiscuous activation in response to multiple stimuli (Smale 2010).

Initial experiments on a limited set of inflammatory genes in macrophages indicated the possibility of a higher frequency of CpG islands at PRGs than at SRGs (Hargreaves et al. 2009; Ramirez-Carrozzi et al. 2009), suggesting that the presence of positioned nucleosomes at SRGs would impose the requirement for chromatin remodelers for gene activation. More recently, a genome-wide analysis of LPS-induced gene expression in macrophages revealed that, among PRGs, CpG-island promoters are most closely associated with transient induction, with more variable promoter architectures among PRGs that show sustained induction. A mechanistic explanation for the link between CpG islands and transient induction has not been obtained. Many SRG promoters have also been found to contain CpG islands (Bhatt et al. 2012). The CpG island–containing SRGs were not apparent in the earlier study because that study focused only on the most potently induced genes, which lack CpG-island promoters. Related to this observation, the most obvious distinction between genes containing and lacking CpG-island promoters is that genes lacking CpG islands, especially among SRGs, are generally induced by a larger magnitude (Bhatt et al. 2012). Therefore, although the presence or absence of a CpG island affects nucleosome occupancy and accessibility of promoters, it is not sufficient to distinguish between PRGs and SRGs. However, CpG island genes differ from non-CpG island genes because of the lower fold induction after stimulation. Mechanistically stable and strategically positioned nucleosomes that reduce basal transcriptional activity impose a tight regulation onto SRGs, and, thus, enable a high dynamic range compared with genes that are basally associated with the transcriptional machinery because of the presence of a CpG island in their promoters.

GENOMIC REGULATORY ELEMENTS THAT CONTROL INFLAMMATORY GENE EXPRESSION

Identification of Genomic *cis*-Regulatory Elements

The recent development of genomic approaches for the study of transcriptional regulation and, specifically, the combination of chromatin immunoprecipitation with multiparallel (or high-throughput) sequencing (ChIP-Seq) has enabled the identification of general principles of transcriptional control that were beyond the reach of single-gene studies (Barski et al. 2007).

The production of antibodies specific for different posttranslationally modified histones allowed the discovery that genomic regions with different functions are also marked with different sets of histone modifications (Bernstein et al. 2007). For instance, TSSs and adjacent regions of active genes or genes poised for activation are associated with trimethylation of histone H3 at lysine 4 (H3K4me3), whereas transcribed genic regions are associated with H3K36me2/3 and H3K79me2. A great advance in the field has been the characterization of a loose yet specific signature associated with distal transcriptional enhancers, whose identification has represented a major challenge for many years because of their variable distance from the target genes they regulate. The enhancer chromatin signature consists of high levels of monomethylated H3K4 (H3K4me1) (Heintzman et al. 2007) associated or not with histone acetylation (Creyghton et al. 2011; Rada-Iglesias et al. 2011). H3K4me1 is also detected at active or poised TSSs, but in those cases it is found on both sides of a central region marked by H3K4me3, thus leading to an overall high H3K4me3/me1 ratio. Conversely, this ratio is completely inverted at enhancers because of the higher H3K4me1 level. Enhancers are also associated with transcriptional coregulators, such as the histone acetyltransferases (HAT) p300, CBP, and Tip60 (Heintzman et al. 2007; Visel et al. 2009a), and, particularly when very active, they are subjected to RNA Pol II–mediated transcription (De Santa et al. 2010; Kim et al. 2010; Natoli and Andrau 2012).

Enhancers Controlling Inflammatory Gene Expression

Enhancer mapping in multiple cell types allowed determination of some basic principles of their usage in mammalian cells (Visel et al. 2009b). In general, an average of 40,000–50,000 regions bearing the chromatin signature of enhancers can be detected in any cell type, although such numbers vary depending on the statistical thresholds used in ChIP-Seq experiments. Although it is clear that the mere presence of a chromatin signature need not imply functionality, it is also clear that most or all enhancers that were previously characterized using genetic and functional approaches can be retrieved in ChIP-Seq experiments.

The initial characterization of enhancers involved in LPS-inducible gene expression in macrophages was based on the ability of stimulus-activated TFs, such as NF-κB and IRFs, to promote the recruitment of the p300 HAT (Ghisletti et al. 2010). LPS-inducible p300 recruitment unveiled thousands of enhancers and revealed their underlying sequence features. In addition to binding sites for LPS-activated TFs such as NF-κB, AP-1, and IRF, these enhancers were almost invariably associated with binding sites for Pu.1, an Ets family protein that controls myeloid development and is expressed at very high levels in terminally differentiated macrophages (Scott et al. 1994; Nerlov and Graf 1998). Pu.1 is constitutively bound to macrophage enhancers and is directly involved, together with partner TFs acting in different phases of macrophage development (Lichtinger et al. 2012), in determining both the deposition of H3K4me1 and the displacement of nucleosomes to generate accessible DNA sequences (via TF-mediated recruitment of histone methyltransferases and ATP-dependent nucleosome remodeling complexes, respectively) (Fig. 2) (Ghisletti et al. 2010; Heinz et al. 2010). More generally, it is now clear that the specificity of the enhancer

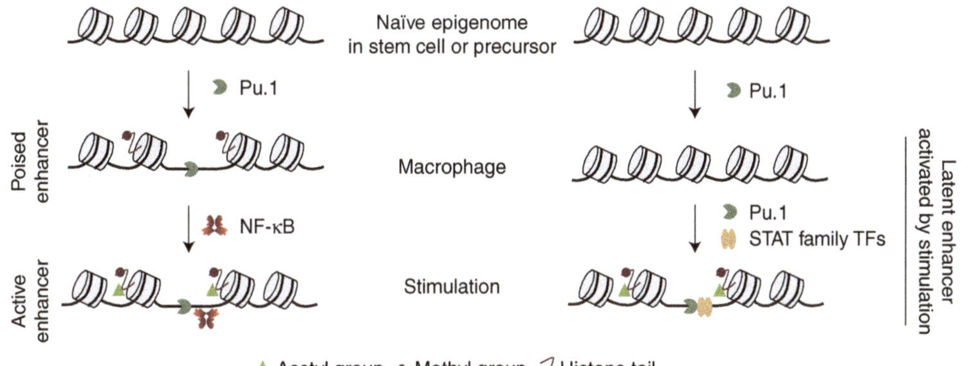

Figure 2. The relationship between nucleosomes, lineage-determining TFs, and TFs activated by inflammatory stimuli. (*Left*) A poised enhancer in which nucleosome depletion and deposition of H3K4me1 is induced by Pu.1 during differentiation. In response to stimulation TFs, such as NF-κB, are recruited to the nucleosome-depleted area maintained by Pu.1. (*Right*) Latent enhancers are not detectable in terminally differentiated cells until a stimulus is delivered that activates TFs (such as Stat TFs) collaborating with Pu.1 to reconfigure local chromatin.

landscape in different cell types is because of the direct activity of TFs involved in lineage specification and in maintenance of terminal differentiation (Natoli 2010). Indeed, at the basic molecular level, an essential function of these master regulators is to select the fraction of the enormous *cis*-regulatory repertoire available in mammalian genomes that will be used to regulate gene expression in that particular cell type. The functional role of H3K4me1 at enhancers is still unclear, although it may be involved in stabilizing the association of some HATs with enhancer nucleosomes (Jeong et al. 2011). Conversely, as discussed above, nucleosome displacement has the obvious consequence of making the underlying binding sites accessible to TFs (such as NF-κB) that are unable to invade nucleosomal DNA. Therefore, the role of Pu.1 in macrophages as well as in other myeloid cells (Garber et al. 2012) is to make a broad and highly cell-type-specific repertoire of regulatory sequences available to other TFs, including those activated in response to inflammatory stimuli (Fig. 2, left).

Clearly, TFs activated by inflammatory stimuli have little, if any, cell-type specificity, being similarly expressed and inducible in most cell types. Nevertheless, inflammatory gene expression is strictly cell-type specific both from a qualitative and a quantitative point of view. For instance, IL12b is expressed only by macrophages and DCs, and TNF-α is produced at much higher levels by macrophages than by any other cell type in the body. The lack of specificity of inflammatory TFs is easily reconciled with the cell-type specificity of inflammatory gene expression when taking into consideration the uniqueness of the preexisting regulatory landscape generated by master regulators (such as Pu.1) that show restricted expression and divergent DNA-binding specificity.

The Impact of Inflammatory Stimuli on the Enhancer Repertoire

In general, active enhancers are associated with acetylated histones, whereas enhancers showing H3K4me1 in the absence of histone acetylation are assumed to be in a poised or in an actively repressed state. In macrophages, the vast majority of enhancers with constitutive H3K4me1 are in a poised state and only a small fraction (<10%) of them can be inducibly acetylated in response to LPS (Ostuni et al. 2013). Different stimuli are able to activate distinct and only partially overlapping fractions of the large repertoire of poised enhancers. As discussed above, this inducible acetylation reflects recruitment to enhancers of TFs activated by stimulation and usually is reverted to almost baseline levels a few hours after activation. A different situation occurs at a smaller subset of enhancers, representing ~15% of the enhancers activated by individual stimuli; at these enhancers, histone H3K4me1 and other enhancer marks are not identifiable in unstimulated cells by ChIP-Seq or related approaches. These latent enhancers (Fig. 2, right) (Ostuni et al. 2013) acquire H3K4me1 and H3K27Ac and undergo an increase in accessibility over several hours after stimulation, thus reflecting a slow process of chromatin reorganization that depends on the functional cooperation between stimulus-activated TFs (such as Stat1 and Stat6 induced in response to IFN-γ and IL-4, respectively) and Pu.1. The repertoire of latent enhancers is very specific to the stimulus used, with very little overlap between unrelated stimuli. Importantly, after stimulation has ceased, while acetylation and binding of Pu.1 and stimulus-activated TFs are rapidly reversed, H3K4me1 persists for days and this correlates with a faster and higher acetylation if macrophages are restimulated. The priming effect on reacetylation may relate to the ability of H3K4me1 to promote and stabilize the recruitment of specific HATs (such as Tip60) to enhancers (Jeong et al. 2011). Irrespective of the underlying mechanism, the persistence of histone marks at latent enhancers after stimulus withdrawal suggests the existence of a sort of short-term memory of the initial stimulation and may help explain the conditioning effect of primary stimulation on the response of macrophages to a subsequent stimulus.

CONCLUDING REMARKS

Recent technological advances have enabled a great and rapid expansion of the knowledge of mechanisms controlling transcriptional responses to inflammatory stimuli. Availability of genome-scale expression data will allow a more and more refined classification of inflammatory genes based on their properties, such as selective induction in response to specific agonists, cell-type-restricted expression and kinetics of activation. At the same time, unraveling transcriptional and chromatin-regulated mechanisms involved in the activation and fine-tuning of inflammatory gene transcription will allow linking regulatory elements and combinations of TFs acting on them to distinct groups of genes, thus providing a systems-level and mechanism-based understanding of the molecular basis of transcriptional control of inflammation. It would be tempting to hypothesize that in the future such knowledge may allow precisely predicting the impact of chemicals and drugs that selectively act on specific signal-transduction pathways (e.g., c-Jun amino-terminal kinase [JNK] inhibitors), TFs (e.g., glucocorticoids), and chromatin regulators (e.g., bromodomain and extraterminal [BET] inhibitors) (Nicodeme et al. 2010).

ACKNOWLEDGMENTS

Research in the S.T.S. group is supported by the U.S. National Institutes of Health. Research on this topic in the G.N. group is mainly supported by the European Research Council Grant NORM (Nuclear Organization of Macrophages).

REFERENCES

Angelov D, Lenouvel F, Hans F, Muller CW, Bouvet P, Bednar J, Moudrianakis EN, Cadet J, Dimitrov S. 2004. The histone octamer is invisible when NF-κB binds to the nucleosome. *J Biol Chem* **279:** 42374–42382.

Barski A, Cuddapah S, Cui K, Roh TY, Schones DE, Wang Z, Wei G, Chepelev I, Zhao K. 2007. High-resolution profiling of histone methylations in the human genome. *Cell* **129:** 823–837.

Bernstein BE, Meissner A, Lander ES. 2007. The mammalian epigenome. *Cell* **128:** 669–681.

Bhatt DM, Pandya-Jones A, Tong AJ, Barozzi I, Lissner MM, Natoli G, Black DL, Smale ST. 2012. Transcript dynamics of proinflammatory genes revealed by sequence analysis of subcellular RNA fractions. *Cell* **150:** 279–290.

Creyghton MP, Cheng AW, Welstead GG, Kooistra T, Carey BW, Steine EJ, Hanna J, Lodato MA, Frampton GM, Sharp PA, et al. 2011. Histone H3K27ac separates active from poised enhancers and predicts developmental state. *Proc Natl Acad Sci* **107:** 21931–21936.

Cui K, Tailor P, Liu H, Chen X, Ozato K, Zhao K. 2004. The chromatin-remodeling BAF complex mediates cellular antiviral activities by promoter priming. *Mol Cell Biol* **24:** 4476–4486.

De Santa F, Barozzi I, Mietton F, Ghisletti S, Polletti S, Tusi BK, Muller H, Ragoussis J, Wei CL, Natoli G. 2010. A large fraction of extragenic RNA pol II transcription sites overlap enhancers. *PLoS Biol* **8:** e1000384.

Doyle S, Vaidya S, O'Connell R, Dadgostar H, Dempsey P, Wu T, Rao G, Sun R, Haberland M, Modlin R, et al. 2002. IRF3 mediates a TLR3/TLR4-specific antiviral gene program. *Immunity* **17:** 251–263.

Fenouil R, Cauchy P, Koch F, Descostes N, Cabeza J, Innocenti Cn, Ferrier P, Spicuglia S, Gut M, Gut I, et al. 2012. CpG islands and GC content dictate nucleosome depletion in a transcription-independent manner at mammalian promoters. *Genome Res* **22:** 2399–2408.

Garber M, Yosef N, Goren A, Raychowdhury R, Thielke A, Guttman M, Robinson J, Minie B, Chevrier N, Itzhaki Z, et al. 2012. A high-throughput chromatin immunoprecipitation approach reveals principles of dynamic gene regulation in mammals. *Mol Cell* **47:** 810–822.

Garvie C, Wolberger C. 2001. Recognition of specific DNA sequences. *Mol Cell* **8:** 937–946.

Ghisletti S, Barozzi I, Mietton F, Polletti S, De Santa F, Venturini E, Gregory L, Lonie L, Chew A, Wei CL, et al. 2010. Identification and characterization of enhancers controlling the inflammatory gene expression program in macrophages. *Immunity* **32:** 317–328.

Hao S, Baltimore D. 2009. The stability of mRNA influences the temporal order of the induction of genes encoding inflammatory molecules. *Nat Immunol* **10:** 281–288.

Hargreaves DC, Horng T, Medzhitov R. 2009. Control of inducible gene expression by signal-dependent transcriptional elongation. *Cell* **138:** 129–145.

Hayden M, Ghosh S. 2012. NF-κB, the first quarter-century: Remarkable progress and outstanding questions. *Genes Dev* **26:** 203–234.

Heintzman N, Stuart R, Hon G, Fu Y, Ching C, Hawkins R, Barrera L, Van Calcar S, Qu C, Ching K, et al. 2007. Distinct and predictive chromatin signatures of transcriptional promoters and enhancers in the human genome. *Nat Genet* **39:** 311–318.

Heinz S, Benner C, Spann N, Bertolino E, Lin YC, Laslo P, Cheng JX, Murre C, Singh H, Glass CK. 2010. Simple combinations of lineage-determining transcription factors prime *cis*-regulatory elements required for macrophage and B cell identities. *Mol Cell* **38:** 576–589.

Huang M, Qian F, Hu Y, Ang C, Li Z, Wen Z. 2002. Chromatin-remodelling factor BRG1 selectively activates a subset of interferon-α-inducible genes. *Nat Cell Biol* **4:** 774–781.

Huxford T, Malek S, Ghosh G. 1999. Structure and mechanism in NF-κB/IκB signaling. *Cold Spring Harb Symp Quant Biol* **64:** 533–540.

Imbalzano AN, Kwon H, Green MR, Kingston RE. 1994. Facilitated binding of TATA-binding protein to nucleosomal DNA. *Nature* **370:** 481–485.

Jeong KW, Kim K, Situ AJ, Ulmer TS, An W, Stallcup MR. 2011. Recognition of enhancer element-specific histone methylation by TIP60 in transcriptional activation. *Nat Struct Mol Biol* **18:** 1358–1365.

Kim T-K, Hemberg M, Gray J, Costa A, Bear D, Wu J, Harmin D, Laptewicz M, Barbara-Haley K, Kuersten S, et al. 2010. Widespread transcription at neuronal activity-regulated enhancers. *Nature* **465:** 182–187.

Lichtinger M, Ingram R, Hannah R, Moller D, Clarke D, Assi S, Lie-A-Ling M, Noailles L, Vijayabaskar M, Wu M, et al. 2012. RUNX1 reshapes the epigenetic landscape at the onset of haematopoiesis. *EMBO J* **31:** 4318–4333.

Litvak V, Ramsey S, Rust A, Zak D, Kennedy K, Lampano A, Nykter M, Shmulevich I, Aderem A. 2009. Function of C/EBPδ in a regulatory circuit that discriminates between transient and persistent TLR4-induced signals. *Nat Immunol* **10:** 437–443.

Liu H, Kang H, Liu R, Chen X, Zhao K. 2002. Maximal induction of a subset of interferon target genes requires the chromatin-remodeling activity of the BAF complex. *Mol Cell Biol* **22:** 6471–6479.

Lone IN, Shukla MS, Charles Richard JL, Peshev ZY, Dimitrov S, Angelov D. 2013. Binding of NF-κB to nucleosomes: Effect of translational positioning, nucleosome remodeling and linker histone H1. *PLoS Genet* **9:** e1003830.

Medzhitov R, Horng T. 2009. Transcriptional control of the inflammatory response. *Nat Rev Immunol* **9:** 692–703.

Metzker M. 2010. Sequencing technologies—The next generation. *Nat Rev Genet* **11:** 31–46.

Natoli G. 2010. Maintaining cell identity through global control of genomic organization. *Immunity* **33:** 12–24.

Natoli G, Andrau JC. 2012. Noncoding transcription at enhancers: General principles and functional models. *Annu Rev Genet* **46:** 1–19.

Natoli G, Saccani S, Bosisio D, Marazzi I. 2005. Interactions of NF-κB with chromatin: The art of being at the right place at the right time. *Nat Immunol* **6:** 439–445.

Nerlov C, Graf T. 1998. PU.1 induces myeloid lineage commitment in multipotent hematopoietic progenitors. *Genes Dev* **12:** 2403–2412.

Nicodeme E, Jeffrey KL, Schaefer U, Beinke S, Dewell S, Chung CW, Chandwani R, Marazzi I, Wilson P, Coste H, et al. 2010. Suppression of inflammation by a synthetic histone mimic. *Nature* **468:** 1119–1123.

O'Shea J, Plenge R. 2012. JAK and STAT signaling molecules in immunoregulation and immune-mediated disease. *Immunity* **36:** 542–550.

Ostuni R, Piccolo V, Barozzi I, Polletti S, Termanini A, Bonifacio S, Curina A, Prosperini E, Ghisletti S, Natoli G. 2013. Latent enhancers activated by stimulation in differentiated cells. *Cell* **152:** 157–171.

Rabani M, Levin JZ, Fan L, Adiconis X, Raychowdhury R, Garber M, Gnirke A, Nusbaum C, Hacohen N, Friedman N, et al. 2011. Metabolic labeling of RNA uncovers principles of RNA production and degradation dynamics in mammalian cells. *Nat Biotechnol* **29:** 436–442.

Rada-Iglesias A, Bajpai R, Swigut T, Brugmann SA, Flynn RA, Wysocka J. 2011. A unique chromatin signature uncovers early developmental enhancers in humans. *Nature* **470:** 279–283.

Ramirez-Carrozzi VR, Braas D, Bhatt DM, Cheng CS, Hong C, Doty KR, Black JC, Hoffmann A, Carey M, Smale ST. 2009. A unifying model for the selective regulation of inducible transcription by CpG islands and nucleosome remodeling. *Cell* **138:** 114–128.

Saccani S, Pantano S, Natoli G. 2001. Two waves of nuclear factor κB recruitment to target promoters. *J Exp Med* **193:** 1351–1359.

Scott EW, Simon MC, Anastasi J, Singh H. 1994. Requirement of transcription factor PU.1 in the development of multiple hematopoietic lineages. *Science* **265:** 1573–1577.

Segal E, Fondufe-Mittendorf Y, Chen L, Thåström A, Field Y, Moore I, Wang J-PZ, Widom J. 2006. A genomic code for nucleosome positioning. *Nature* **442:** 772–778.

Shalek AK, Satija R, Adiconis X, Gertner RS, Gaublomme JT, Raychowdhury R, Schwartz S, Yosef N, Malboeuf C, Lu D, et al. 2013. Single-cell transcriptomics reveals bimodality in expression and splicing in immune cells. *Nature* **498:** 236–240.

Smale ST. 2010. Selective transcription in response to an inflammatory stimulus. *Cell* **140:** 833–844.

Stark G, Darnell J. 2012. The JAK-STAT pathway at twenty. *Immunity* **36:** 503–514.

Struhl K, Segal E. 2013. Determinants of nucleosome positioning. *Nat Struct Mol Biol* **20:** 267–273.

Tamura T, Yanai H, Savitsky D, Taniguchi T. 2008. The IRF family transcription factors in immunity and oncogenesis. *Annu Rev Immunol* **26:** 535–584.

Thomas KE, Galligan CL, Newman RD, Fish EN, Vogel SN. 2006. Contribution of interferon-β to the murine macrophage response to the Toll-like receptor 4 agonist, lipopolysaccharide. *J Biol Chem* **281:** 31119–31130.

Tillo D, Hughes T. 2009. G+C content dominates intrinsic nucleosome occupancy. *BMC Bioinformatics* **10:** 442.

Toshchakov V, Jones B, Perera P-Y, Thomas K, Cody M, Zhang S, Williams B, Major J, Hamilton T, Fenton M, et al. 2002. TLR4, but not TLR2, mediates IFN-β-induced STAT1α/β-dependent gene expression in macrophages. *Nat Immunol* **3:** 392–398.

Visel A, Blow MJ, Li Z, Zhang T, Akiyama JA, Holt A, Plajzer-Frick I, Shoukry M, Wright C, Chen F, et al. 2009a. ChIP-seq accurately predicts tissue-specific activity of enhancers. *Nature* **457:** 854–858.

Visel A, Rubin EM, Pennacchio LA. 2009b. Genomic views of distant-acting enhancers. *Nature* **461:** 199–205.

Wagner E, Eferl R. 2005. Fos/AP-1 proteins in bone and the immune system. *Immunol Rev* **208:** 126–140.

Yamamoto M, Yamazaki S, Uematsu S, Sato S, Hemmi H, Hoshino K, Kaisho T, Kuwata H, Takeuchi O, Takeshige K, et al. 2004. Regulation of Toll/IL-1-receptor-mediated gene expression by the inducible nuclear protein IκBζ. *Nature* **430:** 218–222.

Zaret KS, Carroll JS. 2011. Pioneer transcription factors: Establishing competence for gene expression. *Genes Dev* **25:** 2227–2241.

Inflammasomes

Marcel R. de Zoete[1,3], Noah W. Palm[1,3], Shu Zhu[1,3], and Richard A. Flavell[1,2]

[1]Department of Immunobiology, Yale University School of Medicine, New Haven, Connecticut 06520

[2]Howard Hughes Medical Institute, Yale University, New Haven, Connecticut 06520

Correspondence: richard.flavell@yale.edu

Inflammasomes are large cytosolic multiprotein complexes that assemble in response to detection of infection- or stress-associated stimuli and lead to the activation of caspase-1-mediated inflammatory responses, including cleavage and unconventional secretion of the leaderless proinflammatory cytokines IL-1β and IL-18, and initiation of an inflammatory form of cell death referred to as pyroptosis. Inflammasome activation can be induced by a wide variety of microbial pathogens and generally mediates host defense through activation of rapid inflammatory responses and restriction of pathogen replication. In addition to its role in defense against pathogens, recent studies have suggested that the inflammasome is also a critical regulator of the commensal microbiota in the intestine. Finally, inflammasomes have been widely implicated in the development and progression of various chronic diseases, such as gout, atherosclerosis, and metabolic syndrome. In this perspective, we discuss the role of inflammasomes in infectious and noninfectious inflammation and highlight areas of interest for future studies of inflammasomes in host defense and chronic disease.

Inflammasomes are cytosolic molecular factories that typically consist of a sensor protein, the adaptor protein apoptosis-associated speck-like protein containing a caspase-recruitment domain (ASC), and the proinflammatory caspase, caspase-1 (Schroder and Tschopp 2010). Inflammasome assembly can be triggered by sensing of a variety of stimuli that are associated with infection or cellular stress and culminates in the activation of caspase-1 (Fig. 1) (Strowig et al. 2012; de Zoete and Flavell 2013; Latz et al. 2013). Active caspase-1 subsequently processes the leaderless proinflammatory cytokines pro-IL-1β and pro-IL-18, which are unconventionally secreted on caspase-1 cleavage. Therefore, inflammasome-mediated processing and secretion of pro-IL-1β and pro-IL-18 enables a rapid, yet tightly regulated and highly inducible proinflammatory response. In addition to cytokine secretion, inflammasome activation can also trigger an inflammatory cell death dubbed pyroptosis, which serves to blunt intracellular pathogen replication (Miao et al. 2011).

The NOD-like receptor (or nucleotide-binding domain and leucine-rich repeat containing receptor; NLR) family was the first family of sensor proteins discovered to form inflammasomes and is comprised of 22 genes in humans and 33 genes in mice (Reed et al. 2003; Ting et al. 2008). NLRs are classified

[3]These authors contributed equally to this work.

Cite this article as *Cold Spring Harb Perspect Biol* doi: 10.1101/cshperspect.a016287

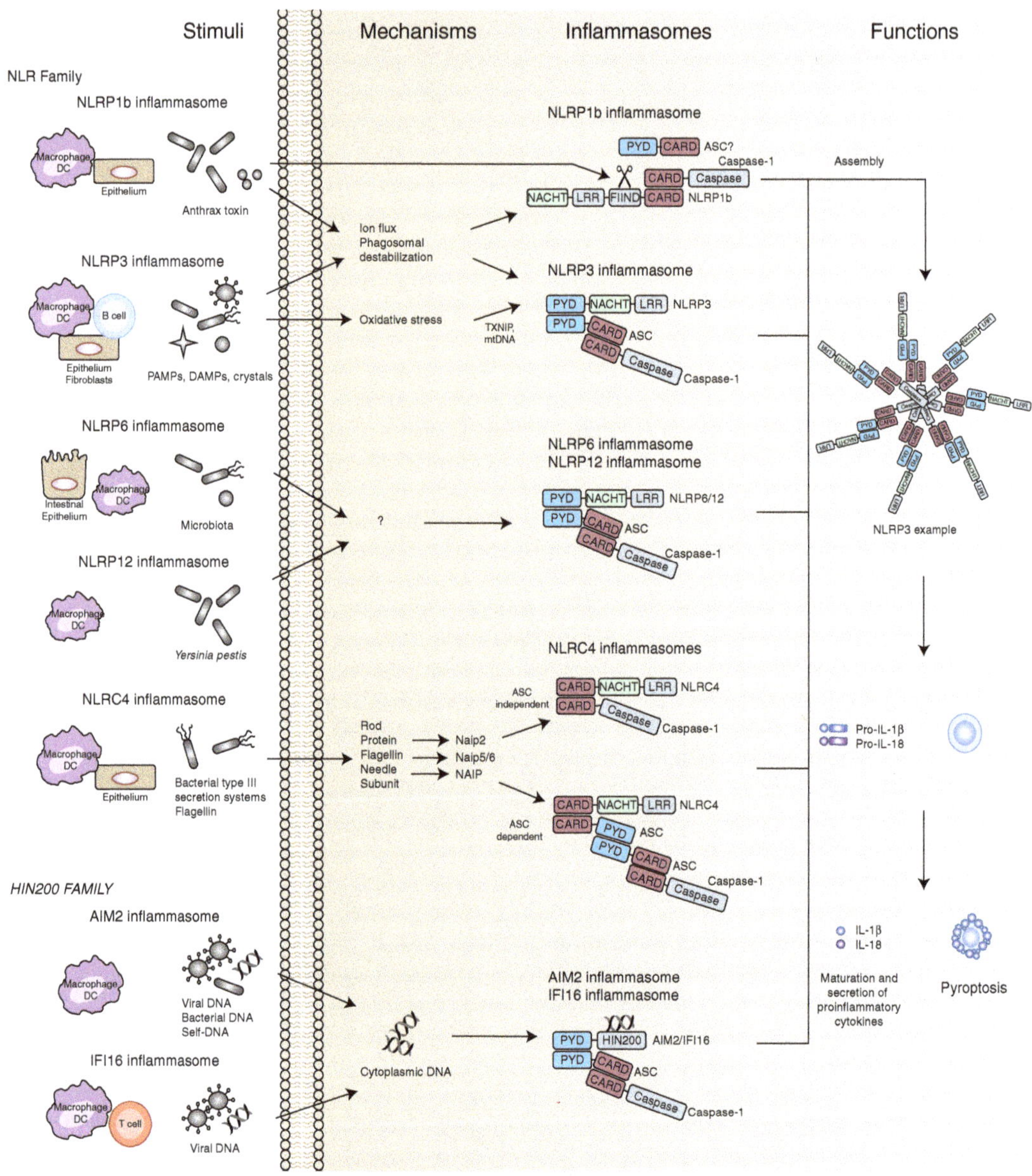

Figure 1. Inflammasome assembly can be triggered by sensing of a variety of stimuli that are associated with infection or cellular stress and culminates in the activation of caspase-1. DC, dendritic cell.

and named according to their domain structure; all NLRs (except NLRP10) contain a leucine-rich repeat (LRR) domain, which is believed to mediate ligand binding, a nucleotide-binding domain (NBD), and a signaling domain (Ting et al. 2008). This signaling domain enables the recruitment of caspase-1 either directly, through a CARD, or indirectly, through a PYRIN domain, which can bind the PYRIN-CARD-containing adaptor ASC. In addition to the NLRs, recent studies have revealed additional families of genes that can nucleate the assembly of inflammasomes: the AIM2-like receptor (ALR) family, which possesses a HIN200 DNA-binding domain instead of an LRR (Schattgen and Fitzgerald 2011), and the RIG-I-like receptor (RLR) family (Ireton and Gale 2011). In this perspective, we will discuss the mechanisms of inflammasome activation, as well as the role of inflammasomes in host defense and disease pathology. In addition, we will attempt to highlight areas in which additional studies will be necessary to clarify or expand our current understanding of inflammasome biology.

MECHANISMS OF INFLAMMASOME ACTIVATION

Many of the most interesting and important questions in the inflammasome field pertain to the identities of the specific signals that lead to the assembly of the different NLRs, ALRs, and RLRs into active inflammasome complexes. Because the expression patterns, molecular structures, and stimuli that lead to activation of the different inflammasomes are highly variable, we will introduce and discuss each one separately.

The NLRP1 Inflammasome

The NLRP1 inflammasome was the first inflammasome to be discovered (Martinon et al. 2002). Although initial studies showed that the NLRP1 inflammasome could assemble spontaneously in cell lysates, more recent studies have described two natural ligands for NLRP1: muramyl dipeptide (MDP), a peptidoglycan fragment from both Gram positive and negative bacteria, and the *Bacillus anthracis* lethal toxin (Boyden and Dietrich 2006; Faustin et al. 2007). These ligands appear to be species-specific because MDP can activate human, but not mouse, NLRP1 (Faustin et al. 2007; Kovarova et al. 2012), whereas anthrax toxin has been shown to activate mouse NLRP1. NLRP1 also shows species-specific differences at the genetic level. Humans possess a single *NLRP1* gene containing both a PYRIN and a CARD domain, whereas mice have a cluster of three homologous genes, *Nlrp1a*, *Nlrp1b*, and *Nlrp1c*, that contain a CARD but lack PYRIN domains. *Nlrp1b* was found to confer responsiveness to anthrax toxin in certain mouse strains, including Balb/c and 129S1; however, C57Bl/6 and SJL/J mice encode *Nlrp1b* variants that are not activated by the toxin (Boyden and Dietrich 2006).

The active fragment of anthrax lethal toxin consists of a zinc metalloproteinase that gains access to the cytosol of infected host cells. Nlrp1b does not directly bind anthrax toxin through its LRR motif, but instead appears to sense the proteolytic activity of the toxin because cleavage of NLRP1b itself proved sufficient for activation of the receptor (Levinsohn et al. 2012; Chavarria-Smith and Vance 2013). In contrast, human NLRP1 was shown to directly bind MDP. This binding induces a conformational change in NLRP1 that allows binding of ATP. ATP hydrolysis presumably then induces NLRP1 oligomerization and provides a platform to recruit and activate caspase-1 (Faustin et al. 2007). MDP-mediated caspase-1 activation by NLRP1 was enhanced by, but did not strictly require, the adaptor protein ASC. In addition to caspase-1, caspase-5 has also been implicated in binding the human NLRP1 complex (Martinon et al. 2002). Further studies will be necessary to reveal the precise mechanisms of NLPR1 activation, to identify any additional NLRP1 ligands, and to elucidate the functions of mouse NLRP1a and NLRP1c.

The NLRP3 Inflammasome

NLRP3 is by far the most thoroughly studied NLR. NLRP3 is expressed at low levels in myeloid cells and can be transcriptionally induced by Toll-like receptor agonists, such as LPS, and by inflammatory cytokines, such as TNF-α, in an NF-κB-dependent manner (Bauernfeind et al. 2009). This is referred to as "signal one" and is necessary to induce transcription of pro-IL-1β and for the NLRP3 inflammasome to become responsive to "signal two," which consists of a broad spectrum of infection and stress-associated signals (Schroder and Tschopp 2010). NLRP3-activating signals include pathogen-associated molecular patterns (PAMPs) and toxins from bacterial (Mariathasan et al. 2006; Duncan et al. 2009; He et al. 2010; Toma et al. 2010; Shimada et al. 2011), viral (Allen et al. 2009; Thomas et al. 2009; Ichinohe et al. 2010; Rajan et al. 2011), fungal (Gross et al. 2009; Hise et al.

2009; Joly et al. 2009), and protozoan pathogens (Shio et al. 2009), as well as host-derived damage-associated molecular patterns (DAMPs) such as ATP, uric acid crystals, and amyloid-β fibrils (Mariathasan et al. 2006; Martinon et al. 2006; Halle et al. 2008). Because many diverse pathogens and pathogen-associated molecules can activate NLRP3, it is unlikely that all of these different stimuli are sensed directly. Instead, it is generally believed that all of these signals converge on a shared molecular event that specifically activates NLRP3.

Three main models for NLRP3 activation have been proposed: the ion flux model, the reactive oxygen species (ROS) model, and the lysosome rupture model. In the ion flux model, changes in cytosolic levels of specific cations, such as K^+, Ca^{2+}, and H^+, are proposed to play a critical role in NLRP3 activation. Several NLRP3 activators were shown to directly induce potent ion fluxes. Extracellular ATP activates the ATP-gated ion channel P2X7 and triggers rapid K^+ efflux (Franchi et al. 2007; Petrilli et al. 2007); Nigericin creates a K^+ pore in the cell membrane (Mariathasan et al. 2006); the influenza M2 protein triggers export of H^+ ions from the Golgi complex into the cytosol (Ichinohe et al. 2010); and high concentrations of extracellular Ca^{2+}, increase cytosolic Ca^{2+}, and cAMP (Lee et al. 2012; Murakami et al. 2012; Rossol et al. 2012). However, it is important to note that ion fluxes also activate other inflammasomes, such as NLRP1b (Fink et al. 2008; Wickliffe et al. 2008; Newman et al. 2009; Ali et al. 2011) and NLRC4 (Arlehamn et al. 2010). This suggests that ion fluxes might only modulate the threshold of caspase-1 activation but not serve as specific signals that trigger NLRP3 inflammasome activation (Lamkanfi and Dixit 2012).

Oxidative stress in the form of ROS has been widely implicated in NLRP3 activation and many NLRP3-activating stimuli, including ATP, alum, uric acid, and Nigericin, all induce ROS production (Schroder and Tschopp 2010). However, the precise role of ROS in NLRP3 inflammasome activation remains somewhat controversial. Initially, intracellular ROS produced via the NADPH oxidase system were thought to activate NLRP3; however, both mouse and human cells defective in NADPH oxidase show normal NLRP3 activation (Latz 2010; van Bruggen et al. 2010). In addition, although ROS inhibitors were shown to inhibit NLRP3 activation, this effect was likely because of the role of ROS in NF-κB mediated up-regulation of NLRP3 and pro-IL-1β transcription (signal one) rather than its role in NLRP3 inflammasome activation itself (Bauernfeind et al. 2011). Recently, a report suggested that increased amounts of ROS were sensed by a complex of Thioredoxin and Thioredoxin-interacting protein (TXNIP), with the latter binding to NLRP3 in response to oxidative stress (Zhou et al. 2010). Two other studies suggest an alternative role for ROS in inflammasome activation, in which oxidized mitochondrial DNA that is released from dysfunctional mitochondria can directly bind to and activate the NLRP3 inflammasome (Nakahira et al. 2011; Shimada et al. 2012). In contrast, ROS-independent mitochondria-derived cardiolipin was shown to bind and activate NLRP3 (Iyer et al. 2013). Although both ROS and mitochondria clearly play a role in inflammasome formation and activation, their precise function has yet to be fully clarified.

Finally, the NLRP3 inflammasome was proposed to sense lysosomal rupture during "frustrated" phagocytosis of crystalline or large particulate molecules, such as uric acid crystals, alum, silica, malarial hemozoin, hydroxyapatite, and amyloid-β (Schroder and Tschopp 2010; Jin et al. 2011). Inhibitors of the lysosomal protease Cathepsin B reduced NLRP3 activation, suggesting a role for proteolytic activity in the activation of NLRP3, similar to what was observed for NLRP1b (Hornung et al. 2008). However, Cathepsin B–deficient mice did not show impaired inflammasome responses (Dostert et al. 2009), and it was recently shown that particulate triggers of the NLRP3 inflammasome also require K^+ efflux (Munoz-Planillo et al. 2013).

Despite considerable efforts, there is currently no agreement in the field on a universal mechanism by which the NLRP3 inflammasome is activated. One likely reason for this is the diversity of stimuli that can lead to NLRP3 activation, many of which have multiple effects on cell physiology. Recently, NLRP3 activation was shown to be even more complex with the discovery of noncanonical inflammasome activation (Kayagaki et al. 2011a). Although the canonical pathway activates NLRP3 directly, noncanonical activation involves the activation of caspase-11 by intracellular LPS released by rupture of bacteria-loaded phagolysosomes or by bacteria that actively enter the cytosol (Hagar et al. 2013; Kayagaki et al. 2013). Notably, the initial discovery of caspase-11-dependent inflammasome activation resulted from the realization that the caspase-1-deficient mice used by most researchers are in fact double knockouts, as they also lack a functional caspase-11 owing to a passenger mutation in the 129 mouse strain, from which the embryonic stem cells to generate the mice were derived (Kayagaki et al. 2011a). Although its nature is currently unknown, presumably the formation of an inflammasome-like complex containing a cytosolic LPS-receptor precedes caspase-11 activation. Also, the mechanism through which caspase-11 interplays with the NLRP3 pathway remains to be elucidated. A possible pathway is that activated caspase-11, which initiates pyroptosis in a similar manner as caspase-1, induces signals that are subsequently sensed by NLRP3. Future research will hopefully clarify the specific mechanisms of both canonical and noncanonical NLRP3 inflammasome activation.

The NLRP6 and NLRP12 Inflammasomes

Unlike NLRP3, which plays a major role in hematopoietic cells such as macrophages, NLRP6 is highly expressed in nonhematopoietic cells, including epithelial cells and goblet cells in the intestine (Elinav et al. 2011; Wlodarska et al. 2014), although there is a claim that it is mainly expressed in hematopoietic cells (Chen et al. 2011). Accordingly, NLRP6 plays an important role in maintaining intestinal homeostasis (Chen et al. 2011; Elinav et al. 2011; Normand et al. 2011; Hu et al. 2013). Despite the clear role of NLRP6 in intestinal homeostasis, many basic questions regarding NLRP6 biology remain unanswered. Early studies suggested that coexpression of NLRP6 with ASC results in a synergistic activation of caspase-1-dependent cytokine processing (Grenier et al. 2002), indicating that NLRP6 might indeed form an inflammasome. Furthermore, alterations in the microbiota associated with NLRP6 deficiency also occurred in mice lacking IL-18, suggesting that NLRP6 activation might trigger IL-18 release (Elinav et al. 2011). However, NLRP6 was recently shown to contribute to mucus secretion by goblet cells through the regulation of autophagy in an IL-1β- and IL-18-independent manner (Wlodarska et al. 2014). In addition, NLRP6 has been shown to be a negative regulator of NF-κB and MAPK signaling, leading to uncontrolled proinflammatory responses in NLRP6-deficient mice during bacterial challenge (Anand and Kanneganti 2013). Additional studies will be necessary to fully understand the role of NLRP6 in immune defense; in particular, identification of the signals that lead to NLRP6 activation, which remain completely unknown, will be critical.

NLRP12 possesses several features that resemble NLRP6. Similar to NLRP6, NLRP12 plays an important role in protecting against DSS-induced colitis and AOM/DSS-induced colon cancer (Zaki et al. 2011; Allen et al. 2012b). Furthermore, NLRP12 also appears to maintain intestinal homeostasis by negatively regulating inflammatory signaling pathways such as NF-κB and MAPK (Zaki et al. 2011; Allen et al. 2012b), and forced coexpression of NLRP12 and ASC results in synergistic activation of caspase-1 and secretion of IL-1β (Wang et al. 2002). In addition, a recent study showed that the NLRP12 inflammasome regulates IL-18 and IL-1β production after *Yersinia pestis* infection, and NLRP12-deficient mice were more susceptible to bacterial challenge (Vladimer et al. 2012). Like NLRP6, the specific PAMPs or DAMPs that can activate NLPR12, and the precise mechanism of activation, await identification.

The NLRC4 Inflammasome

The mechanism of NLRC4 activation appears to be relatively clear in comparison to the other inflammasomes. NLRC4 is activated in response to many different bacterial pathogens, including

Legionella pneumophila, *Pseudomonas aeruginosa*, *Salmonella typhimurium*, and *Shigella flexneri* (Mariathasan et al. 2004; Amer et al. 2006; Franchi et al. 2006; Miao et al. 2006; Lamkanfi et al. 2007; Sutterwala et al. 2007; Suzuki et al. 2007; Miao et al. 2008). NLRC4 senses bacterial flagellin or structural components of the bacterial type III secretion system (T3SS) that are injected or leaked into the host cell (Miao et al. 2010c). These bacterial proteins are directly bound by NLR-family apoptosis-inhibiting proteins (NAIPs) in the cytosol. In mice, NAIP2 binds the T3SS rod protein, whereas NAIP5 and NAIP6 bind flagellin; in humans, the sole NAIP binds the T3SS needle subunit (Kofoed and Vance 2011; Zhao et al. 2011). NAIPs then interact with NLRC4 and trigger assembly of the NLRC4 inflammasome complex, resulting in activation of caspase-1, release of inflammatory cytokines and pyroptosis. NLRC4 inflammasome assembly was recently shown to require phosphorylation of NLRC4, which may drive additional conformational changes that are necessary for inflammasome assembly (Qu et al. 2012).

AIM2-Like Receptors and RIG-I-Like Receptors

Absent in melanoma 2 (AIM2) and the ALRs are members of the PYHIN family, which consists of proteins that contain a PYRIN domain and the conserved DNA-binding domain hematopoietic IFN-inducible nuclear protein with 200-amino acids (HIN-200) domain (Schattgen and Fitzgerald 2011). Therefore, these proteins can theoretically bind nucleic acids and recruit ASC to trigger the formation of an inflammasome. Indeed, AIM2 can form an inflammasome whose assembly is stimulated by recognition of cytosolic DNA of bacterial or viral origin (Fernandes-Alnemri et al. 2010; Jones et al. 2010; Rathinam et al. 2010; Sauer et al. 2010), or self-DNA from apoptotic cells (Choubey 2012; Zhang et al. 2013). Recent crystal structures of AIM2 complexed with DNA have provided particular insight into the mechanism of AIM2 inflammasome activation. Binding of DNA to the HIN domain of AIM2 results in a conformational change and AIM2 oligomerization around the DNA molecule, which then allows for the recruitment of ASC and caspase-1 and inflammasome assembly (Jin et al. 2012, 2013).

In addition to AIM2, humans have three additional ALRs, IFI16, IFIX, and MNDA, whereas mice have an expanded repertoire of ALRs that includes at least 13 members (Brunette et al. 2012). Most of these ALRs remain poorly characterized; however, a number of murine ALRs were found to colocalize with ASC and trigger IL-1β production, suggesting that they can form inflammasomes (Brunette et al. 2012). Furthermore, human IFI16 can form an inflammasome in response to Kaposi's sarcoma–associated herpesvirus infection, and activation of IFI16 in quiescent "bystander" CD4 T cells during HIV infection was found to trigger massive pyroptosis of T cells and is proposed to be a main driver of depletion of CD4 T cells during progression to AIDS (Monroe et al. 2014).

Finally, the RIG-I like receptor (RLR) family member RIG-I, which is best known as an inducer of type I IFN production in response to recognition of viral RNA, was also shown to form an inflammasome (Poeck et al. 2010). However, it remains unclear what determines when RIG-I forms an inflammasome versus when it simply triggers type I IFN production.

INFLAMMASOMES IN HOST DEFENSE AGAINST INFECTIONS

Considering that the main effect of inflammasome activation is pyroptosis and/or the secretion of IL-1β and IL-18, protection against invading microorganisms is likely the primary function of this arm of the innate immune system (Franchi et al. 2012). This is further demonstrated by the abundance of microbe-sensing NLRs, ALRs, and RLRs described so far, which detect microorganisms both directly (e.g., NLRC4, AIM2, NLRP1b, IFI16, RIG-I) and indirectly (e.g., NLRP3). In general, inflammasome-activating microbes can be divided into three categories: intracellular pathogens, extracellular pathogens that secrete toxins or inject virulence proteins into the host cell, and passively "invading" commensals. All three types of inflammasome-activating microbes are frequently found at mucosal and nonmucosal surfaces throughout the body. Not surprisingly, inflammasome proteins are expressed at these surfaces, most prominently by macrophages and dendritic cells (Kummer et al. 2007; Lech et al. 2010; Guarda et al. 2011). Nonetheless, other cell types at the host–microbe interface are increasingly recognized to respond to invading microbes through inflammasome activation. Below, we discuss the main tissues and cells involved in inflammasome-mediated defense against microbial infections.

Intestine

The intestinal tract is continuously colonized with trillions of microbes and continuously exposed to new microbes that are ingested daily. Despite the abundance of commensal bacteria that can potentially activate inflammasomes, the function of inflammasomes in the intestine has been most clearly demonstrated in the context of pathogenic bacterial infections. In the majority of these infections, inflammasomes initiate protective responses: *Salmonella typhimurium*, *Citrobacter rodentium*, *Listeria monocytogenes*, and *Clostridium difficile* infections in mice are all exacerbated in the absence of caspase-1/11 and/or ASC, resulting in increased bacterial burdens, inflammatory responses, tissue damage, and death (Tsuji et al. 2004; Lara-Tejero et al. 2006; Raupach et al. 2006; Ng et al. 2010; Liu et al. 2012b).

Several NLRs have been implicated in the recognition of these pathogens. Both NLRP3 and NLRC4 were shown to detect *S. typhimurium* (Broz et al. 2010, 2012; Carvalho et al. 2012), and IL-1β and IL-18 were found to be important for controlling *S. typhimurium* infection (Raupach et al. 2006). Similar to *S. typhimurium*, *C. rodentium* infection results in NLRP3- and NLRC4-dependent inflammatory responses that limit bacterial burden and tissue pathology, which was proposed to be IL-18 dependent (Kayagaki et al. 2011b; Gurung et al. 2012b; Liu et al. 2012a; Alipour et al. 2013). Unlike for *S. typhimurium*, NLRP3 and NLRC4 were shown to sense *C. rodentium* in nonhematopoietic cells, most likely in intestinal epithelial cells (Gurung et al. 2012a; Nordlander et al. 2013; Song-Zhao et al. 2013). This matches the localization of *C. rodentium*, which does not invade host cells and instead remains firmly attached to intestinal epithelial cells. Other intestinal pathogens that were shown to activate inflammasomes are: (1) *C. difficile*, which secretes two NLRP3-activating toxins resulting in proinflammatory responses that restrict infection; (2) *Yersinia*, which can trigger both the NLRP3 and NLRC4 inflammasomes (Brodsky et al. 2010); (3) and the yeast *Candida albicans*, which has also been shown to activate both NLRP3 and NLRC4, leading to protection in mice (Hise et al. 2009; Joly et al. 2009; Tomalka et al. 2011). Because *C. albicans* is not known to express any of the currently known NLRC4 ligands, this suggests a novel pathway to activate this NLR.

The importance of inflammasome activation in mediating bacterial control is also highlighted by inflammasome-evasion strategies employed by several pathogens. This is nicely demonstrated by the food-borne pathogen *L. monocytogenes*; in vitro, *L. monocytogenes* induces inflammasome responses via NLRP3 through listeriolysin (Warren et al. 2008; Meixenberger et al. 2010; Wu et al. 2010), via NLRC4 through flagellin (Warren et al. 2008), and via AIM2 and RIG-I through released DNA (Kim et al. 2010; Sauer et al. 2010; Tsuchiya et al. 2010; Wu et al. 2010; Abdullah et al. 2012). However, DNA release is kept at a minimum in wild-type bacteria, and flagellin is strongly down-regulated during murine infection (Sauer et al. 2010, 2011). Flagellin-down-regulation is also observed for *S. typhimurium*, which limits NLRC4-mediated pyroptosis (Miao et al. 2010a). When intracellular flagellin expression is enforced, both *S. typhimurium* and *L. monocytogenes* are strongly attenuated because of the potent induction of NLRC4-mediated pyroptosis, showing the potential of this type of cell death to limit intracellular bacterial growth and spread (Miao et al. 2010b; Sauer et al. 2011). Inflammasome-evasion is also displayed by *Yersinia*, which secretes an effector protein that limits inflammasome activation by the *Yersinia* T3SS (Brodsky et al. 2010). The function of caspase-11-mediated pyroptosis remains more enigmatic. Although potentially limiting bacterial intracellular replication, *S. typhimurium* was shown to exploit this pathway as an "exit strategy" for further dissemination as caspase-11-deficient mice showed decreased infection severity (Broz et al. 2012).

Lung

A second major interface between the host and the environment is the lung. Although not as abundantly colonized with microbes as the intestinal tract, the surface of the lung, particularly in the upper airways, is continuously exposed to a wide variety of commensal and pathogenic bacteria, viruses, and fungi. As in the intestine, several different inflammasomes have been shown to detect invading bacteria in the lung: the NLRP3 inflammasome senses infection with *Klebsiella pneumoniae*, *Chlamydia pneumoniae*, *Haemophilus influenzae*, *Streptococcus pneumoniae*, and *Staphylococcus aureus*, the latter two through the detection of pore-forming toxins (Willingham et al. 2009; He et al. 2010; McNeela et al. 2010; Kebaier et al. 2012; Rotta Detto Loria et al. 2013); the NLRC4 inflammasome responds to a number of flagellated bacteria including *Legionella pneumoniae* and *Burkholderia pseudomallei* (Case et al. 2009; Ceballos-Olvera et al. 2011); the NLRP1b inflammasome is activated by secreted toxins from *B. anthracis* (Kovarova et al. 2012); and both the NLRP3 and AIM2 inflammasomes are proposed to play a role during *Mycobacterium tuberculosis* infection (Dorhoi et al. 2012; Saiga et al. 2012). Despite the induction of potent proinflammatory immune responses, which might not always be desirable in the lung, the majority of inflammasome activation helps to fight and resolve bacterial lung infections through pyroptosis, IL-1β and IL-18 secretion.

The role of inflammasomes in protection against viral infections in the lung has been clearly demonstrated by multiple studies. Influenza A virus was shown to strongly activate NLRP3 in the lung, resulting in caspase-1-mediated inflammatory responses that provided protection against and healing after infection (Kanneganti et al. 2006; Allen et al. 2009; Ichinohe et al. 2009; Thomas et al. 2009). Similarly, several other lung viruses were shown to induce NLRP3-mediated immune responses, including Rhinovirus (Triantafilou et al. 2013b), human respiratory syncytial virus (RSV) (Segovia et al. 2012; Triantafilou et al. 2013a), and Varicella-zoster virus (Nour et al. 2011). Whereas NLRP3 activation seems to occur mainly in macrophages, RIG-I was demonstrated to sense influenza A virus through detection of viral RNA in lung epithelial cells (Pothlichet et al. 2013), resulting in a protective type I interferon response. Interestingly, influenza A simultaneously uses the viral protein NS1 to inhibit RIG-I (Gack et al. 2009).

Several viral proteins have been shown to be responsible for NLRP3 activation. Interestingly, all of these proteins target membranes of intracellular organelles. The influenza A M2 and rhinovirus 2B ion channel proteins induce ion fluxes from the Golgi and/or endoplasmic reticulum (Ichinohe et al. 2010; Triantafilou et al. 2013b); the viroporin SH protein from RSV forms pores in Golgi membranes; and the influenza A protein PB1-F2 is incorporated in the membrane of phagolysosomes, in which it has an as-yet unknown function (McAuley et al. 2013). Although these viruses deliberately express these inflammasome-activating proteins, the subsequent initiation of proinflammatory immune responses is usually detrimental to the virus. From the host side, this highlights the strength of "ligand selection" by NLRP3; from the viral side, this suggests an evolutionary difficulty to avoid NLRP3 activation. To circumvent this problem, some viruses, like measles virus and several herpes viruses, have likely developed mechanisms to directly inhibit inflammasomes to remain "under the radar" of the immune system (Gregory et al. 2011; Komune et al. 2011; Johnson et al. 2013).

Although many lung pathogens can directly activate the inflammasome, an additional proinflammatory stimulus comes from host-derived DAMPs following infection-associated tissue damage. Sensing of this type of damage seems to be restricted to the NLRP3 inflammasome, which has been shown to be activated in response to damage-associated molecules like uric acid (Gasse et al. 2009), extracellular ATP (Riteau et al. 2010), and serum amyloid A (Ather et al. 2011). Whether this additional inflammasome activation is detrimental or beneficial depends on the type and severity of the infection.

Other Examples of Inflammasome Activation in Host–Microbial Interactions

Many other surfaces and cells that come in contact with microbes depend on inflammasome-mediated defense strategies. For instance, *Francisella tularensis* induces AIM2-mediated inflammatory responses in the skin during subcutaneous infection (Fernandes-Alnemri et al. 2010) and NLRP3 is activated by the skin commensal *Propionibacterium acnes* (Kistowska et al. 2014), by hepatitis C virus in hepatic macrophages (Burdette et al. 2012), and by encephalomyocarditis virus through the viroporin 2B (Ito et al. 2012). Systemically, CD4 T cells are subject to HIV-induced IFI16 activation, which leads to caspase-1-mediated pyroptosis and subsequent CD4 T-cell depletion (Doitsh et al. 2014; Monroe et al. 2014).

INFLAMMASOMES AND THE COMMENSAL MICROBIOTA

In addition to the many studies linking the inflammasome to defense against pathogenic microbes, recent studies have suggested that the inflammasome may also play a role in shaping the composition of the intestinal microbiota. Our group found that mice lacking NLRP6 harbored an altered intestinal microbiota that was characterized by the presence of *Prevotellaceae* species and conferred increased susceptibility to DSS-induced colitis. This alteration in microbial composition and susceptibility to colitis was communicable to wild-type mice through cohousing or flora transfer, which demonstrates that this phenotype is mediated by alterations in the intestinal microbiota (Chen et al. 2011; Elinav et al. 2011; Normand et al. 2011; Hu et al. 2013). A similar phenotype was observed in mice lacking other inflammasome components or effectors, including $ASC^{-/-}$, caspase-$1^{-/-}$, and IL-$18^{-/-}$ mice. This suggests that the assembly of an NLRP6 inflammasome and secretion of IL-18 are required for maintenance of intestinal hemostasis through regulation of the composition of the microbiota, and that the absence of the NLRP6 inflammasome leads to acquisition or expansion of potentially pathogenic members of the microbiota (Elinav et al. 2011). In addition to affecting susceptibility to DSS colitis, inflammasome-dependent alterations in the microbiota also exacerbated carcinogenesis in an inflammation-dependent model of colon cancer (Hu et al. 2013). Furthermore, alterations in the microbiota in mice deficient in NLRP3, NLRP6, or IL-18 conferred susceptibility to diet-induced metabolic syndrome, including nonalcoholic fatty liver disease and progression to nonalcoholic steatohepatitis and obesity (Henao-Mejia et al. 2012). The precise mechanisms by which NLRP6 regulates microbial composition in the intestine remain unclear, especially because the stimuli for NLRP6 activation are still completely unknown. However, it was recently found that NLRP6 mice display a defect in mucus secretion by goblet cells in the intestine, which leads to impaired mucosal defense (Wlodarska et al. 2014). Therefore, an attractive possibility is that defective mucus secretion in NLRP6 mice may favor outgrowth of potentially pathogenic members of the microbiota and invasion of the gut epithelial environment by these organisms.

Examination of the role of the inflammasome in regulation of the microbiota promises to be a particularly interesting area of research in the future and elucidation of the mechanisms by which inflammasomes are activated in response to members of the intestinal microbiota will be of particular importance.

NONMICROBIAL INFLAMMASOME ACTIVATORS AND DISEASE

Chronic, low-grade inflammation has been linked to a variety of diseases of both obvious and nonobvious inflammatory nature. The triggers of such inflammation can be either microbial derived (e.g., members of the microbiota, as discussed above) or nonmicrobial (e.g., undegraded self-nucleic acids). Nonmicrobial triggers of inflammation can be separated into two categories: endogenous (i.e., self-derived) and exogenous (i.e., environmental). In this section, we will focus on the role of nonmicrobial triggers of the inflammasome in a variety of diseases. Where possible, we will also highlight cases in which treatment with IL-1 blocking therapies has been attempted and assess its relative success or failure.

Cryopyrin-Associated Periodic Syndromes

The genesis of the inflammasome field lies in the discovery that the hereditary diseases Familial Cold Autoinflammatory syndrome (FCAS) and Muckle–Wells syndrome are both caused by mutations in a single novel gene containing a PYRIN domain, which is now known as NLRP3 (Hoffman et al. 2001). Shortly thereafter, it was discovered that de novo mutations in NLRP3 also cause neonatal-onset multisystem inflammatory disease (NOMID), which is also known as chronic infantile neurological cutaneous and articular syndrome (CINCA) (Aksentijevich et al. 2002; Feldmann et al. 2002). These disorders are collectively referred to as Cryopyrin-Associated Periodic syndromes (CAPS) and are considered to be autoinflammatory rather than autoimmune because they are mediated primarily by cytokines of the innate immune system, most notably IL-1β (Kastner et al. 2010; Dinarello et al. 2012). Mice engineered to contain human mutant NLRP3 also show a CAPS-like disorder, and macrophages from these mice activate caspase-1 and secrete IL-1β directly in response to microbial PAMPs without the need for signal 2 (e.g., ATP) (Meng et al. 2009). This suggests that CAPS results from the proclivity of NLRP3 mutants to assemble into active inflammasomes in response to stimuli that are normally insufficient to trigger inflammasome activation.

CAPS patients are highly responsive IL-1 blockade, which suggests that dysregulated production of IL-1β by the NLRP3 inflammasome is the major driver of pathology in CAPS (Hoffman et al. 2004; Goldbach-Mansky et al. 2006; Dinarello et al. 2012; Jesus and Goldbach-Mansky 2014). However, a recent study looking at the role of IL-18 in mice with CAPS-associated NLRP3 mutations found that IL-18 deficiency, like IL-1β deficiency, also dramatically ameliorated disease (Brydges et al. 2013). Furthermore, mice lacking both IL-1β and IL-18 also maintained some residual caspase-1-dependent disease, suggesting that cytokine-independent effects of the inflammasome, such as pyroptosis, may also contribute to CAPS pathology.

Gout

Gout is a form of inflammatory arthritis that most commonly affects the joint at the base of the big toe and results from the pathological accumulation of uric acid, which is the end product of purine catabolism, and the subsequent formation and deposition of uric acid crystals (Rock et al. 2013). In 2006, it was discovered that uric acid crystals can activate the NLRP3 inflammasome and trigger IL-1β secretion, which suggested that inflammasome activation and IL-1β secretion may be a critical driver of arthritis in gout (Martinon et al. 2006). Accordingly, IL-1 blockade was subsequently found to be a highly effective treatment for gout (So et al. 2007; Jesus and Goldbach-Mansky 2014). Osteoarthritis, which is characterized by the degeneration of cartilage, has also been shown to involve NLRP3 activation. Synovial uric acid correlated strongly with synovial fluid IL-1β and IL-18 in patients, and mice deficient in NLRP3 had significantly reduced pathogenesis in osteoarthritis mouse models (Denoble et al. 2011; Jin et al. 2011). Although IL-1 blocking therapy has so far shown less promising results as compared to the treatment of gout, the above results suggests that a detailed investigation of therapies targeting the NLRP3 inflammasome or its mediators in treating osteoarthritis might be fruitful.

Atherosclerosis

Atherosclerosis is a thickening of the artery wall caused by the accumulation and deposition of cholesterol, calcium, and cellular debris in atherosclerotic plaques (Grebe and Latz 2013). Because the accumulation of cholesterol crystals in atherosclerotic plaques is a common feature of atherosclerosis, and various crystals are known to trigger activation of the NLRP3 inflammasome, the Latz group examined whether the inflammasome might be involved in atherosclerosis (Duewell et al. 2010). They found that cholesterol crystals could trigger NLRP3 inflammasome activation, and that the inflammasome contributed to the development of atherosclerosis in mice lacking the low-density lipoprotein (LDL) receptor. Cholesterol crystals were also found to trigger NLRP3 activation in human macrophages (Rajamaki et al. 2010). Thus, inflammasome activation in response to cholesterol crystals may be a critical early trigger of inflammation in atherosclerosis. Notably, recent studies showed that the endocytic PRR CD36 is critical for NLRP3 activation by oxidized LDL, but not in vitro–generated cholesterol crystals (Sheedy et al. 2013). These studies suggest that receptor-mediated uptake of LDL is a primary event in crystal formation and, therefore, inflammasome activation during atherosclerosis (Sheedy et al. 2013). Despite these compelling findings, the importance of the NLRP3 inflammasome in atherosclerosis remains unclear because NLRP3 deficiency had no effect on atherosclerosis development in ApoE-deficient mice fed a high fat diet, which is another common model of atherosclerosis (Menu et al. 2011). However, caspase-1 and CD36 deficiency significantly reduced the severity of atherosclerosis in this model (Chi et al. 2004; Gage et al. 2012; Usui et al. 2012; Sheedy et al. 2013).

Although the role of the NLRP3 inflammasome in atherosclerosis is controversial, the role of IL-1 appears to be clear: both IL-1β deficiency and IL-1 blockade can ameliorate atherosclerosis in mice (Elhage et al. 1998; Kirii et al. 2003). Clinical trials testing the effect of IL-1 blockade on atherosclerosis in humans are in progress (Sheedy and Moore 2013).

Alzheimer's Disease and Amyotrophic Lateral Sclerosis

Recent studies have suggested that misfolded protein aggregates lead to activation of the NLRP3 inflammasome in two neurodegenerative diseases: Alzheimer's disease and amyotrophic lateral sclerosis (ALS) (Masters and O'Neill 2011; Walsh et al. 2014). Alzheimer's disease is a chronic neurodegenerative disease that mainly affects cognitive functioning and is the most common cause of dementia. ALS, which is also known as Lou Gehrig's disease, is a neurodegenerative disease that results from the progressive death of motor neurons, which eventually leads to loss of control of voluntary muscles and, finally, paralysis and death. Both Alzheimer's and ALS are associated with the accumulation of protein aggregates; in Alzheimer's, the amyloid-β protein (Aβ) forms extracellular plaques that are thought to contribute to disease development, whereas mutant forms of superoxide dismutase 1 (SOD1) that form toxic aggregates are the best characterized cause of ALS (Walsh et al. 2014). Fibrilar Aβ was found to trigger activation of the NLRP3 inflammasome in LPS-primed microglial cells, and inflammasome activation and IL-1β secretion contribute to the inflammatory response to Aβ in the brain (Halle et al. 2008). Furthermore, ALS-linked mutant SOD1 was also found to trigger inflammasome activation in microglia in a model of ALS and caspase-1 or IL-1β deficiency significantly ameliorated neurodegenerative disease in SOD1 mutant mice (Meissner et al. 2010a). Aggregates of prion proteins have also been shown to activate the NLRP3 inflammasome, suggesting that aggregated proteins may represent a common class of sterile triggers of inflammasome activation (Hafner-Bratkovic et al. 2012; Shi et al. 2012). Finally, although most studies have focused on the role of the NLRP3 inflammasome, SNPs in NLRP1 have also been linked to Alzheimer's disease (Pontillo et al. 2012a). This suggests that multiple NLRs may play important roles in neurodegenerative disorders.

Taken together, these data suggest that the inflammasome may contribute to the pathogenesis of neurodegenerative disorders that are associated with the accumulation of protein aggregates and where inflammation contributes to disease development and progression. Furthermore, they highlight inflammasome or IL-1 blocking therapies as potential treatments for neurodegenerative disease.

Metabolic Syndrome and Type 2 Diabetes

The role of the inflammasome in metabolic syndrome and type 2 diabetes (T2D) can be separated into two major mechanistic categories: (1) direct roles mediated by sensing of endogenous inflammasome activators; and (2) indirect roles mediated by inflammasome-associated alterations in the composition of the intestinal microbiota. As the latter has been described above (see the section Inflammasomes and the Commensal Microbiota), we will focus on the potential direct role of the inflammasome in metabolic syndrome in this section; however, it is important to note that alterations in the microbiota could also contribute to phenotypes discussed below because these studies did not consider or control for potential effects of the microbiota.

A number of studies have shown that mice deficient in NLRP3, caspase-1, and ASC are protected from high fat diet (HFD)–induced insulin resistance, glucose intolerance, inflammation, and, in some studies, obesity (Stienstra et al. 2010, 2011; Zhou et al. 2010; Vandanmagsar et al. 2011; Wen et al. 2011). Multiple potential mechanisms by which the NLRP3 inflammasome may become activated in HFD-induced metabolic syndrome have been proposed: through binding of TXNIP, which may respond to endoplasmic reticulum–mediated cell stress (Zhou et al. 2010; Lerner et al. 2012; Oslowski et al. 2012); by oligomers of islet amyloid polypeptide (IAPP), a pancreatic hormone that is cosecreted with insulin and triggers IL-1β secretion by islet macrophages (Masters et al. 2010); by ceramide, a lipid that accumulates in adipose tissues in response to HFD (Vandanmagsar et al. 2011); by palmitate, a saturated fatty acid whose concentration in serum is elevated by HFD (Wen et al. 2011); and by endocannabinoids that, in a different model of T2D, activate the NLRP3 inflammasome in macrophages in a cannabinoid receptor-dependent manner (Jourdan et al. 2013).

Where tested, the effects of NLRP3 deficiency on diabetes in mice are partially phenocopied by IL-1R deficiency (Masters et al. 2010). Furthermore, inhibition of NLRP3 inflammasome activation ameliorated metabolic changes associated with HFD (Stienstra et al. 2010; Zhou et al. 2010). In humans, proof-of-concept studies have suggested that IL-1 blockade may significantly ameliorate T2D (Larsen et al. 2007), and larger-scale clinical trials are underway to definitively determine the potential of this treatment strategy (Boni-Schnetzler and Donath 2013).

Autoimmunity

IL-1β and IL-18 are critical for the initiation and control of the adaptive immune response; for example, IL-1β is intimately involved in the instruction of Th17 responses, which are implicated in many autoimmune disorders, and IL-18 supports Th1 responses (Garlanda et al. 2013). For this reason, one might expect that the inflammasome would also play an important role in autoimmunity. Indeed, a variety of studies in both humans and mice have suggested that multiple NLRs play important roles in autoimmune disease. The clearest connection between autoimmunity and the inflammasome is in the case of NLRP1. SNPs in NLRP1 have been linked to vitiligo and vitiligo-associated Addison's disease (Jin et al. 2007), Type 1 diabetes (Magitta et al. 2009), systemic lupus erythematosus (Pontillo et al. 2012b), Kawasaki disease (Onoyama et al. 2012), Addison's disease (Magitta et al. 2009; Zurawek et al. 2010), celiac disease (Pontillo et al. 2011), rheumatoid arthritis (Sui et al. 2012), systemic sclerosis (Dieude et al. 2011), and autoimmune thyroid disease (Alkhateeb et al. 2013). Despite these findings, confirmation of the role of NLRP1 in autoimmunity in mice has remained challenging because mice have three orthologs of NLRP1, whereas humans have only one. Therefore, the mechanism by which NLRP1 SNPs influence autoimmunity remains largely unclear; however, a recent study found that autoimmunity-associated NLRP1 SNPs may lead to greater processing of IL-1β in response to inflammatory stimuli, such as TLR ligands (Levandowski et al. 2013).

Although most findings connecting NLRs to autoimmunity have focused on NLRP1, a few SNPs in NLRP3 have also been linked to autoimmunity. Mutations in NLRP3 have been linked to celiac disease (Pontillo et al. 2010, 2011), and type 1 diabetes (Pontillo et al. 2010). Furthermore, NLRP3-, and IL-1R-deficient mice show reduced disease in a mouse model of multiple sclerosis; this effect is likely a result of defective Th17 differentiation (Sutton et al. 2006; Gris et al. 2010).

Finally, AIM2 has also been suggested to play a role in autoimmunity and autoinflammation. AIM2 blockade was found to ameliorate development of autoimmunity in a model of lupus that is induced via immunization with DNA from apoptotic cells (Zhang et al. 2013). Furthermore, AIM2 has been implicated in the chronic skin disorder psoriasis (Dombrowski et al. 2011).

Environmental Inflammasome Activators: Allergens and Particulates

The role of the inflammasome in the allergic response remains less well studied than its role in antimicrobial defense or in chronic inflammatory diseases. However, a number of studies have shown that certain allergens can indeed trigger inflammasome activation. House dust mite allergens can trigger NLRP3 inflammasome activation in keratinocytes, and honey bee venom can trigger inflammasome activation in both keratinocytes and macrophages (Dai et al. 2011; Dombrowski et al. 2012; Palm and Medzhitov 2013). However, the role of the inflammasome in allergic responses to HDM and venoms remains unclear. Notably, in the case of bee venom, components of the inflammasome were dispensable for the allergic response and instead the inflammasome was critical for the early inflammatory response to envenomation (Palm and Medzhitov 2013). Also, NLRP3 was found to be dispensable for the allergic response to intranasal HDM allergen (Allen et al. 2012a). Aluminum hydroxide, which is an adjuvant notable for its ability to induce allergic responses under certain conditions, can also trigger NLRP3 inflammasome activation (Li et al. 2007, 2008; Eisenbarth et al. 2008; Franchi and Nunez 2008; Kool et al. 2008; McKee et al. 2009). Finally, the inflammasome has been linked to allergic contact dermatitis (ACD) because contact hypersensitivity, which is a model for ACD, was found to require the NLRP3 inflammasome (Sutterwala et al. 2006).

Particulate environmental substances have also been shown to activate the inflammasome. For example, both silica and asbestos can trigger activation of the NLRP3 inflammasome and the inflammasome plays a critical role in the progressive pulmonary fibrotic disorders silicosis and asbestosis, which result from the inhalation of silica and asbestos, respectively (Cassel et al. 2008; Dostert et al. 2008; Hornung et al. 2008). Finally, necrosis triggered by pressure disruption, hypoxia, complement lysis, or chemically induced epithelial cell injury has also been shown to lead to activation of the NLRP3 inflammasome (Iyer et al. 2009; Li et al. 2009). This activation was mediated by ATP release from mitochondria from necrotic cells, and NLRP3 deficiency reduced mortality in a renal ischemic tubular necrosis model (Iyer et al. 2009).

PERSPECTIVE AND FUTURE PROSPECTS

It is abundantly clear that one of the major beneficial roles of the inflammasome is to sense microbial infection and mediate a rapid program of host defense through the immediate secretion of pre-made cytokines and triggering of cell death. These responses generally are highly efficient in fighting off infectious agents of various origins. However, many nonmicrobial activators of the inflammasome are also known, suggesting that inflammasomes can also function as sensors of nonmicrobial signals (e.g., sterile mediators of membrane damage or cellular stress). Most of these nonmicrobial triggers of the inflammasome have been studied mainly because of their pathological roles in disease, whereas examples of beneficial effects of inflammasome activation by nonmicrobial triggers are few and far between. Therefore, current studies support the idea that inflammasome activation by noninfectious triggers is largely unintentional. It is, however, possible that beneficial effects of inflammasome activation by nonmicrobial triggers have been overlooked. It is also notable that almost all known noninfectious triggers of the inflammasome mediate activation through NLRP3, which seems to be uniquely able to respond to a wide range of stimuli. So far it remains unclear whether other NLRs can also sense nonmicrobial signals of physiological stress.

Finally, there remain many NLRs whose respective roles in host defense and/or physiology are just beginning to be appreciated, such as NLRP10 (Eisenbarth et al. 2012), NLRC5 (Cui et al. 2010; Meissner et al. 2010b), NLRC3 (Schneider et al. 2012; Zhang et al. 2014), and many more whose functions are still completely unknown. With these NLRs, one of the major hurdles to overcome seems to be identifying the activating signals. The reason that it has been so difficult to identify ligands and activities for these NLRs is unclear. However, one intriguing possibility is that these NLRs do not respond to the expected signals (e.g., microbial infection), do not engage traditional inflammasome effector responses (e.g., caspase-1 activation, IL-1β or IL-18 secretion, and pyroptosis), or trigger inflammasome activation in unexpected cell types (e.g., lymphocytes or goblet cells). Indeed, the majority of studies of the effects of inflammasome activation have focused on the roles of IL-1β, IL-18, and pyroptosis; furthermore, most studies have looked at inflammasome activation in macrophages. However, a few studies have highlighted potential cytokine- and pyroptosis-independent roles for the inflammasome; for example, inflammasome activation by pore-forming toxins triggers caspase-1-dependent activation of membrane repair (Gurcel 2006), and NLRC4 activation induced by cytosolic delivery of flagellin was shown to trigger rapid production of inflammatory lipid mediators in a caspase-1-dependent manner (von Moltke et al. 2012). Furthermore, it is clear that inflammasomes exist in multiple cell types, including both hematopoetic and nonhematopoetic cells. Additional examples of cytokine- and pyroptosis-independent effects of the inflammasome, and of roles for the inflammasome in additional cell types will presumably be uncovered in future studies. Such studies should even further expand our view of the role of the inflammasome in host physiology and antimicrobial defense.

ACKNOWLEDGMENTS

This work was supported by a Rubicon Fellowship from the Netherlands Organization of Scientific Research (NWO) (M.R.d.Z.), the Cancer Research Institute Irvington Fellowship Program (N.W.P.), a fellowship from Howard Hughes Medical Institute-The Helen Hay Whitney Foundation (S.Z.), Department of Defense Grant (W81XWH-11-1-0745) (R.A.F.), and the Howard Hughes Medical Institute (R.A.F. and M.R.d.Z.).

REFERENCES

Abdullah Z, Schlee M, Roth S, Mraheil MA, Barchet W, Bottcher J, Hain T, Geiger S, Hayakawa Y, Fritz JH, et al. 2012. RIG-I detects infection with live *Listeria* by sensing secreted bacterial nucleic acids. *EMBO J* **31:** 4153–4164.

Aksentijevich I, Nowak M, Mallah M, Chae JJ, Watford WT, Hofmann SR, Stein L, Russo R, Goldsmith D, Dent P, et al. 2002. De novo *CIAS1* mutations, cytokine activation, and evidence for genetic heterogeneity in patients with neonatal-onset multisystem inflammatory disease (NOMID): A new member of the expanding family of pyrin-associated autoinflammatory diseases. *Arthritis Rheum* **46:** 3340–3348.

Ali SR, Timmer AM, Bilgrami S, Park EJ, Eckmann L, Nizet V, Karin M. 2011. Anthrax toxin induces macrophage death by p38 MAPK inhibition but leads to inflammasome activation via ATP leakage. *Immunity* **35:** 34–44.

Alipour M, Lou Y, Zimmerman D, Bording-Jorgensen MW, Sergi C, Liu JJ, Wine E. 2013. A balanced IL-1β activity is required for host response to *Citrobacter rodentium* infection. *PLoS ONE* **8:** e80656.

Alkhateeb A, Jarun Y, Tashtoush R. 2013. Polymorphisms in *NLRP1* gene and susceptibility to autoimmune thyroid disease. *Autoimmunity* **46:** 215–221.

Allen IC, Scull MA, Moore CB, Holl EK, McElvania-TeKippe E, Taxman DJ, Guthrie EH, Pickles RJ, Ting JP. 2009. The NLRP3 inflammasome mediates in vivo innate immunity to influenza A virus through recognition of viral RNA. *Immunity* **30:** 556–565.

Allen IC, Jania CM, Wilson JE, Tekeppe EM, Hua X, Brickey WJ, Kwan M, Koller BH, Tilley SL, Ting JP. 2012a. Analysis of NLRP3 in the development of allergic airway disease in mice. *J Immunol* **188:** 2884–2893.

Allen IC, Wilson JE, Schneider M, Lich JD, Roberts RA, Arthur JC, Woodford RM, Davis BK, Uronis JM, Herfarth HH, et al. 2012b. NLRP12 suppresses colon inflammation and tumorigenesis through the negative regulation of noncanonical NF-κB signaling. *Immunity* **36:** 742–754.

Amer A, Franchi L, Kanneganti TD, Body-Malapel M, Ozoren N, Brady G, Meshinchi S, Jagirdar R, Gewirtz A, Akira S, et al. 2006. Regulation of *Legionella* phagosome maturation and infection through flagellin and host Ipaf. *J Biol Chem* **281:** 35217–35223.

Anand PK, Kanneganti TD. 2013. NLRP6 in infection and inflammation. *Microbes Infect* **15:** 661–668.

Arlehamn CS, Petrilli V, Gross O, Tschopp J, Evans TJ. 2010. The role of potassium in inflammasome activation by bacteria. *J Biol Chem* **285:** 10508–10518.

Ather JL, Ckless K, Martin R, Foley KL, Suratt BT, Boyson JE, Fitzgerald KA, Flavell RA, Eisenbarth SC, Poynter ME. 2011. Serum amyloid A activates the NLRP3 inflammasome and promotes Th17 allergic asthma in mice. *J Immunol* **187:** 64–63.

Bauernfeind FG, Horvath G, Stutz A, Alnemri ES, MacDonald K, Speert D, Fernandes-Alnemri T, Wu J, Monks BG, Fitzgerald KA, et al. 2009. Cutting edge: NF-κB activating pattern recognition and cytokine receptors license NLRP3 inflammasome activation by regulating NLRP3 expression. *J Immunol* **183:** 787–791.

Bauernfeind F, Bartok E, Rieger A, Franchi L, Nunez G, Hornung V. 2011. Cutting edge: Reactive oxygen species inhibitors block priming, but not activation, of the NLRP3 inflammasome. *J Immunol* **187:** 613–617.

Boni-Schnetzler M, Donath MY. 2013. How biologics targeting the IL-1 system are being considered for the treatment of type 2 diabetes. *Br J Clin Pharmacol* **76:** 263–268.

Boyden ED, Dietrich WF. 2006. *Nalp1b* controls mouse macrophage susceptibility to anthrax lethal toxin. *Nat Genet* **38:** 240–244.

Brodsky IE, Palm NW, Sadanand S, Ryndak MB, Sutterwala FS, Flavell RA, Bliska JB, Medzhitov R. 2010. A *Yersinia* effector protein promotes virulence by preventing inflammasome recognition of the type III secretion system. *Cell Host Microbe* **7:** 376–387.

Broz P, Newton K, Lamkanfi M, Mariathasan S, Dixit VM, Monack DM. 2010. Redundant roles for inflammasome receptors NLRP3 and NLRC4 in host defense against *Salmonella*. *J Exp Med* **207:** 1745–1755.

Broz P, Ruby T, Belhocine K, Bouley DM, Kayagaki N, Dixit VM, Monack DM. 2012. Caspase-11 increases susceptibility to *Salmonella* infection in the absence of caspase-1. *Nature* **490:** 288–291.

Brunette RL, Young JM, Whitley DG, Brodsky IE, Malik HS, Stetson DB. 2012. Extensive evolutionary and functional diversity among mammalian AIM2-like receptors. *J Exp Med* **209:** 1969–1983.

Brydges SD, Broderick L, McGeough MD, Pena CA, Mueller JL, Hoffman HM. 2013. Divergence of IL-1, IL-18, and cell death in NLRP3 inflammasomopathies. *J Clin Invest* **123:** 4695–4705.

Burdette D, Haskett A, Presser L, McRae S, Iqbal J, Waris G. 2012. Hepatitis C virus activates interleukin-1β via caspase-1-inflammasome complex. *J Gen Virol* **93:** 235–246.

Carvalho FA, Nalbantoglu I, Aitken JD, Uchiyama R, Su Y, Doho GH, Vijay-Kumar M, Gewirtz AT. 2012. Cytosolic flagellin receptor NLRC4 protects mice against mucosal and systemic challenges. *Mucosal Immunol* **5:** 288–298.

Case CL, Shin S, Roy CR. 2009. Asc and Ipaf Inflammasomes direct distinct pathways for caspase-1 activation in response to *Legionella pneumophila*. *Infect Immun* **77:** 1981–1991.

Cassel SL, Eisenbarth SC, Iyer SS, Sadler JJ, Colegio OR, Tephly LA, Carter AB, Rothman PB, Flavell RA, Sutterwala FS. 2008. The Nalp3 inflammasome is essential for the development of silicosis. *Proc Natl Acad Sci* **105:** 9035–9040.

Ceballos-Olvera I, Sahoo M, Miller MA, Del Barrio L, Re F. 2011. Inflammasome-dependent pyroptosis and IL-18 protect against *Burkholderia pseudomallei* lung infection while IL-1β is deleterious. *PLoS Pathog* **7:** e1002452.

Chavarria-Smith J, Vance RE. 2013. Direct proteolytic cleavage of NLRP1B is necessary and sufficient for inflammasome activation by anthrax lethal factor. *PLoS Pathog* **9:** e1003452.

Chen GY, Liu M, Wang F, Bertin J, Nunez G. 2011. A functional role for Nlrp6 in intestinal inflammation and tumorigenesis. *J Immunol* **186:** 7187–7194.

Chi H, Messas E, Levine RA, Graves DT, Amar S. 2004. Interleukin-1 receptor signaling mediates atherosclerosis associated with bacterial exposure and/or a high-fat diet in a murine apolipoprotein E heterozygote model: Pharmacotherapeutic implications. *Circulation* **110:** 1678–1685.

Choubey D. 2012. DNA-responsive inflammasomes and their regulators in autoimmunity. *Clin Immunol* **142:** 223–231.

Cui J, Zhu L, Xia X, Wang HY, Legras X, Hong J, Ji J, Shen P, Zheng S, Chen ZJ, et al. 2010. NLRC5 negatively regulates the NF-κB and type I interferon signaling pathways. *Cell* **141:** 483–496.

Dai X, Sayama K, Tohyama M, Shirakata Y, Hanakawa Y, Tokumaru S, Yang L, Hirakawa S, Hashimoto K. 2011. Mite allergen is a danger signal for the skin via activation of inflammasome in keratinocytes. *J Allergy Clin Immunol* **127:** 806–814.

Denoble AE, Huffman KM, Stabler TV, Kelly SJ, Hershfield MS, McDaniel GE, Coleman RE, Kraus VB. 2011. Uric acid is a danger signal of increasing risk for osteoarthritis through inflammasome activation. *Proc Natl Acad Sci* **108:** 2088–2093.

de Zoete MR, Flavell RA. 2013. Interactions between Nod-like receptors and intestinal bacteria. *Front Immunol* **4:** 462.

Dieude P, Guedj M, Wipff J, Ruiz B, Riemekasten G, Airo P, Melchers I, Hachulla E, Cerinic MM, Diot E, et al. 2011. *NLRP1* influences the systemic sclerosis phenotype: A new clue for the contribution of innate immunity in systemic sclerosis-related fibrosing alveolitis pathogenesis. *Ann Rheum Dis* **70:** 668–674.

Dinarello CA, Simon A, van der Meer JW. 2012. Treating inflammation by blocking interleukin-1 in a broad spectrum of diseases. *Nat Rev Drug Discov* **11:** 633–652.

Doitsh G, Galloway NL, Geng X, Yang Z, Monroe KM, Zepeda O, Hunt PW, Hatano H, Sowinski S, Munoz-Arias I, et al. 2014. Cell death by pyroptosis drives CD4 T-cell depletion in HIV-1 infection. *Nature* **505:** 509–514.

Dombrowski Y, Peric M, Koglin S, Kammerbauer C, Goss C, Anz D, Simanski M, Glaser R, Harder J, Hornung V, et al. 2011. Cytosolic DNA triggers inflammasome activation in keratinocytes in psoriatic lesions. *Sci Transl Med* **3:** 82ra38.

Dombrowski Y, Peric M, Koglin S, Kaymakanov N, Schmezer V, Reinholz M, Ruzicka T, Schauber J. 2012. Honey bee (*Apis mellifera*) venom induces AIM2 inflammasome activation in human keratinocytes. *Allergy* **67:** 1400–1407.

Dorhoi A, Nouailles G, Jorg S, Hagens K, Heinemann E, Pradl L, Oberbeck-Muller D, Duque-Correa MA, Reece ST, Ruland J, et al. 2012. Activation of the NLRP3 inflammasome by *Mycobacterium tuberculosis* is uncoupled from susceptibility to active tuberculosis. *Eur J Immunol* **42:** 374–384.

Dostert C, Petrilli V, Van Bruggen R, Steele C, Mossman BT, Tschopp J. 2008. Innate immune activation through Nalp3 inflammasome sensing of asbestos and silica. *Science* **320:** 674–677.

Dostert C, Guarda G, Romero JF, Menu P, Gross O, Tardivel A, Suva ML, Stehle JC, Kopf M, Stamenkovic I, et al. 2009. Malarial hemozoin is a Nalp3 inflammasome activating danger signal. *PLoS ONE* **4:** e6510.

Duewell P, Kono H, Rayner KJ, Sirois CM, Vladimer G, Bauernfeind FG, Abela GS, Franchi L, Nunez G, Schnurr M, et al. 2010. NLRP3 inflammasomes are required for atherogenesis and activated by cholesterol crystals. *Nature* **464:** 1357–1361.

Duncan JA, Gao X, Huang MT, O'Connor BP, Thomas CE, Willingham SB, Bergstralh DT, Jarvis GA, Sparling PF, Ting JP. 2009. *Neisseria gonorrhoeae* activates the proteinase cathepsin B to mediate the signaling activities of the NLRP3 and ASC-containing inflammasome. *J Immunol* **182:** 6460–6469.

Eisenbarth SC, Colegio OR, O'Connor W, Sutterwala FS, Flavell RA. 2008. Crucial role for the Nalp3 inflammasome in the immunostimulatory properties of aluminium adjuvants. *Nature* **453:** 1122–1126.

Eisenbarth SC, Williams A, Colegio OR, Meng H, Strowig T, Rongvaux A, Henao-Mejia J, Thaiss CA, Joly S, Gonzalez DG, et al. 2012. NLRP10 is a NOD-like receptor essential to initiate adaptive immunity by dendritic cells. *Nature* **484:** 510–513.

Elhage R, Maret A, Pieraggi MT, Thiers JC, Arnal JF, Bayard F. 1998. Differential effects of interleukin-1 receptor antagonist and tumor necrosis factor binding protein on fatty-streak formation in apolipoprotein E-deficient mice. *Circulation* **97:** 242–244.

Elinav E, Strowig T, Kau AL, Henao-Mejia J, Thaiss CA, Booth CJ, Peaper DR, Bertin J, Eisenbarth SC, Gordon JI, et al. 2011. NLRP6 inflammasome regulates colonic microbial ecology and risk for colitis. *Cell* **145:** 745–757.

Faustin B, Lartigue L, Bruey JM, Luciano F, Sergienko E, Bailly-Maitre B, Volkmann N, Hanein D, Rouiller I, Reed JC. 2007. Reconstituted NALP1 inflammasome reveals two-step mechanism of caspase-1 activation. *Mol Cell* **25:** 713–724.

Feldmann J, Prieur AM, Quartier P, Berquin P, Certain S, Cortis E, Teillac-Hamel D, Fischer A, de Saint Basile G. 2002. Chronic infantile neurological cutaneous and articular syndrome is caused by mutations in *CIAS1*, a gene highly expressed in polymorphonuclear cells and chondrocytes. *Am J Hum Genet* **71:** 198–203.

Fernandes-Alnemri T, Yu JW, Juliana C, Solorzano L, Kang S, Wu J, Datta P, McCormick M, Huang L, McDermott E, et al. 2010. The AIM2 inflammasome is critical for innate immunity to *Francisella tularensis*. *Nat Immunol* **11:** 385–393.

Fink SL, Bergsbaken T, Cookson BT. 2008. Anthrax lethal toxin and *Salmonella* elicit the common cell death pathway of caspase-1-dependent pyroptosis via distinct mechanisms. *Proc Natl Acad Sci* **105:** 4312–4317.

Franchi L, Nunez G. 2008. The Nlrp3 inflammasome is critical for aluminium hydroxide-mediated IL-1β secretion but dispensable for adjuvant activity. *Eur J Immunol* **38:** 2085–2089.

Franchi L, Amer A, Body-Malapel M, Kanneganti TD, Ozoren N, Jagirdar R, Inohara N, Vandenabeele P, Bertin J, Coyle A, et al. 2006. Cytosolic flagellin requires Ipaf for activation of caspase-1 and interleukin 1β in salmonella-infected macrophages. *Nat Immunol* **7:** 576–582.

Franchi L, Kanneganti TD, Dubyak GR, Nunez G. 2007. Differential requirement of P2X7 receptor and intracellular K^+ for caspase-1 activation induced by intracellular and extracellular bacteria. *J Biol Chem* **282:** 18810–18818.

Franchi L, Munoz-Planillo R, Nunez G. 2012. Sensing and reacting to microbes through the inflammasomes. *Nat Immunol* **13:** 325–332.

Gack MU, Albrecht RA, Urano T, Inn KS, Huang IC, Carnero E, Farzan M, Inoue S, Jung JU, Garcia-Sastre A. 2009. Influenza A virus NS1 targets the ubiquitin ligase TRIM25 to evade recognition by the host viral RNA sensor RIG-I. *Cell Host Microbe* **5:** 439–449.

Gage J, Hasu M, Thabet M, Whitman SC. 2012. Caspase-1 deficiency decreases atherosclerosis in apolipoprotein E-null mice. *Can J Cardiol* **28:** 222–229.

Garlanda C, Dinarello CA, Mantovani A. 2013. The interleukin-1 family: Back to the future. *Immunity* **39:** 1003–1018.

Gasse P, Riteau N, Charron S, Girre S, Fick L, Petrilli V, Tschopp J, Lagente V, Quesniaux VF, Ryffel B, et al. 2009. Uric acid is a danger signal activating NALP3 inflammasome in lung injury inflammation and fibrosis. *Am J Respir Crit Care Med* **179:** 903–913.

Goldbach-Mansky R, Dailey NJ, Canna SW, Gelabert A, Jones J, Rubin BI, Kim HJ, Brewer C, Zalewski C, Wiggs E, et al. 2006. Neonatal-onset multisystem inflammatory disease responsive to interleukin-1β inhibition. *N Engl J Med* **355:** 581–592.

Grebe A, Latz E. 2013. Cholesterol crystals and inflammation. *Curr Rheum Rep* **15:** 313.

Gregory SM, Davis BK, West JA, Taxman DJ, Matsuzawa S, Reed JC, Ting JP, Damania B. 2011. Discovery of a viral NLR homolog that inhibits the inflammasome. *Science* **331:** 330–334.

Grenier JM, Wang L, Manji GA, Huang WJ, Al-Garawi A, Kelly R, Carlson A, Merriam S, Lora JM, Briskin M, et al. 2002. Functional screening of five PYPAF family members identifies PYPAF5 as a novel regulator of NF-κB and caspase-1. *FEBS Lett* **530:** 73–78.

Gris D, Ye Z, Iocca HA, Wen H, Craven RR, Gris P, Huang M, Schneider M, Miller SD, Ting JP. 2010. NLRP3 plays a critical role in the development of experimental autoimmune encephalomyelitis by mediating Th1 and Th17 responses. *J Immunol* **185:** 974–981.

Gross O, Poeck H, Bscheider M, Dostert C, Hannesschlager N, Endres S, Hartmann G, Tardivel A, Schweighoffer E, Tybulewicz V, et al. 2009. Syk kinase signalling couples to the Nlrp3 inflammasome for anti-fungal host defence. *Nature* **459:** 433–436.

Guarda G, Zenger M, Yazdi AS, Schroder K, Ferrero I, Menu P, Tardivel A, Mattmann C, Tschopp J. 2011. Differential expression of NLRP3 among hematopoietic cells. *J Immunol* **186:** 2529–2534.

Gurcel L, Abrami L, Girardin S, Tschopp J, van der Goot FG. 2006. Caspase-1 activation of lipid metabolic pathways in response to bacterial pore-forming toxins promotes cell survival. *Cell* **126:** 1135–1145.

Gurung P, Malireddi RK, Anand PK, Demon D, Vande Walle L, Liu Z, Vogel P, Lamkanfi M, Kanneganti TD. 2012a. Toll or interleukin-1 receptor (TIR) domain-containing adaptor inducing interferon-β (TRIF)-mediated caspase-11 protease production integrates toll-like receptor 4 (TLR4) protein- and Nlrp3 inflammasome-mediated host defense against enteropathogens. *J Biol Chem* **287:** 34474–34483.

Gurung P, Malireddi RS, Anand PK, Demon D, Walle LV, Liu Z, Vogel P, Lamkanfi M, Kanneganti T-D. 2012b. Toll or interleukin-1 receptor (TIR) domain-containing adaptor inducing interferon-β (TRIF)-mediated caspase-11 protease production integrates Toll-like receptor 4 (TLR4) protein-and Nlrp3 inflammasome-mediated host defense against enteropathogens. *J Biol Chem* **287:** 34474–34483.

Hafner-Bratkovic I, Bencina M, Fitzgerald KA, Golenbock D, Jerala R. 2012. NLRP3 inflammasome activation in macrophage cell lines by prion protein fibrils as the source of IL-1β and neuronal toxicity. *Cell Mol Life Sci* **69:** 4215–4228.

Hagar JA, Powell DA, Aachoui Y, Ernst RK, Miao EA. 2013. Cytoplasmic LPS activates caspase-11: Implications in TLR4-independent endotoxic shock. *Science* **341:** 1250–1253.

Halle A, Hornung V, Petzold GC, Stewart CR, Monks BG, Reinheckel T, Fitzgerald KA, Latz E, Moore KJ, Golenbock DT. 2008. The NALP3 inflammasome is involved in the innate immune response to amyloid-β. *Nat Immunol* **9:** 857–865.

He X, Mekasha S, Mavrogiorgos N, Fitzgerald KA, Lien E, Ingalls RR. 2010. Inflammation and fibrosis during *Chlamydia pneumoniae* infection is regulated by IL-1 and the NLRP3/ASC inflammasome. *J Immunol* **184:** 5743–5754.

Henao-Mejia J, Elinav E, Jin C, Hao L, Mehal WZ, Strowig T, Thaiss CA, Kau AL, Eisenbarth SC, Jurczak MJ, et al. 2012. Inflammasome-mediated dysbiosis regulates progression of NAFLD and obesity. *Nature* **482:** 179–185.

Hise AG, Tomalka J, Ganesan S, Patel K, Hall BA, Brown GD, Fitzgerald KA. 2009. An essential role for the NLRP3 inflammasome in host defense against the human fungal pathogen *Candida albicans*. *Cell Host Microbe* **5:** 487–497.

Hoffman HM, Mueller JL, Broide DH, Wanderer AA, Kolodner RD. 2001. Mutation of a new gene encoding a putative pyrin-like protein causes familial cold autoinflammatory syndrome and Muckle–Wells syndrome. *Nat Genet* **29:** 301–305.

Hoffman HM, Rosengren S, Boyle DL, Cho JY, Nayar J, Mueller JL, Anderson JP, Wanderer AA, Firestein GS. 2004. Prevention of cold-associated acute inflammation in familial cold autoinflammatory syndrome by interleukin-1 receptor antagonist. *Lancet* **364:** 1779–1785.

Hornung V, Bauernfeind F, Halle A, Samstad EO, Kono H, Rock KL, Fitzgerald KA, Latz E. 2008. Silica crystals and aluminum salts activate the NALP3 inflammasome through phagosomal destabilization. *Nat Immunol* **9:** 847–856.

Hu B, Elinav E, Huber S, Strowig T, Hao L, Hafemann A, Jin C, Wunderlich C, Wunderlich T, Eisenbarth SC, et al. 2013. Microbiota-induced activation of epithelial IL-6 signaling links inflammasome-driven inflammation with transmissible cancer. *Proc Natl Acad Sci* **110:** 9862–9867.

Ichinohe T, Lee HK, Ogura Y, Flavell R, Iwasaki A. 2009. Inflammasome recognition of influenza virus is essential for adaptive immune responses. *J Exp Med* **206:** 79–87.

Ichinohe T, Pang IK, Iwasaki A. 2010. Influenza virus activates inflammasomes via its intracellular M2 ion channel. *Nat Immunol* **11:** 404–410.

Ireton RC, Gale M Jr, 2011. RIG-I like receptors in antiviral immunity and therapeutic applications. *Viruses* **3:** 906–919.

Ito M, Yanagi Y, Ichinohe T. 2012. Encephalomyocarditis virus viroporin 2B activates NLRP3 inflammasome. *PLoS Pathog* **8:** e1002857.

Iyer SS, Pulskens WP, Sadler JJ, Butter LM, Teske GJ, Ulland TK, Eisenbarth SC, Florquin S, Flavell RA, Leemans JC, et al. 2009. Necrotic cells trigger a sterile inflammatory response through the Nlrp3 inflammasome. *Proc Natl Acad Sci* **106:** 20388–20393.

Iyer SS, He Q, Janczy JR, Elliott EI, Zhong Z, Olivier AK, Sadler JJ, Knepper-Adrian V, Han R, Qiao L, et al. 2013. Mitochondrial cardiolipin is required for Nlrp3 inflammasome activation. *Immunity* **39:** 311–323.

Jesus AA, Goldbach-Mansky R. 2014. IL-1 blockade in autoinflammatory syndromes. *Annu Rev Med* **65:** 223–244.

Jin Y, Mailloux CM, Gowan K, Riccardi SL, LaBerge G, Bennett DC, Fain PR, Spritz RA. 2007. *NALP1* in vitiligo-associated multiple autoimmune disease. *N Engl J Med* **356:** 1216–1225.

Jin C, Frayssinet P, Pelker R, Cwirka D, Hu B, Vignery A, Eisenbarth SC, Flavell RA. 2011. NLRP3 inflammasome plays a critical role in the pathogenesis of hydroxyapatite-associated arthropathy. *Proc Natl Acad Sci* **108:** 14867–14872.

Jin T, Perry A, Jiang J, Smith P, Curry JA, Unterholzner L, Jiang Z, Horvath G, Rathinam VA, Johnstone RW, et al. 2012. Structures of the HIN domain:DNA complexes reveal ligand binding and activation mechanisms of the AIM2 inflammasome and IFI16 receptor. *Immunity* **36:** 561–571.

Jin T, Perry A, Smith P, Jiang J, Xiao TS. 2013. Structure of the absent in melanoma 2 (AIM2) pyrin domain provides insights into the mechanisms of AIM2 autoinhibition and inflammasome assembly. *J Biol Chem* **288:** 13225–13235.

Johnson KE, Chikoti L, Chandran B. 2013. Herpes simplex virus 1 infection induces activation and subsequent inhibition of the IFI16 and NLRP3 inflammasomes. *J Virol* **87:** 5005–5018.

Joly S, Ma N, Sadler JJ, Soll DR, Cassel SL, Sutterwala FS. 2009. Cutting edge: *Candida albicans* hyphae formation triggers activation of the Nlrp3 inflammasome. *J Immunol* **183:** 3578–3581.

Jones JW, Kayagaki N, Broz P, Henry T, Newton K, O'Rourke K, Chan S, Dong J, Qu Y, Roose-Girma M, et al. 2010. Absent in melanoma 2 is required for innate immune recognition of *Francisella tularensis. Proc Natl Acad Sci* **107:** 9771–9776.

Jourdan T, Godlewski G, Cinar R, Bertola A, Szanda G, Liu J, Tam J, Han T, Mukhopadhyay B, Skarulis MC, et al. 2013. Activation of the Nlrp3 inflammasome in infiltrating macrophages by endocannabinoids mediates β cell loss in type 2 diabetes. *Nat Med* **19:** 1132–1140.

Kanneganti TD, Body-Malapel M, Amer A, Park JH, Whitfield J, Franchi L, Taraporewala ZF, Miller D, Patton JT, Inohara N, et al. 2006. Critical role for Cryopyrin/Nalp3 in activation of caspase-1 in response to viral infection and double-stranded RNA. *J Biol Chem* **281:** 36560–36568.

Kastner DL, Aksentijevich I, Goldbach-Mansky R. 2010. Autoinflammatory disease reloaded: A clinical perspective. *Cell* **140:** 784–790.

Kayagaki N, Warming S, Lamkanfi M, Vande Walle L, Louie S, Dong J, Newton K, Qu Y, Liu J, Heldens S, et al. 2011a. Non-canonical inflammasome activation targets caspase-11. *Nature* **479:** 117–121.

Kayagaki N, Warming S, Lamkanfi M, Walle LV, Louie S, Dong J, Newton K, Qu Y, Liu J, Heldens S. 2011b. Non-canonical inflammasome activation targets caspase-11. *Nature* **479:** 117–121.

Kayagaki N, Wong MT, Stowe IB, Ramani SR, Gonzalez LC, Akashi-Takamura S, Miyake K, Zhang J, Lee WP, Muszynski A, et al. 2013. Noncanonical inflammasome activation by intracellular LPS independent of TLR4. *Science* **341:** 1246–1249.

Kebaier C, Chamberland RR, Allen IC, Gao X, Broglie PM, Hall JD, Jania C, Doerschuk CM, Tilley SL, Duncan JA. 2012. *Staphylococcus aureus* α-hemolysin mediates virulence in a murine model of severe pneumonia through activation of the NLRP3 inflammasome. *J Infect Dis* **205:** 807–817.

Kim S, Bauernfeind F, Ablasser A, Hartmann G, Fitzgerald KA, Latz E, Hornung V. 2010. *Listeria monocytogenes* is sensed by the NLRP3 and AIM2 inflammasome. *Eur J Immunol* **40:** 1545–1551.

Kirii H, Niwa T, Yamada Y, Wada H, Saito K, Iwakura Y, Asano M, Moriwaki H, Seishima M. 2003. Lack of interleukin-1β decreases the severity of atherosclerosis in ApoE-deficient mice. *Arterioscler Thromb Vasc Biol* **23:** 656–660.

Kistowska M, Gehrke S, Jankovic D, Kerl K, Fettelschoss A, Feldmeyer L, Fenini G, Kolios A, Navarini A, Ganceviciene R, et al. 2014. IL-1β drives inflammatory responses to Propionibacterium acnes in vitro and in vivo. *J Invest Dermatol* **134:** 677–685.

Kofoed EM, Vance RE. 2011. Innate immune recognition of bacterial ligands by NAIPs determines inflammasome specificity. *Nature* **477:** 592–595.

Komune N, Ichinohe T, Ito M, Yanagi Y. 2011. Measles virus V protein inhibits NLRP3 inflammasome-mediated interleukin-1β secretion. *J Virol* **85:** 13019–13026.

Kool M, Petrilli V, De Smedt T, Rolaz A, Hammad H, van Nimwegen M, Bergen IM, Castillo R, Lambrecht BN, Tschopp J. 2008. Cutting edge: Alum adjuvant stimulates inflammatory dendritic cells through activation of the NALP3 inflammasome. *J Immunol* **181:** 3755–3759.

Kovarova M, Hesker PR, Jania L, Nguyen M, Snouwaert JN, Xiang Z, Lommatzsch SE, Huang MT, Ting JP, Koller BH. 2012. NLRP1-dependent pyroptosis leads to acute lung injury and morbidity in mice. *J Immunol* **189:** 2006–2016.

Kummer JA, Broekhuizen R, Everett H, Agostini L, Kuijk L, Martinon F, van Bruggen R, Tschopp J. 2007. Inflammasome components NALP 1 and 3 show distinct but separate expression profiles in human tissues suggesting a site-specific role in the inflammatory response. *J Histochem Cytochem* **55:** 443–452.

Lamkanfi M, Dixit VM. 2012. Inflammasomes and their roles in health and disease. *Ann Rev Cell Dev Biol* **28:** 137–161.

Lamkanfi M, Amer A, Kanneganti TD, Munoz-Planillo R, Chen G, Vandenabeele P, Fortier A, Gros P, Nunez G. 2007. The Nod-like receptor family member Naip5/Birc1e restricts *Legionella pneumophila* growth independently of caspase-1 activation. *J Immunol* **178:** 8022–8027.

Lara-Tejero M, Sutterwala FS, Ogura Y, Grant EP, Bertin J, Coyle AJ, Flavell RA, Galán JE. 2006. Role of the caspase-1 inflammasome in *Salmonella typhimurium* pathogenesis. *J Exp Med* **203:** 1407–1412.

Larsen CM, Faulenbach M, Vaag A, Volund A, Ehses JA, Seifert B, Mandrup-Poulsen T, Donath MY. 2007. Interleukin-1-receptor antagonist in type 2 diabetes mellitus. *N Engl J Med* **356:** 1517–1526.

Latz E. 2010. NOX-free inflammasome activation. *Blood* **116:** 1393–1394.

Latz E, Xiao TS, Stutz A. 2013. Activation and regulation of the inflammasomes. *Nat Rev Immunol* **13:** 397–411.

Lech M, Avila-Ferrufino A, Skuginna V, Susanti HE, Anders HJ. 2010. Quantitative expression of RIG-like helicase, NOD-like receptor and inflammasome-related mRNAs in humans and mice. *Int Immunol* **22:** 717–728.

Lee GS, Subramanian N, Kim AI, Aksentijevich I, Goldbach-Mansky R, Sacks DB, Germain RN, Kastner DL, Chae JJ. 2012. The calcium-sensing receptor regulates the NLRP3 inflammasome through Ca^{2+} and cAMP. *Nature* **492:** 123–127.

Lerner AG, Upton JP, Praveen PV, Ghosh R, Nakagawa Y, Igbaria A, Shen S, Nguyen V, Backes BJ, Heiman M, et al. 2012. IRE1α induces thioredoxin-interacting protein to activate the NLRP3 inflammasome and promote programmed cell death under irremediable ER stress. *Cell Metab* **16:** 250–264.

Levandowski CB, Mailloux CM, Ferrara TM, Gowan K, Ben S, Jin Y, McFann KK, Holland PJ, Fain PR, Dinarello CA, et al. 2013. *NLRP1* haplotypes associated with vitiligo and autoimmunity increase interleukin-1β processing via the NLRP1 inflammasome. *Proc Natl Acad Sci* **110:** 2952–2956.

Levinsohn JL, Newman ZL, Hellmich KA, Fattah R, Getz MA, Liu S, Sastalla I, Leppla SH, Moayeri M. 2012. Anthrax lethal factor cleavage of Nlrp1 is required for activation of the inflammasome. *PLoS Pathog* **8:** e1002638.

Li H, Nookala S, Re F. 2007. Aluminum hydroxide adjuvants activate caspase-1 and induce IL-1β and IL-18 release. *J Immunol* **178:** 5271–5276.

Li H, Willingham SB, Ting JP, Re F. 2008. Cutting edge: Inflammasome activation by alum and alum's adjuvant effect are mediated by NLRP3. *J Immunol* **181:** 17–21.

Li H, Ambade A, Re F. 2009. Cutting edge: Necrosis activates the NLRP3 inflammasome. *J Immunol* **183:** 1528–1532.

Liu Z, Zaki MH, Vogel P, Gurung P, Finlay BB, Deng W, Lamkanfi M, Kanneganti T.-D. 2012a. Role of inflammasomes in host defense against *Citrobacter rodentium* infection. *J Biol Chem* **287:** 16955–16964.

Liu Z, Zaki MH, Vogel P, Gurung P, Finlay BB, Deng W, Lamkanfi M, Kanneganti TD. 2012b. Role of inflammasomes in host defense against *Citrobacter rodentium* infection. *J Biol Chem* **287:** 16955–16964.

Magitta NF, Boe Wolff AS, Johansson S, Skinningsrud B, Lie BA, Myhr KM, Undlien DE, Joner G, Njolstad PR, Kvien TK, et al. 2009. A coding polymorphism in NALP1 confers risk for autoimmune Addison's disease and type 1 diabetes. *Genes Immun* **10:** 120–124.

Mariathasan S, Newton K, Monack DM, Vucic D, French DM, Lee WP, Roose-Girma M, Erickson S, Dixit VM. 2004. Differential activation of the inflammasome by caspase-1 adaptors ASC and Ipaf. *Nature* **430:** 213–218.

Mariathasan S, Weiss DS, Newton K, McBride J, O'Rourke K, Roose-Girma M, Lee WP, Weinrauch Y, Monack DM, Dixit VM. 2006. Cryopyrin activates the inflammasome in response to toxins and ATP. *Nature* **440:** 228–232.

Martinon F, Burns K, Tschopp J. 2002. The inflammasome: A molecular platform triggering activation of inflammatory caspases and processing of pro-IL-β. *Mol Cell* **10:** 417–426.

Martinon F, Petrilli V, Mayor A, Tardivel A, Tschopp J. 2006. Gout-associated uric acid crystals activate the NALP3 inflammasome. *Nature* **440:** 237–241.

Masters SL, O'Neill LA. 2011. Disease-associated amyloid and misfolded protein aggregates activate the inflammasome. *Trends Mol Med* **17:** 276–282.

Masters SL, Dunne A, Subramanian SL, Hull RL, Tannahill GM, Sharp FA, Becker C, Franchi L, Yoshihara E, Chen Z, et al. 2010. Activation of the NLRP3 inflammasome by islet amyloid polypeptide provides a mechanism for enhanced IL-1β in type 2 diabetes. *Nat Immunol* **11:** 897–904.

McAuley JL, Tate MD, MacKenzie-Kludas CJ, Pinar A, Zeng W, Stutz A, Latz E, Brown LE, Mansell A. 2013. Activation of the NLRP3 inflammasome by IAV virulence protein PB1-F2 contributes to severe pathophysiology and disease. *PLoS Pathog* **9:** e1003392.

McKee AS, Munks MW, MacLeod MK, Fleenor CJ, Van Rooijen N, Kappler JW, Marrack P. 2009. Alum induces innate immune responses through macrophage and mast cell sensors, but these sensors are not required for alum to act as an adjuvant for specific immunity. *J Immunol* **183:** 4403–4414.

McNeela EA, Burke A, Neill DR, Baxter C, Fernandes VE, Ferreira D, Smeaton S, El-Rachkidy R, McLoughlin RM, Mori A, et al. 2010. Pneumolysin activates the NLRP3 inflammasome and promotes proinflammatory cytokines independently of TLR4. *PLoS Pathog* **6:** e1001191.

Meissner F, Seger RA, Moshous D, Fischer A, Reichenbach J, Zychlinsky A. 2010a. Inflammasome activation in NADPH oxidase defective mononuclear phagocytes from patients with chronic granulomatous disease. *Blood* **116:** 1570–1573.

Meissner TB, Li A, Biswas A, Lee KH, Liu YJ, Bayir E, Iliopoulos D, van den Elsen PJ, Kobayashi KS. 2010b. NLR family member NLRC5 is a transcriptional regulator of MHC class I genes. *Proc Natl Acad Sci* **107:** 13794–13799.

Meixenberger K, Pache F, Eitel J, Schmeck B, Hippenstiel S, Slevogt H, N'Guessan P, Witzenrath M, Netea MG, Chakraborty T, et al. 2010. *Listeria monocytogenes*–infected human peripheral blood mononuclear cells produce IL-1β, depending on listeriolysin O and NLRP3. *J Immunol* **184:** 922–930.

Meng G, Zhang F, Fuss I, Kitani A, Strober W. 2009. A mutation in the *Nlrp3* gene causing inflammasome hyperactivation potentiates Th17 cell-dominant immune responses. *Immunity* **30:** 860–874.

Menu P, Pellegrin M, Aubert JF, Bouzourene K, Tardivel A, Mazzolai L, Tschopp J. 2011. Atherosclerosis in ApoE-deficient mice progresses independently of the NLRP3 inflammasome. *Cell Death Dis* **2:** e137.

Miao EA, Alpuche-Aranda CM, Dors M, Clark AE, Bader MW, Miller SI, Aderem A. 2006. Cytoplasmic flagellin activates caspase-1 and secretion of interleukin 1β via Ipaf. *Nat Immunol* **7:** 569–575.

Miao EA, Ernst RK, Dors M, Mao DP, Aderem A. 2008. *Pseudomonas aeruginosa* activates caspase 1 through Ipaf. *Proc Natl Acad Sci* **105:** 2562–2567.

Miao EA, Leaf IA, Treuting PM, Mao DP, Dors M, Sarkar A, Warren SE, Wewers MD, Aderem A. 2010a. Caspase-1-induced pyroptosis is an innate immune effector mechanism against intracellular bacteria. *Nat Immunol* **11:** 1136–1142.

Miao EA, Leaf IA, Treuting PM, Mao DP, Dors M, Sarkar A, Warren SE, Wewers MD, Aderem A. 2010b. Caspase-1-induced pyroptosis is an innate immune effector mechanism against intracellular bacteria. *Nat Immunol* **11:** 1136–1142.

Miao EA, Mao DP, Yudkovsky N, Bonneau R, Lorang CG, Warren SE, Leaf IA, Aderem A. 2010c. Innate immune detection of the type III secretion apparatus through the NLRC4 inflammasome. *Proc Natl Acad Sci* **107:** 3076–3080.

Miao EA, Rajan JV, Aderem A. 2011. Caspase-1-induced pyroptotic cell death. *Immunol Rev* **243:** 206–214.

Monroe KM, Yang Z, Johnson JR, Geng X, Doitsh G, Krogan NJ, Greene WC. 2014. IFI16 DNA sensor is required for death of lymphoid CD4 T cells abortively infected with HIV. *Science* **343:** 428–432.

Munoz-Planillo R, Kuffa P, Martinez-Colon G, Smith BL, Rajendiran TM, Nunez G. 2013. K^+efflux is the common trigger of NLRP3 inflammasome activation by bacterial toxins and particulate matter. *Immunity* **38:** 1142–1153.

Murakami T, Ockinger J, Yu J, Byles V, McColl A, Hofer AM, Horng T. 2012. Critical role for calcium mobilization in activation of the NLRP3 inflammasome. *Proc Natl Acad Sci* **109:** 11282–11287.

Nakahira K, Haspel JA, Rathinam VA, Lee SJ, Dolinay T, Lam HC, Englert JA, Rabinovitch M, Cernadas M, Kim HP, et al. 2011. Autophagy proteins regulate innate immune responses by inhibiting the release of mitochondrial DNA mediated by the NALP3 inflammasome. *Nat Immunol* **12:** 222–230.

Newman ZL, Leppla SH, Moayeri M. 2009. CA-074Me protection against anthrax lethal toxin. *Infect Immun* **77:** 4327–4336.

Ng J, Hirota SA, Gross O, Li Y, Ulke-Lemee A, Potentier MS, Schenck LP, Vilaysane A, Seamone ME, Feng H, et al. 2010. *Clostridium difficile* toxin-induced inflammation and intestinal injury are mediated by the inflammasome. *Gastroenterology* **139:** 542–552.

Nordlander S, Pott J, Maloy KJ. 2013. NLRC4 expression in intestinal epithelial cells mediates protection against an enteric pathogen. *Mucosal Immunol* **7:** 775–785.

Normand S, Delanoye-Crespin A, Bressenot A, Huot L, Grandjean T, Peyrin-Biroulet L, Lemoine Y, Hot D, Chamaillard M. 2011. Nod-like receptor pyrin domain-containing protein 6 (NLRP6) controls epithelial self-renewal and colorectal carcinogenesis upon injury. *Proc Natl Acad Sci* **108:** 9601–9606.

Nour AM, Reichelt M, Ku CC, Ho MY, Heineman TC, Arvin AM. 2011. Varicella-zoster virus infection triggers formation of an interleukin-1β (IL-1β)-processing inflammasome complex. *J Biol Chem* **286:** 17921–17933.

Onoyama S, Ihara K, Yamaguchi Y, Ikeda K, Yamaguchi K, Yamamura K, Hoshina T, Mizuno Y, Hara T. 2012. Genetic susceptibility to Kawasaki disease: Analysis of pattern recognition receptor genes. *Hum Immunol* **73:** 654–660.

Oslowski CM, Hara T, O'Sullivan-Murphy B, Kanekura K, Lu S, Hara M, Ishigaki S, Zhu LJ, Hayashi E, Hui ST, et al. 2012. Thioredoxin-interacting protein mediates ER stress-induced β-cell death through initiation of the inflammasome. *Cell Metab* **16:** 265–273.

Palm NW, Medzhitov R. 2013. Role of the inflammasome in defense against venoms. *Proc Natl Acad Sci* **110:** 1809–1814.

Petrilli V, Papin S, Dostert C, Mayor A, Martinon F, Tschopp J. 2007. Activation of the NALP3 inflammasome is triggered by low intracellular potassium concentration. *Cell Death Differ* **14:** 1583–1589.

Poeck H, Bscheider M, Gross O, Finger K, Roth S, Rebsamen M, Hannesschlager N, Schlee M, Rothenfusser S, Barchet W, et al. 2010. Recognition of RNA virus by RIG-I results in activation of CARD9 and inflammasome signaling for interleukin 1β production. *Nat Immunol* **11:** 63–69.

Pontillo A, Brandao L, Guimaraes R, Segat L, Araujo J, Crovella S. 2010. Two SNPs in *NLRP3* gene are involved in the predisposition to type-1 diabetes and celiac disease in a pediatric population from northeast Brazil. *Autoimmunity* **43:** 583–589.

Pontillo A, Vendramin A, Catamo E, Fabris A, Crovella S. 2011. The missense variation Q705K in CIAS1/*NALP3*/*NLRP3* gene and an NLRP1 haplotype are associated with celiac disease. *Am J Gastroenterol* **106:** 539–544.

Pontillo A, Catamo E, Arosio B, Mari D, Crovella S. 2012a. NALP1/*NLRP1* genetic variants are associated with Alzheimer disease. *Alzheimer Dis Assoc Disord* **26:** 277–281.

Pontillo A, Girardelli M, Kamada AJ, Pancotto JA, Donadi EA, Crovella S, Sandrin-Garcia P. 2012b. Polimorphisms in inflammasome genes are involved in the predisposition to systemic lupus erythematosus. *Autoimmunity* **45:** 271–278.

Pothlichet J, Meunier I, Davis BK, Ting JP, Skamene E, von Messling V, Vidal SM. 2013. Type I IFN triggers RIG-I/TLR3/NLRP3-dependent inflammasome activation in influenza A virus infected cells. *PLoS Pathog* **9:** e1003256.

Qu Y, Misaghi S, Izrael-Tomasevic A, Newton K, Gilmour LL, Lamkanfi M, Louie S, Kayagaki N, Liu J, Komuves L, et al. 2012. Phosphorylation of NLRC4 is critical for inflammasome activation. *Nature* **490:** 539–542.

Rajamaki K, Lappalainen J, Oorni K, Valimaki E, Matikainen S, Kovanen PT, Eklund KK. 2010. Cholesterol crystals activate the NLRP3 inflammasome in human macrophages: A novel link between cholesterol metabolism and inflammation. *PLoS ONE* **5:** e11765.

Rajan JV, Rodriguez D, Miao EA, Aderem A. 2011. The NLRP3 inflammasome detects encephalomyocarditis virus and vesicular stomatitis virus infection. *J Virol* **85:** 4167–4172.

Rathinam VA, Jiang Z, Waggoner SN, Sharma S, Cole LE, Waggoner L, Vanaja SK, Monks BG, Ganesan S, Latz E, et al. 2010. The AIM2 inflammasome is essential for host defense against cytosolic bacteria and DNA viruses. *Nat Immunol* **11:** 395–402.

Raupach B, Peuschel S-K, Monack DM, Zychlinsky A. 2006. Caspase-1-mediated activation of interleukin-1β (IL-1β) and IL-18 contributes to innate immune defenses against *Salmonella enterica* serovar typhimurium infection. *Infect Immun* **74:** 4922–4926.

Reed JC, Doctor K, Rojas A, Zapata JM, Stehlik C, Fiorentino L, Damiano J, Roth W, Matsuzawa S, Newman R, et al. 2003. Comparative analysis of apoptosis and inflammation genes of mice and humans. *Genome Res* **13:** 1376–1388.

Riteau N, Gasse P, Fauconnier L, Gombault A, Couegnat M, Fick L, Kanellopoulos J, Quesniaux VF, Marchand-Adam S, Crestani B, et al. 2010. Extracellular ATP is a danger signal activating P2X7 receptor in lung inflammation and fibrosis. *Am J Respir Crit Care Med* **182:** 774–783.

Rock KL, Kataoka H, Lai JJ. 2013. Uric acid as a danger signal in gout and its comorbidities. *Nat Rev Rheumatol* **9:** 13–23.

Rossol M, Pierer M, Raulien N, Quandt D, Meusch U, Rothe K, Schubert K, Schoneberg T, Schaefer M, Krugel U, et al. 2012. Extracellular Ca^{2+} is a danger signal activating the NLRP3 inflammasome through G protein-coupled calcium sensing receptors. *Nat Commun* **3:** 1329.

Rotta Detto Loria J, Rohmann K, Droemann D, Kujath P, Rupp J, Goldmann T, Dalhoff K. 2013. Nontypeable *Haemophilus influenzae* infection upregulates the NLRP3 inflammasome and leads to caspase-1-dependent secretion of interleukin-1β—A possible pathway of exacerbations in COPD. *PLoS ONE* **8:** e66818.

Saiga H, Kitada S, Shimada Y, Kamiyama N, Okuyama M, Makino M, Yamamoto M, Takeda K. 2012. Critical role of AIM2 in *Mycobacterium tuberculosis* infection. *Int Immunol* **24:** 637–644.

Sauer JD, Witte CE, Zemansky J, Hanson B, Lauer P, Portnoy DA. 2010. *Listeria monocytogenes* triggers AIM2-mediated pyroptosis upon infrequent bacteriolysis in the macrophage cytosol. *Cell Host Microbe* **7:** 412–419.

Sauer JD, Pereyre S, Archer KA, Burke TP, Hanson B, Lauer P, Portnoy DA. 2011. *Listeria monocytogenes* engineered to activate the Nlrc4 inflammasome are severely attenuated and are poor inducers of protective immunity. *Proc Natl Acad Sci* **108:** 12419–12424.

Schattgen SA, Fitzgerald KA. 2011. The PYHIN protein family as mediators of host defenses. *Immunol Rev* **243:** 109–118.

Schneider M, Zimmermann AG, Roberts RA, Zhang L, Swanson KV, Wen H, Davis BK, Allen IC, Holl EK, Ye Z, et al. 2012. The innate immune sensor NLRC3 attenuates Toll-like receptor signaling via modification of the signaling adaptor TRAF6 and transcription factor NF-κB. *Nat Immunol* **13:** 823–831.

Schroder K, Tschopp J. 2010. The inflammasomes. *Cell* **140:** 821–832.

Segovia J, Sabbah A, Mgbemena V, Tsai SY, Chang TH, Berton MT, Morris IR, Allen IC, Ting JP, Bose S. 2012. TLR2/MyD88/NF-κB pathway, reactive oxygen species, potassium efflux activates NLRP3/ASC inflammasome during respiratory syncytial virus infection. *PLoS ONE* **7:** e29695.

Sheedy FJ, Moore KJ. 2013. IL-1 signaling in atherosclerosis: Sibling rivalry. *Nat Immunol* **14:** 1030–1032.

Sheedy FJ, Grebe A, Rayner KJ, Kalantari P, Ramkhelawon B, Carpenter SB, Becker CE, Ediriweera HN, Mullick AE, Golenbock DT, et al. 2013. CD36 coordinates NLRP3 inflammasome activation by facilitating intracellular nucleation of soluble ligands into particulate ligands in sterile inflammation. *Nat Immunol* **14:** 812–820.

Shi F, Yang L, Kouadir M, Yang Y, Wang J, Zhou X, Yin X, Zhao D. 2012. The NALP3 inflammasome is involved in neurotoxic prion peptide-induced microglial activation. *J Neuroinflammation* **9:** 73.

Shimada K, Crother TR, Karlin J, Chen S, Chiba N, Ramanujan VK, Vergnes L, Ojcius DM, Arditi M. 2011. Caspase-1 dependent IL-1β secretion is critical for host defense in a mouse model of *Chlamydia pneumoniae* lung infection. *PLoS ONE* **6:** e21477.

Shimada K, Crother TR, Karlin J, Dagvadorj J, Chiba N, Chen S, Ramanujan VK, Wolf AJ, Vergnes L, Ojcius DM, et al. 2012. Oxidized mitochondrial DNA activates the NLRP3 inflammasome during apoptosis. *Immunity* **36:** 401–414.

Shio MT, Eisenbarth SC, Savaria M, Vinet AF, Bellemare MJ, Harder KW, Sutterwala FS, Bohle DS, Descoteaux A, Flavell RA, et al. 2009. Malarial hemozoin activates the NLRP3 inflammasome through Lyn and Syk kinases. *PLoS Pathog* **5:** e1000559.

So A, De Smedt T, Revaz S, Tschopp J. 2007. A pilot study of IL-1 inhibition by anakinra in acute gout. *Arthritis Res Ther* **9:** R28.

Song-Zhao GX, Srinivasan N, Pott J, Baban D, Frankel G, Maloy KJ. 2013. Nlrp3 activation in the intestinal epithelium protects against a mucosal pathogen. *Mucosal Immunol* **7:** 763–774.

Stienstra R, Joosten LA, Koenen T, van Tits B, van Diepen JA, van den Berg SA, Rensen PC, Voshol PJ, Fantuzzi G, Hijmans A, et al. 2010. The inflammasome-mediated caspase-1 activation controls adipocyte differentiation and insulin sensitivity. *Cell Metabol* **12:** 593–605.

Stienstra R, van Diepen JA, Tack CJ, Zaki MH, van de Veerdonk FL, Perera D, Neale GA, Hooiveld GJ, Hijmans A, Vroegrijk I, et al. 2011. Inflammasome is a central player in the induction of obesity and insulin resistance. *Proc Natl Acad Sci* **108:** 15324–15329.

Strowig T, Henao-Mejia J, Elinav E, Flavell R. 2012. Inflammasomes in health and disease. *Nature* **481:** 278–286.

Sui J, Li H, Fang Y, Liu Y, Li M, Zhong B, Yang F, Zou Q, Wu Y. 2012. *NLRP1* gene polymorphism influences gene transcription and is a risk factor for rheumatoid arthritis in Han Chinese. *Arthritis Rheum* **64:** 647–654.

Sutterwala FS, Ogura Y, Szczepanik M, Lara-Tejero M, Lichtenberger GS, Grant EP, Bertin J, Coyle AJ, Galan JE, Askenase PW, et al. 2006. Critical role for NALP3/CIAS1/Cryopyrin in innate and adaptive immunity through its regulation of caspase-1. *Immunity* **24:** 317–327.

Sutterwala FS, Mijares LA, Li L, Ogura Y, Kazmierczak BI, Flavell RA. 2007. Immune recognition of *Pseudomonas aeruginosa* mediated by the IPAF/NLRC4 inflammasome. *J Exp Med* **204:** 3235–3245.

Sutton C, Brereton C, Keogh B, Mills KH, Lavelle EC. 2006. A crucial role for interleukin (IL)-1 in the induction of IL-17-producing T cells that mediate autoimmune encephalomyelitis. *J Exp Med* **203:** 1685–1691.

Suzuki T, Franchi L, Toma C, Ashida H, Ogawa M, Yoshikawa Y, Mimuro H, Inohara N, Sasakawa C, Nunez G. 2007. Differential regulation of caspase-1 activation, pyroptosis, and autophagy via Ipaf and ASC in *Shigella*-infected macrophages. *PLoS Pathog* **3:** e111.

Thomas PG, Dash P, Aldridge JR Jr., Ellebedy AH, Reynolds C, Funk AJ, Martin WJ, Lamkanfi M, Webby RJ, Boyd KL, et al. 2009. The intracellular sensor NLRP3 mediates key innate and healing responses to influenza A virus via the regulation of caspase-1. *Immunity* **30:** 566–575.

Ting JP, Lovering RC, Alnemri ES, Bertin J, Boss JM, Davis BK, Flavell RA, Girardin SE, Godzik A, Harton JA, et al. 2008. The NLR gene family: A standard nomenclature. *Immunity* **28:** 285–287.

Toma C, Higa N, Koizumi Y, Nakasone N, Ogura Y, McCoy AJ, Franchi L, Uematsu S, Sagara J, Taniguchi S, et al. 2010. Pathogenic *Vibrio* activate NLRP3 inflammasome via cytotoxins and TLR/nucleotide-binding oligomerization domain-mediated NF-κB signaling. *J Immunol* **184:** 5287–5297.

Tomalka J, Ganesan S, Azodi E, Patel K, Majmudar P, Hall BA, Fitzgerald KA, Hise AG. 2011. A novel role for the NLRC4 inflammasome in mucosal defenses against the fungal pathogen *Candida albicans*. *PLoS Pathog* **7:** e1002379.

Triantafilou K, Kar S, Vakakis E, Kotecha S, Triantafilou M. 2013a. Human respiratory syncytial virus viroporin SH: A viral recognition pathway used by the host to signal inflammasome activation. *Thorax* **68:** 66–75.

Triantafilou K, Kar S, van Kuppeveld FJ, Triantafilou M. 2013b. Rhinovirus-induced calcium flux triggers NLRP3 and NLRC5 activation in bronchial cells. *Am J Respir Cell Mol Biol* **49:** 923–934.

Tsuchiya K, Hara H, Kawamura I, Nomura T, Yamamoto T, Daim S, Dewamitta SR, Shen Y, Fang R, Mitsuyama M. 2010. Involvement of absent in melanoma 2 in inflammasome activation in macrophages infected with Listeria monocytogenes. *J Immunol* **185:** 1186–1195.

Tsuji NM, Tsutsui H, Seki E, Kuida K, Okamura H, Nakanishi K, Flavell RA. 2004. Roles of caspase-1 in *Listeria* infection in mice. *Int Immunol* **16:** 335–343.

Usui F, Shirasuna K, Kimura H, Tatsumi K, Kawashima A, Karasawa T, Hida S, Sagara J, Taniguchi S, Takahashi M. 2012. Critical role of caspase-1 in vascular inflammation and development of atherosclerosis in Western diet-fed apolipoprotein E-deficient mice. *Biochem Biophys Res Commun* **425:** 162–168.

van Bruggen R, Koker MY, Jansen M, van Houdt M, Roos D, Kuijpers TW, van den Berg TK. 2010. Human NLRP3 inflammasome activation is Nox1-4 independent. *Blood* **115:** 5398–5400.

Vandanmagsar B, Youm YH, Ravussin A, Galgani JE, Stadler K, Mynatt RL, Ravussin E, Stephens JM, Dixit VD. 2011. The NLRP3 inflammasome instigates obesity-induced inflammation and insulin resistance. *Nat Med* **17:** 179–188.

Vladimer GI, Weng D, Paquette SW, Vanaja SK, Rathinam VA, Aune MH, Conlon JE, Burbage JJ, Proulx MK, Liu Q, et al. 2012. The NLRP12 inflammasome recognizes *Yersinia pestis*. *Immunity* **37:** 96–107.

von Moltke J, Trinidad NJ, Moayeri M, Kintzer AF, Wang SB, van Rooijen N, Brown CR, Krantz BA, Leppla SH, Gronert K, et al. 2012. Rapid induction of inflammatory lipid mediators by the inflammasome in vivo. *Nature* **490:** 107–111.

Walsh JG, Muruve DA, Power C. 2014. Inflammasomes in the CNS. *Nat Rev Neurosci* **15:** 84–97.

Wang L, Manji GA, Grenier JM, Al-Garawi A, Merriam S, Lora JM, Geddes BJ, Briskin M, DiStefano PS, Bertin J. 2002. PYPAF7, a novel PYRIN-containing Apaf1-like protein that regulates activation of NF-κB and caspase-1-dependent cytokine processing. *J Biol Chem* **277:** 29874–29880.

Warren SE, Mao DP, Rodriguez AE, Miao EA, Aderem A. 2008. Multiple Nod-like receptors activate caspase 1 during *Listeria monocytogenes* infection. *J Immunol* **180:** 7558–7564.

Wen H, Gris D, Lei Y, Jha S, Zhang L, Huang MT, Brickey WJ, Ting JP. 2011. Fatty acid-induced NLRP3-ASC inflammasome activation interferes with insulin signaling. *Nat Immunol* **12:** 408–415.

Wickliffe KE, Leppla SH, Moayeri M. 2008. Anthrax lethal toxin-induced inflammasome formation and caspase-1 activation are late events dependent on ion fluxes and the proteasome. *Cell Microbiol* **10:** 332–343.

Willingham SB, Allen IC, Bergstralh DT, Brickey WJ, Huang MT, Taxman DJ, Duncan JA, Ting JP. 2009. NLRP3 (NALP3, Cryopyrin) facilitates in vivo caspase-1 activation, necrosis, and HMGB1 release via inflammasome-dependent and -independent pathways. *J Immunol* **183:** 2008–2015.

Wlodarska M, Thaiss CA, Nowarski R, Henao-Mejia J, Zhang JP, Brown EM, Frankel G, Levy M, Katz MN, Philbrick WM, et al. 2014. NLRP6 inflammasome orchestrates the colonic host-microbial interface by regulating goblet cell mucus secretion. *Cell* **156:** 1045–1059.

Wu J, Fernandes-Alnemri T, Alnemri ES. 2010. Involvement of the AIM2, NLRC4, and NLRP3 inflammasomes in caspase-1 activation by *Listeria monocytogenes*. *J Clin Immunol* **30:** 693–702.

Zaki MH, Vogel P, Malireddi RK, Body-Malapel M, Anand PK, Bertin J, Green DR, Lamkanfi M, Kanneganti TD. 2011. The NOD-like receptor NLRP12 attenuates colon inflammation and tumorigenesis. *Cancer Cell* **20:** 649–660.

Zhang W, Cai Y, Xu W, Yin Z, Gao X, Xiong S. 2013. AIM2 facilitates the apoptotic DNA-induced systemic lupus erythematosus via arbitrating macrophage functional maturation. *J Clin Immunol* **33:** 925–937.

Zhang L, Mo J, Swanson KV, Wen H, Petrucelli A, Gregory SM, Zhang Z, Schneider M, Jiang Y, Fitzgerald KA, et al. 2014. NLRC3, a member of the NLR family of proteins, is a negative regulator of innate immune signaling induced by the DNA sensor STING. *Immunity* **40:** 329–341.

Zhao Y, Yang J, Shi J, Gong YN, Lu Q, Xu H, Liu L, Shao F. 2011. The NLRC4 inflammasome receptors for bacterial flagellin and type III secretion apparatus. *Nature* **477:** 596–600.

Zhou R, Tardivel A, Thorens B, Choi I, Tschopp J. 2010. Thioredoxin-interacting protein links oxidative stress to inflammasome activation. *Nat Immunol* **11:** 136–140.

Zurawek M, Fichna M, Januszkiewicz-Lewandowska D, Gryczynska M, Fichna P, Nowak J. 2010. A coding variant in *NLRP1* is associated with autoimmune Addison's disease. *Hum Immunol* **71:** 530–534.

Tumor Necrosis Factor Superfamily in Innate Immunity and Inflammation

John Šedý, Vasileios Bekiaris, and Carl F. Ware

Laboratory of Molecular Immunology, Infectious and Inflammatory Disease Center, Sanford Burnham Medical Research Institute, La Jolla, California 92037

Correspondence: cware@sanfordburnham.org

The tumor necrosis factor superfamily (TNFSF) and its corresponding receptor superfamily (TNFRSF) form communication pathways required for developmental, homeostatic, and stimulus-responsive processes in vivo. Although this receptor–ligand system operates between many different cell types and organ systems, many of these proteins play specific roles in immune system function. The TNFSF and TNFRSF proteins lymphotoxins, LIGHT (homologous to lymphotoxins, exhibits inducible expression, and competes with HSV glycoprotein D for herpes virus entry mediator [HVEM], a receptor expressed by T lymphocytes), lymphotoxin-β receptor (LT-βR), and HVEM are used by embryonic and adult innate lymphocytes to promote the development and homeostasis of lymphoid organs. Lymphotoxin-expressing innate-acting B cells construct microenvironments in lymphoid organs that restrict pathogen spread and initiate interferon defenses. Recent results illustrate how the communication networks formed among these cytokines and the coreceptors B and T lymphocyte attenuator (BTLA) and CD160 both inhibit and activate innate lymphoid cells (ILCs), innate γδ T cells, and natural killer (NK) cells. Understanding the role of TNFSF/TNFRSF and interacting proteins in innate cells will likely reveal avenues for future therapeutics for human disease.

The tumor necrosis factor superfamily (TNFSF) and its corresponding receptor superfamily (TNFRSF) mediate developmental, homeostatic, and stimulus-responsive processes in many organ systems (Locksley et al. 2001; Wiens and Glenney 2011). The ligands and receptors in the tumor necrosis factor (TNF) superfamilies form communication pathways between many different cell types (Ware 2005). The focus of this review is on the lymphotoxin and LIGHT (TNFSF-14)-related subset of TNFSF, their interacting receptors, and in their roles as effectors and regulators of innate immunity and inflammation. In particular, we focus on recent results that illustrate the key roles of these cytokines in the ability of both conventional innate lymphoid cells, such as natural killer (NK) cells, and unconventional innate-acting cells, such as B lymphocytes and γδ T cells, to initiate and attenuate inflammation.

TNF is well known as a critical factor in eliciting rapid inflammatory events acting through two distinct receptors, TNFR1 and TNFR2 (for review, see Walczak 2011). In general, ligation of these receptors results in activation of caspases, E3:ubiquitin ligases, or both. Death do-

Cite this article as *Cold Spring Harb Perspect Biol* doi: 10.1101/cshperspect.a016279

main containing receptors, such as TNFR1, recruit caspase 8, whereas lymphotoxin-β receptor (LT-βR) forms an E3 ligase liberating the nuclear factor κ-light-chain-enhancer of activated B-cell (NF-κB)-inducing serine kinase (NIK) from ubiquitination and degradation (Sanjo et al. 2010). LT-βR signaling plays a key role in lymphoid organogenesis and homeostasis of lymphoid tissue microarchitecture. Herpes virus entry mediator (HVEM), TNFRSF14 acts as a molecular switch between proinflammatory and inhibitory signaling by serving as both ligand and receptor for multiple ligands (or coreceptors), creating a network for cellular communication (Kaye 2008; Cai and Freeman 2009; Murphy and Murphy 2010; Ware and Šedý 2011). The cellular ligands for HVEM come from two distinct families: the TNF-related cytokines LIGHT (TNFSF14) and lymphotoxin-α (LT-α) (Mauri et al. 1998), and the Ig superfamily members B and T lymphocyte attenuator (BTLA) (Šedý et al. 2005) and CD160 (Cai et al. 2008). The cross-utilization of ligands by HVEM, the LT-β receptor, and the two receptors for TNF (Fig. 1) create a network of signaling systems that together form a broader network

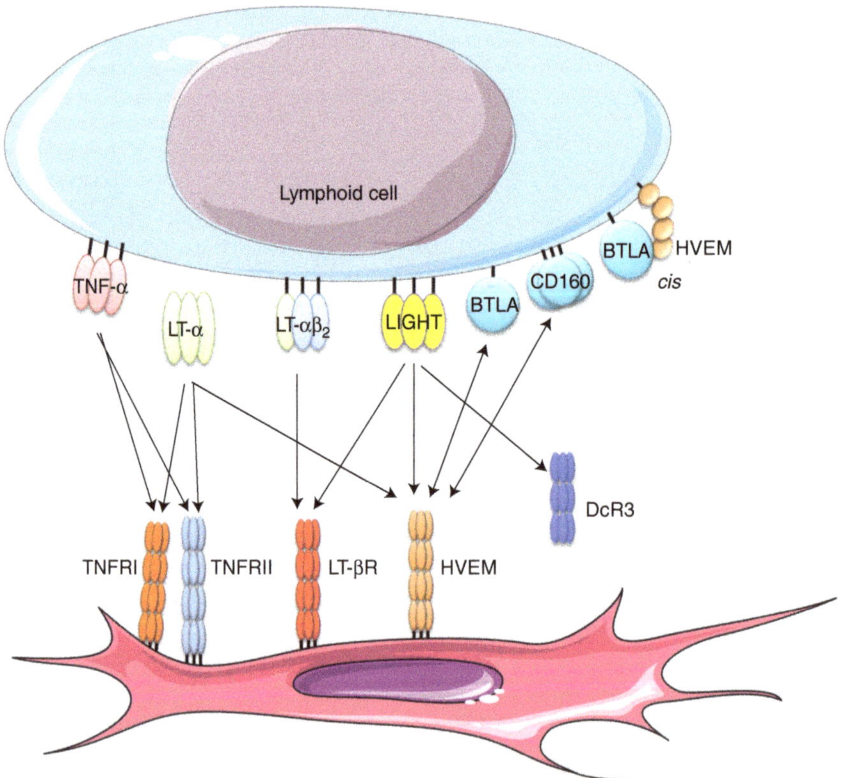

Figure 1. The lymphotoxin LIGHT-related network. The diagram depicts the binding interactions between cytokines and receptors related to lymphotoxins. The arrows define the specificity of the ligand–receptor interactions. Arrowheads define the directionality of signaling, with dual arrowheads defining bidirectional signaling. The TNF-related ligands include TNF-α, LT-α, LT-α1β2, and LIGHT (TNFSF14) and are shown as trimers in their membrane-bound form and expressed in lymphoid cells. Their cognate receptors, TNFRI, TNFRII, LT-βR, and HVEM (TNFRSF14), are expressed in stromal and myeloid cells. Decoy receptor-3 is secreted and also binds Fas ligand and TL1A (TNFSF25) (not shown). HVEM binds the Ig superfamily members BTLA and CD160, which form bidirectional-signaling pathways. BTLA and HVEM are coexpressed in lymphocytes forming a complex in *cis*. Not shown in this diagram are herpesvirus proteins gD and UL144 that signal via HVEM or BTLA. DcR3, decoy receptor 3. (From Bjordahl et al. 2013; reprinted, with permission, from Elsevier © 2013.)

of pathways regulating inflammation, and innate and adaptive immune responses (Ware 2005; Šedý et al. 2008; Ware and Šedý 2011). The critical issue currently being addressed is interpreting these molecular pathways with physiological processes, particularly in the context of host defense. The recent examples we provide in this review hopefully help address how to interpret inflammatory models to aid in designing new approaches to alter real-world disease processes in patients.

LT-αβ AND TNFR-DEPENDENT IMMUNITY IN INNATE LYMPHOID CELLS

Peripheral lymphoid organs form during embryonic development requiring lymphoid tissue inducer cells (LTi) that engage stromal organizing cells. Although often thought of as the critical sites for generating adaptive immune responses, the spleen, lymph nodes, and Peyer's patches also serve to centralize innate defenses, particularly the initial type 1 interferon (IFN-I) response to virus infections (see below). LT-αβ–LT-βR signaling is the key pathway in the formation of lymph nodes and Peyer's patches. The LT-αβ that activates stromal cell LT-βR to initiate lymph node (LN) development is expressed in LTi cells, a highly specialized type of hematopoietic cell (Mebius et al. 1997; Cupedo et al. 2009). Embryonic LTi cells originate in the fetal liver and respond to CXCL13 to migrate to the nascent lymphoid tissues (van de Pavert et al. 2009). LTi lack LT-αβ before entry into newly formed LN, but express surface LT-α1β2 on activation through receptor activator of NF-κB (RANK) (TNFRSF11a) by its ligand (Vondenhoff et al. 2009). Following the expression of LT-α1β2, lymphoid tissue formation initiates. Expression of LT-α1β2 by LTi is also induced and sustained by IL-7 (Yoshida et al. 2002), whereas stromal cell-produced chemokines recruit additional LTi in a feed-forward loop. Successful completion of the lymphoid developmental program in rodents is largely concluded during the first week of neonatal life (Nolte et al. 2003).

Although unique during embryonic life, in the adult, LTi cells constitute a subset of a recently appreciated family of lymphocytes that are collectively known as innate lymphoid cells (ILCs) (Fig. 2). ILCs have been categorized into three main groups depending on their functional capacity. Thus, there are group 1 ILCs that secrete mainly IFN-γ (ILC1), group 2 ILCs that specialize in the production of type 2 cytokines, such as IL-5 and IL-13 (ILC2), and group 3 ILCs, which primarily produce IL-17 and IL-22 (ILC3) (Spits et al. 2013). LTi cells have been classified as ILC3 because they express IL-22 within the embryo and share a common progenitor that depends on constitutive expression of the transcription factor retinoid-related orphan receptor γt (RORγt) (Sawa et al. 2010; Vonarbourg et al. 2010). ILCs play very important roles in many inflammatory and infectious responses (reviewed elsewhere, see Bernink et al. 2013). Interestingly, it appears that LT-αβ and LT-α are also important for the function of ILC3 by regulating the IL-22-dependent clearance of intestinal bacterial infections (Wang et al. 2010) and inducing immunoglobulin A (IgA) production (Kruglov et al. 2013).

In addition to LT-αβ, LTi, and ILC (particularly ILC1 and ILC3) express high levels of an array of TNFSF ligands including but not limited to TNF-α, LIGHT (TNFSF14), OX40-ligand (TNFSF4), CD30-ligand (TNFSF8), RANK (TNFSF11a), RANK-ligand (TNFSF11), and receptors including HVEM, TNFR2 (TNF RSF1B), and death receptor (DR) 3 (TNFRSF 25) (Kim et al. 2003, 2006b). With the exception of CD30-ligand and the provision of survival signals to NK cells during viral infection (Bekiaris et al. 2009), the individual contribution of each receptor or ligand in ILC-driven innate immune responses has not been formally addressed. However, given their shared signaling properties with LT-αβ, TNF, and LIGHT may contribute compensatory mechanisms for ILC function during immune development and infection.

ILC are not the only lymphocytes that have been linked with TNFR-driven innate immunity. There are at least two subpopulations of γδ T cells that have lost T-cell-receptor responsiveness, yet display bona fide innate functionality (Wencker et al. 2014). These two γδ T-cell sub-

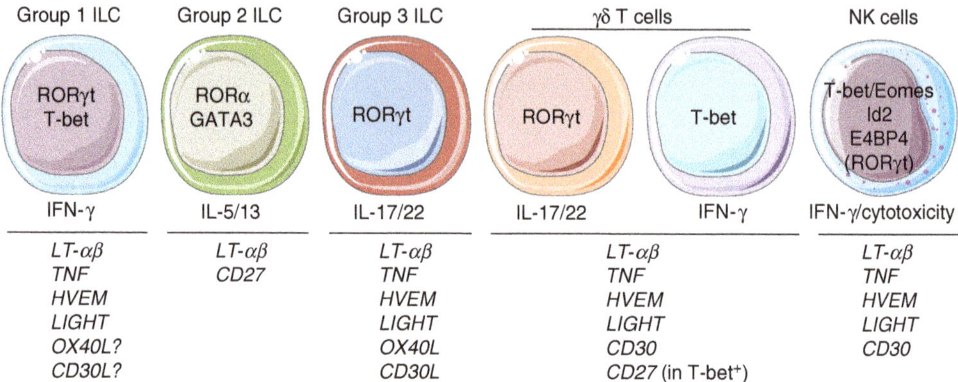

Figure 2. TNFRSF proteins associated with innate cells. Groups 1, 2, and 3 ILC, γδ T cells, and NK cells are displayed showing the transcription factors required for specific differentiation of each innate lineage (within the cell). Differentiation of ILC and γδ T cells is discussed within the text. NK cell differentiation has been shown to depend on a number of factors including both the T-box factors T-bet and Eomes, the basic leucine zipper factor E4BP4, and the E protein repressor Id2 for maintenance of mature NK cells (Boos et al. 2007; Gascoyne et al. 2009; Kamizono et al. 2009; Gordon et al. 2012). Cytokine production and effector function are displayed directly *below* the cells, and the TNFRSF and TNFSF proteins discussed are shown at the *bottom*.

sets can be broadly identified by the expression of the TNFRSF CD27 as CD27^{+}IFN-γ^{+} and CD27^{-}IL-17^{+} (Ribot et al. 2009). Embryonic thymic signaling from CD27 and LT-βR in the stromal cells have been shown to regulate the balanced development of IFN-γ- versus IL-17-producing γδ T cells (Silva-Santos et al. 2005; Ribot et al. 2009), whereas LT-βR is critical for the full maturation and activation of CD27^{-}IL-17^{+} γδ T cells (Powolny-Budnicka et al. 2011). It is likely that the LT-βR requirement is indirect because LT-βR expression is restricted to stromal or cells of myeloid lineage (Murphy et al. 1998). Furthermore, CD30 has also been shown to provide survival and activation signals to γδ T cells at mucosal sites (Sun et al. 2013).

In humans, we recently reported the presence of a CD4^{+}CD3^{-} innate-like T cells that are characterized by high levels of constitutive TNF and inducible surface LT-αβ that are prevalent in the blood of rheumatoid arthritis patients (Bekiaris et al. 2013b). The functional role of TNF and LT-αβ in CD4^{+}CD3^{-} innate T cells has not been elucidated.

Although not required for the formation of the spleen, the LT-αβ–LT-βR pathway provides key signals that drive the maturation of the spleen and its microarchitecture. Maintenance of mature splenic organization and homeostasis is initially dependent on embryonic LTi cells, whereas B lymphocytes provide homeostatic signals at later times (Cyster 2003). In contrast, innate lymphocytes continue to provide key homeostatic signals via the LT-αβ–LT-βR pathway in lymph nodes. The LT-αβ–LT-βR system plays multiple roles in forming and maintaining the microarchitecture providing key differentiation signals to stromal cells (Cyster 2003), subsets of dendritic cells (DC) (CD11c^{hi}CD8^{-}) (Kabashima et al. 2005; De Trez et al. 2008), follicular dendritic cells (Endres et al. 1999), and marginal sinus macrophages (Kraal and Mebius 2006). In addition to regulating antiviral responses via type I IFN, engagement of the LT-βR on the embryonic lymphoid tissue stromal cells initiates a signaling cascade that promotes the differentiation of these cells into a specialized population known as stromal organizer cells (White et al. 2007). Stromal organizing cells then express the homeostatic chemokines CXCL13, CCL19, and CCL21 (Dejardin et al. 2002) that will attract B and T lymphocytes after birth and organize them into their distinct T-cell zones and B follicles the regions found in mature lymph nodes. In addition, stromal organizing cells express the cytokine interleukin

(IL)-7, which is the major survival factor for naïve lymphocytes within the developing and mature lymphoid organ (Link et al. 2007; Meier et al. 2007). Recent work has shown that within lymph nodes the specific progenitor to stromal organizing cell is a preadipocyte, which differentiates using an LT-β R-dependent mechanism (Benezech et al. 2012).

INNATE B CELLS AND SENTINEL PATHOGEN PERMISSIVE CELLS

The sinuses in the spleen and lymph nodes provide the essential filtering mechanism that capture blood- and lymph-borne pathogens (Junt et al. 2008). Within these structures, the sinus-lining macrophages in the spleen and lymph nodes are critically important for capturing pathogens. Their differentiation requires B-cell expression of LT-αβ to produce the initial wave of type 1 interferons (IFN-I including both IFN-β and IFN-α). Studies of two distinct viruses in their natural hosts, mouse cytomegalovirus (CMV) a β-herpesvirus with a large DNA genome (Schneider et al. 2008) and vesicular stomatitis virus (VSV), a Rhabdovirus with a small RNA genome (Moseman et al. 2012) revealed the innate role of B cells in providing LT-αβ that drives the initial IFN-I response. Mice deficient in *Lt-β*, specifically in B cells fail to produce IFN-I in response to CMV or VSV infection. The "innateness" of B cells in IFN-I-mediated defense against VSV was clearly defined using the D_HLMP2A mice (Casola et al. 2004). Activating signals from the herpesvirus LMP2A gene retain B cells and have normal lymphoid tissue architecture, yet are devoid of surface or secreted antibodies.

The initial IFN-I defense to CMV occurs rapidly, peaking within 8–12 h after infection and accounts for >80% of the IFN-I in the circulation. Mouse CMV infects the CCL21 expressing LT-βR-dependent stromal cells in the splenic marginal sinus and T-cell zone (Benedict et al. 2006). IFN-I production requires active signaling by the LT-βR within these infected cells (Banks et al. 2005). IFN-I induction measured as IFN-β or IFN-α mRNA increases by several orders of magnitude in LT-βR^+ stromal cells paralleling viral gene expression, yet is independent of TLR signaling (Schneider et al. 2008). In contrast, LT-βR signaling in DC produces IFN-β at levels reflecting homeostasis (Summers deLuca et al. 2011). Thus, communication between innate B cells and stromal cells initiates and amplifies the earliest IFN-I response to mouse CMV infection (Fig. 3) (Schneider et al. 2008).

Infection with VSV differs from CMV in that macrophages are the primary target following subcutaneous infection and macrophages provide the source of IFN-I. Innate B cells via LT-αβ–LT-βR pathway differentiate $CD169^+$ subcapsular macrophages into a permissive state for VSV replication. In macrophages, the permissive state of these cells is controlled by *Usp18*, which encodes an ISG-15 deconjugating enzyme that also restricts IFN-I receptor signaling, destabilizing the effects of IFN-I (Kim et al. 2006a, 2008). The *Usp18* gene is needed to mount an effective adaptive (antibody) response to VSV (Honke et al. 2012).

The consequences of infection are severe when the LT-IFN axis fails to function. In the absence of LT-βR signaling, IFN-I production and VSV replication in lymph nodes is blocked; however, neurons innervating the nodes become a virus target leading to spread of virus to the central nervous system with ensuing paralytic disease. In contrast, CMV replicates in the absence of IFN-I or LT-βR signaling with disastrous consequences for lymphoid tissues, causing a massive apoptosis of T and B cells in the absence of direct infection (splenic necrosis). IFN-I is a key factor in promoting T- and B-cell viability during antigen challenge (Marrack et al. 1999) indicating that the LT-IFN axis promotes adaptive defenses.

The conceptual outcome of these results suggests that the LT-IFN axis creates a restricted microenvironment containing sentinel pathogen permissive cells located in the sinuses of lymphoid organs as the first line of innate defense to viral pathogens (Fig. 3) (Khanna and Lefrancois 2012; Ware and Benedict 2012). The LT-βR system controls expression of CXCL13 and CCL21 that position $CD169^+$ macrophages within lymph node and splenic microenvironment. The membrane-anchored position of the

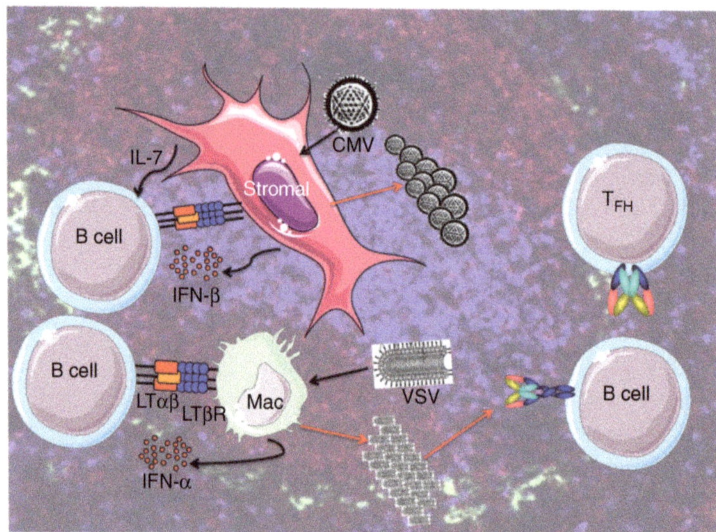

Figure 3. Innate B cells and sentinel pathogen permissive cells. Innate B cells in lymphoid organs express the LT-αβ that specifically engages the LT-βR expressed in lymphoid tissue stromal cells and myeloid lineage cells including marginal sinus macrophages. LT-αβ signals differentiate stromal cells macrophages into sentinel IFN-I producing cells and allow virus production. The B cell to stromal cell interaction creates microenvironments of lymphocytes through secretion of chemokines and IL-7. CMV infects stromal cells in the splenic marginal sinus from which IFN-I is rapidly expressed and secreted. B-cell expression of LT-αβ is also required for the differentiation and recruitment of Siglec1-expressing macrophages in lymph organs. Usp18 creates a permissive state for virus replication in macrophages. Vesicular stomatitis virus (VSV) infects macrophages inducing production of IFN-I. Virus replication and progeny are produced (red arrows) in the permissive stromal cells or $CD169^+$ macrophages. IFN-αβ protects uninfected cells in the surrounding microenvironment. High virus production allows antigen-specific B cells to capture antigen, migrate, and engage T follicular helper (T_{FH}) cells for antibody production. Mac, macrophage.

LT-αβ complex indicates that only those macrophages in cell contact with innate B cells will differentiate into a permissive state, and, consequently, creating a restricted microenvironment for pathogen replication. Usp18 is an example of a gene that induces unresponsiveness to IFN-I signaling, thus restricting virus replication in the sinus-lining macrophages. It is not clear whether this mechanism operates in stromal cells. This IFN-I unresponsive state allows for high virus replication and, thus, antigen production in a restricted microenvironment. High antigen concentration would promote efficient B-cell recognition for transport to follicular regions to generate high affinity antibody responses. Additionally, it is predicted that IFN-I resistant mutations would not be selected during the initial rounds of virus replication. Additional pathogens show strong dependence on sinus-lining macrophages and the LT-IFN axis suggesting a broader host defense strategy. It is not surprising then that some successful pathogens have evolved strategies to evade these pathways (Šedý et al. 2008).

ROLE OF THE HVEM-BTLA-CD160 SYSTEM IN INNATE IMMUNITY

HVEM is one of the most evolutionary conserved TNFRSF members with orthologs expressed in many species including humans and lamprey (Guo et al. 2009). Diverse binding modalities of HVEM with LIGHT, LT-α3, and the immunoglobulin (Ig) superfamily members BTLA and CD160 in *trans* and in *cis* have opened up the TNFR signaling pathways to further regulation and unique immunological properties. HVEM was originally identified as an entry receptor for herpes simplex virus and is a widely expressed in hematopoietic cells and in epithelia cells, and binds the inducible TNF family members LT-α3 and LIGHT (Murphy et al. 2006; Ware and Šedý 2011).

BTLA is an inhibitory receptor that binds HVEM at a distinct surface from TNF ligands through its Ig domain, resulting in bidirectional signaling in cells expressing both proteins (Šedý et al. 2005; Cheung et al. 2009; Murphy and Murphy 2010). BTLA was originally identified as a Th1-specific transcript, but is expressed in many cells in the immune system including B cells, αβ and γδ T cells, NK cells, DC, and macrophages (Watanabe et al. 2003; Murphy and Murphy 2010; Bekiaris et al. 2013a; Šedý et al. 2013). HVEM activates BTLA resulting in phosphorylation of its cytoplasmic domain and recruitment of the Src homology-2 (SH2) containing phosphatase (SHP)-1 and 2 to its carboxy-terminal motif (Watanabe et al. 2003; Šedý et al. 2005). The hematopoietic and epithelial cell–specific SHP-1 protein functions predominantly to dephosphorylate tyrosine within activated kinases or adaptor proteins, thus limiting signal transduction pathways (Van Vactor et al. 1998; Pao et al. 2007). In contrast, a major function of the more ubiquitously expressed SHP-2 protein is to activate RAS/mitogen-activated protein kinase (MAPK) signaling, although SHP-2 has been associated with activation of several inhibitory receptors (Tiganis and Bennett 2007). Direct targets of BTLA-associated SHP-1 and -2 phosphatase activity are thought to include the CD3ζ chain in T cells and Syk in B cells (Wu et al. 2007; Vendel et al. 2009), but little is known about BTLA signaling in other cell subsets. A peptide derived from an additional tyrosine-containing motif within the BTLA cytoplasmic tail could recruit growth factor receptor bound-2 (Grb2) and the p85 regulatory subunit of phosphoinositide-3-kinase (PI3K), although a function of these associations has not been ascribed in vivo (Gavrieli and Murphy 2006). HVEM and LIGHT are each expressed on a variety of cell types including T cells, DC, macrophages, and NK cells, and activation of HVEM by LIGHT or BTLA leads to NF-κB-driven gene activation and inflammatory signaling (Murphy et al. 2006).

HVEM was also identified as a ligand for CD160, a glycosylphosphoinosiol (GPI)-linked receptor expressed at low levels in many cells and showing greater expression in NK cells (Cai et al. 2008; Le Bouteiller et al. 2011). CD160 has been shown to interact MHC-I proteins in mouse and human, including HLA-C (Le Bouteiller et al. 2002; Maeda et al. 2005). HLA-C binds CD160 in NK cells to activate cytolysis, and the expression of IFN-γ, TNF-α, and IL-6 (Le Bouteiller et al. 2002; Barakonyi et al. 2004). CD160 may function as a costimulatory receptor in the absence of CD28 on T cells to activate PI3K signaling (Nikolova et al. 2002; Rabot et al. 2007). The mechanism of GPI-linked CD160 signaling is unclear, although it may transmit signals with cell surface–associated CD2 in NK cells (Rabot et al. 2006). Additionally, in $CD4^+$ T cells, CD160 was shown to inhibit T-cell activation in response to HVEM ligation (Cai et al. 2008). In contrast, activated NK cells can express an alternate spliced form of CD160 that encodes a transmembrane domain and cytoplasmic tail. This form of CD160 costimulates protein kinase B (PKB; AKT) and extracellular signal-regulated kinase (ERK) signaling in response to HVEM activation.

CD160 AND BTLA CONTROL OF NK-CELL ANTIVIRAL ACTIVATION

NK cells mediate clearance of infected cells largely through the secretion of inflammatory cytokines, such as TNF-α or IFN-γ, or through direct lysis of infected cells (Vivier et al. 2011). Many pathogens that establish chronic or latent infections use a variety of immunoregulatory mechanisms to evade initial clearance by innate effector cells, such as NK cells (Lanier 2008). The role of HVEM as a receptor for HSV raised the possibility that viruses may alter the activity of HVEM or its ligands during infection as an immunoregulatory mechanism. We previously showed that human CMV expresses a viral mimic of HVEM (orf UL144) that binds and activates BTLA to inhibit T-cell proliferation (Benedict et al. 1999; Cheung et al. 2005). We and others found that the orf UL144 protein could not bind LT-α3 or LIGHT, likely because of its limited domain structure (Benedict et al. 1999; Poole et al. 2006). However, we also found that the orf UL144 protein could not bind CD160, which binds an overlapping site on HVEM with BTLA (Kojima et al. 2011; Šedý et al. 2013). HVEM binding to CD160 enhanced NK-cell activation in response to CMV and costimulated NK cytolytic function and cytokine release, whereas NK lines expressing high levels of BTLA had reduced cytolytic function.

Together, these data showed that BTLA and CD160 counter-regulate NK-cell activation in response to HVEM binding (Fig. 4). Expression of BTLA following NKG2D stimulation likely serves to limit CD160 costimulatory signals as a means to shut down activation (Šedý et al. 2013). The expression of a BTLA-binding protein in CMV that avoids CD160 activation may have evolved to promote infection by selectively engaging inhibitory pathways and tip the balance in favor of reduced NK-cell activation. As an additional measure against NK killing, CMV expresses a second protein, orf UL141, which binds and down-regulates expression of two apoptotic TNF receptors (DR4 and 5) in infected cells (Smith et al. 2013).

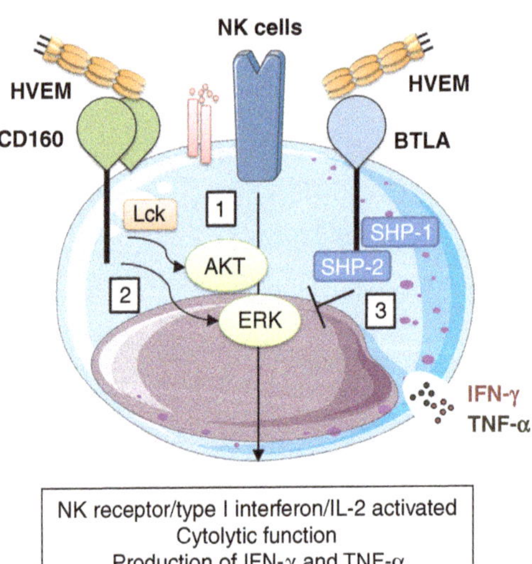

Figure 4. Dual regulation of NK-cell activation by CD160 and BTLA. (*1*) NK cells are activated by NK receptors (e.g., NKG2D), and cytokine receptors (e.g., IL-2, type I interferon), to signal through the ERK and Akt pathways; (*2*) CD160 costimulates ERK and Akt activation, enhancing NK cell effector function; and (*3*) BTLA inhibits NK-cell activation resulting in reduced target cell lysis.

HVEM costimulation of NK cells is greatest in donors that had not been previously infected with human CMV. In CMV seropositive individuals, an NKG2C^{+} population of NK cells expands that is suggested to acquire memory cell properties and, thus, may be less reliant on costimulatory signals (Lopez-Verges et al. 2011; Foley et al. 2012a,b; Muntasell et al. 2013). We did not detect differences in CD160 expression between NKG2C^{+} or NKG2C^{-} cells. Thus, the response of NKG2C^{-} cells to HVEM may reflect a requirement for CD160 signaling to initiate transcriptional programs already active in NKG2C^{+} cells. Activating NKG2C and inhibitory NKG2A receptors bind to human leukocyte antigen (HLA)-E, which presents leader peptides from MHC class I. In accordance with the missing-self hypothesis of NK-cell activation, the absence of NKG2A inhibitory signaling through SHP-1 allows for NK activation. The role for lower affinity interactions between NKG2C and HLA-E-peptide complexes is less clear, although it has been proposed that a stable complex of HLA-E with a CMV nonamer peptide from the orf UL40 protein may activate NKG2C (Heatley et al. 2013; Muntasell et al. 2013). Intriguingly, HLA-E polymorphisms have been associated with the development of psoriasis lesions and decreased frequencies of NKG2C^{+} NK cells (Batista et al. 2013; Patel et al. 2013; Zeng et al. 2013). Future studies will be required to clarify how these receptors are involved in the development of disease.

BTLA INHIBITION OF γδ T-CELL HOMEOSTASIS AND ACTIVATION

We and others recently observed the inhibitory effect of BTLA in innate γδ T cells (Bekiaris et al. 2013a; Gertner-Dardenne et al. 2013). Mice deficient in BTLA had significantly increased IL-17-producing RORγt^{+}CD27^{-} γδ T cells that are implicated in autoimmune diseases (Sutton et al. 2012; Ribot and Silva-Santos 2013). Additionally, we found that BTLA deficiency conferred a competitive advantage to γδ T cells in mixed bone marrow chimera. γδ T-cell homeostatic expansion was previously shown to be dependent on common γ chain signaling through IL-7 and IL-15 (Baccala et al. 2005). BTLA-deficient γδ T cells also show increased IL-17 and TNF-α production in response to IL-7. Together these data indicate that BTLA may directly regulate cytokine signals in innate cells. SHP-1 has been shown to regulate signaling of several cytokines, including those using the common γ chain, whereas Janus kinase (JAK) and signal transducer and activator of transcription (STAT) proteins have been described as substrates for both SHP-1 and SHP-2 (Rakesh and Agrawal 2005; Pao et al. 2007; Tiganis and Bennett 2007). Recently, SHP-1 was reported to limit TH17 development through inhibition of IL-21 signaling, and T-cell homeostasis through inhibition of IL-4 (Mauldin et al. 2012; Johnson et al. 2013). Notably, it is unclear how SHP phosphatases are recruited to cytokine receptors, or whether BTLA participates in recruitment to receptors. We observed induction of BTLA expression in response to IL-7 signaling, providing a mechanism through which cytokine signals can be regulated (Fig. 5). This cytokine signaling circuit was also present in ILC3, which also gained a competitive advantage by BTLA deficiency in mixed bone marrow reconstitution experiments, and which also induced BTLA expression. Thus, in ILC and γδ T cells, IL-7 is regulated by negative feedback through BTLA.

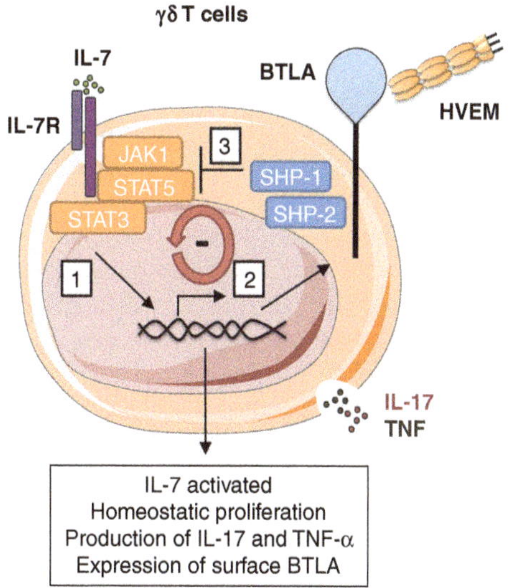

Figure 5. BTLA regulation of IL-7 responses in γδ T cells. (*1*) IL-7 binding to IL-7 receptor activates JAK1 inducing STAT3 and STAT5 activation and nuclear localization; (*2*) IL-7 signaling in part drives homeostatic proliferation of γδ T cells (and ILC), IL-17 and TNF-α expression in $CD27^-$ γδ T cells, and increased levels of surface BTLA in γδ T cells (and ILC); and (*3*) BTLA expression inhibits IL-7 responses in γδ T cells.

Using the Aldara/Imiquimod (IMQ) cream–induced psoriasis model (van der Fits et al. 2009), we further show the necessity of BTLA to suppress skin inflammation by regulating γδ T-cell expansion and cytokine secretion. IMQ cream–induced psoriasis is strongly dependent on skin resident nonlymphoid cells (Cai et al. 2011) and is largely TLR7-dependent, although it may also induce TLR7-independent inflammation (Pantelyushin et al. 2012; Walter et al. 2013). Whether BTLA or other inhibitory receptors act very early to suppress TLR signaling in the skin is currently unknown. Interestingly, whereas IMQ can induce IFN-αβ in a TLR7/MyD88-dependent mechanism (Hemmi et al. 2002), it was recently shown that signaling downstream from the type I IFN receptor is not required for the development of psoriasis or the activation of IL-17-producing γδ T cells (Walter et al. 2013; Wohn et al. 2013). Because of the importance of BTLA in inhibiting $CD27^-IL\text{-}17^+$ γδ T-cell responses and the key role of innate IL-17 during bacterial infection, it would make evolutionary sense to maintain low levels of BTLA at steady state. In this regard, we show that RORγt repressed BTLA transcription through direct binding to the *Btla* promoter (Bekiaris et al. 2013a).

In humans, proliferation of γδ T cells expressing Vγ9- and Vδ2-encoding T-cell receptors was reduced by HVEM binding to BTLA, and enhanced by antibodies or decoy receptors that blocked the HVEM-BTLA interaction (Gertner-Dardenne et al. 2013). $V\gamma9V\delta2^+$ T cells are the predominant population of γδ T cells circulating within human blood, and have potent cytolytic activity against a wide variety of tumors and infected cells (Caccamo et al. 2010; Kalyan and Kabelitz 2013). Although the cognate receptor for Vγ9 Vδ2^{+} T cells remains unclear, they are activated by nonpeptide phosphorylated isoprenoid pathway metabolites whose presentation is facilitated by stress-induced proteins and CD277 (Harly et al. 2012). Recently, much attention has been focused on evaluating the effectiveness of phosphoantigens or activated Vγ9Vδ2^{+} T cells as anticancer therapies (reviewed in Caccamo et al. 2010). Blocking the BTLA-HVEM interaction has been proposed as a further measure to enhance antitumor responses of Vγ9Vδ2^{+} T cells (Lopez 2013).

LT-βR LIGHT HVEM-BTLA COUNTER-REGULATION OF DC HOMEOSTASIS

The role of TNFRSF in balancing DC homeostasis is well documented. In this regard, LT-βR is necessary for the proliferation of lymphoid tissue resident CD8^{-} DC (Kabashima et al. 2005). In contrast, although mice deficient in LT-βR have reduced numbers of CD8^{-} DCs, their numbers in BTLA- or HVEM-deficient animals show the opposite phenotype, with significantly increased CD8^{-} DC (De Trez et al. 2008), suggesting that ITIM-dependent signals suppress the impact of the LT-βR pathway to balance DC homeostasis. Similarly, recent data have shown that along with Notch2, LT-βR instructs the normal development of tissue resident and migratory intestinal DC subsets (Satpathy et al. 2013). Moreover, BTLA expression in DC and macrophages is necessary to suppress LPS-induced TNF and IL-12 production, and prevent endotoxin shock (Kobayashi et al. 2013). The importance of BTLA in DC biology has been recently reemphasized by the discovery that it is one of the most highly expressed genes in mouse and human CD103^{+}CD11b^{-} DC (Watchmaker et al. 2014).

ASSOCIATION OF HVEM NETWORK WITH DISEASE

A wide variety of tumors express HVEM and, thus, have the potential to inhibit the proliferation of Vγ9Vδ2^{+} T cells. Interestingly, several groups have reported that a high percentage of adult onset and pediatric follicular lymphoma and diffuse large B-cell lymphoma contain deletions in the gene-encoding HVEM (*TNFRSF14*), and are associated with worse prognosis (Cheung et al. 2010; Launay et al. 2012; Lohr et al. 2012; Bjordahl et al. 2013; Martin-Guerrero et al. 2013). We previously argued that *TNFRSF14* deletions may be acquired in more aggressive tumors as an adaptation to prevent NK cell costimulation through CD160 (Šedý et al. 2013). CD160 is also present in human γδ T cells, and identifies the recently characterized lytic ILC1 population, although it is not clear how this receptor signals in these cells (Maiza et al. 1993; Fuchs et al. 2013). Another possible selective pressure for follicular lymphoma to delete *TNFRSF14* is to prevent interactions with BTLA-expressing T_{FH} cells (Chtanova et al. 2004; Nurieva et al. 2008; M'Hidi et al. 2009). Follicular lymphomas containing a higher content of T cells with T_{FH} markers are associated with increased survival (Byers et al. 2008; Carreras et al. 2009; Pangault et al. 2010). Further studies will be required to determine the relative contributions of innate and adaptive compartments in shaping the tumor microenvironment and the development of more aggressive tumors.

BTLA has also been shown to control inflammation induced by a number of infections in mice including malaria and listeria (Lepenies et al. 2007; Sun et al. 2009). In humans, increased expression of BTLA in lymphocytes is associated with the presence of chronic infections, such as cytomegalovirus and hepatitis B (Serriari et al. 2010; Cai et al. 2013). In contrast, decreased BTLA expression in lymphocytes is observed during infection with human immunodeficiency virus, possibly owing to regulation by type I interferon (Xu et al. 2009; Zhang et al. 2011; Boliar et al. 2012; Larsson et al. 2013). The HVEM-BTLA-CD160 signaling complex is also critical for the innate immune properties of intestinal intraepithelial lymphocytes. Thus, following bacterial infection CD160 in intraepithelial lymphocytes engages HVEM on the epithelium to increase levels of the IL-22 receptor and initiate a STAT3-dependent pathway necessary for clearing the infection (Shui et al. 2012). Whether these are CD160-expressing ILC1 (see above) remains to be determined.

Table 1. Gene polymorphisms or altered cellular expression of HVEM, LIGHT, BTLA, and CD160 in human disease

	Human disease			
Gene	Autoimmunity	Cancer	Pathogen responses	Other
TNFRSF14 (*HVEM*)	Asthma (Jung et al. 2003) Dermatitis (Jung et al. 2003) Rheumatoid arthritis (Jung et al. 2003; Raychaudhuri et al. 2008; Perdigones et al. 2010; Shang et al. 2012a; Herraez et al. 2013) Multiple sclerosis (Blanco-Kelly et al. 2011) Ulcerative colitis (Anderson et al. 2011; Jostins et al. 2012) Celiac disease (Dubois et al. 2010)	Follicular lymphoma (Cheung et al. 2010; Launay et al. 2012)	CMV (Cheung et al. 2005)	Atherosclerosis (Lee et al. 2001) Obesity (Bassols et al. 2010)
TNFSF14 (*LIGHT*)	Multiple sclerosis (Sawcer et al. 2011) Rheumatoid arthritis (Celik et al. 2009) Dermatitis (Kotani et al. 2012)	Melanoma (Mortarini et al. 2005) Breast cancer (Gantsev et al. 2013)	Hepatitis C (Celik et al. 2009)	Obesity (Bassols et al. 2010) Atherosclerosis (Scholz et al. 2005) Vascular dementia (Kong et al. 2008) Stroke (Liu et al. 2008) Ulcerative interstitial cystitis (Ogawa et al. 2010) Amyotrophic lateral sclerosis (Aebischer et al. 2012) Sickle cell disease (Garrido et al. 2012)
BTLA	Rheumatoid arthritis (Lin et al. 2006; Oki et al. 2011; Shang et al. 2012a,b)	Breast cancer (Fu et al. 2009) Chronic lymphocytic leukemia (CLL)/small lymphocytic lymphoma (M'Hidi et al. 2009)	HIV (Zhang et al. 2011) CMV (Cheung et al. 2005; Serriari et al. 2010)	
CD160	Dermatitis (Abecassis et al. 2007)	CLL/hairy cell leukemia/mantle cell lymphoma (Farren et al. 2011)	HIV (Peretz et al. 2012)	Paroxysmal nocturnal hemoglobinuria (Giustiniani et al. 2012)

The prevention of autoimmune recognition is thought to be controlled by lymphocyte inhibitory receptors, such as CTLA4, PD-1, BTLA, and their ligands (Watanabe and Nakajima 2012). BTLA has been shown to be important in regulating autoimmunity in several disease models, such as experimental autoimmune encephalitis, autoimmune cardiomyopathy, dermatitis, and airway hypersensitivity (Watanabe et al. 2003; Tao et al. 2005; Deppong et al. 2006; Bekiaris et al. 2013a). In a model of $CD4^+$ T-cell-driven colitis, HVEM expression in recipient animals lacking T or B cells was shown to prevent the development of disease (Steinberg et al. 2008). A SNP in the human *BTLA* locus is associated with rheumatoid arthritis (RA) (Table 1) (Lin et al. 2006; Oki et al. 2011). SNPs in *TNFRSF14* have also been associated with RA as well as multiple sclerosis, ulcerative colitis, and celiac disease, possibly owing to the role of HVEM as a focal point of network interactions (Raychaudhuri et al. 2008; Dubois et al. 2010; Perdigones et al. 2010; Anderson et al. 2011; Blanco-Kelly et al. 2011; Jostins et al. 2012; Herraez et al. 2013).

CONCLUDING REMARKS

The association of innate cells in diseased tissues and their role in promoting inflammation in autoimmune disease is well established. Blockade of TNF has proven effective in clinical settings of inflammation (rheumatoid arthritis psoriasis and inflammatory bowel diseases), although significant subsets of patients are refractory to treatment with TNF inhibitors, and in some diseases (e.g., multiple sclerosis), TNF inhibitors are contraindicated. One must consider whether other pathways are active in these pathologies, and indeed whether these pathologies may arise from alterations in inhibitory signaling pathways that may, at least in part, be because of defects in innate cell function.

ACKNOWLEDGMENTS

The authors thank Lisa Marie Bellovich for figures and editing and the National Institutes of Health (AI-033068, AI48073, and CA164679) and Jean Perkins Family Foundation for support.

REFERENCES

Abecassis S, Giustiniani J, Meyer N, Schiavon V, Ortonne N, Campillo JA, Bagot M, Bensussan A. 2007. Identification of a novel $CD160^+$ $CD4^+$ T-lymphocyte subset in the skin: A possible role for CD160 in skin inflammation. *J Invest Dermatol* **127:** 1161–1166.

Aebischer J, Moumen A, Sazdovitch V, Seilhean D, Meininger V, Raoul C. 2012. Elevated levels of IFN-γ and LIGHT in the spinal cord of patients with sporadic amyotrophic lateral sclerosis. *Eur J Neurol* **19:** 752–759, e745–756.

Anderson CA, Boucher G, Lees CW, Franke A, D'Amato M, Taylor KD, Lee JC, Goyette P, Imielinski M, Latiano A, et al. 2011. Meta-analysis identifies 29 additional ulcerative colitis risk loci, increasing the number of confirmed associations to 47. *Nat Genet* **43:** 246–252.

Baccala R, Witherden D, Gonzalez-Quintial R, Dummer W, Surh CD, Havran WL, Theofilopoulos AN. 2005. γδ T cell homeostasis is controlled by IL-7 and IL-15 together with subset-specific factors. *J Immunol* **174:** 4606–4612.

Banks TA, Rickert S, Benedict CA, Ma L, Ko M, Meier J, Ha W, Schneider K, Granger SW, Turovskaya O, et al. 2005. A lymphotoxin-IFN-β axis essential for lymphocyte survival revealed during cytomegalovirus infection. *J Immunol* **174:** 7217–7225.

Barakonyi A, Rabot M, Marie-Cardine A, Aguerre-Girr M, Polgar B, Schiavon V, Bensussan A, Le Bouteiller P. 2004. Cutting edge: Engagement of CD160 by its HLA-C physiological ligand triggers a unique cytokine profile secretion in the cytotoxic peripheral blood NK cell subset. *J Immunol* **173:** 5349–5354.

Bassols J, Moreno JM, Ortega F, Ricart W, Fernandez-Real JM. 2010. Characterization of herpes virus entry mediator as a factor linked to obesity. *Obesity (Silver Spring)* **18:** 239–246.

Batista MD, Ho EL, Kuebler PJ, Milush JM, Lanier LL, Kallas EG, York VA, Chang D, Liao W, Unemori P, et al. 2013. Skewed distribution of natural killer cells in psoriasis skin lesions. *Exp Dermatol* **22:** 64–66.

Bekiaris V, Gaspal F, McConnell FM, Kim MY, Withers DR, Sweet C, Anderson G, Lane PJ. 2009. NK cells protect secondary lymphoid tissue from cytomegalovirus via a CD30-dependent mechanism. *Eur J Immunol* **39:** 2800–2808.

Bekiaris V, Šedý JR, Macauley MG, Rhode-Kurnow A, Ware CF. 2013a. The inhibitory receptor BTLA controls γδ T cell homeostasis and inflammatory responses. *Immunity* **12:** 1082–1094.

Bekiaris V, Šedý JR, Rossetti M, Spreafico R, Sharma S, Rhode-Kurnow A, Ware BC, Huang N, Macauley MG, Norris PS, et al. 2013b. Human $CD4^+CD3^-$ innate-like T cells provide a source of TNF and lymphotoxin-αβ and are elevated in rheumatoid arthritis. *J Immunol* **191:** 4611–4618.

Benedict C, Butrovich K, Lurain N, Corbeil J, Rooney I, Schenider P, Tschopp J, Ware C. 1999. Cutting Edge: A novel viral TNF receptor superfamily member in virulent strains of human cytomegalovirus. *J Immunol* **162:** 6967–6970.

Benedict CA, De Trez C, Schneider K, Ha S, Patterson G, Ware CF. 2006. Specific remodeling of splenic architecture by cytomegalovirus. *PLoS Pathog* **2:** e16.

Benezech C, Mader E, Desanti G, Khan M, Nakamura K, White A, Ware CF, Anderson G, Caamano JH. 2012. Lymphotoxin-β receptor signaling through NF-κB2-RelB pathway reprograms adipocyte precursors as lymph node stromal cells. *Immunity* **37:** 721–734.

Bernink J, Mjosberg J, Spits H. 2013. Th1- and Th2-like subsets of innate lymphoid cells. *Immunol Rev* **252:** 133–138.

Bjordahl RL, Steidl C, Gascoyne RD, Ware CF. 2013. Lymphotoxin network pathways shape the tumor microenvironment. *Curr Opin Immunol* **25:** 222–229.

Blanco-Kelly F, Alvarez-Lafuente R, Alcina A, Abad-Grau MM, de Las Heras V, Lucas M, de la Concha EG, Fernandez O, Arroyo R, Matesanz F, et al. 2011. Members 6B and 14 of the TNF receptor superfamily in multiple sclerosis predisposition. *Genes Immun* **12:** 145–148.

Boliar S, Murphy MK, Tran TC, Carnathan DG, Armstrong WS, Silvestri G, Derdeyn CA. 2012. B-lymphocyte dysfunction in chronic HIV-1 infection does not prevent cross-clade neutralization breadth. *J Virol* **86:** 8031–8040.

Boos MD, Yokota Y, Eberl G, Kee BL. 2007. Mature natural killer cell and lymphoid tissue-inducing cell development requires Id2-mediated suppression of E protein activity. *J Exp Med* **204:** 1119–1130.

Byers RJ, Sakhinia E, Joseph P, Glennie C, Hoyland JA, Menasce LP, Radford JA, Illidge T. 2008. Clinical quantitation of immune signature in follicular lymphoma by RT-PCR-based gene expression profiling. *Blood* **111:** 4764–4770.

Caccamo N, Dieli F, Meraviglia S, Guggino G, Salerno A. 2010. γδ T cell modulation in anticancer treatment. *Curr Cancer Drug Targets* **10:** 27–36.

Cai G, Freeman GJ. 2009. The CD160, BTLA, LIGHT/HVEM pathway: A bidirectional switch regulating T-cell activation. *Immunol Rev* **229:** 244–258.

Cai G, Anumanthan A, Brown JA, Greenfield EA, Zhu B, Freeman GJ. 2008. CD160 inhibits activation of human $CD4^+$ T cells through interaction with herpesvirus entry mediator. *Nat Immunol* **9:** 176–185.

Cai Y, Shen X, Ding C, Qi C, Li K, Li X, Jala VR, Zhang HG, Wang T, Zheng J, et al. 2011. Pivotal role of dermal IL-17-producing γδ T cells in skin inflammation. *Immunity* **35:** 596–610.

Cai G, Nie X, Li L, Hu L, Wu B, Lin J, Jiang C, Wang H, Wang X, Shen Q. 2013. B and T lymphocyte attenuator is highly expressed on intrahepatic T cells during chronic HBV infection and regulates their function. *J Gastroenterol* **48:** 1362–1372.

Carreras J, Lopez-Guillermo A, Roncador G, Villamor N, Colomo L, Martinez A, Hamoudi R, Howat WJ, Montserrat E, Campo E. 2009. High numbers of tumor-infiltrating programmed cell death 1–positive regulatory lymphocytes are associated with improved overall survival in follicular lymphoma. *J Clin Oncol* **27:** 1470–1476.

Casola S, Otipoby KL, Alimzhanov M, Humme S, Uyttersprot N, Kutok JL, Carroll MC, Rajewsky K. 2004. B cell receptor signal strength determines B cell fate. *Nat Immunol* **5:** 317–327.

Celik S, Shankar V, Richter A, Hippe HJ, Akhavanpoor M, Bea F, Erbel C, Urban S, Blank N, Wambsganss N, et al. 2009. Proinflammatory and prothrombotic effects on human vascular endothelial cells of immune-cell-derived LIGHT. *Eur J Med Res* **14:** 147–156.

Cheung TC, Humphreys IR, Potter KG, Norris PS, Shumway HM, Tran BR, Patterson G, Jean-Jacques R, Yoon M, Spear PG, et al. 2005. Evolutionarily divergent herpesviruses modulate T cell activation by targeting the herpesvirus entry mediator cosignaling pathway. *Proc Natl Acad Sci* **102:** 13218–13223.

Cheung TC, Steinberg MW, Oborne LM, Macauley MG, Fukuyama S, Sanjo H, D'Souza C, Norris PS, Pfeffer K, Murphy KM, et al. 2009. Unconventional ligand activation of herpesvirus entry mediator signals cell survival. *Proc Natl Acad Sci* **106:** 6244–6249.

Cheung KJ, Johnson NA, Affleck JG, Severson T, Steidl C, Ben-Neriah S, Schein J, Morin RD, Moore R, Shah SP, et al. 2010. Acquired TNFRSF14 mutations in follicular lymphoma are associated with worse prognosis. *Cancer Res* **70:** 9166–9174.

Chtanova T, Tangye SG, Newton R, Frank N, Hodge MR, Rolph MS, Mackay CR. 2004. T follicular helper cells express a distinctive transcriptional profile, reflecting their role as non-Th1/Th2 effector cells that provide help for B cells. *J Immunol* **173:** 68–78.

Cupedo T, Crellin NK, Papazian N, Rombouts EJ, Weijer K, Grogan JL, Fibbe WE, Cornelissen JJ, Spits H. 2009. Human fetal lymphoid tissue-inducer cells are interleukin 17-producing precursors to $RORC^+$ $CD127^+$ natural killer-like cells. *Nat Immunol* **10:** 66–74.

Cyster JG. 2003. Lymphoid organ development and cell migration. *Immunol Rev* **195:** 5–14.

Dejardin E, Droin NM, Delhase M, Haas E, Cao Y, Makris C, Li ZW, Karin M, Ware CF, Green DR. 2002. The lymphotoxin-β receptor induces different patterns of gene expression via two NF-κB pathways. *Immunity* **17:** 525–535.

Deppong C, Juehne TI, Hurchla M, Friend LD, Shah DD, Rose CM, Bricker TL, Shornick LP, Crouch EC, Murphy TL, et al. 2006. Cutting edge: B and T lymphocyte attenuator and programmed death receptor-1 inhibitory receptors are required for termination of acute allergic airway inflammation. *J Immunol* **176:** 3909–3913.

De Trez C, Schneider K, Potter K, Droin N, Fulton J, Norris PS, Ha SW, Fu YX, Murphy T, Murphy KM, et al. 2008. The inhibitory HVEM-BTLA pathway counter regulates lymphotoxin receptor signaling to achieve homeostasis of dendritic cells. *J Immunol* **180:** 238–248.

Dubois PC, Trynka G, Franke L, Hunt KA, Romanos J, Curtotti A, Zhernakova A, Heap GA, Adany R, Aromaa A, et al. 2010. Multiple common variants for celiac disease influencing immune gene expression. *Nat Genet* **42:** 295–302.

Endres R, Alimzhanov MB, Plitz T, Futterer A, Kosco-Vilbois MH, Nedospasov SA, Rajewsky K, Pfeffer K. 1999. Mature follicular dendritic cell networks depend on expression of lymphotoxin β receptor by radioresistant stromal cells and of lymphotoxin β and tumor necrosis factor by B cells. *J Exp Med* **189:** 159–168.

Farren TW, Giustiniani J, Liu FT, Tsitsikas DA, Macey MG, Cavenagh JD, Oakervee HE, Taussig D, Newland AC, Calaminici M, et al. 2011. Differential and tumor-specific expression of CD160 in B-cell malignancies. *Blood* **118:** 2174–2183.

Foley B, Cooley S, Verneris MR, Curtsinger J, Luo X, Waller EK, Anasetti C, Weisdorf D, Miller JS. 2012a. Human cytomegalovirus (CMV)-induced memory-like NKG2C$^+$ NK cells are transplantable and expand in vivo in response to recipient CMV antigen. *J Immunol* **189:** 5082–5088.

Foley B, Cooley S, Verneris MR, Pitt M, Curtsinger J, Luo X, Lopez-Verges S, Lanier LL, Weisdorf D, Miller JS. 2012b. Cytomegalovirus reactivation after allogeneic transplantation promotes a lasting increase in educated NKG2C$^+$ natural killer cells with potent function. *Blood* **119:** 2665–2674.

Fu Z, Li D, Jiang W, Wang L, Zhang J, Xu F, Pang D. 2009. Association of BTLA gene polymorphisms with the risk of malignant breast cancer in Chinese women of Heilongjiang Province. *Breast Cancer Res Treat* **120:** 195–202.

Fuchs A, Vermi W, Lee JS, Lonardi S, Gilfillan S, Newberry RD, Cella M, Colonna M. 2013. Intraepithelial type 1 innate lymphoid cells are a unique subset of IL-12- and IL-15-responsive IFN-γ-producing cells. *Immunity* **38:** 769–781.

Gantsev SK, Umezawa K, Islamgulov DV, Khusnutdinova EK, Ishmuratova RS, Frolova VY, Kzyrgalin SR. 2013. The role of inflammatory chemokines in lymphoid neoorganogenesis in breast cancer. *Biomed Pharmacother* **67:** 363–366.

Garrido VT, Proenca-Ferreira R, Dominical VM, Traina F, Bezerra MA, de Mello MR, Colella MP, Araujo AS, Saad ST, Costa FF, et al. 2012. Elevated plasma levels and platelet-associated expression of the pro-thrombotic and pro-inflammatory protein, TNFSF14 (LIGHT), in sickle cell disease. *Br J Haematol* **158:** 788–797.

Gascoyne DM, Long E, Veiga-Fernandes H, de Boer J, Williams O, Seddon B, Coles M, Kioussis D, Brady HJ. 2009. The basic leucine zipper transcription factor E4BP4 is essential for natural killer cell development. *Nat Immunol* **10:** 1118–1124.

Gavrieli M, Murphy KM. 2006. Association of Grb-2 and PI3K p85 with phosphotyrosile peptides derived from BTLA. *Biochem Biophys Res Commun* **345:** 1440–1445.

Gertner-Dardenne J, Fauriat C, Orlanducci F, Thibult ML, Pastor S, Fitzgibbon J, Bouabdallah R, Xerri L, Olive D. 2013. The co-receptor BTLA negatively regulates human Vγ9Vδ2 T-cell proliferation: A potential way of immune escape for lymphoma cells. *Blood* **122:** 922–931.

Giustiniani J, Alaoui SS, Marie-Cardine A, Bernard J, Olive D, Bos C, Razafindratsita A, Petropoulou A, de Latour RP, Le Bouteiller P, et al. 2012. Possible pathogenic role of the transmembrane isoform of CD160 NK lymphocyte receptor in paroxysmal nocturnal hemoglobinuria. *Curr Mol Med* **12:** 188–198.

Gordon SM, Chaix J, Rupp LJ, Wu J, Madera S, Sun JC, Lindsten T, Reiner SL. 2012. The transcription factors T-bet and Eomes control key checkpoints of natural killer cell maturation. *Immunity* **36:** 55–67.

Guo P, Hirano M, Herrin BR, Li J, Yu C, Sadlonova A, Cooper MD. 2009. Dual nature of the adaptive immune system in lampreys. *Nature* **459:** 796–801.

Harly C, Guillaume Y, Nedellec S, Peigne CM, Monkkonen H, Monkkonen J, Li J, Kuball J, Adams EJ, Netzer S, et al. 2012. Key implication of CD277/butyrophilin-3 (BTN3A) in cellular stress sensing by a major human γδ T-cell subset. *Blood* **120:** 2269–2279.

Heatley SL, Pietra G, Lin J, Widjaja JM, Harpur CM, Lester S, Rossjohn J, Szer J, Schwarer A, Bradstock K, et al. 2013. Polymorphism in human cytomegalovirus UL40 impacts on recognition of human leukocyte antigen-E (HLA-E) by natural killer cells. *J Biol Chem* **288:** 8679–8690.

Hemmi H, Kaisho T, Takeuchi O, Sato S, Sanjo H, Hoshino K, Horiuchi T, Tomizawa H, Takeda K, Akira S. 2002. Small anti-viral compounds activate immune cells via the TLR7 MyD88-dependent signaling pathway. *Nat Immunol* **3:** 196–200.

Herraez DL, Martinez-Bueno M, Riba L, de la Torre IG, Sacnun M, Goni M, Berbotto GA, Paira S, Musuruana JL, Graf CE, et al. 2013. Rheumatoid arthritis in Latin Americans enriched for Amerindian ancestry is associated with loci in chromosomes 1, 12, and 13, and the HLA class II region. *Arthritis Rheum* **65:** 1457–1467.

Honke N, Shaabani N, Cadeddu G, Sorg UR, Zhang DE, Trilling M, Klingel K, Sauter M, Kandolf R, Gailus N, et al. 2012. Enforced viral replication activates adaptive immunity and is essential for the control of a cytopathic virus. *Nat Immunol* **13:** 51–57.

Johnson DJ, Pao LI, Dhanji S, Murakami K, Ohashi PS, Neel BG. 2013. Shp1 regulates T cell homeostasis by limiting IL-4 signals. *J Exp Med* **210:** 1419–1431.

Jostins L, Ripke S, Weersma RK, Duerr RH, McGovern DP, Hui KY, Lee JC, Schumm LP, Sharma Y, Anderson CA, et al. 2012. Host-microbe interactions have shaped the genetic architecture of inflammatory bowel disease. *Nature* **491:** 119–124.

Jung HW, La SJ, Kim JY, Heo SK, Wang S, Kim KK, Lee KM, Cho HR, Lee HW, Kwon B, et al. 2003. High levels of soluble herpes virus entry mediator in sera of patients with allergic and autoimmune diseases. *Exp Mol Med* **35:** 501–508.

Junt T, Scandella E, Ludewig B. 2008. Form follows function: Lymphoid tissue microarchitecture in antimicrobial immune defence. *Nat Rev Immunol* **8:** 764–775.

Kabashima K, Banks TA, Ansel KM, Lu TT, Ware CF, Cyster JG. 2005. Intrinsic lymphotoxin-β receptor requirement for homeostasis of lymphoid tissue dendritic cells. *Immunity* **22:** 439–450.

Kalyan S, Kabelitz D. 2013. Defining the nature of human γδ T cells: A biographical sketch of the highly empathetic. *Cell Mol Immunol* **10:** 21–29.

Kamizono S, Duncan GS, Seidel MG, Morimoto A, Hamada K, Grosveld G, Akashi K, Lind EF, Haight JP, Ohashi PS, et al. 2009. Nfil3/E4bp4 is required for the development and maturation of NK cells in vivo. *J Exp Med* **206:** 2977–2986.

Kaye J. 2008. CD160 and BTLA: LIGHTs out for CD4$^+$ T cells. *Nat Immunol* **9:** 122–124.

Khanna KM, Lefrancois L. 2012. B cells, not just for antibody anymore. *Immunity* **36:** 315–317.

Kim MY, Gaspal FM, Wiggett HE, McConnell FM, Gulbranson-Judge A, Raykundalia C, Walker LS, Goodall MD, Lane PJ. 2003. $CD4^+CD3^-$ accessory cells costimulate primed CD4 T cells through OX40 and CD30 at sites where T cells collaborate with B cells. *Immunity* **18:** 643–654.

Kim KI, Yan M, Malakhova O, Luo JK, Shen MF, Zou W, de la Torre JC, Zhang DE. 2006a. Ube1L and protein ISGylation are not essential for α/β interferon signaling. *Mol Cell Biol* **26:** 472–479.

Kim MY, Toellner KM, White A, McConnell FM, Gaspal FM, Parnell SM, Jenkinson E, Anderson G, Lane PJ. 2006b. Neonatal and adult $CD4^+$ $CD3^-$ cells share similar gene expression profile, and neonatal cells up-regulate OX40 ligand in response to TL1A (TNFSF15). *J Immunol* **177:** 3074–3081.

Kim JH, Luo JK, Zhang DE. 2008. The level of hepatitis B virus replication is not affected by protein ISG15 modification but is reduced by inhibition of UBP43 (USP18) expression. *J Immunol* **181:** 6467–6472.

Kobayashi Y, Iwata A, Suzuki K, Suto A, Kawashima S, Saito Y, Owada T, Kobayashi M, Watanabe N, Nakajima H. 2013. B and T lymphocyte attenuator inhibits LPS-induced endotoxic shock by suppressing Toll-like receptor 4 signaling in innate immune cells. *Proc Natl Acad Sci* **110:** 5121–5126.

Kojima R, Kajikawa M, Shiroishi M, Kuroki K, Maenaka K. 2011. Molecular basis for herpesvirus entry mediator recognition by the human immune inhibitory receptor CD160 and its relationship to the cosignaling molecules BTLA and LIGHT. *J Mol Biol* **413:** 762–772.

Kong M, Kim Y, Lee C. 2008. Promoter sequence variants of LIGHT are associated with female vascular dementia. *J Biomed Sci* **15:** 545–552.

Kotani H, Masuda K, Tamagawa-Mineoka R, Nomiyama T, Soga F, Nin M, Asai J, Kishimoto S, Katoh N. 2012. Increased plasma LIGHT levels in patients with atopic dermatitis. *Clin Exp Immunol* **168:** 318–324.

Kraal G, Mebius R. 2006. New insights into the cell biology of the marginal zone of the spleen. *Int Rev Cytol* **250:** 175–215.

Kruglov AA, Grivennikov SI, Kuprash DV, Winsauer C, Prepens S, Seleznik GM, Eberl G, Littman DR, Heikenwalder M, Tumanov AV, et al. 2013. Nonredundant function of soluble LT-α3 produced by innate lymphoid cells in intestinal homeostasis. *Science* **342:** 1243–1246.

Lanier LL. 2008. Evolutionary struggles between NK cells and viruses. *Nat Rev Immunol* **8:** 259–268.

Larsson M, Shankar EM, Che KF, Saeidi A, Ellegard R, Barathan M, Velu V, Kamarulzaman A. 2013. Molecular signatures of T-cell inhibition in HIV-1 infection. *Retrovirology* **10:** 31.

Launay E, Pangault C, Bertrand P, Jardin F, Lamy T, Tilly H, Tarte K, Bastard C, Fest T. 2012. High rate of TNFRSF14 gene alterations related to 1p36 region in de novo follicular lymphoma and impact on prognosis. *Leukemia* **26:** 559–562.

Le Bouteiller P, Barakonyi A, Giustiniani J, Lenfant F, Marie-Cardine A, Aguerre-Girr M, Rabot M, Hilgert I, Mami-Chouaib F, Tabiasco J, et al. 2002. Engagement of CD160 receptor by HLA-C is a triggering mechanism used by circulating natural killer (NK) cells to mediate cytotoxicity. *Proc Natl Acad Sci* **99:** 16963–16968.

Le Bouteiller P, Tabiasco J, Polgar B, Kozma N, Giustiniani J, Siewiera J, Berrebi A, Aguerre-Girr M, Bensussan A, Jabrane-Ferrat N. 2011. CD160: A unique activating NK cell receptor. *Immunol Lett* **138:** 93–96.

Lee WH, Kim SH, Lee Y, Lee BB, Kwon B, Song H, Kwon BS, Park JE. 2001. Tumor necrosis factor receptor superfamily 14 is involved in atherogenesis by inducing proinflammatory cytokines and matrix metalloproteinases. *Arterioscler Thromb Vasc Biol* **21:** 2004–2010.

Lepenies B, Pfeffer K, Hurchla MA, Murphy TL, Murphy KM, Oetzel J, Fleischer B, Jacobs T. 2007. Ligation of B and T lymphocyte attenuator prevents the genesis of experimental cerebral malaria. *J Immunol* **179:** 4093–4100.

Lin SC, Kuo CC, Chan CH. 2006. Association of a BTLA gene polymorphism with the risk of rheumatoid arthritis. *J Biomed Sci* **13:** 853–860.

Link A, Vogt TK, Favre S, Britschgi MR, Acha-Orbea H, Hinz B, Cyster JG, Luther SA. 2007. Fibroblastic reticular cells in lymph nodes regulate the homeostasis of naive T cells. *Nat Immunol* **8:** 1255–1265.

Liu GZ, Fang LB, Hjelmstrom P, Gao XG. 2008. Enhanced plasma levels of LIGHT in patients with acute atherothrombotic stroke. *Acta Neurol Scand* **118:** 256–259.

Locksley RM, Killeen N, Lenardo MJ. 2001. The TNF and TNF receptor superfamilies: Integrating mammalian biology. *Cell* **104:** 487–501.

Lohr JG, Stojanov P, Lawrence MS, Auclair D, Chapuy B, Sougnez C, Cruz-Gordillo P, Knoechel B, Asmann YW, Slager SL, et al. 2012. Discovery and prioritization of somatic mutations in diffuse large B-cell lymphoma (DLBCL) by whole-exome sequencing. *Proc Natl Acad Sci* **109:** 3879–3884.

Lopez RD. 2013. Inhibiting inhibitory pathways in human $\gamma\delta$ T cells. *Blood* **122:** 857–858.

Lopez-Verges S, Milush JM, Schwartz BS, Pando MJ, Jarjoura J, York VA, Houchins JP, Miller S, Kang SM, Norris PJ, et al. 2011. Expansion of a unique $CD57^+NKG2C^{hi}$ natural killer cell subset during acute human cytomegalovirus infection. *Proc Natl Acad Sci* **108:** 14725–14732.

Maeda M, Carpenito C, Russell RC, Dasanjh J, Veinotte LL, Ohta H, Yamamura T, Tan R, Takei F. 2005. Murine CD160, Ig-like receptor on NK cells and NKT cells, recognizes classical and nonclassical MHC class I and regulates NK cell activation. *J Immunol* **175:** 4426–4432.

Maiza H, Leca G, Mansur IG, Schiavon V, Boumsell L, Bensussan A. 1993. A novel 80-kD cell surface structure identifies human circulating lymphocytes with natural killer activity. *J Exp Med* **178:** 1121–1126.

Marrack P, Kappler J, Mitchell T. 1999. Type I interferons keep activated T cells alive. *J Exp Med* **189:** 521–530.

Martin-Guerrero I, Salaverria I, Burkhardt B, Szczepanowski M, Baudis M, Bens S, de Leval L, Garcia-Orad A, Horn H, Lisfeld J, et al. 2013. Recurrent loss of heterozygosity in 1p36 associated with *TNFRSF14* mutations in *IRF4* translocation negative pediatric follicular lymphomas. *Haematologica* **98:** 1237–1241.

Mauldin IS, Tung KS, Lorenz UM. 2012. The tyrosine phosphatase SHP-1 dampens murine Th17 development. *Blood* **119:** 4419–4429.

Mauri DN, Ebner R, Montgomery RI, Kochel KD, Cheung TC, Yu GL, Ruben S, Murphy M, Eisenberg RJ, Cohen GH, et al. 1998. LIGHT, a new member of the TNF superfamily, and lymphotoxin α are ligands for herpesvirus entry mediator. *Immunity* **8:** 21–30.

Mebius RE, Rennert P, Weissman IL. 1997. Developing lymph nodes collect $CD4^+CD3^-$ LT-β^+ cells that can differentiate to APC, NK cells, and follicular cells but not T or B cells. *Immunity* **7:** 493–504.

Meier D, Bornmann C, Chappaz S, Schmutz S, Otten LA, Ceredig R, Acha-Orbea H, Finke D. 2007. Ectopic lymphoid-organ development occurs through interleukin 7-mediated enhanced survival of lymphoid-tissue-inducer cells. *Immunity* **26:** 643–654.

M'Hidi H, Thibult ML, Chetaille B, Rey F, Bouadallah R, Nicollas R, Olive D, Xerri L. 2009. High expression of the inhibitory receptor BTLA in T-follicular helper cells and in B-cell small lymphocytic lymphoma/chronic lymphocytic leukemia. *Am J Clin Pathol* **132:** 589–596.

Mortarini R, Scarito A, Nonaka D, Zanon M, Bersani I, Montaldi E, Pennacchioli E, Patuzzo R, Santinami M, Anichini A. 2005. Constitutive expression and costimulatory function of LIGHT/TNFSF14 on human melanoma cells and melanoma-derived microvesicles. *Cancer Res* **65:** 3428–3436.

Moseman EA, Iannacone M, Bosurgi L, Tonti E, Chevrier N, Tumanov A, Fu YX, Hacohen N, von Andrian UH. 2012. B cell maintenance of subcapsular sinus macrophages protects against a fatal viral infection independent of adaptive immunity. *Immunity* **36:** 415–426.

Muntasell A, Vilches C, Angulo A, Lopez-Botet M. 2013. Adaptive reconfiguration of the human NK-cell compartment in response to cytomegalovirus: A different perspective of the host-pathogen interaction. *Eur J Immunol* **43:** 1133–1141.

Murphy TL, Murphy KM. 2010. Slow down and survive: Enigmatic immunoregulation by BTLA and HVEM. *Annu Rev Immunol* **28:** 389–411.

Murphy M, Walter BN, Pike-Nobile L, Fanger NA, Guyre PM, Browning JL, Ware CF, Epstein LB. 1998. Expression of the lymphotoxin β-receptor on follicular stromal cells in human lymphoid tissue. *Cell Death Differ* **5:** 497–505.

Murphy KM, Nelson CA, Šedý JR. 2006. Balancing co-stimulation and inhibition with BTLA and HVEM. *Nat Rev Immunol* **6:** 671–681.

Nikolova M, Marie-Cardine A, Boumsell L, Bensussan A. 2002. BY55/CD160 acts as a co-receptor in TCR signal transduction of a human circulating cytotoxic effector T lymphocyte subset lacking CD28 expression. *Int Immunol* **14:** 445–451.

Nolte MA, Belien JA, Schadee-Eestermans I, Jansen W, Unger WW, van Rooijen N, Kraal G, Mebius RE. 2003. A conduit system distributes chemokines and small blood-borne molecules through the splenic white pulp. *J Exp Med* **198:** 505–512.

Nurieva RI, Chung Y, Hwang D, Yang XO, Kang HS, Ma L, Wang YH, Watowich SS, Jetten AM, Tian Q, et al. 2008. Generation of T follicular helper cells is mediated by interleukin-21 but independent of T helper 1, 2, or 17 cell lineages. *Immunity* **29:** 138–149.

Ogawa T, Homma T, Igawa Y, Seki S, Ishizuka O, Imamura T, Akahane S, Homma Y, Nishizawa O. 2010. CXCR3 binding chemokine and TNFSF14 over expression in bladder urothelium of patients with ulcerative interstitial cystitis. *J Urol* **183:** 1206–1212.

Oki M, Watanabe N, Owada T, Oya Y, Ikeda K, Saito Y, Matsumura R, Seto Y, Iwamoto I, Nakajima H. 2011. A functional polymorphism in B and T lymphocyte attenuator is associated with susceptibility to rheumatoid arthritis. *Clin Dev Immunol* **2011:** 305656.

Pangault C, Ame-Thomas P, Ruminy P, Rossille D, Caron G, Baia M, De Vos J, Roussel M, Monvoisin C, Lamy T, et al. 2010. Follicular lymphoma cell niche: Identification of a preeminent IL-4-dependent T_{FH}–B cell axis. *Leukemia* **24:** 2080–2089.

Pantelyushin S, Haak S, Ingold B, Kulig P, Heppner FL, Navarini AA, Becher B. 2012. Rorγt^+ innate lymphocytes and γδ T cells initiate psoriasiform plaque formation in mice. *J Clin Invest* **122:** 2252–2256.

Pao LI, Badour K, Siminovitch KA, Neel BG. 2007. Nonreceptor protein-tyrosine phosphatases in immune cell signaling. *Annu Rev Immunol* **25:** 473–523.

Patel F, Marusina AI, Duong C, Adamopoulos IE, Maverakis E. 2013. NKG2C, HLA-E and their association with psoriasis. *Exp Dermatol* **22:** 797–799.

Perdigones N, Vigo AG, Lamas JR, Martinez A, Balsa A, Pascual-Salcedo D, de la Concha EG, Fernandez-Gutierrez B, Urcelay E. 2010. Evidence of epistasis between *TNFRSF14* and *TNFRSF6B* polymorphisms in patients with rheumatoid arthritis. *Arthritis Rheum* **62:** 705–710.

Peretz Y, He Z, Shi Y, Yassine-Diab B, Goulet JP, Bordi R, Filali-Mouhim A, Loubert JB, El-Far M, Dupuy FP, et al. 2012. CD160 and PD-1 co-expression on HIV-specific CD8 T cells defines a subset with advanced dysfunction. *PLoS Pathog* **8:** e1002840.

Poole E, King CA, Sinclair JH, Alcami A. 2006. The UL144 gene product of human cytomegalovirus activates NF-κB via a TRAF6-dependent mechanism. *EMBO J* **25:** 4390–4399.

Powolny-Budnicka I, Riemann M, Tanzer S, Schmid RM, Hehlgans T, Weih F. 2011. RelA and RelB transcription factors in distinct thymocyte populations control lymphotoxin-dependent interleukin-17 production in γδ T cells. *Immunity* **34:** 364–374.

Rabot M, Bensussan A, Le Bouteiller P. 2006. Engagement of the CD160 activating NK cell receptor leads to its association with CD2 in circulating human NK cells. *Transpl Immunol* **17:** 36–38.

Rabot M, El Costa H, Polgar B, Marie-Cardine A, Aguerre-Girr M, Barakonyi A, Valitutti S, Bensussan A, Le Bouteiller P. 2007. CD160-activating NK cell effector functions depend on the phosphatidylinositol-3-kinase recruitment. *Int Immunol* **19:** 401–409.

Rakesh K, Agrawal DK. 2005. Controlling cytokine signaling by constitutive inhibitors. *Biochem Pharmacol* **70:** 649–657.

Raychaudhuri S, Remmers EF, Lee AT, Hackett R, Guiducci C, Burtt NP, Gianniny L, Korman BD, Padyukov L, Kurreeman FA, et al. 2008. Common variants at CD40 and other loci confer risk of rheumatoid arthritis. *Nat Genet* **40:** 1216–1223.

Ribot JC, Silva-Santos B. 2013. Differentiation and activation of γδ T Lymphocytes: Focus on CD27 and CD28 costimulatory receptors. *Adv Exp Med Biol* **785:** 95–105.

Ribot JC, deBarros A, Pang DJ, Neves JF, Peperzak V, Roberts SJ, Girardi M, Borst J, Hayday AC, Pennington DJ, et al. 2009. CD27 is a thymic determinant of the balance between interferon-γ- and interleukin-17-producing γδ T cell subsets. *Nat Immunol* **10:** 427–436.

Sanjo H, Zajonc DM, Braden R, Norris PS, Ware CF. 2010. Allosteric regulation of the ubiquitin:NIK and ubiquitin:TRAF3 E3 ligases by the lymphotoxin-β receptor. *J Biol Chem* **285:** 17148–17155.

Satpathy AT, Briseno CG, Lee JS, Ng D, Manieri NA, Kc W, Wu X, Thomas SR, Lee WL, Turkoz M, et al. 2013. Notch2-dependent classical dendritic cells orchestrate intestinal immunity to attaching-and-effacing bacterial pathogens. *Nat Immunol* **14:** 937–948.

Sawa S, Cherrier M, Lochner M, Satoh-Takayama N, Fehling HJ, Langa F, Di Santo JP, Eberl G. 2010. Lineage relationship analysis of RORγt^{+} innate lymphoid cells. *Science* **330:** 665–669.

Sawcer S, Hellenthal G, Pirinen M, Spencer CC, Patsopoulos NA, Moutsianas L, Dilthey A, Su Z, Freeman C, Hunt SE, et al. 2011. Genetic risk and a primary role for cell-mediated immune mechanisms in multiple sclerosis. *Nature* **476:** 214–219.

Schneider K, Loewendorf A, De Trez C, Fulton J, Rhode A, Shumway H, Ha S, Patterson G, Pfeffer K, Nedospasov SA, et al. 2008. Lymphotoxin-mediated crosstalk between B cells and splenic stroma promotes the initial type I interferon response to cytomegalovirus. *Cell Host Microbe* **3:** 67–76.

Scholz H, Sandberg W, Damas JK, Smith C, Andreassen AK, Gullestad L, Froland SS, Yndestad A, Aukrust P, Halvorsen B. 2005. Enhanced plasma levels of LIGHT in unstable angina: Possible pathogenic role in foam cell formation and thrombosis. *Circulation* **112:** 2121–2129.

Šedý JR, Gavrieli M, Potter KG, Hurchla MA, Lindsley RC, Hildner K, Scheu S, Pfeffer K, Ware CF, Murphy TL, et al. 2005. B and T lymphocyte attenuator regulates T cell activation through interaction with herpesvirus entry mediator. *Nat Immunol* **6:** 90–98.

Šedý JR, Spear PG, Ware CF. 2008. Cross-regulation between herpesviruses and the TNF superfamily members. *Nat Rev Immunol* **8:** 861–873.

Šedý JR, Bjordahl RL, Bekiaris V, Macauley MG, Ware BC, Norris PS, Lurain NS, Benedict CA, Ware CF. 2013. CD160 Activation by herpesvirus entry mediator augments inflammatory cytokine production and cytolytic function by NK cells. *J Immunol* **15:** 828–836.

Serriari NE, Gondois-Rey F, Guillaume Y, Remmerswaal EB, Pastor S, Messal N, Truneh A, Hirsch I, van Lier RA, Olive D. 2010. B and T lymphocyte attenuator is highly expressed on CMV-specific T cells during infection and regulates their function. *J Immunol* **185:** 3140–3148.

Shang Y, Guo G, Cui Q, Li J, Ruan Z, Chen Y. 2012a. The expression and anatomical distribution of BTLA and its ligand HVEM in rheumatoid synovium. *Inflammation* **35:** 1102–1112.

Shang YJ, Cui QF, Li JL, Guo GN, Zhu WY. 2012b. Expression of the costimulatory molecule BTLA in synovial tissues from rheumatoid arthritis patients. *Xi Bao Yu Fen Zi Mian Yi Xue Za Zhi* **28:** 643–646.

Shui JW, Larange A, Kim G, Vela JL, Zahner S, Cheroutre H, Kronenberg M. 2012. HVEM signalling at mucosal barriers provides host defence against pathogenic bacteria. *Nature* **488:** 222–225.

Silva-Santos B, Pennington DJ, Hayday AC. 2005. Lymphotoxin-mediated regulation of γδ cell differentiation by αβ T cell progenitors. *Science* **307:** 925–928.

Smith W, Tomasec P, Aicheler R, Loewendorf A, Nemcovicova I, Wang EC, Stanton RJ, Macauley M, Norris P, Willen L, et al. 2013. Human cytomegalovirus glycoprotein UL141 targets the TRAIL death receptors to thwart host innate antiviral defenses. *Cell Host Microbe* **13:** 324–335.

Spits H, Artis D, Colonna M, Diefenbach A, Di Santo JP, Eberl G, Koyasu S, Locksley RM, McKenzie AN, Mebius RE, et al. 2013. Innate lymphoid cells—A proposal for uniform nomenclature. *Nat Rev Immunol* **13:** 145–149.

Steinberg MW, Turovskaya O, Shaikh RB, Kim G, McCole DF, Pfeffer K, Murphy KM, Ware CF, Kronenberg M. 2008. A crucial role for HVEM and BTLA in preventing intestinal inflammation. *J Exp Med* **205:** 1463–1476.

Summers deLuca L, Ng D, Gao Y, Wortzman ME, Watts TH, Gommerman JL. 2011. LT-βR signaling in dendritic cells induces a type I IFN response that is required for optimal clonal expansion of CD8^{+} T cells. *Proc Natl Acad Sci* **108:** 2046–2051.

Sun Y, Brown NK, Ruddy MJ, Miller ML, Lee Y, Wang Y, Murphy KM, Pfeffer K, Chen L, Kaye J, et al. 2009. B and T lymphocyte attenuator tempers early infection immunity. *J Immunol* **183:** 1946–1951.

Sun X, Shibata K, Yamada H, Guo Y, Muta H, Podack ER, Yoshikai Y. 2013. CD30L/CD30 is critical for maintenance of IL-17A-producing γδ T cells bearing Vγ6 in mucosa-associated tissues in mice. *Mucosal Immunol* **6:** 1191–1201.

Sutton CE, Mielke LA, Mills KH. 2012. IL-17-producing γδ T cells and innate lymphoid cells. *Eur J Immunol* **42:** 2221–2231.

Tao R, Wang L, Han R, Wang T, Ye Q, Honjo T, Murphy TL, Murphy KM, Hancock WW. 2005. Differential effects of B and T lymphocyte attenuator and programmed death-1 on acceptance of partially versus fully MHC-mismatched cardiac allografts. *J Immunol* **175:** 5774–5782.

Tiganis T, Bennett AM. 2007. Protein tyrosine phosphatase function: The substrate perspective. *Biochem J* **402:** 1–15.

van de Pavert SA, Olivier BJ, Goverse G, Vondenhoff MF, Greuter M, Beke P, Kusser K, Hopken UE, Lipp M, Niederreither K, et al. 2009. Chemokine CXCL13 is essential for lymph node initiation and is induced by retinoic acid and neuronal stimulation. *Nat Immunol* **10:** 1193–1199.

van der Fits L, Mourits S, Voerman JS, Kant M, Boon L, Laman JD, Cornelissen F, Mus AM, Florencia E, Prens EP, et al. 2009. Imiquimod-induced psoriasis-like skin inflammation in mice is mediated via the IL-23/IL-17 axis. *J Immunol* **182:** 5836–5845.

Van Vactor D, O'Reilly AM, Neel BG. 1998. Genetic analysis of protein tyrosine phosphatases. *Curr Opin Genet Dev* **8:** 112–126.

Vendel AC, Calemine-Fenaux J, Izrael-Tomasevic A, Chauhan V, Arnott D, Eaton DL. 2009. B and T lymphocyte attenuator regulates B cell receptor signaling by targeting Syk and BLNK. *J Immunol* **182:** 1509–1517.

Vivier E, Raulet DH, Moretta A, Caligiuri MA, Zitvogel L, Lanier LL, Yokoyama WM, Ugolini S. 2011. Innate or adaptive immunity? The example of natural killer cells. *Science* **331:** 44–49.

Vonarbourg C, Mortha A, Bui VL, Hernandez PP, Kiss EA, Hoyler T, Flach M, Bengsch B, Thimme R, Holscher C, et al. 2010. Regulated expression of nuclear receptor RORγt confers distinct functional fates to NK cell receptor-expressing RORγt$^+$ innate lymphocytes. *Immunity* **33:** 736–751.

Vondenhoff MF, Greuter M, Goverse G, Elewaut D, Dewint P, Ware CF, Hoorweg K, Kraal G, Mebius RE. 2009. LT-βR signaling induces cytokine expression and up-regulates lymphangiogenic factors in lymph node anlagen. *J Immunol* **182:** 5439–5445.

Walczak H. 2011. TNF and ubiquitin at the crossroads of gene activation, cell death, inflammation, and cancer. *Immunol Rev* **244:** 9–28.

Walter A, Schafer M, Cecconi V, Matter C, Urosevic-Maiwald M, Belloni B, Schonewolf N, Dummer R, Bloch W, Werner S, et al. 2013. Aldara activates TLR7-independent immune defence. *Nat Commun* **4:** 1560.

Wang Y, Koroleva EP, Kruglov AA, Kuprash DV, Nedospasov SA, Fu YX, Tumanov AV. 2010. Lymphotoxin β receptor signaling in intestinal epithelial cells orchestrates innate immune responses against mucosal bacterial infection. *Immunity* **32:** 403–413.

Ware CF. 2005. Network communications: Lymphotoxins, LIGHT, and TNF. *Annu Rev Immunol* **23:** 787–819.

Ware CF, Benedict C. 2012. Innate B cells: Oxymoron or validated concept?. *F1000Res* **1:** 8.

Ware CF, Šedý JR. 2011. TNF superfamily networks: Bidirectional and interference pathways of the herpesvirus entry mediator (TNFSF14). *Curr Opin Immunol* **23:** 627–631.

Watanabe N, Nakajima H. 2012. Coinhibitory molecules in autoimmune diseases. *Clin Dev Immunol* **2012:** 269756.

Watanabe N, Gavrieli M, Šedý JR, Yang J, Fallarino F, Loftin SK, Hurchla MA, Zimmerman N, Sim J, Zang X, et al. 2003. BTLA is a lymphocyte inhibitory receptor with similarities to CTLA-4 and PD-1. *Nat Immunol* **4:** 670–679.

Watchmaker PB, Lahl K, Lee M, Baumjohann D, Morton J, Kim SJ, Zeng R, Dent A, Ansel KM, Diamond B, et al. 2014. Comparative transcriptional and functional profiling defines conserved programs of intestinal DC differentiation in humans and mice. *Nat Immunol* **15:** 98–108.

Wencker M, Turchinovich G, Di Marco Barros R, Deban L, Jandke A, Cope A, Hayday AC. 2014. Innate-like T cells straddle innate and adaptive immunity by altering antigen-receptor responsiveness. *Nat Immunol* **15:** 80–87.

White A, Carragher D, Parnell S, Msaki A, Perkins N, Lane P, Jenkinson E, Anderson G, Caamano JH. 2007. Lymphotoxin α-dependent and -independent signals regulate stromal organiser cell homeostasis during lymph node organogenesis. *Blood* **15:** 1950–1959.

Wiens GD, Glenney GW. 2011. Origin and evolution of TNF and TNF receptor superfamilies. *Dev Comp Immunol* **35:** 1324–1335.

Wohn C, Ober-Blobaum JL, Haak S, Pantelyushin S, Cheong C, Zahner SP, Onderwater S, Kant M, Weighardt H, Holzmann B, et al. 2013. Langerin^{neg} conventional dendritic cells produce IL-23 to drive psoriatic plaque formation in mice. *Proc Natl Acad Sci* **110:** 10723–10728.

Wu TH, Zhen Y, Zeng C, Yi HF, Zhao Y. 2007. B and T lymphocyte attenuator interacts with CD3ζ and inhibits tyrosine phosphorylation of TCRζ complex during T-cell activation. *Immunol Cell Biol* **85:** 590–595.

Xu XS, Zhang Z, Gu LL, Wang FS. 2009. BTLA Characterization and its association with disease progression in patients with chronic HIV-1 infection. *Xi Bao Yu Fen Zi Mian Yi Xue Za Zhi* **25:** 1158–1160.

Yoshida H, Naito A, Inoue J, Satoh M, Santee-Cooper SM, Ware CF, Togawa A, Nishikawa S, Nishikawa S. 2002. Different cytokines induce surface lymphotoxin-αβ on IL-7 receptor-α cells that differentially engender lymph nodes and Peyer's patches. *Immunity* **17:** 823–833.

Zeng X, Chen H, Gupta R, Paz-Altschul O, Bowcock AM, Liao W. 2013. Deletion of the activating NKG2C receptor and a functional polymorphism in its ligand HLA-E in psoriasis susceptibility. *Exp Dermatol* **22:** 679–681.

Zhang Z, Xu X, Lu J, Zhang S, Gu L, Fu J, Jin L, Li H, Zhao M, Zhang J, et al. 2011. B and T lymphocyte attenuator down-regulation by HIV-1 depends on type I interferon and contributes to T-cell hyperactivation. *J Infect Dis* **203:** 1668–1678.

IL-6 in Inflammation, Immunity, and Disease

Toshio Tanaka[1,2], Masashi Narazaki[3], and Tadamitsu Kishimoto[4]

[1]Department of Clinical Application of Biologics, Osaka University Graduate School of Medicine, Osaka University, Osaka 565-0871, Japan

[2]Department of Immunopathology, World Premier International Immunology Frontier Research Center, Osaka University, Osaka 565-0871, Japan

[3]Department of Respiratory Medicine, Allergy and Rheumatic Diseases, Osaka University Graduate School of Medicine, Osaka University, Osaka 565-0871, Japan

[4]Laboratory of Immune Regulation, World Premier International Immunology Frontier Research Center, Osaka University, Osaka 565-0871, Japan

Correspondence: kishimoto@ifrec.osaka-u.ac.jp

Interleukin 6 (IL-6), promptly and transiently produced in response to infections and tissue injuries, contributes to host defense through the stimulation of acute phase responses, hematopoiesis, and immune reactions. Although its expression is strictly controlled by transcriptional and posttranscriptional mechanisms, dysregulated continual synthesis of IL-6 plays a pathological effect on chronic inflammation and autoimmunity. For this reason, tocilizumab, a humanized anti-IL-6 receptor antibody, was developed. Various clinical trials have since shown the exceptional efficacy of tocilizumab, which resulted in its approval for the treatment of rheumatoid arthritis and juvenile idiopathic arthritis. Moreover, tocilizumab is expected to be effective for other intractable immune-mediated diseases. In this context, the mechanism for the continual synthesis of IL-6 needs to be elucidated to facilitate the development of more specific therapeutic approaches and analysis of the pathogenesis of specific diseases.

IL-6 is a soluble mediator with a pleiotropic effect on inflammation, immune response, and hematopoiesis. At first, distinct functions of IL-6 were studied and given distinct names based on their biological activity. For example, the name B-cell stimulatory factor 2 (BSF-2) was based on the ability to induce differentiation of activated B cells into antibody (Ab)-producing cells (Kishimoto 1985), the name hepatocyte-stimulating factor (HSF) on the effect of acute phase protein synthesis on hepatocytes, the name hybridoma growth factor (HGF) on the enhancement of growth of fusion cells between plasma cells and myeloma cells, or the name interferon (IFN)-β2 owing to its IFN antiviral activity. When the BSF-2 cDNA was successfully cloned in 1986 (Hirano et al. 1986), however, it was found that the molecules with different names studied by various groups were in fact identical, resulting in the single name IL-6 (Kishimoto 1989). Human IL-6 is made up of 212 amino acids, including a 28-amino-acid

Cite this article as *Cold Spring Harb Perspect Biol* doi: 10.1101/cshperspect.a016295

signal peptide, and its gene has been mapped to chromosome 7p21. Although the core protein is ~20 kDa, glycosylation accounts for the size of 21–26 kDa of natural IL-6.

BIOLOGICAL EFFECT OF IL-6 ON INFLAMMATION AND IMMUNITY

After IL-6 is synthesized in a local lesion in the initial stage of inflammation, it moves to the liver through the bloodstream, followed by the rapid induction of an extensive range of acute phase proteins such as C-reactive protein (CRP), serum amyloid A (SAA), fibrinogen, haptoglobin, and α1-antichymotrypsin (Fig. 1) (Heinrich et al. 1990). On the other hand, IL-6 reduces the production of fibronectin, albumin, and transferrin. These biological effects on hepatocytes were at first studied as belonging to HSF. When high-level concentrations of SAA persist for a long time, it leads to a serious complication of several chronic inflammatory diseases through the generation of amyloid A amyloidosis (Gillmore et al. 2001). This results in amyloid fibril deposition, which causes progressive deterioration in various organs. IL-6 is also involved in the regulation of serum iron and zinc levels via control of their transporters. As for serum iron, IL-6 induces hepcidin production, which blocks the action of iron transporter ferroportin 1 on gut and, thus, reduces serum iron levels (Nemeth et al. 2004). This means that the IL-6-hepcidin axis is responsible for hypoferremia and anemia associated with chronic inflammation. IL-6 also enhances zinc importer ZIP14 expression on hepatocytes and so induces hypozincemia seen in inflammation (Liuzzi et al. 2005). When IL-6 reaches the bone marrow, it promotes megakaryocyte maturation, thus leading to the release of platelets (Ishibashi et al. 1989). These changes in acute phase protein levels and red blood cell and platelet counts are used for the evaluation of inflammatory severity in routine clinical laboratory examinations.

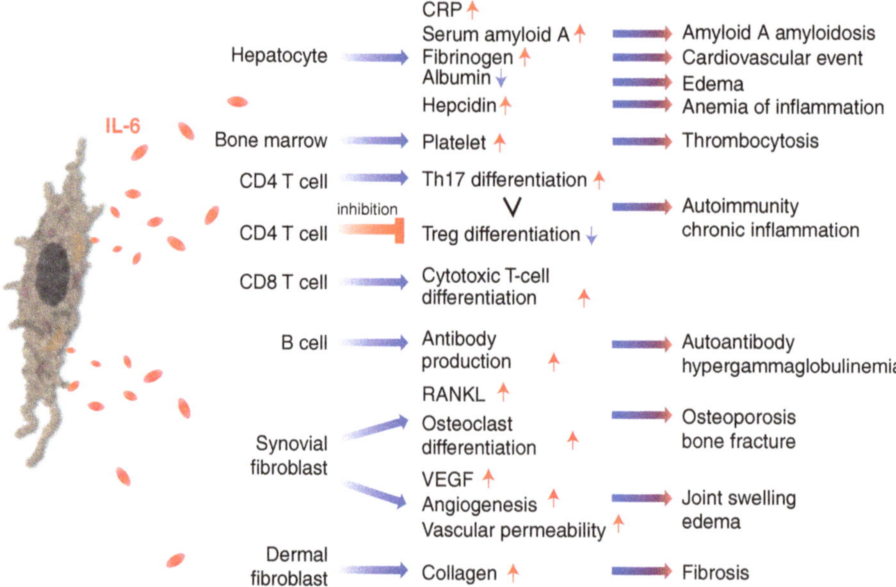

Figure 1. IL-6 in inflammation, immunity, and disease. IL-6 is a cytokine featuring pleiotropic activity; it induces synthesis of acute phase proteins such as CRP, serum amyloid A, fibrinogen, and hepcidin in hepatocytes, whereas it inhibits production of albumin. IL-6 also plays an important role on acquired immune response by stimulation of antibody production and of effector T-cell development. Moreover, IL-6 can promote differentiation or proliferation of several nonimmune cells. Because of the pleiotropic activity, dysregulated continual production of IL-6 leads to the onset or development of various diseases. Treg, regulatory T cell; RANKL, receptor activator of nuclear factor κB (NF-κB) ligand; VEGF, vascular endothelial growth factor.

Cite this article as *Cold Spring Harb Perspect Biol* doi: 10.1101/cshperspect.a016295

Furthermore, IL-6 promotes specific differentiation of naïve $CD4^+$ T cells, thus performing an important function in the linking of innate to acquired immune response. It has been shown that IL-6, in combination with transforming growth factor (TGF)-β, is indispensable for Th17 differentiation from naïve $CD4^+$ T cells (Korn et al. 2009), but that IL-6 also inhibits TGF-β-induced Treg differentiation (Bettelli et al. 2006). Up-regulation of the Th17/Treg balance is considered to be responsible for the disruption of immunological tolerance, and is thus pathologically involved in the development of autoimmune and chronic inflammatory diseases (Kimura and Kishimoto 2010). It has been further shown that IL-6 also promotes T-follicular helper-cell differentiation as well as production of IL-21 (Ma et al. 2012), which regulates immunoglobulin (Ig) synthesis and IgG4 production in particular. IL-6 also induces the differentiation of $CD8^+$ T cells into cytotoxic T cells (Okada et al. 1988). Under one of its previous names, BSF-2, IL-6 was found to be able to induce the differentiation of activated B cells into Ab-producing plasma cells, so that continuous oversynthesis of IL-6 results in hypergammaglobulinemia and autoantibody production.

IL-6 exerts various effects other than those on hepatocytes and lymphocytes and these are frequently detected in chronic inflammatory diseases (Kishimoto 1989; Hirano et al. 1990; Akira et al. 1993). One of these effects is that, when IL-6 is generated in bone marrow stromal cells, it stimulates the RANKL (Hashizume et al. 2008), which is indispensable for the differentiation and activation of osteoclasts (Kotake et al. 1996), and this leads to bone resorption and osteoporosis (Poli et al. 1994). IL-6 also induces excess production of VEGF, leading to enhanced angiogenesis and increased vascular permeability, which are pathological features of inflammatory lesions and are seen in, for example, synovial tissues of rheumatoid arthritis (RA) or edema of remitting seronegative symmetrical synovitis with pitting edema (RS3PE) syndrome (Nakahara et al. 2003; Hashizume et al. 2009). Finally, it has been reported that IL-6 aids keratinocyte proliferation (Grossman et al. 1989) or the generation of collagen in dermal fibroblasts that may account for changes in the skin of patients with systemic sclerosis (Duncan and Berman 1991).

REGULATION OF IL-6 SYNTHESIS

IL-6 functions as a mediator for notification of the occurrence of some emergent event. IL-6 is generated in an infectious lesion and sends out a warning signal to the entire body. The signature of exogenous pathogens, known as pathogen-associated molecular patterns, is recognized in the infected lesion by pathogen-recognition receptors (PRRs) of immune cells such as monocytes and macrophages (Kumar et al. 2011). These PRRs comprise Toll-like receptors (TLRs), retinoic acid-inducible gene-1-like receptors, nucleotide-binding oligomerization domain-like receptors, and DNA receptors. They stimulate a range of signaling pathways including NF-κB, and enhance the transcription of the mRNA of inflammatory cytokines such as IL-6, tumor necrosis factor (TNF)-α, and IL-1β. TNF-α and IL-1β also activate transcription factors to produce IL-6.

IL-6 also issues a warning signal in the event of tissue damage. Damage-associated molecular patterns (DAMPs), which are released from damaged or dying cells in noninfectious inflammations such as burn or trauma, directly or indirectly promote inflammation. During sterile surgical operations, an increase in serum IL-6 levels precedes elevation of body temperature and serum acute phase protein concentration (Nishimoto et al. 1989). DAMPs from injured cells contain a variety of molecules such as mitochondrial (mt) DNA, high mobility group box 1 (HMGB1), and S100 proteins (Bianchi 2007). Serum mtDNA levels in trauma patients are thousands of times higher than in controls and this elevation leads to TLR9 stimulation and NF-κB activation (Zhang et al. 2010), whereas binding of HMGB1 to TLR2, TLR4, and the receptor of advanced glycation end products (RAGE) can promote inflammation. The S100 family of proteins comprises more than 25 members, some of which also interact with RAGE to evoke sterile inflammation (Sims et al. 2010).

In addition to immune-mediated cells, mesenchymal cells, endothelial cells, fibroblasts, and many other cells are involved in the production of IL-6 in response to various stimuli (Akira et al. 1993). The fact that IL-6 issues a warning signal to indicate occurrence of an emergency accounts for the strict regulation of IL-6 synthesis both gene transcriptionally and posttranscriptionally. A number of transcription factors have been shown to regulate the IL-6 gene transcription (Fig. 2). The functional *cis*-regulatory elements in the human IL-6 gene 5′ flanking region are found binding sites for NF-κB, specificity protein 1 (SP1), nuclear factor IL-6 (NF-IL-6) (also known as CAAT/enhancer-binding protein β), activator protein 1 (AP-1), and interferon regulatory factor 1 (Libermann and Baltimore 1990; Akira and Kishimoto 1992; Matsusaka et al. 1993). Activation of *cis*-regulatory elements by stimulation with IL-1, TNF, TLR-mediated signal, and forskolin lead to activation of the IL-6 promoter.

A polymorphism at position -174 of the IL-6 promoter region is reportedly associated with systemic onset juvenile idiopathic arthritis (Fishman et al. 1998) and susceptibility to RA in Europeans (Lee et al. 2012). Stimulation with lipopolysaccharide (LPS) and IL-1 did not evoke any response in a reporter assay using -174 C construct. A -174 G construct, on the other hand, was found to promote transcription of the reporter gene, suggesting that a genetic back-

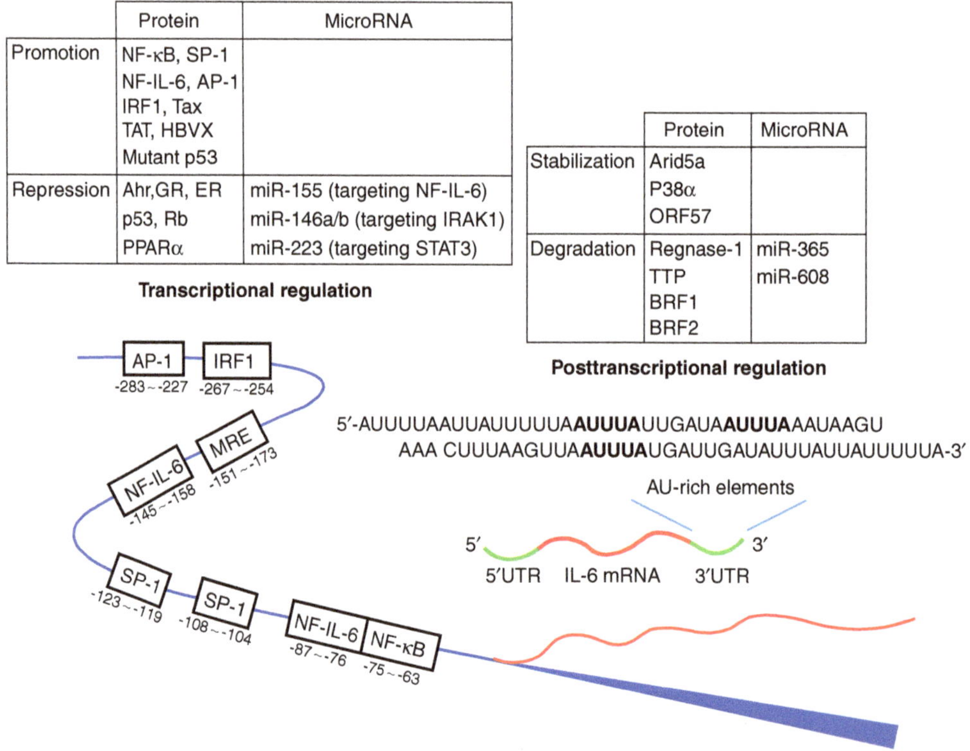

Figure 2. Transcriptional and posttranscriptional regulation of IL-6 gene. The expression and degradation of IL-6 mRNA is regulated transcriptionally and posttranscriptionally by several proteins and microRNAs. Activation of these proteins and microRNAs determines the fate of IL-6 mRNA. NF-IL-6, nuclear factor of IL-6; Tax, transactivator protein; TAT, transactivator of the transcription; HBVX, hepatitis B virus X protein; Ahr, aryl hydrocarbon receptor; GR, glucocorticoid receptor; ER, estrogen receptor; Rb, retinoblastoma; PPARα, peroxisome proliferator–activated receptor α; miR, microRNA; IRAK1, IL-1 receptor–associated kinase 1; STAT3, signal transducer and activator of transcription 3; ORF, open reading frame; TTP, tristetraprolin; BRF1, butyrate response factor 1.

 Cite this article as *Cold Spring Harb Perspect Biol* doi: 10.1101/cshperspect.a016295

ground of excess IL-6 production constitutes a risk factor for juvenile idiopathic arthritis and RA.

An interesting finding is that some viral products enhance the DNA-binding activity of NF-κB and NF-IL-6, resulting in an increase in IL-6 mRNA transcription. An instance of this phenomenon is that interaction with NF-κB of the Tax derived from the human T lymphotropic virus 1 enhances IL-6 production (Ballard et al. 1988; Leung and Nabel 1988). Another example is the enhancement of both NF-κB and NF-IL-6 DNA-binding activity by the transactivator of the TAT protein of the human immunodeficiency virus 1 (Scala et al. 1994; Ambrosino et al. 1997). Moreover, it has been shown that DNA binding of NF-IL-6 can be enhanced by the human hepatitis B virus X protein (Mahe et al. 1991; Ohno et al. 1999).

On the other hand, some transcription factors suppress IL-6 expression. Peroxisome proliferator–activated receptors (PPARs) are ligand-activated transcription factors consisting of three subtypes: α, β, and γ. Among three PPARs, fibrates-activated PPARα interacts with c-Jun and p65 NF-κB subunits, which negatively regulate IL-6 transcription (Delerive et al. 1999). In addition, some hormone receptors have been identified as repressors of IL-6 expression. The increase in serum IL-6 after menopause or ovarectomy is reportedly associated with suppression of IL-6 expression by estrogen receptors (Jilka et al. 1992), whereas activation of the glucocorticoid receptor can repress IL-6 expression, and this is thought to be one of mechanisms responsible for the anti-inflammatory effects of corticosteroids (Ray and Prefontaine 1994). It has further been shown that retinoblastoma protein and p53 repress the IL-6 gene promoter, whereas it is up-regulated by mutant p53 (Santhanam et al. 1991).

In addition, some microRNAs directly or indirectly regulate transcription activity. Interaction of microRNA-155 with the 3′ untranslated regions (UTR) of NF-IL-6 results in suppression of NF-IL-6 expression (He et al. 2009), whereas microRNA-146a/b and -223 indirectly suppress transcription of IL-6 by respectively targeting IL-1 receptor–associated kinase 1 and STAT3 (Chen et al. 2012; Zilahi et al. 2012).

PRODUCTION AND FUNCTION OF IL-6 AND ARYL HYDROCARBON RECEPTOR

Aryl hydrocarbon receptor (Ahr) not only affects IL-6 transcription, but also regulates innate and acquired immune response. Ahr, also known as the dioxin receptor, is a ligand-activated transcription factor that belongs to the basic helix-loop-helix PER-ARNT-SIM family (Burbach et al. 1992; Ema et al. 1992). Ahr is present in the cytoplasm, where it forms a complex with Ahr-interacting protein (Bell and Poland 2000). On binding with a ligand, Ahr moves to the nucleus and dimerizes with the Ahr nuclear translocator (Arnt). Within the nucleus, the Ahr/Arnt heterodimer then binds to the xenobiotic response element (XRE), which leads to various toxicological effects (Fujii-Kuriyama et al. 1994; Dragan and Schrenk 2000; Ohtake et al. 2003; Puga et al. 2005). Although the physiological ligands for Ahr are not well known, indoleamine 2,3-dioxygenase (IDO), which catalyzes tryptophan into kynurenine, is induced by Ahr signaling and kynurenine is one of the ligands of Ahr (Vogel et al. 2008; Jux et al. 2009).

An animal model of RA was used to show the essential role of Ahr in the induction of Th17 cells and Th17-dependent collagen-induced arthritis (CIA) (Kimura et al. 2008; Nakahama et al. 2011). Stimulation of naïve T cells with IL-6 plus TGF-β (Th17 cell–inducing condition) induced Ahr expression and deletion of the Ahr gene nullified the induction of Th17 cells.

Ahr interacted with and inhibited the activities of STAT1 or STAT5, which mediate the anti-inflammatory signals of IL-27 and IFN-γ, or IL-2, respectively (Harrington et al. 2005; Stumhofer et al. 2006; Laurence et al. 2007; Kimura et al. 2008), thus suppressing the inhibitory signals for the induction of Th17 cells. Retinoid-related orphan receptors (ROR) γ and α, which are activated by STAT3, are essential transcription factors for Th17-cell induction (Ivanov et al. 2006; Yang et al. 2008) and Ahr was found not to affect the ROR-γ and -α expression. In the Ahr gene–deficient mice, no arthritis developed in CIA. Moreover, with T-cell-specific deletion of the Ahr gene, no development of CIA was observed (Nakahama et al. 2011). These results clearly show that CIA is a T-cell-dependent disease and the presence of Ahr is essential for its development. In these Ahr-deficient mice, the number of Th17 cells decreased and that of Th1 cells increased but no significant changes were observed in Foxp3-expressing Treg cells.

Ahr also regulates Th17-cell induction through regulation of microRNAs. In our study, Ahr-induced microRNA-132/212 cluster under Th17 cell–inducing conditions and transfection of microRNA-212 into naïve T cells under these circumstances augmented the expression of IL-17-related genes such as IL-17A, IL-22, and IL-23R (Nakahama et al. 2013). One of the target genes of this microRNA is B-cell lymphoma 6, which is known as an inhibitor of Th17-cell induction (Yu et al. 2009). All of these findings show that Ahr accelerates inflammation through the enhancement of Th17-cell induction by several mechanisms.

Interestingly, Ahr showed a negative regulatory effect on peritoneal macrophages and bone marrow–derived dendritic cells (BMDC) (Nguyen et al. 2010). In the absence of Ahr, LPS-induced production of inflammatory cytokines such as IL-6, TNF, and IL-12 showed major increases in macrophages, indicating that Ahr negatively regulates inflammatory cytokine production. Ahr interacts with STAT1 and NF-κB and the resultant complex of Ahr/STAT1 and NF-κB leads to inhibition of the promoter activity of IL-6 and other inflammatory cytokines (Kimura et al. 2009). In BMDC, Ahr is required for the activation of IDO leading to kynurenine production because the deletion of Ahr in BMDC leads to loss of IL-10 and kynurenine production. Coculture of naïve T cells with Ahr-deficient BMDC in the presence of LPS resulted in reduction of Treg-cell induction, whereas addition of kynurenine rescued the induction of Treg cells by BMDC (Nguyen et al. 2010). These findings indicate that Ahr is required for the regulatory BMDC cells through the induction of IDO.

STABILIZATION AND DEGRADATION OF IL-6 MRNA (ARID5A AND REGNASE-1)

As for posttranscriptional regulation of cytokine expression, cytokine mRNA is controlled through both the 5′ and 3′UTR (Chen and Shyu 1995; Anderson 2008). Initiation of mRNA translation is determined by the 5′UTR, and the stability of mRNA by the 3′UTR. IL-6 mRNA is regulated by modulation of AU-rich elements located in the 3′UTR region, whereas a number of RNA-binding proteins and microRNAs bind to the 3′UTRs and regulate the stability of IL-6 mRNA (Fig. 2). For example, IL-6 mRNA stabilization is promoted by mitogen-activated protein kinase (MAPK) p38α via 3′UTRs of IL-6 (Zhao et al. 2008), and the stabilization of both viral and human IL-6 mRNA by the Kaposi's sarcoma–associated herpesvirus (KSHV) ORF-57 by competing with the binding of microRNA-1293 to the viral or of microRNA-608 to the human IL-6 mRNA (Kang et al. 2011). RNA-binding proteins, such as TTP and BRF1 and 2, on the other hand, promote IL-6 mRNA degradation (Palanisamy et al. 2012), whereas IL-6 mRNA levels are reduced by microRNAs such as microRNA-365 and -608 through direct interaction with IL-6 3′UTR (Kang et al. 2011; Xu et al. 2011).

It was recently found that a nuclease known as regulatory RNase-1 (regnase-1) (also known as Zc3h12a) plays a part in the destabilization of IL-6 mRNA, and that the relevant knockout mice spontaneously develop autoimmune diseases accompanied by splenomegaly and lymphadenopathy (Matsushita et al. 2009). The inhibitor of NF-κB (IκB) kinase (IKK) complex controls IL-6 mRNA stability by phosphorylating regnase-1 in response to IL-1R/TLR stimulation (Iwasaki et al. 2011). Phosphorylated regnase-1 underwent ubiquitination and degradation. Regnase-1 reexpressed in IL-1R/TLR-activated cells was found to feature delayed kinetics, and regnase-1 mRNA to be negatively regulated by regnase-1 itself via a stem-loop region present in the regnase-1 3′UTR. These findings show that IKK complex phosphorylates not only IκBα, activating transcription, but also regnase-1, releasing the brake on IL-6 mRNA expression. Regnase-1 also regulates the mRNAs of a set of genes, including c-Rel, Ox40, and IL-2 through cleavage of their 3′UTRs in T cells. T-cell receptor engagement then leads to cleavage of regnase-1, which frees T cells from regnase-mediated suppression, thus indicating that regnase-1 may play a crucial role in T-cell activation (Uehata et al. 2013).

We have recently identified a novel RNA-binding protein, AT-rich interactive domain-containing protein 5a (Arid5a), which binds to the 3′UTR of IL-6 mRNA, resulting in the selective stabilization of IL-6 but not of TNF-α or IL-12 mRNA (Masuda et al. 2013). Arid5a expression was found to be enhanced in macrophages in response to LPS, IL-1β, and IL-6, and also to be induced under Th17-polarizing conditions in T cells. We also found that Arid5a gene deficiency inhibited elevation of IL-6 levels in LPS-injected mice and preferential Th17-cell development in experimental autoimmune encephalomyelitis. Moreover, Arid5a counteracted the destabilizing function of regnase-1 on IL-6 mRNA (Fig. 3), indicating that the balance between Arid5a and regnase-1 plays an important role in IL-6 mRNA stability. All of these results suggest that posttranscriptional regulation of IL-6 mRNA by Arid5a and regnase-1 may play an important role in the expression of IL-6 and that the predominance of Arid5a over regnase-1 promotes inflammatory processes and possibly induces the development of autoimmune inflammatory diseases.

During the so-called "cytokine storm," a potentially fatal immune reaction induced by hyperactivation of T cells, a major boost in IL-6 production is observed but without comparable production of other inflammatory cytokines. A recent study showed that the cytokine storm induced by cancer immunotherapy using T-cell transfection was counteracted by the anti-IL-6 receptor antibody, tocilizumab (Grupp et al. 2013). Experimentally, inhalation by mice of peroxidized phospholipids induced a cytokine storm resulting from a greatly marked increase in the production of IL-6 but not TNF (Imai et al. 2008).

These results showing the IL-6-specific elevation without any effect on the other inflammatory cytokines strongly suggest the impor-

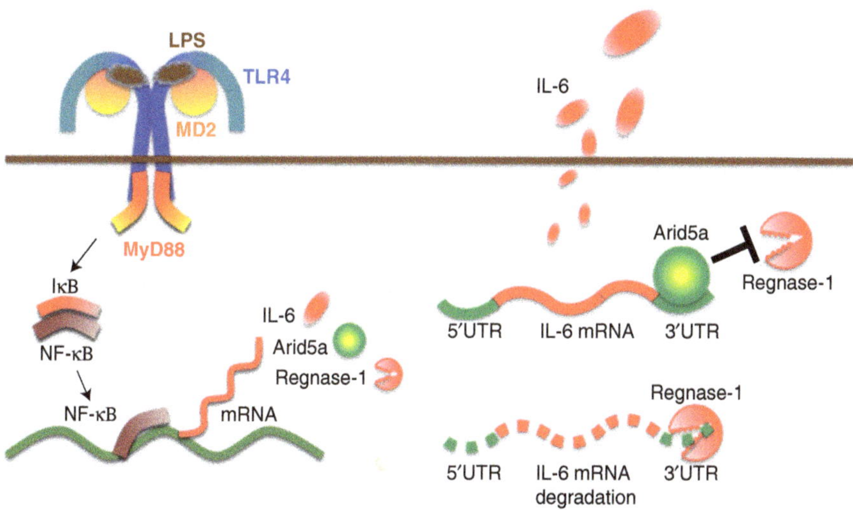

Figure 3. IL-6 synthesis and regulation of IL-6 mRNA stability by Arid5a. Pathogen-associated molecular patterns are recognized by pathogen-recognition receptors to induce proinflammatory cytokines; in this figure, TLR4 recognizes LPS and induces IL-6 mRNA via activation of the NF-κB signaling pathway. Regnase-1 promotes IL-6 mRNA degradation, whereas Arid5a inhibits destabilizing effects of regnase-1. The balance between Arid5a and regnase-1 is important for the regulation of IL-6 mRNA. MD2, myeloid differentiation protein 2; MyD88, myeloid differentiation primary response 88; IκB, inhibitor of NF-κB.

tance of posttranscriptional modification of IL-6 mRNA by Arid5a and regnase-1. The balance between Arid5a and regnase-1 may determine the pathological increase of IL-6 in various diseases including autoimmunity and even cytokine storm.

IL-6 RECEPTOR–MEDIATED SIGNALING SYSTEM

The IL-6 receptor–signaling system is made up of two receptor chains and downstream signaling molecules (Kishimoto et al. 1992). The IL-6 receptor (IL-6R) constitutes the IL-6-binding chain, which occurs in two forms, 80-kDa transmembrane and 50–55-kDa–soluble IL-6R (sIL-6R), whereas 130-kDa gp130 constitutes the signal-transducing chain. Both of these proteins belong to the cytokine receptor family with a Trp-Ser-X-Trp-Ser motif (Yamasaki et al. 1988; Hibi et al. 1990). sIL-6R without the cytoplasmic region is present in human serum and after IL-6 binding to sIL-6R; the resultant complex induces the IL-6 signal on gp130-expressing cells (Narazaki et al. 1993). The pleiotropic effect of IL-6 on various cells derives from the broad range of gp130 expression observed on cells (Taga and Kishimoto 1997). After binding of IL-6 to IL-6R, the IL-6/IL-6R complex in turn induces homodimerization of gp130 and triggers a downstream signal cascade (Fig. 4) (Murakami et al. 1993). The activated IL-6 receptor complex is generated in the form of a hexameric structure comprising two molecules each of IL-6, IL-6R, and gp130 (Boulanger et al. 2003). Of these components, IL-6R is a unique binding-receptor for IL-6, whereas the signal-transducing chain gp130 is shared by members of the IL-6 family of cytokines, that is, leukemia inhibitory factor, oncostatin M, ciliary neurotrophic factor, IL-11, cardiotrophin 1, cardiotrophin-like cytokine, IL-27, and IL-35. Although all of these cytokines thus bind to their specific binding receptors, they use the same gp130 for their signals (Kishimoto et al. 1994, 1995). The only exception is virus-encoded IL-6, which is the product of KSHV (also known as human herpesvirus 8), and directly binds to and activates gp130 (Aoki et al. 2001). The mechanism that the IL-6 cytokine family members use to employ the common signal transducer makes it clear why members of the IL-6 family show functional redundancy. The molecular elucidation of the IL-6-signaling system finally solved the long-standing mystery of why cytokines featured pleiotropy and redundancy.

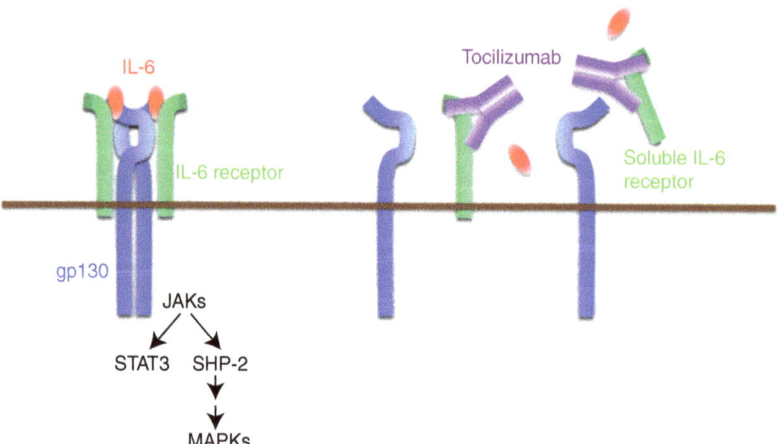

Figure 4. IL-6 receptor system and IL-6 blocker, a humanized anti-IL-6 receptor antibody tocilizumab. IL-6 binds to soluble and transmembrane IL-6R and the complex, then induces homodimerization of gp130, leading to activation of the signaling system. A humanized anti-IL-6R antibody, tocilizumab, blocks IL-6-mediated signaling pathway by its inhibition of IL-6 binding to both receptors. JAKs, Janus kinases; SHP-2, SH2-domain containing protein tyrosine phosphatase-2.

 Cite this article as *Cold Spring Harb Perspect Biol* doi: 10.1101/cshperspect.a016295

Activation of gp130 in turn triggers activation of downstream signaling molecules, that is, the Janus kinase (JAK)-STAT3 pathway and the JAK-SHP-2-mitogen-activated protein (MAP) kinase pathway. The regulation of various sets of IL-6 responsive genes, including acute phase proteins, is accounted for by the transcription factor STAT3, which also induces the suppressor of cytokine signaling 1 (SOCS1) and SOCS3, which share the SH2-domain. In this context, SOCS1 binds to tyrosine-phosphorylated JAK (Naka et al. 1997), whereas SOCS3 binds to tyrosine-phosphorylated gp130 (Schmitz et al. 2000) to stop IL-6 signaling by means of a negative feedback loop.

IL-6 AND DISEASE

An immediate and transient expression of IL-6 is generated in response to environmental stress factors such as infections and tissue injuries. This expression triggers an alarm signal and activates host defense mechanisms against stress. Removal of the source of stress from the host is followed by cessation of IL-6-mediated activation of the signal-transduction cascade by negative regulatory systems such as ligand-induced internalization and degradation of gp130 and recruitment of SOCS (Naka et al. 1997), as well as degradation of IL-6 mRNA by regnase-1 leading to termination of IL-6 production. However, dysregulated and persistent IL-6 production of mostly unknown etiology, one of which may be the unbalance between Arid5a and regnase-1, in certain cell populations leads to the development of various diseases. This association of IL-6 with disease development was first shown in a case of cardiac myxoma. The culture of fluid obtained from the myxoma tissues of a patient who presented with fever, polyarthritis with positivity for antinuclear factor, elevated CRP level, and hypergammaglobulinemia, contained a large quantity of IL-6, which suggested that IL-6 may contribute to chronic inflammation and autoimmunity (Hirano et al. 1987). Subsequent studies have shown that dysregulation of IL-6 production occurs in the synovial cells of RA (Hirano et al. 1988), swollen lymph nodes of Castleman's disease (Yoshizaki et al. 1989), myeloma cells (Kawano et al. 1988), and peripheral blood cells or involved tissues in various other autoimmune and chronic inflammatory diseases and even malignant cells in cancers (Nishimoto et al. 1989, 2005).

Moreover, the pathological role of IL-6 in disease development has been shown in numerous animal models of diseases as well as the fact that IL-6 blockade by means of gene knockout or administration of anti-IL-6 or anti-IL-6R Ab can result in the preventive or therapeutic suppression of disease development. For example, IL-6 blockade resulted in a noticeable reduction in susceptibility to Castleman's disease–like symptoms in IL-6 transgenic mice (Katsume et al. 2002). Similar effects were observed in models of RA (Alonzi et al. 1998; Ohshima et al. 1998; Fujimoto et al. 2008), systemic lupus erythematosus (Mihara et al. 1998), systemic sclerosis (Kitaba et al. 2012), inflammatory myopathies (Okiyama et al. 2009), experimental autoimmune uveoretinitis (Haruta et al. 2011), experimental autoimmune encephalomyelitis (Serada et al. 2008), and many other diseases.

IL-6 TARGETING AS STRATEGY FOR TREATMENT OF IMMUNE-MEDIATED DISEASES

In view of the range of biological activities of IL-6 and its pathological role in various diseases described above, it was anticipated that IL-6 targeting would constitute a novel treatment strategy for various immune-mediated diseases. The development of tocilizumab was a direct result of this hypothesis. Tocilizumab is a humanized anti-IL-6R monoclonal Ab of the IgG1 class that was generated by grafting the complementarity determining regions of a mouse antihuman IL-6R Ab onto human IgG1 (Sato et al. 1993), and it blocks IL-6-mediated signal transduction by inhibiting IL-6 binding to transmembrane and soluble IL-6R (Fig. 4). The outstanding efficacy, tolerability, and safety of tocilizumab were verified in numerous worldwide clinical trials initiated in the late 1990s. This has resulted in the approval of this biologic for the treatment of RA in more than 100 countries (Tanaka et al. 2013), as well as for systemic and polyarticular juvenile idiopathic arthritis (Yokota et al. 2008, 2012; De Benedetti et al. 2012) and Castleman's disease (Nishimoto et al. 2005) in several countries. Although other biologics including TNF inhibitors, T-cell stimulator blocker, B-cell depletory, or IL-1R antagonist are currently used for RA, tocilizumab has proved its superior efficacy as monotherapy for moderate-to-severe active RA (Tanaka and Kishimoto 2011; Emery et al. 2013) and is recommended as a first-line biologic. Moreover, the outstanding efficacy of tocilizumab for systemic juvenile idiopathic arthritis has led to the recognition of the start of a new era in the treatment of this disease, which was long considered to be one of the most intractable juvenile diseases (Sandborg and Mellins 2012).

Furthermore, there are strong indications based on favorable results detailed in numerous recent case reports, series, and pilot studies of the off-label application of tocilizumab that it can be used for the treatment of various immune-mediated diseases (Tanaka and Kishimoto 2012; Tanaka et al. 2012). These comprise autoimmune, chronic inflammatory, autoinflammatory, and other diseases. The first group includes systemic sclerosis, inflammatory myopathies, large vessel vasculitis, systemic lupus erythematosus, relapsing polychondritis, autoimmune hemolytic anemia, acquired hemophilia A, neuromyelitis optica, and Cogan's syndrome. The second group includes adult-onset Still's disease, amyloid A amyloidosis, polymyalgia rheumatica, RS3PE, Bechet's disease, uveitis, Crohn's disease, graft-versus-host disease, pulmonary arterial hypertension, and IgG4-related diseases. The third group comprises such autoinflammatory diseases as TNF-receptor-associated periodic syndrome and chronic inflammatory neurological cutaneous articular syndrome, whereas other diseases include atherosclerosis, type 2 diabetes mellitus, atopic dermatitis, sciatica, and cancer-associated cachexia. Clinical trials are in progress to identify additional indications for tocilizumab (Table 1).

Table 1. Ongoing clinical trials of tocilizumab

Targeted diseases	Status	Identifier
Diabetes mellitus (type 2), obesity	Phase 2	NCT01073826
Graves' ophthalmopathy	Phase 3	NCT01297699
Cardiovascular disease in RA (vs. etanercept)	Phase 4	NCT01331837
Polymyalgia rheumatica	Phase 2	NCT01396317
	Phase 2	NCT01713842
Giant-cell arteritis	Phase 2	NCT01450137
	Phase 3	NCT01791153
Steroid refractory acute GVHD	Phase 1/2	NCT01475162
After HSCT	Phase 2	NCT01757197
Non-ST elevation myocardial infarction	Phase 2	NCT01491074
Noninfectious uveitis	Phase 1/2	NCT01717170
Systemic sclerosis	Phase 2/3	NCT01532869
Transplant rates awaiting kidney transplantation	Phase 1/2	NCT01594424
JIA-associated uveitis	Phase 1/2	NCT01603355
Recurrent ovarian cancer	Phase 1/2	NCT01637532
Behcet's syndrome	Phase 2	NCT01693653
Schizophrenia	Phase 1	NCT01696929
Erdheim–Chester disease	Phase 2	NCT01727206
Primary Sjögren's syndrome	Phase 2/3	NCT01782235
Fibrous dysplasia of bone	Phase 2	NCT01791842

Registered with ClinicalTrials.gov for diseases other than rheumatoid arthritis, juvenile idiopathic arthritis, and Castleman's disease.

GVHD, graft-versus-host disease; HSCT, hematopoietic stem cell transplant; JIA, juvenile idiopathic arthritis.

CONCLUDING REMARKS AND FUTURE PROSPECTS

The first report of the existence of soluble factors for the enhancement of IgG and IgE Ab responses was published by Kishimoto and Ishizaka in 1973, and it took 13 years until the actual cloning of the IL-6 gene (Hirano et al. 1986). After this success, the fundamental research progressed rapidly and the whole picture of the IL-6 signaling system was completed in the early 1990s (Kishimoto et al. 1994). In parallel with this development, the pathological involvement of IL-6 in various diseases was also established (Nishimoto et al. 1989, 2005; Yoshizaki et al. 1989). It was found that if free serum concentration of tocilizumab is maintained at more than 1 µg/ml, CRP remains negative (Nishimoto et al. 2008), indicating that IL-6 plays a major role in the induction of CRP expression and that IL-6 may be involved in the development of almost all chronic inflammatory diseases with CRP elevation. Clinical trials of tocilizumab started in the late 1990s and this humanized monoclonal Ab was approved for the treatment of Castleman's disease in 2005 in Japan, nearly 20 years after the successful molecular cloning of the IL-6 gene (Fig. 5) (Kishimoto 2005). During the following years, tocilizumab has become a first-line biologic for the treatment of moderate-to-severe active RA and the only approved first-line biologic for systemic juvenile idiopathic arthritis. It is anticipated that during the next decade this IL-6 blocker will be widely used for

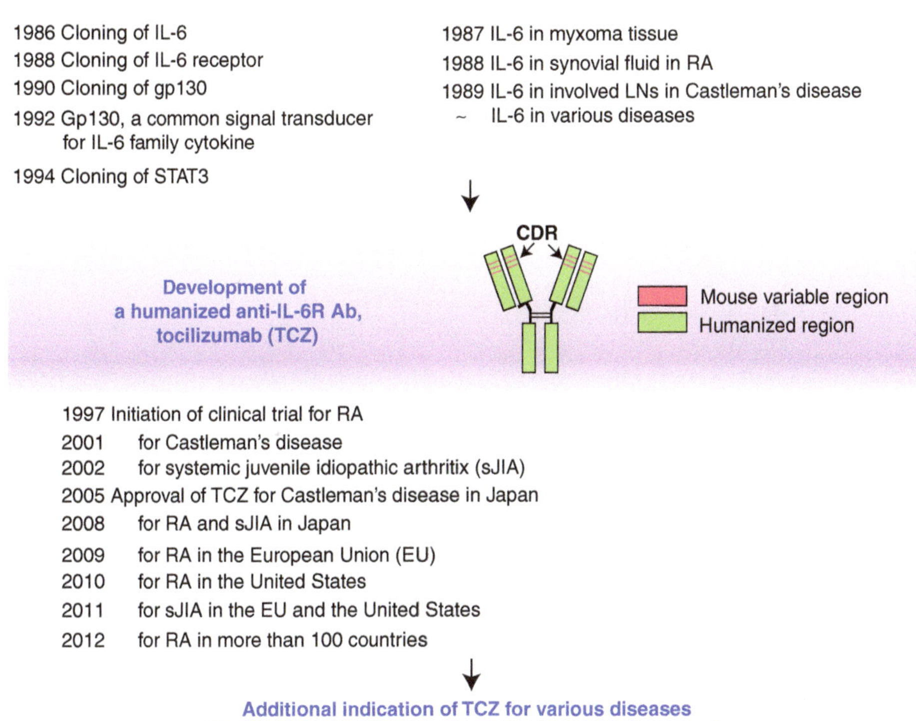

Figure 5. Major discoveries in IL-6-related research and establishment of IL-6 targeting strategy for immune-mediated diseases. Basic research regarding IL-6 clarified the molecular basis of the characteristics of cytokine, redundancy, and pleiotropy, whereas clinical research revealed its pathological significance in disease development. These findings led to the concept that IL-6 targeting would constitute a novel therapeutic strategy for immune-mediated diseases, and indeed, tocilizumab, a humanized anti-IL-6R antibody, became an innovative biologic for the treatment of intractable diseases such as RA, systemic juvenile idiopathic arthritis (sJIA), and Castleman's disease. It will be expected that this strategy would be widely applicable for other immune-mediated diseases. LNs, lymph nodes; TCZ, tocilizumab; CDR, complementarity determining region.

the treatment of various as-yet intractable diseases including cytokine storm and its application will overcome the refractoriness of such diseases.

To achieve this goal, however, there are several hurdles to overcome. First, additional clinical trials will be needed to evaluate the efficacy and safety of tocilizumab for various diseases. The second, and also important, hurdle is to clarify the mechanisms that render tocilizumab efficacious for phenotypically different diseases. In the case of amyloid A amyloidosis, the ameliorative effect of tocilizumab can be explained by its potent suppression of serum amyloid A protein synthesis (Tanaka et al. 2011). In the case of RA, it has been shown that tocilizumab treatment led to improvement in systemic and joint inflammatory markers (Garnero et al. 2010; Terpos et al. 2011; Kanbe et al. 2012), but it remained to be determined whether it can rectify fundamental immunological abnormalities, such as autoantibody production or imbalance of effector $CD4^+$ T-cell subsets (Tanaka 2013). However, recent preliminary results indicate that tocilizumab treatment can correct the imbalance between Th17 and Treg in peripheral blood $CD4^+$ T cells (Samson et al. 2012; Pesce et al. 2013). Moreover, it was shown that the treatment caused a reduction in the level of serum IgG4-class anticyclic citrullinated peptide Ab in RA (Carbone et al. 2013). Anti-aquaporin 4 (AQP4) Ab plays a pathological role in neuromyelitis optica, and tocilizumab treatment was found to improve clinical symptoms and reduce serum anti-AQP4 Ab titers, perhaps by inhibiting cell survival of the plasmablasts secreting this Ab (Chihara et al. 2011; Araki et al. 2013). If IL-6 blockade can actually correct these immunological abnormalities, it will in fact be possible to use tocilizumab for the treatment of a wide variety of immune-mediated diseases.

Finally, the mystery remains why IL-6 is persistently expressed in distinct cell populations in various diseases. Accurate and detailed analyses of proteins such as Arid5a and regnase-1 and of microRNAs that regulate IL-6 synthesis will be helpful for solving this mystery, whereas clarification of the mechanism(s) involved will facilitate the identification of more specific target molecules and investigations into the pathogenesis of specific diseases.

ACKNOWLEDGMENTS

T.K. holds a patent of tocilizumab and has received royalties for Actemra. T.T. received grants and payment for lectures including service on speaker's bureaus from Chugai Pharmaceutical Co., Ltd. M.N. received payment for lectures including service on speaker's bureaus from Chugai Pharmaceutical Co., Ltd.

REFERENCES

Akira S, Kishimoto T. 1992. IL-6 and NF-IL6 in acute-phase response and viral infection. *Immunol Rev* **127:** 25–50.

Akira S, Taga T, Kishimoto T. 1993. Interleukin-6 in biology and medicine. *Adv Immunol* **54:** 1–78.

Alonzi T, Fattori E, Lazzaro D, Costa P, Probert L, Kollias G, De Benedetti F, Poli V, Ciliberto G. 1998. Interleukin 6 is required for the development of collagen-induced arthritis. *J Exp Med* **187:** 461–468.

Ambrosino C, Ruocco MR, Chen X, Mallardo M, Baudi F, Trematerra S, Quinto I, Venuta S, Scala G. 1997. HIV-1 Tat induces the expression of the interleukin-6 (IL6) gene by binding to the IL6 leader RNA and by interacting with CAAT enhancer-binding protein β (NF-IL6) transcription factors. *J Biol Chem* **272:** 14883–14892.

Anderson P. 2008. Post-transcriptional control of cytokine production. *Nature Immunol* **9:** 353–359.

Aoki Y, Narazaki M, Kishimoto T, Tosato G. 2001. Receptor engagement by viral interleukin-6 encoded by Kaposi sarcoma-associated herpesvirus. *Blood* **98:** 3042–3049.

Araki M, Aranami T, Matsuoka T, Nakamura M, Miyake S, Yamamura T. 2013. Clinical improvement in a patient with neuromyelitis optica following therapy with the anti-IL-6 receptor monoclonal antibody tocilizumab. *Mod Rheumatol* **23:** 827–831.

Ballard DW, Bohnlein E, Lowenthal JW, Wano Y, Franza BR, Greene WC. 1988. HTLV-I tax induces cellular proteins that activate the κB element in the IL-2 receptor α gene. *Science* **241:** 1652–1655.

Bell DR, Poland A. 2000. A binding of aryl hydrocarbon receptor (AhR) to AhR-interacting protein. The role of hsp90. *J Biol Chem* **275:** 36407–36414.

Bettelli E, Carrier Y, Gao W, Korn T, Strom TB, Oukka M, Weiner HL, Kuchroo VK. 2006. Reciprocal developmental pathways for the generation of pathogenic effector TH17 and regulatory T cells. *Nature* **441:** 235–238.

Bianchi ME. 2007. DAMPs, PAMPs and alarmins: All we need to know about danger. *J Leukoc Biol* **81:** 1–5.

Boulanger MJ, Chow DC, Brevnova EE, Garcia KC. 2003. Hexameric structure and assembly of the interleukin-6/IL-6 α-receptor/gp130 complex. *Science* **300:** 2101–2104.

Burbach KM, Poland A, Bradfield CA. 1992. Cloning of the Ah-receptor cDNA reveals a distinctive ligand-activated transcription factor. *Proc Natl Acad Sci* **89:** 8185–8189.

Carbone G, Wilson A, Diehl SA, Bunn J, Cooper SM, Rincon M. 2013. Interleukin-6 receptor blockade selectively reduces IL-21 production by CD4 T cells and IgG4 autoantibodies in rheumatoid arthritis. *Int J Biol Sci* **9:** 279–288.

Chen CY, Shyu AB. 1995. AU-rich elements: Characterization and importance in mRNA degradation. *Trends Biochem Sci* **20:** 465–470.

Chen Q, Wang H, Liu Y, Song Y, Lai L, Han Q, Cao X, Wang Q. 2012. Inducible microRNA-223 down-regulation promotes TLR-triggered IL-6 and IL-1β production in macrophages by targeting STAT3. *PLoS ONE* **7:** e42971.

Chihara N, Aranami T, Sato W, Miyazaki Y, Miyake S, Okamoto T, Ogawa M, Toda T, Yamamura T. 2011. Interleukin 6 signaling promotes anti-aquaporin 4 autoantibody production from plasmablasts in neuromyelitis optica. *Proc Natl Acad Sci* **108:** 3701–3706.

De Benedetti F, Brunner HI, Ruperto N, Kenwright A, Wright S, Calvo I, Cuttica R, Ravelli A, Schneider R, Woo P, et al. 2012. Randomized trial of tocilizumab in systemic juvenile idiopathic arthritis. *N Engl J Med* **367:** 2385–2395.

Delerive P, De Bosscher K, Besnard S, Vanden Berghe W, Peters JM, Gonzalez FJ, Fruchart JC, Tedgui A, Haegeman G, Staels B. 1999. Peroxisome proliferator-activated receptor α negatively regulates the vascular inflammatory gene response by negative cross-talk with transcription factors NF-κB and AP-1. *J Biol Chem* **274:** 32048–32054.

Dragan YP, Schrenk D. 2000. Animal studies addressing the carcinogenicity of TCDD (or related compounds) with an emphasis in tumour promotion. *Food Addit Contam* **17:** 289–302.

Duncan MR, Berman B. 1991. Stimulation of collagen and glycosaminoglycan production in cultured human adult dermal fibroblasts by recombinant human interleukin 6. *J Invest Dermatol* **97:** 686–692.

Ema M, Sogawa K, Watanabe N, Chujoh Y, Matsushita N, Gotoh O, Funae Y, Fujii-Kuriyama Y. 1992. cDNA cloning and structure of mouse putative Ah receptor. *Biochem Biophys Res Commun* **184:** 246–253.

Emery P, Sebba A, Huizinga TW. 2013. Biologic and oral disease-modifying antirheumatic drug monotherapy in rheumatoid arthritis. *Ann Rheum Dis* **72:** 1897–1904 .

Fishman D, Faulds G, Jeffery R, Mohamed-Ali V, Yudkin JS, Humphries S, Woo P. 1998. The effect of novel polymorphisms in the interleukin-6 (IL-6) gene on IL-6 transcription and plasma IL-6 levels, and an association with systemic-onset juvenile chronic arthritis. *J Clin Invest* **102:** 1369–1376.

Fujii-Kuriyama Y, Ema M, Miura J, Sogawa K. 1994. Ah receptor: A novel ligand-activated transcription factor. *Exp Clin Immunogenet* **1:** 65–74.

Fujimoto M, Serada S, Mihara M, Uchiyama Y, Yoshida H, Koike N, Ohsugi Y, Nishikawa T, Ripley B, Kimura A, et al. 2008. Interleukin-6 blockade suppresses autoimmune arthritis in mice by the inhibition of inflammatory Th17 responses. *Arthritis Rheum* **58:** 3710–3719.

Garnero P, Thompson E, Woodworth T, Smolen JS. 2010. Rapid and sustained improvement in bone and cartilage turnover markers with the anti-interleukin-6 receptor inhibitor tocilizumab plus methotrexate in rheumatoid arthritis patients with an inadequate response to methotrexate: Results from a substudy of the multicenter double-blind, placebo-controlled trial of tocilizumab in inadequate responders to methotrexate alone. *Arthritis Rheum* **62:** 33–43.

Gillmore JD, Lovat LB, Persey MR, Pepys MB, Hawkins PN. 2001. Amyloid load and clinical outcome in AA amyloidosis in relation to circulating concentration of serum amyloid A protein. *Lancet* **358:** 24–29.

Grossman RM, Krueger J, Yourish D, Granelli-Piperno A, Murphy DP, May LT, Kupper TS, Sehgal PB, Gottlieb AB. 1989. Interleukin 6 is expressed in high levels in psoriatic skin and stimulates proliferation of cultured human keratinocytes. *Proc Natl Acad Sci* **86:** 6367–6371.

Grupp SA, Kalos M, Barrett D, Aplenc R, Porter DL, Rheingold SR, Teachey DT, Chew A, Hauck B, Wright JF, et al. 2013. Chimeric antigen receptor-modified T cells for acute lymphoid leukemia. *N Engl J Med* **368:** 1509–1518.

Harrington LE, Hatton RD, Mangan PR, Turner H, Murphy TL, Murphy KM, Weaver CT. 2005. Interleukin 17-producing $CD4^+$ effector T cells develop via a lineage distinct from the T helper type 1 and 2 lineages. *Nat Immunol* **6:** 1123–1132.

Haruta H, Ohguro N, Fujimoto M, Hohki S, Terabe F, Serada S, Nomura S, Nishida K, Kishimoto T, Naka T. 2011. Blockade of interleukin-6 signaling suppresses not only th17 but also interphotoreceptor retinoid binding protein-specific Th1 by promoting regulatory T cells in experimental autoimmune uveoretinitis. *Invest Ophthalmol Vis Sci* **52:** 3264–3271.

Hashizume M, Hayakawa N, Mihara M. 2008. IL-6 trans-signalling directly induces RANKL on fibroblast-like synovial cells and is involved in RANKL induction by TNF-α and IL-17. *Rheumatology (Oxford)* **47:** 1635–1640.

Hashizume M, Hayakawa N, Suzuki M, Mihara M. 2009. IL-6/sIL-6R trans-signalling, but not TNF-α induced angiogenesis in a HUVEC and synovial cell co-culture system. *Rheumatol Int* **29:** 1449–1454.

He M, Xu Z, Ding T, Kuang DM, Zheng L. 2009. MicroRNA-155 regulates inflammatory cytokine production in tumor-associated macrophages via targeting C/EBPβ. *Cell Mol Immunol* **6:** 343–352.

Heinrich PC, Castell JV, Andus T. 1990. Interleukin-6 and the acute phase response. *Biochem J* **265:** 621–636.

Hibi M, Murakami M, Saito M, Hirano T, Taga T, Kishimoto T. 1990. Molecular cloning and expression of an IL-6 signal transducer, gp130. *Cell* **63:** 1149–1157.

Hirano T, Yasukawa K, Harada H, Taga T, Watanabe Y, Matsuda T, Kashiwamura S, Nakajima K, Koyama K, Iwamatsu A, et al. 1986. Complementary DNA for a novel human interleukin (BSF-2) that induces B lymphocytes to produce immunoglobulin. *Nature* **324:** 73–76.

Hirano T, Taga T, Yasukawa K, Nakajima K, Nakano N, Takatsuki F, Shimizu M, Murashima A, Tsunasawa S, Sakiyama F, et al. 1987. Human B-cell differentiation factor defined by an anti-peptide antibody and its possible role in autoantibody production. *Proc Natl Acad Sci* **84:** 228–231.

Hirano T, Matsuda T, Turner M, Miyasaka N, Buchan G, Tang B, Sato K, Shimizu M, Maini R, Feldmann M, et al. 1988. Excessive production of interleukin 6/B cell stimulatory factor-2 in rheumatoid arthritis. *Eur J Immunol* **18:** 1797–1801.

Hirano T, Akira S, Taga T, Kishimoto T. 1990. Biological and clinical aspects of interleukin 6. *Immunol Today* **11:** 443–449.

Imai Y, Kuba K, Neely GG, Yaghubian-Malhami R, Perkmann T, van Loo G, Ermolaeva M, Veldhuizen R, Leung YH, Wang H, et al. 2008. Identification of oxidative stress and Toll-like receptor 4 signaling as a key pathway of acute lung injury. *Cell* **133:** 235–249.

Ishibashi T, Kimura H, Shikama Y, Uchida T, Kariyone S, Hirano T, Kishimoto T, Takatsuki F, Akiyama Y. 1989. Interleukin-6 is a potent thrombopoietic factor in vivo in mice. *Blood* **74:** 1241–1244.

Ivanov II, McKenzie BS, Zhou L, Tadokoro CE, Lepelley A, Lafaille JJ, Cua DJ, Littman DR. 2006. The orphan nuclear receptor RORγ directs the differentiation program of proinflammatory IL-17$^+$ T helper cells. *Cell* **126:** 1121–1133.

Iwasaki H, Takeuchi O, Teraguchi S, Matsushita K, Uehata T, Kuniyoshi K, Satoh T, Saitoh T, Matsushita M, Standley DM, et al. 2011. The IκB kinase complex regulates the stability of cytokine-encoding mRNA induced by TLR-IL-1R by controlling degradation of regnase-1. *Nat Immunol* **12:** 1167–1175.

Jilka RL, Hangoc G, Girasole G, Passeri G, Williams DC, Abrams JS, Boyce B, Broxmeyer H, Manolagas SC. 1992. Increased osteoclast development after estrogen loss: mediation by interleukin-6. *Science* **257:** 88–91.

Jux B, Kadow S, Esser C. 2009. Langerhans cell maturation and contact hypersensitivity are impaired in aryl hydrocarbon receptor-null mice. *J Immunol* **182:** 6709–6717.

Kanbe K, Nakamura A, Inoue Y, Hobo K. 2012. Osteoprotegerin expression in bone marrow by treatment with tocilizumab in rheumatoid arthritis. *Rheumatol Int* **32:** 2669–2674.

Kang JG, Pripuzova N, Majerciak V, Kruhlak M, Le SY, Zheng ZM. 2011. Kaposi's sarcoma-associated herpesvirus ORF57 promotes escape of viral and human interleukin-6 from microRNA-mediated suppression. *J Virol* **85:** 2620–2630.

Katsume A, Saito H, Yamada Y, Yorozu K, Ueda O, Akamatsu K, Nishimoto N, Kishimoto T, Yoshizaki K, Ohsugi Y. 2002. Anti-interleukin 6 (IL-6) receptor antibody suppresses Castleman's disease like symptoms emerged in IL-6 transgenic mice. *Cytokine* **20:** 304–311.

Kawano M, Hirano T, Matsuda T, Taga T, Horii Y, Iwato K, Asaoku H, Tang B, Tanabe O, Tanaka H, et al. 1988. Autocrine generation and requirement of BSF-2/IL-6 for human multiple myelomas. *Nature* **332:** 83–85.

Kimura A, Kishimoto T. 2010. IL-6: Regulator of Treg/Th17 balance. *Eur J Immunol* **40:** 1830–1835.

Kimura A, Naka T, Nohara K, Fujii-Kuriyama Y, Kishimoto T. 2008. Aryl hydrocarbon receptor regulates Stat1 activation and participates in the development of Th17 cells. *Proc Natl Acad Sci* **105:** 9721–9726.

Kimura A, Naka T, Nakahama T, Chinen I, Masuda K, Nohara K, Fujii-Kuriyama Y, Kishimoto T. 2009. Aryl hydrocarbon receptor in combination with Stat1 regulates LPS-induced inflammatory responses. *J Exp Med* **206:** 2027–2035.

Kishimoto T. 1985. Factors affecting B-cell growth and differentiation. *Annu Rev Immunol* **3:** 133–157.

Kishimoto T. 1989. The biology of interleukin-6. *Blood* **74:** 1–10.

Kishimoto T. 2005. Interleukin-6: From basic science to medicine—40 years in immunology. *Annu Rev Immunol* **23:** 1–21.

Kishimoto T, Ishizaka K. 1973. Regulation of antibody response in vitro: VII. Enhancing soluble factors for IgG and IgE antibody response. *J Immunol* **111:** 1194–1205.

Kishimoto T, Akira S, Taga T. 1992. Interleukin-6 and its receptor: A paradigm for cytokines. *Science* **258:** 593–597.

Kishimoto T, Taga T, Akira S. 1994. Cytokine signal transduction. *Cell* **76:** 253–262.

Kishimoto T, Akira S, Narazaki M, Taga T. 1995. Interleukin-6 family of cytokines and gp130. *Blood* **86:** 1243–1254.

Kitaba S, Murota H, Terao M, Azukizawa H, Terabe F, Shima Y, Fujimoto M, Tanaka T, Naka T, Kishimoto T, et al. 2012. Blockade of interleukin-6 receptor alleviates disease in mouse model of scleroderma. *Am J Pathol* **180:** 165–176.

Korn T, Bettelli E, Oukka M, Kuchroo VK. 2009. IL-17 and Th17 cells. *Annu Rev Immunol* **27:** 485–517.

Kotake S, Sato K, Kim KJ, Takahashi N, Udagawa N, Nakamura I, Yamaguchi A, Kishimoto T, Suda T, Kashiwazaki S. 1996. Interleukin-6 and soluble interleukin-6 receptors in the synovial fluids from rheumatoid arthritis patients are responsible for osteoclast-like cell formation. *J Bone Miner Res* **11:** 88–95.

Kumar H, Kawai T, Akira S. 2011. Pathogen recognition by the innate immune system. *Int Rev Immunol* **30:** 16–34.

Laurence A, Tato CM, Davidson TS, Kanno Y, Chen Z, Yao Z, Blank RB, Meylan F, Siegel R, Hennighausen L, et al. 2007. Interleukin-2 signaling via STAT5 constrains T helper 17 cell generation. *Immunity* **26:** 371–381.

Lee YH, Bae SC, Choi SJ, Ji JD, Song GG. 2012. The association between interleukin-6 polymorphisms and rheumatoid arthritis: A meta-analysis. *Inflamm Res* **61:** 665–671.

Leung K, Nabel GJ. 1988. HTLV-1 transactivator induces interleukin-2 receptor expression through an NF-κB-like factor. *Nature* **333:** 776–778.

Libermann TA, Baltimore D. 1990. Activation of interleukin-6 gene expression through the NF-κB transcription factor. *Mol Cell Biol* **10:** 2327–2334.

Liuzzi JP, Lichten LA, Rivera S, Blanchard RK, Aydemir TB, Knutson MD, Ganz T, Cousins RJ. 2005. Interleukin-6 regulates the zinc transporter Zip14 in liver and contributes to the hypozincemia of the acute-phase response. *Proc Natl Acad Sci* **102:** 6843–6848.

Ma CS, Deenick EK, Batten M, Tangye SG. 2012. The origins, function, and regulation of T follicular helper cells. *J Exp Med* **209:** 1241–1253.

Mahe Y, Mukaida N, Kuno K, Akiyama M, Ikeda N, Matsushima K, Murakami S. 1991. Hepatitis B virus X protein transactivates human interleukin-8 gene through acting on nuclear factor κB and CCAAT/enhancer-binding protein-like *cis*-elements. *J Biol Chem* **266:** 13759–13763.

Masuda K, Ripley B, Nishimura R, Mino T, Takeuchi O, Shioi G, Kiyonari H, Kishimoto T. 2013. Arid5a controls IL-6 mRNA stability, which contributes to elevation of IL-6 level in vivo. *Proc Natl Acad Sci* **110:** 9409–9414.

Matsusaka T, Fujikawa K, Nishio Y, Mukaida N, Matsushima K, Kishimoto T, Akira S. 1993. Transcription factors NF-IL6 and NF-κB synergistically activate transcription of the inflammatory cytokines, interleukin 6 and interleukin 8. *Proc Natl Acad Sci* **90:** 10193–10197.

Matsushita K, Takeuchi O, Standley DM, Kumagai Y, Kawagoe T, Miyake T, Satoh T, Kato H, Tsujimura T, Nakamura H, et al. 2009. Zc3h12a is an RNase essential for controlling immune responses by regulating mRNA decay. *Nature* **458:** 1185–1190.

Mihara M, Takagi N, Takeda Y, Ohsugi Y. 1998. IL-6 receptor blockage inhibits the onset of autoimmune kidney disease in NZB/W F1 mice. *Clin Exp Immunol* **112:** 397–402.

Murakami M, Hibi M, Nakagawa N, Nakagawa T, Yasukawa K, Yamanishi K, Taga T, Kishimoto T. 1993. IL-6-induced homodimerization of gp130 and associated activation of a tyrosine kinase. *Science* **260:** 1808–1810.

Naka T, Narazaki M, Hirata M, Matsumoto T, Minamoto S, Aono A, Nishimoto N, Kajita T, Taga T, Yoshizaki K, et al. 1997. Structure and function of a new STAT-induced STAT inhibitor. *Nature* **387:** 924–929.

Nakahama T, Kimura A, Nguyen NT, Chinen I, Hanieh H, Nohara K, Fujii-Kuriyama Y, Kishimoto T. 2011. Aryl hydrocarbon receptor deficiency in T cells suppresses the development of collagen-induced arthritis. *Proc Natl Acad Sci* **108:** 14222–14227.

Nakahama T, Hanieh H, Nguyen NT, Chinen I, Ripley B, Millrine D, Lee S, Nyati KK, Dubey PK, Chowdhury K, et al. 2013. Aryl hydrocarbon receptor-mediated induction of the microRNA-132/212 cluster promotes interleukin-17-producing T-helper cell differentiation. *Proc Natl Acad Sci* **110:** 11964–11969.

Nakahara H, Song J, Sugimoto M, Hagihara K, Kishimoto T, Yoshizaki K, Nishimoto N. 2003. Anti-interleukin-6 receptor antibody therapy reduces vascular endothelial growth factor production in rheumatoid arthritis. *Arthritis Rheum* **48:** 1521–1529.

Narazaki M, Yasukawa K, Saito T, Ohsugi Y, Fukui H, Koishihara Y, Yancopoulos GD, Taga T, Kishimoto T. 1993. Soluble forms of the interleukin-6 signal-transducing receptor component gp130 in human serum possessing a potential to inhibit signals through membrane-anchored gp130. *Blood* **82:** 1120–1126.

Nemeth E, Rivera S, Gabayan V, Keller C, Taudorf S, Pedersen BK, Ganz T. 2004. IL-6 mediates hypoferremia of inflammation by inducing the synthesis of the iron regulatory hormone hepcidin. *J Clin Invest* **113:** 1271–1276.

Nguyen NT, Kimura A, Nakahama T, Chinen I, Masuda K, Nohara K, Fujii-Kuriyama Y, Kishimoto T. 2010. Aryl hydrocarbon receptor negatively regulates dendritic cell immunogenicity via a kynurenine-dependent mechanism. *Proc Natl Acad Sci* **107:** 19961–19966.

Nishimoto N, Yoshizaki K, Tagoh H, Monden M, Kishimoto S, Hirano T, Kishimoto T. 1989. Elevation of serum interleukin 6 prior to acute phase proteins on the inflammation by surgical operation. *Clin Immunol Immunopathol* **50:** 399–401.

Nishimoto N, Kanakura Y, Aozasa K, Johkoh T, Nakamura M, Nakano S, Nakano N, Ikeda Y, Sasaki T, Nishioka K, et al. 2005. Humanized anti-interleukin-6 receptor antibody treatment of multicentric Castleman disease. *Blood* **106:** 2627–2632.

Nishimoto N, Terao K, Mima T, Nakahara H, Takagi N, Kakehi T. 2008. Mechanisms and pathologic significances in increase in serum interleukin-6 (IL-6) and soluble IL-6 receptor after administration of an anti-IL-6 receptor antibody, tocilizumab, in patients with rheumatoid arthritis and Castleman disease. *Blood* **112:** 3959–3964.

Ohno H, Kaneko S, Lin Y, Kobayashi K, Murakami S. 1999. Human hepatitis B virus X protein augments the DNA binding of nuclear factor for IL-6 through its basic-leucine zipper domain. *J Med Virol* **58:** 11–18.

Ohshima S, Saeki Y, Mima T, Sasai M, Nishioka K, Nomura S, Kopf M, Katada Y, Tanaka T, Suemura M, et al. 1998. Interleukin 6 plays a key role in the development of antigen-induced arthritis. *Proc Natl Acad Sci* **95:** 8222–8226.

Ohtake F, Takeyama K, Matsumoto T, Kitagawa H, Yamamoto Y, Nohara K, Tohyama C, Krust A, Mimura J, Chambon P, et al. 2003. Modulation of oestrogen receptor signalling by association with the activated dioxin receptor. *Nature* **423:** 545–550.

Okada M, Kitahara M, Kishimoto S, Matsuda T, Hirano T, Kishimoto T. 1988. IL-6/BSF-2 functions as a killer helper factor in the in vitro induction of cytotoxic T cells. *J Immunol* **141:** 1543–1549.

Okiyama N, Sugihara T, Iwakura Y, Yokozeki H, Miyasaka N, Kohsaka H. 2009. Therapeutic effects of interleukin-6 blockade in a murine model of polymyositis that does not require interleukin-17A. *Arthritis Rheum* **60:** 2505–2512.

Palanisamy V, Jakymiw A, Van Tubergen EA, D'Silva NJ, Kirkwood KL. 2012. Control of cytokine mRNA expression by RNA-binding proteins and microRNAs. *J Dent Res* **91:** 651–658.

Pesce B, Soto L, Sabugo F, Wurmann P, Cuchacovich M, Lopez MN, Sotelo PH, Molina MC, Aguillon JC, Catalan D. 2013. Effect of interleukin-6 receptor blockade on the balance between regulatory T cells and T helper type 17 cells in rheumatoid arthritis patients. *Clin Exp Immunol* **171:** 237–242.

Poli V, Balena R, Fattori E, Markatos A, Yamamoto M, Tanaka H, Ciliberto G, Rodan GA, Costantini F. 1994. Interleukin-6 deficient mice are protected from bone loss caused by estrogen depletion. *EMBO J* **13:** 1189–1196.

Puga A, Tomlinson CR, Xia Y. 2005. Ah receptor signals cross-talk with multiple developmental pathways. *Biochem Pharmacol* **69:** 199–207.

Ray A, Prefontaine KE. 1994. Physical association and functional antagonism between the p65 subunit of transcription factor NF-κB and the glucocorticoid receptor. *Proc Natl Acad Sci* **91:** 752–756.

Samson M, Audia S, Janikashvili N, Ciudad M, Trad M, Fraszczak J, Ornetti P, Maillefert JF, Miossec P, Bonnotte B. 2012. Brief report: Inhibition of interleukin-6 function corrects Th17/Treg cell imbalance in patients with rheumatoid arthritis. *Arthritis Rheum* **64:** 2499–2503.

Sandborg C, Mellins ED. 2012. A new era in the treatment of systemic juvenile idiopathic arthritis. *N Engl J Med* **367:** 2439–2440.

Santhanam U, Ray A, Sehgal PB. 1991. Repression of the interleukin 6 gene promoter by p53 and the retinoblastoma susceptibility gene product. *Proc Natl Acad Sci* **88:** 7605–7609.

Sato K, Tsuchiya M, Saldanha J, Koishihara Y, Ohsugi Y, Kishimoto T, Bendig MM. 1993. Reshaping a human antibody to inhibit the interleukin 6-dependent tumor cell growth. *Cancer Res* **53:** 851–856.

Scala G, Ruocco MR, Ambrosino C, Mallardo M, Giordano V, Baldassarre F, Dragonetti E, Quinto I, Venuta S. 1994. The expression of the interleukin 6 gene is induced by the human immunodeficiency virus 1 TAT protein. *J Exp Med* **179:** 961–971.

Schmitz J, Weissenbach M, Haan S, Heinrich PC, Schaper F. 2000. SOCS3 exerts its inhibitory function on interleukin-6 signal transduction through the SHP2 recruitment sites of gp130. *J Biol Chem* **275:** 12848–12856.

Serada S, Fujimoto M, Mihara M, Koike N, Ohsugi Y, Nomura S, Yoshida H, Nishikawa T, Terabe F, Ohkawara T, et al. 2008. IL-6 blockade inhibits the induction of myelin antigen-specific Th17 cells and Th1 cells in experimental autoimmune encephalomyelitis. *Proc Natl Acad Sci* **105:** 9041–9046.

Sims GP, Rowe DC, Rietdijk ST, Herbst R, Coyle AJ. 2010. HMGB1 and RAGE in inflammation and cancer. *Annu Rev Immunol* **28:** 367–388.

Stumhofer JS, Laurence A, Wilson EH, Huang E, Tato CM, Johnson LM, Villarino AV, Huang Q, Yoshimura A, Sehy D, et al. 2006. Interleukin 27 negatively regulates the development of interleukin 17-producing T helper cells during chronic inflammation of the central nervous system. *Nat Immunol* **7:** 937–945.

Taga T, Kishimoto T. 1997. Gp130 and the interleukin-6 family of cytokines. *Annu Rev Immunol* **15:** 797–819.

Tanaka T. 2013. Can IL-6 blockade rectify imbalance between Tregs and Th17 cells? *Immunotherapy* **5:** 695–697.

Tanaka T, Kishimoto T. 2011. Immunotherapy of tocilizumab for rheumatoid arthritis. *J Clin Cell Immunol* **S6:** 001.

Tanaka T, Kishimoto T. 2012. Targeting interleukin-6: All the way to treat autoimmune and inflammatory diseases. *Int J Biol Sci* **8:** 1227–1236.

Tanaka T, Hagihara K, Hishitani Y, Ogata A. 2011. Tocilizumab for the treatment of AA amyloidosis. In *Amyloidosis—An insight to disease of systems and novel therapies* (ed. Isil Adadan Guvenc MD), pp. 155–170. InTech Open Access, Rijeka, Croatia.

Tanaka T, Narazaki M, Kishimoto T. 2012. Therapeutic targeting of the interleukin-6 receptor. *Annu Rev Pharmacol Toxicol* **52:** 199–219.

Tanaka T, Ogata A, Narazaki M. 2013. Tocilizumab: An updated review of its use in the treatment of rheumatoid arthritis and its application for other immune-mediated diseases. *Clin Med Insights Ther* **5:** 33–52.

Terpos E, Fragiadaki K, Konsta M, Bratengeier C, Papatheodorou A, Sfikakis PP. 2011. Early effects of IL-6 receptor inhibition on bone homeostasis: A pilot study in women with rheumatoid arthritis. *Clin Exp Rheumatol* **29:** 921–925.

Uehata T, Iwasaki H, Vandenbon A, Matsushita K, Hernandez-Cuellar E, Kuniyoshi K, Satoh T, Mino T, Suzuki Y, Standley DM, et al. 2013. Malt1-induced cleavage of regnase-1 in $CD4^+$ helper T cells regulates immune activation. *Cell* **153:** 1036–1049.

Vogel CFA, Goth SR, Dong B, Pessah IN, Natsumura F. 2008. Aryl hydrocarbon receptor signaling mediates expression of indoleamine 2,3-dioxygenase. *Biochem Biophys Res Commun* **375:** 331–335.

Xu Z, Xiao SB, Xu P, Xie Q, Cao L, Wang D, Luo R, Zhong Y, Chen HC, Fang LR. 2011. MiR-365, a novel negative regulator of interleukin-6 gene expression, is cooperatively regulated by Sp1 and NF-κB. *J Biol Chem* **286:** 21401–21412.

Yamasaki K, Taga T, Hirata Y, Yawata H, Kawanishi Y, Seed B, Taniguchi T, Hirano T, Kishimoto T. 1988. Cloning and expression of the human interleukin-6 (BSF-2/IFN-β2) receptor. *Science* **241:** 825–828.

Yang XO, Pappu BP, Nurieva R, Akimzhanov A, Kang HS, Chung Y, Ma L, Shah B, Panopoulos AD, Schluns KS, et al. 2008. T helper 17 lineage differentiation is programmed by orphan nuclear receptors RORα and RORγ. *Immunity* **28:** 29–39.

Yokota S, Imagawa T, Mori M, Miyamae T, Aihara Y, Takei S, Iwata N, Umebayashi H, Murata T, Miyoshi M, et al. 2008. Efficacy and safety of tocilizumab in patients with systemic-onset juvenile idiopathic arthritis: A randomised, double-blind, placebo-controlled, withdrawal phase III trial. *Lancet* **371:** 998–1006.

Yokota S, Tanaka T, Kishimoto T. 2012. Efficacy, safety and tolerability of tocilizumab in patients with systemic juvenile idiopathic arthritis. *Ther Adv Musculoskelet Dis* **4:** 387–397.

Yoshizaki K, Matsuda T, Nishimoto N, Kuritani T, Taeho L, Aozasa K, Nakahata T, Kawai H, Tagoh H, Komori T, et al. 1989. Pathogenic significance of interleukin-6 (IL-6/BSF-2) in Castleman's disease. *Blood* **74:** 1360–1367.

Yu D, Rao S, Tsai LM, Lee SK, He Y, Sutcliffe EL, Srivastava M, Linterman M, Zheng L, Simpson N, et al. 2009. The transcriptional repressor Bcl-6 directs T follicular cell lineage commitment. *Immunity* **31:** 457–468.

Zhang Q, Raoof M, Chen Y, Sumi Y, Sursal T, Junger W, Brohi K, Itagaki K, Hauser CJ. 2010. Circulating mitochondrial DAMPs cause inflammatory responses to injury. *Nature* **464:** 104–107.

Zhao W, Liu M, Kirkwood KL. 2008. pp38α stabilizes interleukin-6 mRNA via multiple AU-rich elements. *J Biol Chem* **283:** 1778–1785.

Zilahi E, Tarr T, Papp G, Griger Z, Sipka S, Zeher M. 2012. Increased microRNA-146a/b, TRAF6 gene and decreased IRAK1 gene expressions in the peripheral mononuclear cells of patients with Sjögren's syndrome. *Immunol Lett* **141:** 165–168.

The Chemokine System in Innate Immunity

Caroline L. Sokol and Andrew D. Luster

Center for Immunology & Inflammatory Diseases, Division of Rheumatology, Allergy and Immunology, Massachusetts General Hospital, Harvard Medical School, Boston, Massachusetts 02114

Correspondence: aluster@mgh.harvard.edu

Chemokines are chemotactic cytokines that control the migration and positioning of immune cells in tissues and are critical for the function of the innate immune system. Chemokines control the release of innate immune cells from the bone marrow during homeostasis as well as in response to infection and inflammation. They also recruit innate immune effectors out of the circulation and into the tissue where, in collaboration with other chemoattractants, they guide these cells to the very sites of tissue injury. Chemokine function is also critical for the positioning of innate immune sentinels in peripheral tissue and then, following innate immune activation, guiding these activated cells to the draining lymph node to initiate and imprint an adaptive immune response. In this review, we will highlight recent advances in understanding how chemokine function regulates the movement and positioning of innate immune cells at homeostasis and in response to acute inflammation, and then we will review how chemokine-mediated innate immune cell trafficking plays an essential role in linking the innate and adaptive immune responses.

Chemokines are chemotactic cytokines that control cell migration and cell positioning throughout development, homeostasis, and inflammation. The immune system, which is dependent on the coordinated migration of cells, is particularly dependent on chemokines for its function. Not only do chemokines guide immune effector cells to sites of infection or inflammation, but they also coordinate interactions between immune cells. By doing so, chemokines promote interactions between the innate and adaptive immune systems, thus shaping and providing the necessary context for the development of optimal adaptive immune responses. This review will aim to provide an overview of the function of the chemokine system, with emphasis placed on its role in the innate immune system.

CHEMOKINES AND CHEMOKINE RECEPTORS

The chemokine family consists of approximately 50 endogenous chemokine ligands in humans and mice (Table 1). These small, 8- to 12-kDa protein ligands promote increased motility and directional migration when they bind to their corresponding cell-surface receptor. The chemokine ligands are divided into four groups based on the positioning of their initial cysteine residues: XC, CC, CXC, and CX3C. Thus, CC chemokine ligands (CCLs) have two

Cite this article as *Cold Spring Harb Perspect Biol* doi: 10.1101/cshperspect.a016303

Table 1. Chemokines

Chemokine	Other names	Receptor	Key/main immune function
CXCL1	GRO-α, MGSA, mouse KC	CXCR2	Neutrophil trafficking
CXCL2	GRO-β, MIP-2α, mouse MIP2	CXCR2	Neutrophil trafficking
CXCL3	GRO-γ, MIP-2β	CXCR2	Neutrophil trafficking
CXCL4	PF4	?	Procoagulant
CXCL5	ENA-78, mouse LIX	CXCR2	Neutrophil trafficking
CXCL6	GCP-2	CXCR1, CXCR2	Neutrophil trafficking
CXCL7	NAP-2	CXCR2	Neutrophil trafficking
CXCL8	IL-8 (no mouse)	CXCR1, CXCR2	Neutrophil trafficking
CXCL9	Mig	CXCR3	Th1 response; Th1, CD8, NK trafficking
CXCL10	IP-10	CXCR3	Th1 response; Th1, CD8, NK trafficking
CXCL11	I-TAC	CXCR3	Th1 response; Th1, CD8, NK trafficking
CXCL12	SDF-1	CXCR4	Bone-marrow homing
CXCL13	BLC, BCA-1	CXCR5	B-cell and T_{FH}-positioning LN
CXCL14	BRAK	?	Macrophage skin homing (human)
Cxcl15	Lungkine (mouse only)	?	?
CXCL16		CXCR6	NKT and ILC migration and survival
CXCL17		?	Macrophage and DC chemotaxis
CCL1	I-309, mouse TCA3	CCR8	Th2 cell and Treg trafficking
CCL2	MCP-1, mouse JE	CCR2	Inflammatory monocyte trafficking
CCL3	MIP-1α	CCR1, CCR5	Macrophage and NK-cell migration; T-cell–DC interactions
CCL4	MIP-1β	CCR5	Macrophage and NK-cell migration; T-cell–DC interactions
CCL5	RANTES	CCR1, CCR3, CCR5	Macrophage and NK-cell migration; T-cell–DC interactions
Ccl6	C-10, MRP-1 (mouse only)	Unknown	?
CCL7	MCP-3, mouse Fic or MARC	CCR2, CCR3	Monocyte mobilization
CCL8	MCP-2	CCR1, CCR2, CCR3, CCR5 (human); CCR8 (mouse)	Th2 response, skin homing mouse
Ccl9/10	MIP-1γ, MRP-2 (mouse only)	Unknown	?
CCL11	Eotaxin-1	CCR3	Eosinophil and basophil migration
Ccl12	MCP-5 (mouse only)	CCR2	Inflammatory monocyte trafficking
CCL13	MCP-4	CCR2, CCR3, CCR5	Th2 responses
CCL14	HCC-1	CCR1	?
CCL15	Leukotactin-1, HCC-2, MIP-5	CCR1, CCR3	?
CCL16	HCC-4, NCC-4, LEC	CCR1, CCR2, CCR5	?
CCL17	TARC	CCR4	Th2 responses, Th2-cell migration, Treg, lung, and skin homing
CCL18	PARC, DC-CK1	CCR8	Th2 response, marker AAM, skin homing
CCL19	ELC, MIP-3β	CCR7	T-cell and DC homing to LN
CCL20	MIP-3α, LARC	CCR6	Th17 responses, B-cell, and DC homing to gut-associated lymphoid tissue
CCL21	SLC, 6CKine	CCR6, CCR7	T-cell and DC homing to LN
CCL22	MDC	CCR4	Th2 response, Th2-cell migration, Treg migration
CCL23	MPIF-1, MIP-3	Unknown	?
CCL24	Eotaxin-2, MPIF-2	CCR3	Eosinophil and basophil migration
CCL25	TECK	CCR9	T-cell homing to gut, thymocyte migration

Continued

Table 1. *Continued*

Chemokine	Other names	Receptor	Key/main immune function
CCL26	Eotaxin-3	CCR3, CX3CR1	Eosinophil and basophil migration
CCL27	CTAK	CCR10	T-cell homing to skin
CCL28	MEC	CCR3, CCR10	T-cell and IgA plasma–cell homing to mucosa
XCL1	Lymphotactin, SCM-1α	XCR1	Cross presentation by $CD8^+$ DCs
XCL2	SCM-1β	XCR1	Cross presentation by $CD8^+$ DCs
CX3CL1	Fractalkine	CX3CR1	NK, monocyte, and T-cell migration

This table was created from data adapted from Bachelerie et al. 2013.

AAM, alternatively activated macrophage; ILC, innate lymphoid cell; LN, lymph node; NK, natural killer; NKT, natural killer T; T_{FH}, follicular helper T cell; Treg, regulatory T cell.

adjoining amino-terminal cysteine residues, whereas CX3CL1 has three amino acids separating the two initial cysteine residues. Most chemokines are secreted into the extracellular space where they remain soluble or are bound to extracellular matrix components, thus forming transient or stable concentration gradients, respectively. Chemokines and their gradients are detected by binding to specific chemokine receptors.

Chemokine receptors are a group of approximately 20 rhodopsin-like seven-transmembrane-spanning receptors in humans and mice (Table 2). These receptors are G-protein-coupled and signal via pertussis toxin-sensitive Gi-type G proteins. The chemokine receptors show varying levels of binding specificity and promiscuity. For example, CXCR4 binds only CXCL12, whereas CCR1 can bind to six different chemokine ligands. Despite this promiscuity, chemokine receptors do not bind different groups of chemokines (e.g., CCL and CXCL chemokines), and they are named based on the group that they bind; CCR chemokine receptors bind CCL chemokines, whereas CXCR receptors bind CXCL chemokines. In addition to the signaling receptors, chemokine receptors also include a group of four atypical receptors. These are similar in structure to the signaling receptors, but lack an intracellular motif required for signaling through Gi-type G proteins. These nonsignaling, atypical receptors appear to play primary roles in shaping chemokine gradients by scavenging and promoting transcytosis of chemokine ligands.

EVOLUTION OF CHEMOKINES AND CHEMOKINE RECEPTORS

Chemokines are evolutionarily ancient, having first appeared 700 million years ago in a common ancestor of the vertebrate lineage. Although chemotaxis has been described in invertebrate cells, an analog of the chemokine system does not appear to exist in the invertebrates (DeVries et al. 2006). After their initial appearance, chemokines have undergone extensive and rapid evolution, owing to large gene duplication events and subsequent selection. Human chemokine ligand genes are predominantly found on chromosomes 4 and 17 in regions with significant numbers of psueodogenes, which correlates with a distant gene duplication event with subsequent diversification, selection, and loss (Zlotnik et al. 2006; Nomiyama et al. 2010). However, there is significant variation in chemokine ligands between different evolutionary lineages. The zebrafish genome contains more than 100 discrete chemokine genes (Zlotnik et al. 2006; Nomiyama et al. 2010). Whether this is the result of selection events attributable to different pathogen exposure or the result of random genetic duplication events remains to be determined.

Similar to chemokine ligands, chemokine receptors show evidence of rapid expansion and evolution after their appearance in a common vertebrate ancestor. The sea lamprey, a jawless fish, contains six known chemokine receptors. Concurrent with their ligands, further expansion of chemokine receptors occurred

Table 2. Chemokine receptors

Receptor	Immune cell expression	Key immune function
G-protein-coupled chemokine receptors		
CXCR1	N>Mo, NK, MC, Ba, CD8 T_{EFF}	Neutrophil trafficking
CXCR2	N>Mo, NK, MC, Ba, CD8 T	B-cell lymphopoiesis Neutrophil egress from bone marrow Neutrophil trafficking
CXCR3	Th1, CD8 T_{CM} and T_{EM}, NK, NKT, pDC, B, Treg, T_{FH}	Th1-type adaptive immunity
CXCR4	Most (if not all) leukocytes	Hematopoiesis Organogenesis Bone marrow homing
CXCR5	B, T_{FH}, T_{FR}, CD8 T_{EM}	B- and T-cell trafficking in lymphoid tissue to B-cell zone/follicles
CXCR6	Th1, Th17, γδ T, iLC, NKT, NK, PC	Innate lymphoid cell function Adaptive immunity
CCR1	Mo, MΦ, N, Th1, Ba, DC	Innate immunity Adaptive immunity
CCR2	Mo, MΦ, Th1, iDC, Ba, NK	Monocyte trafficking Th1-type adaptive immunity
CCR3	Eo>Ba, MC	Th2-type adaptive immunity Eosinophil distribution and trafficking
CCR4	Th2, skin- and lung-homing T, Treg>Th17, CD8 T, Mo, B, iDC	Homing of T cells to skin and lung Th2-type immune response
CCR5	Mo, MΦ, Th1, NK, Treg, CD8 T, DC, N	Type-1 adaptive immunity
CCR6	Th17>iDC, γδ T, NKT, NK, Treg, T_{FH}	iDC trafficking, GALT development Th17 adaptive immune responses
CCR7	T_N, T_{CM}, T_{RCM}, mDC, B	mDC, and B- and T-cell trafficking in lymphoid tissue to T-cell zone Egress of DC and T cells from tissue
CCR8	Th2, Treg, skin T_{RM}, γδ T, Mo, MΦ	Immune surveillance in skin type-2 adaptive immunity, thymopoiesis
CCR9	Gut homing T, thymocytes, B, DC, pDC	Homing of T cells to gut GALT development and function, thymopoiesis
CCR10	Skin-homing T, IgA plasmablasts	Humoral immunity at mucosal sites Immune surveillance in skin
XCR1	Cross-presenting $CD8^+$ DC, thymic DC	Ag cross-presentation by $CD8^+$ DCs
CX3CR1	Resident Mo, MΦ, MG, Th1, CD8 T_{EM}, NK, γδ T, DC	Patrolling monocytes in innate immunity Microglial-cell and NK-cell migration type-1 adaptive immunity
Atypical chemokine receptors (non-G-protein-coupled signaling)		
ACKR1 (DARC; Duffy)	RBC, LEC	Chemokine transcytosis Chemokine scavenging
ACKR2 (D6)	LEC, DC, B	Chemokine scavenging
ACKR3 (CXCR7)	Stromal cells, B	Shaping chemokine gradients for CXCR4
ACKR4 (CCRL1; CCX-CKR)	Thymic epithelium	Chemokine scavenging

Data in table is modified from Bachelerie et al. 2013.

B, B cell; Ba, basophil; Eo, eosinophils; GALT, gut-associated lymphoid tissue; iDC, immature DC; LEC, lymphatic endothelium; MΦ, macrophage; MG, microglia; Mo, monocyte; N, neutrophil; pDC, plasmacytoid DC; PC, plasma cell; RBC, red blood cell; T_{CM}, central memory T cell; T_{EFF}, effector T cell; T_{FH}, follicular helper T cell; T_{FR}, follicular regulatory T cell; T_N, naïve T cell; T_{RCM}, recirculating memory T cell.

with species diversification. The elephant shark, a cartilaginous fish, contains 13 putative chemokine receptors, whereas the zebrafish contains at least 32 different chemokine receptors (Zlotnik et al. 2006). Mammals contain less, indicating that this expansion in chemokine receptors occurred following the split of the teleost lineage with the mammalian lineage. Humans and mice contain 17 signaling chemokine receptors, with the majority present on human chromosome 3 (Zlotnik et al. 2006).

THE CHEMOKINE SYSTEM IN INNATE IMMUNE CELL HOMEOSTASIS

Developing Immune Cells

Maintenance of hematopoietic stem cells and developing innate immune cells takes place largely in the bone marrow (BM) and is dependent on CXCL12/CXCR4 interactions. The retention of hematopoietic stem cells within BM niches is dependent on CXCL12, which is produced by CXCL12-abundant reticular cells, and binds to CXCR4 on hematopoietic stem cells (Ara et al. 2003). As immune cell development progresses past the hematopoietic stem cell, CXCL12/CXCR4 interactions remain essential for BM retention and normal development of multiple immune lineages, including B cells, monocytes, macrophages, neutrophils, natural killer (NK) cells, and plasmacytoid dendritic cells (Mercier et al. 2012). Developing neutrophils are retained in the BM by CXCR4 signals, and their maturation coincides with down-regulation of CXCR4, permitting mature neutrophils to enter the peripheral blood (Suratt et al. 2004; Broxmeyer et al. 2005). This process is defective in patients with warts, hypogammaglobulinemia, infections, and myelokathexis (WHIM) syndrome, which is caused by mutations in CXCR4 that enhance responsiveness to CXCL12 (Hernandez et al. 2003; Gulino et al. 2004). Neutrophils in the WHIM syndrome are unable to normally decrease responsiveness to CXCL12 and are therefore trapped within the BM, resulting in peripheral neutropenia.

CXCR4-mediated signaling plays a major role in promoting BM retention of many immune cells. However, exit from the BM may not be entirely passive. In studies examining monocyte development and release from the BM, blockade of CXCR4 induces only a small increase in the number of peripheral blood monocytes (Wang et al. 2009). However, blockade of CCR2, which is uniformly expressed by early monocytes, leads to decreases in circulating monocytes with concomitant increases in BM monocytes (Serbina and Pamer 2006; Wang et al. 2009). However, the source of CCR2 ligands under homeostatic conditions is unclear.

Neutrophils

After release from the BM, neutrophils enter the bloodstream where they await inflammatory stimuli that would promote their migration into peripheral tissue. The peripheral bloodstream is thought to be the major compartment for neutrophils under homeostatic conditions; however, recent reports indicate that CXCR4 signaling may promote the formation of a marginated pool of neutrophils in the lung vasculature that can be rapidly mobilized (Devi et al. 2013). CXCR4 may also play an important role in neutrophil elimination; senescent neutrophils up-regulate expression of CXCR4 and this may promote reentry into the BM where they undergo apoptotic cell death (Martin et al. 2003).

Monocytes

Exit of monocytes from the BM appears to be dependent on CXCR4 and possibly CCR2-mediated signaling under homeostatic conditions. Upon exiting, monocytes differentiate into proinflammatory and anti-inflammatory subsets based on their expression of CCR2 and CX3CR1, respectively. The proinflammatory ($CCR2^+$) monocyte population remains within the peripheral blood and spleen, whereas the anti-inflammatory ($CX3CR1^+$) population is distributed in the peripheral blood and nonlymphoid organs (Fig. 1) (Geissmann et al. 2003). This is likely because of differential homeostatic expression of CCR2 and CX3CR1 ligands; in the absence of inflammation, there is little expres-

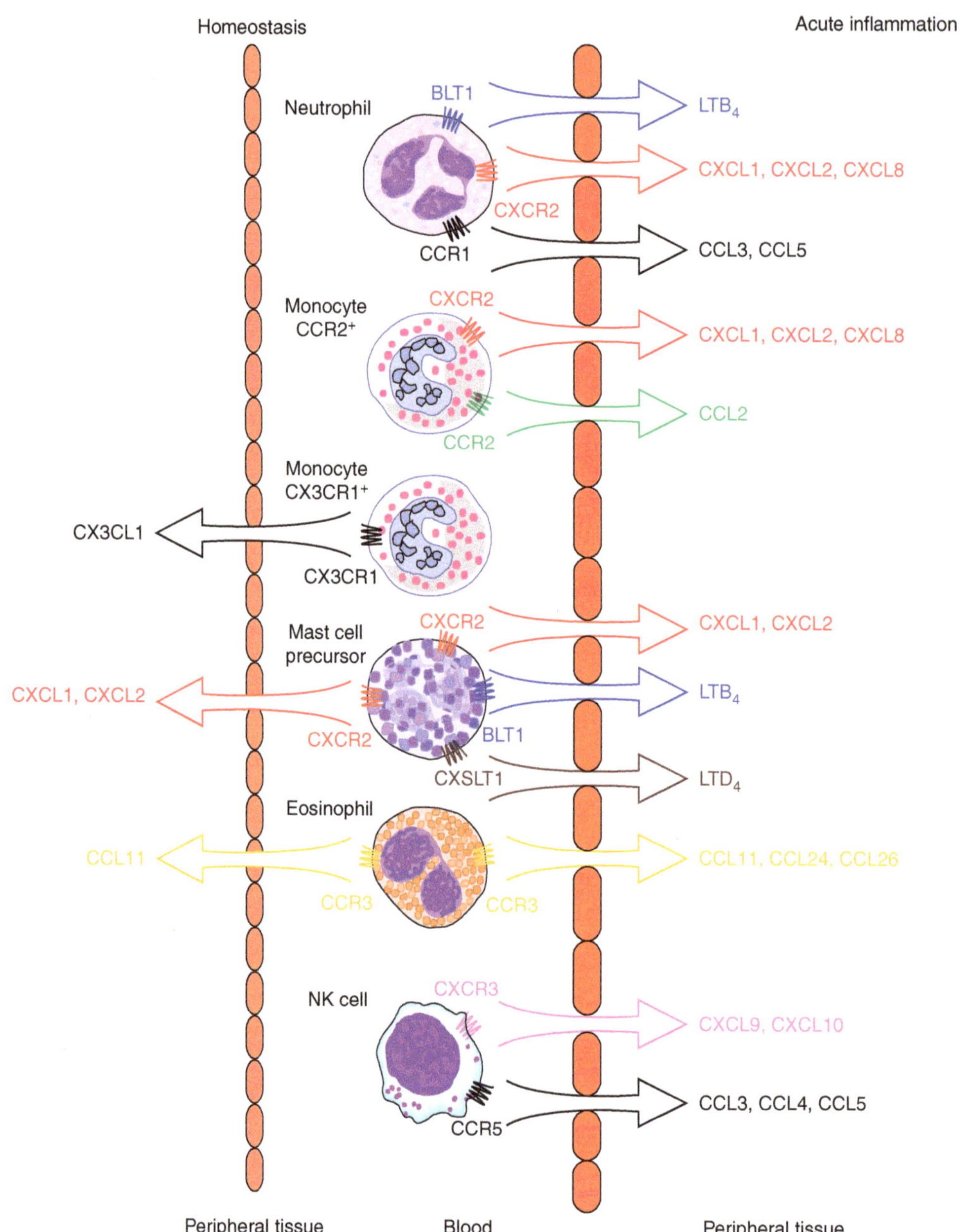

Figure 1. Chemokine control of innate immune cell migration in homeostasis and inflammation. Under homeostatic conditions, CX3CR1$^+$ monocytes presumably migrate into the periphery following CX3CL1 gradients, mast cell precursors migrate into the gastrointestinal tract following CXCL1 and CXCL2 gradients, and eosinophils migrate into tissue following CCL11 gradients. In acute inflammation, neutrophils may leave the bloodstream and migrate into the periphery following gradients of LTB_4, CXCL1, CXCL2, CXCL8, CCL3, and CCL5. CCR2$^+$ monocytes will migrate following CXCL1, CXCL2, CXCL8, and CCL2 gradients. Mast cell precursors also migrate in response to leukotriene B4 (LTB_4) and leukotriene D4 (LTD_4) via activation of their respective G-protein-coupled receptors BLT1 and CYSLT1. Eosinophils will migrate into inflammatory sites in response to CCL11, CCL24, and CCL26 gradients. NK cells will migrate following CXCL9, CXCL10, CCL3, CCL4, and CCL5 gradients.

sion of CCR2 ligands in the periphery, whereas CX3CL1 is homeostatically expressed by endothelium in various tissues (Foussat et al. 2000). However, CX3CR1$^+$ monocytes may follow additional chemokine signals as they enter the periphery and develop into tissue macrophages under homeostatic conditions. CXCL14 is homeostatically produced by human fibroblasts in the skin and lamina propria, and tissue macrophages can be found in close association with CXCL14-producing cells (Kurth et al. 2001). However, the full role of CXCL14 and its receptor, which remains to be identified, is not fully appreciated.

Eosinophils and Basophils

Under homeostatic conditions, eosinophils are found in the blood and the peripheral tissues, with the gastrointestinal tract making up the main reservoir of eosinophils. This baseline migration into the periphery is largely dependent on the production of CCL11 (eotaxin-1)/CCR3 interactions that promote eosinophil release from the BM and entry into peripheral tissues (Fig. 1) (Palframan et al. 1998; Mishra et al. 1999). Basophils express CXCR4 and so it is possible that they are released into the peripheral blood by CXCR4-mediated mechanisms, where they remain in the absence of inflammatory stimuli. Human basophils constitutively express CXCR1, CXCR4, CCR1, CCR2, and a majority express CCR3 (Uguccioni et al. 1997; Marone et al. 2005).

Mast Cells

Mast cells (MCs) are found throughout all vascularized tissues, where they act as innate immune sentinels and effector cells. MC precursors express CXCR4, as well as CXCR2, CCR3, and CCR5, and likely exit the BM following similar pathways to other innate immune cells (Ochi et al. 1999). They then migrate into the periphery via unknown mechanisms, although migration in response to epithelial- and fibroblast-derived stem cell factor (SCF) has been postulated (Nilsson et al. 1996). Migration signals may be tissue dependent; CXCR2 deficiency in MC precursors leads to a specific defect in intestinal homing of MC precursors and loss of intestinal MCs (Fig. 1) (Abonia et al. 2005). Once in the peripheral tissues, MCs further differentiate and alter their chemokine receptor expression; whether this is because of constitutive signaling necessary for tissue retention is unknown (Halova et al. 2012).

Dendritic Cells

Under homeostatic conditions, dendritic cells (DCs) develop in situ from DC precursor cells that first populate the periphery. No common chemokine for DC precursor positioning has been described and the chemokines that guide this transit remain largely unknown, although CCL2, CXCL14, and CCL20 have been implicated. CCR2 deficiency leads to decreases in the skin resident langerin-positive DCs, but not in other DC populations (Bogunovic et al. 2006). CCL20/CCR6 interactions are essential for normal DC migration to the subepithelial dome of the Peyer's patches, but no other DC populations are affected (Cook et al. 2000; Varona et al. 2001). Finally, CXCL14 has been postulated to play a role in DC precursor positioning based on its ability to induce chemotaxis in human DC precursors (Schaerli et al. 2005). However, CXCL14-deficient mice show no defect in DC positioning, casting doubt on the universal role of CXCL14 in DC positioning (Meuter et al. 2007).

DC precursors differentiate into immature resident DCs that survey the periphery for signs of infection or inflammation. Immature, resident DCs express various chemokine receptors (e.g., CCR1, CCR2, CCR5, CCR6, CXCR1, CXCR2, and CXCR4) that allow DCs to migrate to sites of inflammation, but they may also actively promote their maintenance in the periphery (Randolph et al. 2008). However, even in the absence of inflammatory stimuli, DCs can become "semi-mature," a state defined by increased surface expression of major histocompatibility complex class II (MHCII) and CCR7. The up-regulation of CCR7 allows these DCs to sense CCL21, which is produced by the lymphatic vessels in peripheral tissues, thus permitting semi-mature DCs to migrate into the lymphatics (Forster et al. 1999; Weber et al. 2013). CCR7 signaling is a common pathway for homeostatic DC migration from diverse peripheral tissues; CCR7 has been shown to be essential for migration of DCs from the skin, lung, and gastrointestinal tract (Ohl et al. 2004; Worbs et al. 2006). However, there is redundancy in CCR7 ligands; within the peripheral tissues, CCL21 promotes migration to the lymphatic vessels, but once inside the vessel, CCL19 and CCL21 act additively to promote migration to the lymph node (LN).

Because soluble or particulate antigens may have direct access to the secondary lymphoid organs via the lymphatics or bloodstream, proper homeostatic positioning of DCs within the secondary lymphoid organs is essential for immune defense. Within the LN, the fibroblastic reticular cells homeostatically produce CCL19 and CCL21, which promote DC localization to the T-cell area. In the spleen, $CD4^+$ DCs express high levels of the G-protein-coupled receptor EBI2, the receptor for 7α, 25-dihydroxycholesterol, which is necessary for the normal positioning of $CD4^+$ DCs in the bridging channels of the spleen and the subsequent immune response to particulate antigens (Gatto et al. 2013). Finally, although sphingosine-1-phosphate receptor 1 (S1PR1) blockade has no effect on DC localization in the LN, it does lead to the redistribution of immature $CD4^+$ DCs from the bridging channels to the splenic marginal zone (Czeloth et al. 2007). This is specific to DC localization under homeostatic conditions; activated or mature DC positioning is not affected. Thus, although DC migration is a universal pathway governed by CCR7 signaling, positioning in the periphery and secondary lymphoid organs may be specific for different DC subsets.

Innate Lymphoid Cells

Innate lymphoid cells (ILCs) are a broad group of innate lymphocyte-like cells that are notable for not undergoing recombination of antigen receptors and clonal selection. They have been organized into three groups based on cytokine production. Group 1 ILCs includes ILC1 and classical NK cells. Human NK cells can be split into $CD56^{dim}$ and $CD56^{bright}$ populations. Under homeostatic conditions, $CD56^{dim}$ NK cells comprise the majority of blood resident NK cells, whereas $CD56^{bright}$ are preferentially localized to secondary lymphoid organs. As expected, this differential localization is accompanied by differences in chemokine receptor expression. Although all NK cells express CXCR1, CXCR3, and CXCR4, $CD56^{high}$ NK cells also express CCR7, allowing for homeostatic migration to the LNs (Maghazachi 2010). Group 2 ILCs include ILC2, also referred to as natural helper cells or nuocytes. ILC2 express CXCR4 and CCR9, which promote their homeostatic distribution (Walker et al. 2013). Using a CXCR6 reporter mouse, ~50% of ILC2 was also shown to express the CXCR6 reporter (Roediger et al. 2013). However, ILC2 may not actually express CXCR6 and expression of the reporter may simply reflect prior expression by an ILC2 precursor cell (Possot et al. 2011; Constantinides et al. 2014). Finally, group 3 ILCs include ILC3 and lymphoid tissue- inducer (LTi) cells. LTi cells differentiate from a $CXCR6^{high}$ LTi precursor cell and the mature LTi cells express CXCR5 and CCR6 (Constantinides et al. 2014). This CCR6 expression allows LTi cells to migrate into the intestinal epithelium in response to epithelial derived CCL20 and β-defensins, which are produced in response to commensal bacteria (Sawa et al. 2010).

THE CHEMOKINE SYSTEM IN ACUTE INFLAMMATION

Acute inflammation is a complex process that must coordinate the dual goals of providing initial immune protection as well as to initiate the adaptive immune response. This coordination starts with the homeostatic prepositioning of innate immune cells throughout the periphery, where they act as local sensors of infection and inflammation through the activation of pattern recognition receptors (PRRs), the inflammasome, and/or RNA and DNA sensors. Examples of these prepositioned cellular sensors of infection or inflammation include MCs, macrophages, and DCs. Upon activation, these local innate immune cells release inflammatory cyto-

Cite this article as *Cold Spring Harb Perspect Biol* doi: 10.1101/cshperspect.a016303

kines and chemokines that promote the entry of additional, often blood resident, innate effector cells such as neutrophils and monocytes. These cells follow chemokine gradients to the site of inflammation, but can themselves be activated to produce additional inflammatory cytokines and chemokines that promote further inflammatory cell entry. This feed-forward mechanism not only allows for rapid amplification of the effector response, but also allows the innate immune system to shape the inflammatory response. At the same time, activated DCs change their responsiveness to chemokine gradients, allowing for rapid migration to secondary lymphoid organs and initiation of the adaptive immune response. We will discuss the role of the chemokine system in shaping the acute inflammatory response as well as its role in shaping the activation of the adaptive immune response.

Induction of Acute Inflammation by Resident Immune Cells

With the exception of neutrophils, monocytes, and basophils, almost all innate immune cells are present to some extent in the periphery under homeostatic conditions. There they lie in wait as sensors of pathogen invasion, via PRRs, or tissue damage, via the interleukin (IL)-33 pathway as one example. MCs and macrophages are classically described as essential immune sensors, based on their expression of a wide variety of PRRs and their broad localization throughout all vascularized tissues. MCs are uniquely capable of responding immediately to infectious or inflammatory stimuli. Lipopolysaccharide (LPS) stimulation of murine peritoneal MCs leads to immediate release of CXCL1 and CXCL2-containing granules, but not histamine-containing granules, as well as transcriptional activation of CXCL1 and CXCL2. This promotes early (within 2 h of stimulus) neutrophil recruitment that is abolished in MC-depleted mice, but not in macrophage-depleted mice (De Filippo et al. 2013). Additionally, in mouse models of airway hyperreactivity, MCs release preformed CCL1, promoting early migration of CCR8 expressing Th2 effector cells (Gonzalo et al. 2007). However, release of preformed mediators acts only temporarily to promote inflammatory cell entry; sustained recruitment requires transcriptional activation and secretion of additional chemokines.

Although preformed MC chemokines are necessary for immediate neutrophil recruitment in an LPS intraperitoneal injection model, neutrophil migration is normal in MC-deficient mice within 4 h after LPS injection (De Filippo et al. 2013). This is attributable to chemokine production by macrophages, which produce a wide range of chemokines including, but not limited to, CXCL1, CXCL2, CXCL8, CCL2, CCL3, CCL4, and CCL5. Similarly, MCs have been shown to produce CCL2, CCL3, CCL4, CCL5, CCL11, CCL20, CXCL1, CXCL2, CXCL8, CXCL9, CXCL10, and CXCL11 (Marshall 2004). Finally, viral exposure of DCs leads to the production of many of the same chemokines as well as CXCL16 (Piqueras et al. 2006). Group 2 ILCs appear to play an important role in acute inflammation in response to epithelial damage. IL-33 produced in response to epithelial damage by the protease allergen papain activates ILC2, leading to their production of CCL17 and CCL22 (Halim et al. 2014). Depletion of any of these cell types results in impaired inflammatory cell migration, although the degree of impairment is dependent on the model used.

In addition to producing chemokines, resident innate immune cells also produce inflammatory cytokines, such as tumor necrosis factor (TNF) and IL-1. These cytokines can alter the chemokine environment by inducing further production of chemokines, altering the presentation of chemokines by endothelium, or by altering the response to chemokine gradients. Cytokine-activated epithelium can produce a host of chemokines, including CCL2, CCL3, CCL4, CCL5, CXCL1, CXCL2, CXCL3, CXCL5, and CXCL8 (Kagnoff and Eckmann 1997). Likewise, cytokines can alter chemokine presentation by endothelial cells. In a model of experimental autoimmune encephalitis, IL-17 stimulation of brain endothelial cells leads to abluminal expression of CXCR7, which acts to remove CXCL12. In the absence of endothelial CXCL12, leukocytes enter the brain and induce disease pathogenesis (Cruz-Orengo et al. 2011).

Whether this occurs in other disease settings and in other organs remains to be seen. Finally, cytokines can also impact the responsiveness of immune cells to available chemokines. For instance, activated ILC2 releases IL-13, which promotes increased DC responsiveness to CCR7 ligands and thus induces DC migration into the draining LNs (Halim et al. 2014).

ENTRY OF BLOOD-BORNE CELLS INTO SITES OF ACUTE INFLAMMATION

Neutrophils

Although resident innate immune cells are the initial responders to inflammatory cues, circulating innate cells such as neutrophils, monocytes, and eosinophils quickly become the major immune cells during acute inflammation. Neutrophils are stereotypical cells of acute inflammation; they express many chemokine receptors, including CXCR2 and CCR1 in mice and CXCR1 and CXCR2 in humans, which respond to early chemokines released by MCs and macrophages (Fig. 1). Once activated in tissue, neutrophils up-regulate other chemokine receptors, such as CCR5, which has been shown to act as a chemokine scavenger (Ariel et al. 2006). Neutrophils also express chemotactic receptors for complement, lipid mediators, such as leukotriene B_4 (LTB_4), and bacterial products, such as formylated peptides, including formylmethionyl-leucyl-phenylalanine (fMLP). But how do neutrophils respond to distant chemokine gradients that exist beyond the endothelium? As previously discussed, endothelial cells can be activated by inflammatory cytokines produced by innate immune cells. Additionally, the pericyte, a structural support cell of the endothelium, has been shown to be a "hot spot" for neutrophil migration and to up-regulate adhesion molecules and chemokines after activation via PRRs (Proebstl et al. 2012; Stark et al. 2013). This activation induces the endothelial expression of P-selectin, E-selectin, and integrins, which bind to neutrophils, slowing their movement and causing them to roll along the endothelium. Rolling neutrophils are then able to bind to chemokines, such as CXCL1, CXCL2, or CXCL8, which are presented on the luminal surface of the endothelium. There are two established sources for these luminal chemokines: direct production by the activated endothelium or endothelial presentation of distantly produced chemokines. In the second case, the atypical chemokine receptor, known as the Duffy antigen receptor for chemokines (DARC or ACKR1), binds free chemokines on the abluminal surface of the endothelium. The DARC/chemokine complex is then transcytosed across the endothelial cell and ultimately presented on the luminal surface, where the bound chemokine is able to activate and promote immune cell transmigration (Proudfoot et al. 2003; Pruenster et al. 2009). Given the indiscriminate tissue damage seen in acute inflammation, this process is under tight control with rapid endocytosis and destruction of immobilized chemokines (Hillyer and Male 2005).

Once they have transcytosed, neutrophils follow chemoattractant gradients as they migrate through the interstitium to sites of acute inflammation. Along the way, activated neutrophils produce CCL3, CCL4, CCL5, CCL20, CXCL1, CXCL8, CXCL9, and CXCL10. Thus, they can further amplify the initial acute inflammatory response by inducing additional leukocyte entry (Bennouna et al. 2003). Chemokines make up one class of chemoattractants; however, neutrophils can respond to a wide variety of chemoattractant molecules that include lipid mediators, bacterial products, and complement fragments. These different classes of chemoattractants can act simultaneously or sequentially. One example of sequential activity of chemoattractants is in a model of sterile thermal tissue injury. In this model, CXCL2 gradients form around the area of tissue injury and promote neutrophil chemotaxis toward the injury site (McDonald et al. 2010). However, because chemokine production is dependent on live cells, the central site of necrosis lacked a supportive CXCL2 gradient. Instead, endogenous formyl peptides, which are produced by mitochondria and released upon cellular damage, promoted neutrophil chemotaxis to the site of cellular injury (McDonald et al. 2010; Zhang et al. 2010). Thus, CXCL2 gradients promoted initial migration to the area of necrosis, followed by FPR1

Cite this article as *Cold Spring Harb Perspect Biol* doi: 10.1101/cshperspect.a016303

signaling that promoted neutrophil entry into the necrotic center (Fig. 2A).

Beyond providing linear directional information, different chemoattractant classes may play important roles in permitting neutrophils to amplify their own migration. Such interactions between chemoattractant classes have been shown in the case of LTB_4 and chemokines. Using a mouse model of autoimmune arthritis, we and other investigators have shown that LTB_4 acting through the G-protein-coupled receptor BLT1 on neutrophils is necessary for initial neutrophil recruitment into the joint (Chen et al. 2006; Kim et al. 2006). Once inside the joint, these intial "scout" neutrophils are activated by synovial-immune complexes to produce neutrophil-active chemokines, such as CXCL2, as well as the proinflammatory cytokine IL-1β.

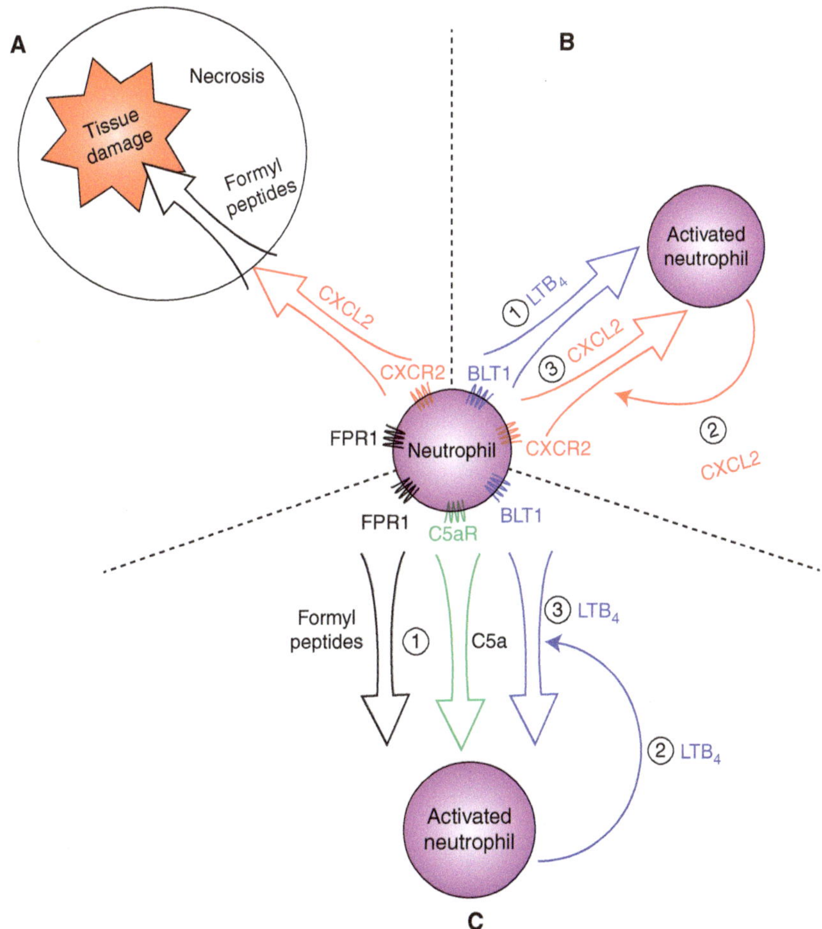

Figure 2. Mechanisms of neutrophil migration in the inflamed periphery. (*A*) Neutrophils migrate to sites of tissue damage by following sequential chemoattractant gradients. Neutrophils first follow CXCL2 gradients, and then in areas of tissue necrosis follow gradients of endogenous formyl peptides via FPR1 signaling. (*B*) Neutrophils can amplify their own recruitment through the production of chemokines. Neutrophils migrate to inflammatory sites by following LTB_4 gradients (1) where they are activated by inflammatory stimuli and produce additional chemokines, such as CXCL2 (2) that promote further neutrophil entry by establishing an additional chemokine gradient (3). (*C*) Neutrophils can be activated by chemotactic agents, such as formyl peptides or C5a, leading to LTB_4 production, which establishes a relay system for neutrophil swarming. Formyl-peptide or C5a-activated neutrophils (1) produce LTB_4 (2), which creates a chemoattactant gradient that induces more neutrophils to enter and swarm to the site of inflammation (3).

This inflammatory cytokine then activates local synoviocytes as well as macrophages and endothelial cells to produce additional acute inflammatory chemokines, such as CCL3, CCL4, CCL5, CXCL1, and CXCL2. These chemokines then rapidly amplify neutrophil entry into the joint by binding to the CCR1 and CXCR2 receptors on neutrophils (Chou et al. 2010). Thus, LTB_4 induces a necessary initial neutrophil migration that then amplifies inflammatory cell influx characteristic of acute inflammation (Fig. 2B).

In addition to creating sequential waves of neutrophil recruitment, different chemoattractant classes may enhance the response to each other or act to amplify the chemotactic gradient. Stimulation of the G-protein-coupled formyl peptide receptor FPR1 on neutrophils leads to neutrophil LTB_4 production, which then feeds back in an autocrine and paracrine fashion and promotes further neutrophil migration toward formyl peptides (Afonso et al. 2012). Likewise, in the murine model of immune-complex-induced arthritis discussed above, activation of the G-protein-coupled complement receptor C5aR was required for the generation of LTB_4 and the induction of arthritis (Sadik et al. 2012). Indeed, production of LTB_4 by neutrophils appears to be an essential mechanism for extending the range and enhancing the stability of chemotactic signals. Activated neutrophils produce LTB_4, which then acts as a neutrophil chemoattractant. Thus, LTB_4 production by activated neutrophils can feed-forward and extend the range of initial chemotactic signals. Additionally, this LTB_4 can activate neutrophils, promoting further LTB_4 production, which acts to further amplify the initial inflammatory stimulus (Afonso et al. 2012). LTB_4-mediated amplification of neutrophil activation and chemotaxis induces the coordinated chemotaxis and clustering (i.e., swarming) of neutrophils at sites of tissue damage in vivo (Fig. 2C) (Lammermann et al. 2013).

Monocytes

Along with neutrophils, blood-borne monocytes are recruited early in the setting of acute inflammation by activated endothelium. As discussed previously, monocytes are divided into $CCR2^+$ and $CX3CR1^+$ groups. $CCR2^+$ monocytes are largely resident in the peripheral blood during homeostasis, but rapidly migrate to areas of acute inflammation. Although they are defined by their expression of CCR2, they also express CXCR2, which may play an important role in initial activation and transmigration. In an atheroma model, CXCL8/CXCR2 interactions were necessary for firm adhesion of monocytes to vascular endothelium (Gerszten et al. 1999). CCR2 signaling then promotes proinflammatory monocyte migration into peripheral tissues in response to stable extracellular matrix–associated CCL2 gradients (Fig. 1) (Kuziel et al. 1997; Proudfoot et al. 2003). Interestingly, although CXCL2 does not appear to promote monocyte migration in response to CXCL8, it does promote migration in response to its atypical ligand macrophage migration inhibitory factor (MIF). MIF binds to CD74 and CXCR2 on monocytes and macrophages, leading to CXCR2 signaling and integrin-dependent chemotaxis of monocytes, which is necessary for maintenance of atherosclerotic plaques (Bernhagen et al. 2007). Additionally, CCR1 and CCR5 have been shown to promote chemotaxis of $CCR2^+$ monocytes in in vitro experiments, but in vivo data are lacking (Weber et al. 2001; Le Borgne et al. 2006; Shi and Pamer 2011).

In addition to promoting $CCR2^+$ monocyte recruitment to sites of acute inflammation, CCL2 may also have long-range effects in draining LNs and BM. Soluble CCL2 has been detected in the afferent lymph, where it drains to the LN, binds to HEVs, and induces $CCR2^+$ monocyte entry into the draining LN (Palframan et al. 2001). Additionally, CCL2 plays an important role in mobilizing inflammatory monocytes from the BM (Fig. 1). The source of this CCL2 is unclear. It is possible that locally produced CCL2 enters the systemic circulation, but given the mechanisms in place to remove circulating chemokines, there may not be sufficient concentrations of CCL2 that reach the BM. Alternatively, small amounts of pathogen-associated molecular patterns may enter the systemic circulation during infection, activate PRRs on BM niche cells, and induce local CCL2 production by BM cells (Shi et al. 2011). Given the heterogeneity of acute inflammatory stimuli that induce monocyte egress, it is likely that multiple mechanisms contribute to the chemokine-dependent mobilization of inflammatory monocytes.

The chemotactic properties of CX3CR1$^+$ monocytes are less understood. Anti-inflammatory monocytes are characterized by their high-level expression of CX3CR1 and lack of CCR2, although CCR2$^+$ monocytes also express intermediate levels of CX3CR1, and CX3CR1 deficiency leads to minimal defects in migration of monocytes. CX3CR1 was shown to play a role in the early migration of inflammatory monocytes into the spleen after *Listeria* infection, but this was a temporary defect (Auffray et al. 2007). Instead, CX3CR1 may fulfill nonchemotactic roles. CX3CR1 has been shown to play an essential role in promoting integrin-mediated adhesion within the vessel, which allows for the patrolling phenotype of CX3CR1$^+$ monocytes (Auffray et al. 2007). Additionally, CX3CR1 may provide a prosurvival signal to CX3CR1$^+$ monocytes, which may underlie the proatherogenic role of CX3CL1/CX3CR1 (Moatti et al. 2001; Landsman et al. 2009).

Eosinophils

Eosinophils express the chemokine receptors CCR1 and CCR3, allowing them to respond to a wide variety of inflammatory stimuli. However, eosinophils are best characterized by their role in allergic and parasitic responses, during which they migrate in response to the eotaxins (CCL11, CCL24, and CCL26). Eotaxins can be released directly by stimulation of innate immune cells, but experiments in murine asthma models have illustrated that IL-4 and IL-13 exposure are necessary for optimal production of CCL11, CCL24, and CCL26 (Menzies-Gow and Robinson 2001). In murine models of airway inflammation, CCL11 and CCL24 play additive roles in promoting eosinophil migration into the lung (Humbles et al. 2002; Mattes et al. 2002; Pope et al. 2005). The role of CCL26 is less understood, but it likely plays an important role in human eosinophil migration as it is the predominant chemokine that drives eosinophil migration across IL-4-activated endothelial and epithelial monolayers (Fig. 1) (Cuvelier and Patel 2001; Yuan et al. 2006). CCL11 activity is regulated by CD26, a cell-surface protease that cleaves CCL11 into a partial antagonist: a protein that cannot induce chemotaxis, but can still bind to CCR3 and desensitize its response to CCL11. This leads to the inhibition of CCR3-mediated chemotaxis, and ultimately to the suppression of eosinophilic inflammation (Struyf et al. 1999; Yan et al. 2012). Finally, analogous to the inflammation-amplification mechanisms used by other acute inflammatory cells, thymic stromal lymphopoietin (TSLP)–stimulated eosinophils also produce CCL2, CXCL1, and CXCL8 promoting further influx of inflammatory monocytes and neutrophils (Wong et al. 2010).

Dendritic Cells

Once activated by inflammatory cytokines or PRR ligation, DCs undergo a maturation process and down-regulate expression of chemokine receptors expressed on immature DCs and up-regulate CCR7 expression (Sallusto et al. 1998). Thus, instead of altering the expression of chemokine ligands, DC migration occurs in response to alterations in chemokine receptor expression. Mature, CCR7^{high}, DCs follow stable gradients of CCL21, which increase the mobility and directional migration of mature DCs toward the lymphatic endothelium (Tal et al. 2011). There, DCs enter the lumen of the lymphatic vessel via gaps in the basement membrane beneath the lymphatic endothelial cells (Pflicke and Sixt 2009). Once in the lumen, they crawl along following the direction of lymphatic flow until they reach the collecting lymphatics, at which point they freely flow with the lymph to the draining LN and then transmigrate through the floor of the subcapsular sinus and into the T-cell zone in a CCR7-dependent process (Braun et al. 2011).

Innate Lymphoid Cells

As discussed above, innate lymphocytes are thought to primarily be tissue resident in order to provide a first line of defense, especially at mucosal sites. However, parabiosis studies have revealed that there are both tissue-resident and recirculating populations of NK and natural killer T (NKT) cells (Thomas et al. 2011; Sojka et al. 2014). Innate lymphocytes constitutively express inflammatory chemokine receptors, which allow them to rapidly migrate and respond to inflammatory stimuli. NK cells use CCR5 signaling to migrate to sites of acute inflammation in murine-infection models, whereas CXCR3 signaling is used in models of cardiac transplant and hepatitis (Hancock et al. 2001; Hokeness et al. 2005; Khan et al. 2006; Wald et al. 2006). The infiltration of activated, IFN-γ-producing NK cells into peripheral tissues then promotes further cell migration via production of the IFN-γ-inducible CXCR3 ligands, CXCL9 and CXCL10 (Fig. 1). NKT cells have similarly been shown to initially migrate into the acutely inflamed liver via a CXCR6-dependent process where they then promote local accumulation of monocytes and macrophages (Wehr et al. 2013). Chemokine-mediated trafficking of innate lymphoid cells also serves to link innate and adaptive immune responses. For example, CXCR3-mediated NK cell migration guides IFN-γ-secreting NK cells into the LN where they help prime the LN for a Th1 response (Martin-Fontecha et al. 2004). Similarly, in the periphery, a blockade of NK cell CX3CR1-mediated migration into the brain in a mouse model of experimental autoimmune encephalitis led to enhanced disease caused by unabated Th17 formation in the absence of NK-cell-derived IFN-γ (Hao et al. 2010).

THE CHEMOKINE SYSTEM IN INNATE CONTROL OF ADAPTIVE IMMUNITY

Once in the draining LN or other secondary lymphoid organ, mature DCs migrate to specific regions and secrete their own chemokines to enhance their interactions with adaptive immune cells. The combination of specific localization and chemoattraction of adaptive immune cells appears to be necessary for optimal activation and instruction of the developing adaptive immune response.

$CD4^+$ Th1 Priming

Th1 cells have been shown to preferentially express the chemokine receptors CCR5, CXCR3, and CXCR6, but CXCR3 may play a specific role in $CD4^+$ Th1 priming (Bromley et al. 2008; Groom et al. 2012). In response to cutaneous immunization with antigen-pulsed mature DCs or immunization with antigen and adjuvant, CXCL9 and CXCL10 were produced in the interfollicular and medullary regions, respectively, in the draining LN. This expression promoted migration of $CD4^+$ T cells to these DC-rich regions, while, at the same time, activated DCs produced CXCL10. Together, this led to clustering and the formation of stable contacts between $CD4^+$ T cells and DCs, which were necessary for optimal Th1 differentiation (Groom et al. 2012). In addition to promoting direct interactions between $CD4^+$ T cells and DCs, this localization may also place developing Th1 cells close to innate sources of IFN-γ, such as CXCR3 expressing NK, NKT, γδ T, and innate-like $CD8^+$ T cells (Bajenoff et al. 2006; Groom and Luster 2011; Kastenmuller et al. 2012; Oghumu et al. 2013).

$CD4^+$ Th2 and Tfh Priming

Like $CD4^+$ Th1 cells, optimal induction of Th2 cell differentiation appears to be dependent on chemokine guidance. Using a helminth infection model, Th2 and T follicular helper (Tfh) cell induction was shown to be dependent on CXCL13/CXCR5 interactions (Leon et al. 2012). CXCR5-mediated signaling in both DCs and $CD4^+$ T cells led to migration to the perifollicular regions of the LN in response to B-cell-derived CXCL13. A blockade of CXCR5 signaling prevented optimal Th2 and Tfh differentiation (Leon et al. 2012). However, it remains unclear why localization to this region is necessary for Th2 differentiation and whether an innate immune cell involved in cytokine production or Th2 skewing cell is similarly localized to this region.

CD8$^+$ T-Cell Priming

Optimal priming of CD8$^+$ T cells requires interactions with DCs that have been "licensed" by CD4$^+$ T cells. This requires the coordinated migration of DCs, CD4$^+$ T cells and CD8$^+$ T cells, which appears to be caused by CCR5- and CCR4-mediated signaling. In response to CD4$^+$ T-cell and DC interactions, both cells secrete CCR5 ligands, which bind to a subset of CD8$^+$ T cells within the LN (Castellino et al. 2006). This promotes sustained contacts between CD8$^+$ T cells and the CD4$^+$ licensed DCs, which is necessary for optimal CD8$^+$ T-cell memory responses. However, the precise role for CCR5 remains unclear as viral infection models using different vaccinia virus strains have shown contradictory results for the role of CCR5 in CD8$^+$ T-cell priming (Hickman et al. 2011; Kastenmuller et al. 2013). Finally, in a model of CD8$^+$ T-cell priming using injection of α-galactosylceramide, CCR4 signaling promoted stable contacts between CD8$^+$ T cells and DCs (Semmling et al. 2010). However, whether this pathway is relevant in the setting of more physiologic immune stimuli remains unknown.

CONCLUDING REMARKS

Chemokines are essential for the positioning of innate immune sentinels at mucosal barriers and for the recruitment of the first line of innate immune effector cells to sites of infection and inflammation. Chemokine function is also essential for maintaining an adequate pool of circulating immune cells at homeostasis and at times of stress by governing their release from the BM. Once in the tissue, chemokines collaborate with other chemoattractants, such as lipid mediators, formylated peptides, and complement components, to guide innate immune effectors to the very site of tissue damage and pathogen replication. Chemokine function is also necessary to translate an innate immune response into an adaptive immune response. Innate immune stimuli—through activation of PRRs—set in motion a genetic program that induces the expression of chemokines from resident tissue innate immune cells and also modulates the expression of chemokine receptors on DCs. The induction of chemokine and chemokine receptor expression orchestrates the movement of antigen-loaded DCs from the tissue into lymphoid tissue to activate T and B cells to initiate the adaptive immune response. Chemokines downstream from PRR activation also help guide the newly activated T cells back into the tissue where the innate immune system first sensed the foreign challenge. In addition, during secondary immune responses, chemokines induced by antigen-specific lymphocyte responses recruit innate immune cells into sites of inflammation, serving to amplify the adaptive response with innate immune effector cells. Thus, chemokines and their receptors serve a critical function in coordinating the interdependent innate and adaptive immune responses.

ACKNOWLEDGMENTS

The writers are supported by grants from the National Institutes of Health, and C.L.S. is also supported by the Asthma and Allergy Foundation of America. We thank Jason Griffith for helpful discussions in the preparation of this manuscript.

REFERENCES

Abonia JP, Austen KF, Rollins BJ, Joshi SK, Flavell RA, Kuziel WA, Koni PA, Gurish MF. 2005. Constitutive homing of mast cell progenitors to the intestine depends on autologous expression of the chemokine receptor CXCR2. *Blood* **105:** 4308–4313.

Afonso PV, Janka-Junttila M, Lee YJ, McCann CP, Oliver CM, Aamer KA, Losert W, Cicerone MT, Parent CA. 2012. LTB4 is a signal-relay molecule during neutrophil chemotaxis. *Dev Cell* **22:** 1079–1091.

Ara T, Tokoyoda K, Sugiyama T, Egawa T, Kawabata K, Nagasawa T. 2003. Long-term hematopoietic stem cells require stromal cell-derived factor-1 for colonizing bone marrow during ontogeny. *Immunity* **19:** 257–267.

Ariel A, Fredman G, Sun YP, Kantarci A, Van Dyke TE, Luster AD, Serhan CN. 2006. Apoptotic neutrophils and T cells sequester chemokines during immune response resolution through modulation of CCR5 expression. *Nat Immunol* **7:** 1209–1216.

Auffray C, Fogg D, Garfa M, Elain G, Join-Lambert O, Kayal S, Sarnacki S, Cumano A, Lauvau G, Geissmann F. 2007. Monitoring of blood vessels and tissues by a population of monocytes with patrolling behavior. *Science* **317:** 666–670.

Bachelerie FB-BA, Burkhardt AM, Combadiere C, Farber JM, Graham GJ, Horuk R, Sparre-Ulrich AH, Locati M, Luster AD, Mantovani A, et al. 2013. International Union of Basic and Clinical Pharmacology. LXXXIX. Update on the extended family of chemokine receptors and introducing a new nomenclature for atypical chemokine receptors. *Pharmacol Rev* **11:** 1–79.

Bajenoff M, Breart B, Huang AY, Qi H, Cazareth J, Braud VM, Germain RN, Glaichenhaus N. 2006. Natural killer cell behavior in lymph nodes revealed by static and real-time imaging. *J Exp Med* **203:** 619–631.

Bennouna S, Bliss SK, Curiel TJ, Denkers EY. 2003. Cross-talk in the innate immune system: Neutrophils instruct recruitment and activation of dendritic cells during microbial infection. *J Immunol* **171:** 6052–6058.

Bernhagen J, Krohn R, Lue H, Gregory JL, Zernecke A, Koenen RR, Dewor M, Georgiev I, Schober A, Leng L, et al. 2007. MIF is a noncognate ligand of CXC chemokine receptors in inflammatory and atherogenic cell recruitment. *Nat Med* **13:** 587–596.

Bogunovic M, Ginhoux F, Wagers A, Loubeau M, Isola LM, Lubrano L, Najfeld V, Phelps RG, Grosskreutz C, Scigliano E, et al. 2006. Identification of a radio-resistant and cycling dermal dendritic cell population in mice and men. *J Exp Med* **203:** 2627–2638.

Braun A, Worbs T, Moschovakis GL, Halle S, Hoffmann K, Bolter J, Munk A, Forster R. 2011. Afferent lymph-derived T cells and DCs use different chemokine receptor CCR7-dependent routes for entry into the lymph node and intranodal migration. *Nat Immunol* **12:** 879–887.

Bromley SK, Mempel TR, Luster AD. 2008. Orchestrating the orchestrators: Chemokines in control of T cell traffic. *Nat Immunol* **9:** 970–980.

Broxmeyer HE, Orschell CM, Clapp DW, Hangoc G, Cooper S, Plett PA, Liles WC, Li X, Graham-Evans B, Campbell TB, et al. 2005. Rapid mobilization of murine and human hematopoietic stem and progenitor cells with AMD3100, a CXCR4 antagonist. *J Exp Med* **201:** 1307–1318.

Castellino F, Huang AY, Altan-Bonnet G, Stoll S, Scheinecker C, Germain RN. 2006. Chemokines enhance immunity by guiding naive $CD8^+$ T cells to sites of $CD4^+$ T cell-dendritic cell interaction. *Nature* **440:** 890–895.

Chen M, Lam BK, Kanaoka Y, Nigrovic PA, Audoly LP, Austen KF, Lee DM. 2006. Neutrophil-derived leukotriene B4 is required for inflammatory arthritis. *J Exp Med* **203:** 837–842.

Chou RC, Kim ND, Sadik CD, Seung E, Lan Y, Byrne MH, Haribabu B, Iwakura Y, Luster AD. 2010. Lipid-cytokine-chemokine cascade drives neutrophil recruitment in a murine model of inflammatory arthritis. *Immunity* **33:** 266–278.

Constantinides MG, McDonald BD, Verhoef PA, Bendelac A. 2014. A committed precursor to innate lymphoid cells. *Nature* **508:** 397–401.

Cook DN, Prosser DM, Forster R, Zhang J, Kuklin NA, Abbondanzo SJ, Niu XD, Chen SC, Manfra DJ, Wiekowski MT, et al. 2000. CCR6 mediates dendritic cell localization, lymphocyte homeostasis, and immune responses in mucosal tissue. *Immunity* **12:** 495–503.

Cruz-Orengo L, Holman DW, Dorsey D, Zhou L, Zhang P, Wright M, McCandless EE, Patel JR, Luker GD, Littman DR, et al. 2011. CXCR7 influences leukocyte entry into the CNS parenchyma by controlling abluminal CXCL12 abundance during autoimmunity. *J Exp Med* **208:** 327–339.

Cuvelier SL, Patel KD. 2001. Shear-dependent eosinophil transmigration on interleukin 4-stimulated endothelial cells: A role for endothelium-associated eotaxin-3. *J Exp Med* **194:** 1699–1709.

Czeloth N, Schippers A, Wagner N, Muller W, Kuster B, Bernhardt G, Forster R. 2007. Sphingosine-1 phosphate signaling regulates positioning of dendritic cells within the spleen. *J Immunol* **179:** 5855–5863.

De Filippo K, Dudeck A, Hasenberg M, Nye E, van Rooijen N, Hartmann K, Gunzer M, Roers A, Hogg N. 2013. Mast cell and macrophage chemokines CXCL1/CXCL2 control the early stage of neutrophil recruitment during tissue inflammation. *Blood* **121:** 4930–4937.

Devi S, Wang Y, Chew WK, Lima R, A-González N, Mattar CN, Chong SZ, Schlitzer A, Bakocevic N, Chew S, et al. 2013. Neutrophil mobilization via plerixafor-mediated CXCR4 inhibition arises from lung demargination and blockade of neutrophil homing to the bone marrow. *J Exp Med* **210:** 2321–2336.

DeVries ME, Kelvin AA, Xu L, Ran L, Robinson J, Kelvin DJ. 2006. Defining the origins and evolution of the chemokine/chemokine receptor system. *J Immunol* **176:** 401–415.

Forster R, Schubel A, Breitfeld D, Kremmer E, Renner-Muller I, Wolf E, Lipp M. 1999. CCR7 coordinates the primary immune response by establishing functional microenvironments in secondary lymphoid organs. *Cell* **99:** 23–33.

Foussat A, Coulomb-L'Hermine A, Gosling J, Krzysiek R, Durand-Gasselin I, Schall T, Balian A, Richard Y, Galanaud P, Emilie D. 2000. Fractalkine receptor expression by T lymphocyte subpopulations and in vivo production of fractalkine in human. *Eur J Immunol* **30:** 87–97.

Gatto D, Wood K, Caminschi I, Murphy-Durland D, Schofield P, Christ D, Karupiah G, Brink R. 2013. The chemotactic receptor EBI2 regulates the homeostasis, localization and immunological function of splenic dendritic cells. *Nat Immunol* **14:** 446–453.

Geissmann F, Jung S, Littman DR. 2003. Blood monocytes consist of two principal subsets with distinct migratory properties. *Immunity* **19:** 71–82.

Gerszten RE, Garcia-Zepeda EA, Lim YC, Yoshida M, Ding HA, Gimbrone MA Jr, Luster AD, Luscinskas FW, Rosenzweig A. 1999. MCP-1 and IL-8 trigger firm adhesion of monocytes to vascular endothelium under flow conditions. *Nature* **398:** 718–723.

Gonzalo JA, Qiu Y, Lora JM, Al-Garawi A, Villeval JL, Boyce JA, Martinez-A C, Marquez G, Goya I, Hamid Q, et al. 2007. Coordinated involvement of mast cells and T cells in allergic mucosal inflammation: Critical role of the CC chemokine ligand 1:CCR8 axis. *J Immunol* **179:** 1740–1750.

Groom JR, Luster AD. 2011. CXCR3 ligands: Redundant, collaborative and antagonistic functions. *Immunol Cell Biol* **89:** 207–215.

Groom JR, Richmond J, Murooka TT, Sorensen EW, Sung JH, Bankert K, von Andrian UH, Moon JJ, Mempel TR, Luster AD. 2012. CXCR3 chemokine receptor-ligand interactions in the lymph node optimize $CD4^+$ T helper 1 cell differentiation. *Immunity* **37:** 1091–1103.

Gulino AV, Moratto D, Sozzani S, Cavadini P, Otero K, Tassone L, Imberti L, Pirovano S, Notarangelo LD, Soresina R, et al. 2004. Altered leukocyte response to CXCL12 in patients with warts hypogammaglobulinemia, infections, myelokathexis (WHIM) syndrome. *Blood* **104:** 444–452.

Halim TY, Steer CA, Matha L, Gold MJ, Martinez-Gonzalez I, McNagny KM, McKenzie AN, Takei F. 2014. Group 2 innate lymphoid cells are critical for the initiation of adaptive T helper 2 cell-mediated allergic lung inflammation. *Immunity* **40:** 425–435.

Halova I, Draberova L, Draber P. 2012. Mast cell chemotaxis—Chemoattractants and signaling pathways. *Front Immunol* **3:** 119.

Hancock WW, Gao W, Csizmadia V, Faia KL, Shemmeri N, Luster AD. 2001. Donor-derived IP-10 initiates development of acute allograft rejection. *J Exp Med* **193:** 975–980.

Hao J, Liu R, Piao W, Zhou Q, Vollmer TL, Campagnolo DI, Xiang R, La Cava A, Van Kaer L, Shi FD. 2010. Central nervous system (CNS)-resident natural killer cells suppress Th17 responses and CNS autoimmune pathology. *J Exp Med* **207:** 1907–1921.

Hernandez PA, Gorlin RJ, Lukens JN, Taniuchi S, Bohinjec J, Francois F, Klotman ME, Diaz GA. 2003. Mutations in the chemokine receptor gene CXCR4 are associated with WHIM syndrome, a combined immunodeficiency disease. *Nat Genet* **34:** 70–74.

Hickman HD, Li L, Reynoso GV, Rubin EJ, Skon CN, Mays JW, Gibbs J, Schwartz O, Bennink JR, Yewdell JW. 2011. Chemokines control naive $CD8^+$ T cell selection of optimal lymph node antigen presenting cells. *J Exp Med* **208:** 2511–2524.

Hillyer P, Male D. 2005. Expression of chemokines on the surface of different human endothelia. *Immunol Cell Biol* **83:** 375–382.

Hokeness KL, Kuziel WA, Biron CA, Salazar-Mather TP. 2005. Monocyte chemoattractant protein-1 and CCR2 interactions are required for IFN-α/β-induced inflammatory responses and antiviral defense in liver. *J Immunol* **174:** 1549–1556.

Humbles AA, Lu B, Friend DS, Okinaga S, Lora J, Al-Garawi A, Martin TR, Gerard NP, Gerard C. 2002. The murine CCR3 receptor regulates both the role of eosinophils and mast cells in allergen-induced airway inflammation and hyperresponsiveness. *Proc Natl Acad Sci* **99:** 1479–1484.

Kagnoff MF, Eckmann L. 1997. Epithelial cells as sensors for microbial infection. *J Clin Invest* **100:** 6–10.

Kastenmuller W, Torabi-Parizi P, Subramanian N, Lammermann T, Germain RN. 2012. A spatially organized multicellular innate immune response in lymph nodes limits systemic pathogen spread. *Cell* **150:** 1235–1248.

Kastenmuller W, Brandes M, Wang Z, Herz J, Egen JG, Germain RN. 2013. Peripheral prepositioning and local CXCL9 chemokine-mediated guidance orchestrate rapid memory $CD8^+$ T cell responses in the lymph node. *Immunity* **38:** 502–513.

Khan IA, Thomas SY, Moretto MM, Lee FS, Islam SA, Combe C, Schwartzman JD, Luster AD. 2006. CCR5 is essential for NK cell trafficking and host survival following *Toxoplasma gondii* infection. *PLoS Pathog* **2:** e49.

Kim ND, Chou RC, Seung E, Tager AM, Luster AD. 2006. A unique requirement for the leukotriene B4 receptor BLT1 for neutrophil recruitment in inflammatory arthritis. *J Exp Med* **203:** 829–835.

Kurth I, Willimann K, Schaerli P, Hunziker T, Clark-Lewis I, Moser B. 2001. Monocyte selectivity and tissue localization suggests a role for breast and kidney-expressed chemokine (BRAK) in macrophage development. *J Exp Med* **194:** 855–861.

Kuziel WA, Morgan SJ, Dawson TC, Griffin S, Smithies O, Ley K, Maeda N. 1997. Severe reduction in leukocyte adhesion and monocyte extravasation in mice deficient in CC chemokine receptor 2. *Proc Natl Acad Sci* **94:** 12053–12058.

Lammermann T, Afonso PV, Angermann BR, Wang JM, Kastenmuller W, Parent CA, Germain RN. 2013. Neutrophil swarms require LTB4 and integrins at sites of cell death in vivo. *Nature* **498:** 371–375.

Landsman L, Bar-On L, Zernecke A, Kim KW, Krauthgamer R, Shagdarsuren E, Lira SA, Weissman IL, Weber C, Jung S. 2009. CX3CR1 is required for monocyte homeostasis and atherogenesis by promoting cell survival. *Blood* **113:** 963–972.

Le Borgne M, Etchart N, Goubier A, Lira SA, Sirard JC, van Rooijen N, Caux C, Ait-Yahia S, Vicari A, Kaiserlian D, et al. 2006. Dendritic cells rapidly recruited into epithelial tissues via CCR6/CCL20 are responsible for $CD8^+$ T cell crosspriming in vivo. *Immunity* **24:** 191–201.

Leon B, Ballesteros-Tato A, Browning JL, Dunn R, Randall TD, Lund FE. 2012. Regulation of T_H2 development by $CXCR5^+$ dendritic cells and lymphotoxin-expressing B cells. *Nat Immunol* **13:** 681–690.

Maghazachi AA. 2010. Role of chemokines in the biology of natural killer cells. *Curr Top Microbiol Immunol* **341:** 37–58.

Marone G, Triggiani M, de Paulis A. 2005. Mast cells and basophils: Friends as well as foes in bronchial asthma? *Trends Immunol* **26:** 25–31.

Marshall JS. 2004. Mast-cell responses to pathogens. *Nat Rev Immunol* **4:** 787–799.

Martin C, Burdon PC, Bridger G, Gutierrez-Ramos JC, Williams TJ, Rankin SM. 2003. Chemokines acting via CXCR2 and CXCR4 control the release of neutrophils from the bone marrow and their return following senescence. *Immunity* **19:** 583–593.

Martin-Fontecha A, Thomsen LL, Brett S, Gerard C, Lipp M, Lanzavecchia A, Sallusto F. 2004. Induced recruitment of NK cells to lymph nodes provides IFN-γ for T_H1 priming. *Nat Immunol* **5:** 1260–1265.

Mattes J, Yang M, Mahalingam S, Kuehr J, Webb DC, Simson L, Hogan SP, Koskinen A, McKenzie AN, Dent LA, et al. 2002. Intrinsic defect in T cell production of interleukin (IL)-13 in the absence of both IL-5 and eotaxin precludes the development of eosinophilia and airways hyperreactivity in experimental asthma. *J Exp Med* **195:** 1433–1444.

McDonald B, Pittman K, Menezes GB, Hirota SA, Slaba I, Waterhouse CC, Beck PL, Muruve DA, Kubes P. 2010. Intravascular danger signals guide neutrophils to sites of sterile inflammation. *Science* **330:** 362–366.

Menzies-Gow A, Robinson DS. 2001. Eosinophil chemokines and chemokine receptors: Their role in eosinophil accumulation and activation in asthma and potential as therapeutic targets. *J Asthma* **38:** 605–613.

Mercier FE, Ragu C, Scadden DT. 2012. The bone marrow at the crossroads of blood and immunity. *Nat Rev Immunol* **12:** 49–60.

Meuter S, Schaerli P, Roos RS, Brandau O, Bosl MR, von Andrian UH, Moser B. 2007. Murine CXCL14 is dispensable for dendritic cell function and localization within peripheral tissues. *Mol Cell Biol* **27:** 983–992.

Mishra A, Hogan SP, Lee JJ, Foster PS, Rothenberg ME. 1999. Fundamental signals that regulate eosinophil homing to the gastrointestinal tract. *J Clin Invest* **103:** 1719–1727.

Moatti D, Faure S, Fumeron F, Amara Mel W, Seknadji P, McDermott DH, Debre P, Aumont MC, Murphy PM, de Prost D, et al. 2001. Polymorphism in the fractalkine receptor CX3CR1 as a genetic risk factor for coronary artery disease. *Blood* **97:** 1925–1928.

Nilsson G, Johnell M, Hammer CH, Tiffany HL, Nilsson K, Metcalfe DD, Siegbahn A, Murphy PM. 1996. C3a and C5a are chemotaxins for human mast cells and act through distinct receptors via a pertussis toxin-sensitive signal transduction pathway. *J Immunol* **157:** 1693–1698.

Nomiyama H, Osada N, Yoshie O. 2010. The evolution of mammalian chemokine genes. *Cytokine Growth Factor Rev* **21:** 253–262.

Ochi H, Hirani WM, Yuan Q, Friend DS, Austen KF, Boyce JA. 1999. T helper cell type 2 cytokine-mediated comitogenic responses and CCR3 expression during differentiation of human mast cells in vitro. *J Exp Med* **190:** 267–280.

Oghumu S, Dong R, Varikuti S, Shawler T, Kampfrath T, Terrazas CA, Lezama-Davila C, Ahmer BM, Whitacre CC, Rajagopalan S, et al. 2013. Distinct populations of innate $CD8^+$ T cells revealed in a CXCR3 reporter mouse. *J Immunol* **190:** 2229–2240.

Ohl L, Mohaupt M, Czeloth N, Hintzen G, Kiafard Z, Zwirner J, Blankenstein T, Henning G, Forster R. 2004. CCR7 governs skin dendritic cell migration under inflammatory and steady-state conditions. *Immunity* **21:** 279–288.

Palframan RT, Collins PD, Williams TJ, Rankin SM. 1998. Eotaxin induces a rapid release of eosinophils and their progenitors from the bone marrow. *Blood* **91:** 2240–2248.

Palframan RT, Jung S, Cheng G, Weninger W, Luo Y, Dorf M, Littman DR, Rollins BJ, Zweerink H, Rot A, et al. 2001. Inflammatory chemokine transport and presentation in HEV: A remote control mechanism for monocyte recruitment to lymph nodes in inflamed tissues. *J Exp Med* **194:** 1361–1373.

Pflicke H, Sixt M. 2009. Preformed portals facilitate dendritic cell entry into afferent lymphatic vessels. *J Exp Med* **206:** 2925–2935.

Piqueras B, Connolly J, Freitas H, Palucka AK, Banchereau J. 2006. Upon viral exposure, myeloid and plasmacytoid dendritic cells produce 3 waves of distinct chemokines to recruit immune effectors. *Blood* **107:** 2613–2618.

Pope SM, Zimmermann N, Stringer KF, Karow ML, Rothenberg ME. 2005. The eotaxin chemokines and CCR3 are fundamental regulators of allergen-induced pulmonary eosinophilia. *J Immunol* **175:** 5341–5350.

Possot C, Schmutz S, Chea S, Boucontet L, Louise A, Cumano A, Golub R. 2011. Notch signaling is necessary for adult, but not fetal, development of $ROR\gamma t^+$ innate lymphoid cells. *Nat Immunol* **12:** 949–958.

Proebstl D, Voisin MB, Woodfin A, Whiteford J, D'Acquisto F, Jones GE, Rowe D, Nourshargh S. 2012. Pericytes support neutrophil subendothelial cell crawling and breaching of venular walls in vivo. *J Exp Med* **209:** 1219–1234.

Proudfoot AE, Handel TM, Johnson Z, Lau EK, LiWang P, Clark-Lewis I, Borlat F, Wells TN, Kosco-Vilbois MH. 2003. Glycosaminoglycan binding and oligomerization are essential for the in vivo activity of certain chemokines. *Proc Natl Acad Sci* **100:** 1885–1890.

Pruenster M, Mudde L, Bombosi P, Dimitrova S, Zsak M, Middleton J, Richmond A, Graham GJ, Segerer S, Nibbs RJ, et al. 2009. The Duffy antigen receptor for chemokines transports chemokines and supports their promigratory activity. *Nat Immunol* **10:** 101–108.

Randolph GJ, Ochando J, Partida-Sanchez S. 2008. Migration of dendritic cell subsets and their precursors. *Annu Rev Immunol* **26:** 293–316.

Roediger B, Kyle R, Yip KH, Sumaria N, Guy TV, Kim BS, Mitchell AJ, Tay SS, Jain R, Forbes-Blom E, et al. 2013. Cutaneous immunosurveillance and regulation of inflammation by group 2 innate lymphoid cells. *Nat Immunol* **14:** 564–573.

Sadik CD, Kim ND, Iwakura Y, Luster AD. 2012. Neutrophils orchestrate their own recruitment in murine arthritis through C5aR and FcγR signaling. *Proc Natl Acad Sci* **109:** E3177–E3185.

Sallusto F, Schaerli P, Loetscher P, Schaniel C, Lenig D, Mackay CR, Qin S, Lanzavecchia A. 1998. Rapid and coordinated switch in chemokine receptor expression during dendritic cell maturation. *Eur J Immunol* **28:** 2760–2769.

Sawa S, Cherrier M, Lochner M, Satoh-Takayama N, Fehling HJ, Langa F, Di Santo JP, Eberl G. 2010. Lineage relationship analysis of $ROR\gamma t^+$ innate lymphoid cells. *Science* **330:** 665–669.

Schaerli P, Willimann K, Ebert LM, Walz A, Moser B. 2005. Cutaneous CXCL14 targets blood precursors to epidermal niches for Langerhans cell differentiation. *Immunity* **23:** 331–342.

Semmling V, Lukacs-Kornek V, Thaiss CA, Quast T, Hochheiser K, Panzer U, Rossjohn J, Perlmutter P, Cao J, Godfrey DI, et al. 2010. Alternative cross-priming through CCL17-CCR4-mediated attraction of CTLs toward NKT cell-licensed DCs. *Nat Immunol* **11:** 313–320.

Serbina NV, Pamer EG. 2006. Monocyte emigration from bone marrow during bacterial infection requires signals mediated by chemokine receptor CCR2. *Nat Immunol* **7:** 311–317.

Shi C, Pamer EG. 2011. Monocyte recruitment during infection and inflammation. *Nat Rev Immunol* **11:** 762–774.

Shi C, Jia T, Mendez-Ferrer S, Hohl TM, Serbina NV, Lipuma L, Leiner I, Li MO, Frenette PS, Pamer EG. 2011. Bone marrow mesenchymal stem and progenitor cells induce monocyte emigration in response to circulating toll-like receptor ligands. *Immunity* **34:** 590–601.

Sojka DK, Plougastel-Douglas B, Yang L, Pak-Wittel MA, Artyomov MN, Ivanova Y, Zhong C, Chase JM, Rothman PB, Yu J, et al. 2014. Tissue-resident natural killer (NK) cells are cell lineages distinct from thymic and conventional splenic NK cells. *eLife* **3:** e01659.

Stark K, Eckart A, Haidari S, Tirniceriu A, Lorenz M, von Bruhl ML, Gartner F, Khandoga AG, Legate KR, Pless R, et al. 2013. Capillary and arteriolar pericytes attract innate leukocytes exiting through venules and "instruct" them with pattern-recognition and motility programs. *Nat Immunol* **14:** 41–51.

Struyf S, Proost P, Schols D, De Clercq E, Opdenakker G, Lenaerts JP, Detheux M, Parmentier M, De Meester I, Scharpe S, et al. 1999. CD26/dipeptidyl-peptidase IV down-regulates the eosinophil chemotactic potency, but not the anti-HIV activity of human eotaxin by affecting its interaction with CC chemokine receptor 3. *J Immunol* **162:** 4903–4909.

Suratt BT, Petty JM, Young SK, Malcolm KC, Lieber JG, Nick JA, Gonzalo JA, Henson PM, Worthen GS. 2004. Role of the CXCR4/SDF-1 chemokine axis in circulating neutrophil homeostasis. *Blood* **104:** 565–571.

Tal O, Lim HY, Gurevich I, Milo I, Shipony Z, Ng LG, Angeli V, Shakhar G. 2011. DC mobilization from the skin requires docking to immobilized CCL21 on lymphatic endothelium and intralymphatic crawling. *J Exp Med* **208:** 2141–2153.

Thomas SY, Scanlon ST, Griewank KG, Constantinides MG, Savage AK, Barr KA, Meng F, Luster AD, Bendelac A. 2011. PLZF induces an intravascular surveillance program mediated by long-lived LFA-1-ICAM-1 interactions. *J Exp Med* **208:** 1179–1188.

Uguccioni M, Mackay CR, Ochensberger B, Loetscher P, Rhis S, LaRosa GJ, Rao P, Ponath PD, Baggiolini M, Dahinden CA. 1997. High expression of the chemokine receptor CCR3 in human blood basophils. Role in activation by eotaxin, MCP-4, and other chemokines. *J Clin Invest* **100:** 1137–1143.

Varona R, Villares R, Carramolino L, Goya I, Zaballos A, Gutierrez J, Torres M, Martinez AC, Marquez G. 2001. CCR6-deficient mice have impaired leukocyte homeostasis and altered contact hypersensitivity and delayed-type hypersensitivity responses. *J Clin Invest* **107:** R37–R45.

Wald O, Weiss ID, Wald H, Shoham H, Bar-Shavit Y, Beider K, Galun E, Weiss L, Flaishon L, Shachar I, et al. 2006. IFN-γ acts on T cells to induce NK cell mobilization and accumulation in target organs. *J Immunol* **176:** 4716–4729.

Walker JA, Barlow JL, McKenzie AN. 2013. Innate lymphoid cells—How did we miss them? *Nat Rev Immunol* **13:** 75–87.

Wang Y, Cui L, Gonsiorek W, Min SH, Anilkumar G, Rosenblum S, Kozlowski J, Lundell D, Fine JS, Grant EP. 2009. CCR2 and CXCR4 regulate peripheral blood monocyte pharmacodynamics and link to efficacy in experimental autoimmune encephalomyelitis. *J Inflamm (Lond)* **6:** 32.

Weber C, Weber KS, Klier C, Gu S, Wank R, Horuk R, Nelson PJ. 2001. Specialized roles of the chemokine receptors CCR1 and CCR5 in the recruitment of monocytes and T_H1-like/CD45RO$^+$ T cells. *Blood* **97:** 1144–1146.

Weber M, Hauschild R, Schwarz J, Moussion C, de Vries I, Legler DF, Luther SA, Bollenbach T, Sixt M. 2013. Interstitial dendritic cell guidance by haptotactic chemokine gradients. *Science* **339:** 328–332.

Wehr A, Baeck C, Heymann F, Niemietz PM, Hammerich L, Martin C, Zimmermann HW, Pack O, Gassler N, Hittatiya K, et al. 2013. Chemokine receptor CXCR6-dependent hepatic NK T cell accumulation promotes inflammation and liver fibrosis. *J Immunol* **190:** 5226–5236.

Wong CK, Hu S, Cheung PF, Lam CW. 2010. Thymic stromal lymphopoietin induces chemotactic and prosurvival effects in eosinophils: Implications in allergic inflammation. *Am J Respir Cell Mol Biol* **43:** 305–315.

Worbs T, Bode U, Yan S, Hoffmann MW, Hintzen G, Bernhardt G, Forster R, Pabst O. 2006. Oral tolerance originates in the intestinal immune system and relies on antigen carriage by dendritic cells. *J Exp Med* **203:** 519–527.

Yan S, Gessner R, Dietel C, Schmiedek U, Fan H. 2012. Enhanced ovalbumin-induced airway inflammation in CD26$^{-/-}$ mice. *Eur J Immunol* **42:** 533–540.

Yuan Q, Campanella GS, Colvin RA, Hamilos DL, Jones KJ, Mathew A, Means TK, Luster AD. 2006. Membrane-bound eotaxin-3 mediates eosinophil transepithelial migration in IL-4-stimulated epithelial cells. *Eur J Immunol* **36:** 2700–2714.

Zhang Q, Raoof M, Chen Y, Sumi Y, Sursal T, Junger W, Brohi K, Itagaki K, Hauser CJ. 2010. Circulating mitochondrial DAMPs cause inflammatory responses to injury. *Nature* **464:** 104–107.

Zlotnik A, Yoshie O, Nomiyama H. 2006. The chemokine and chemokine receptor superfamilies and their molecular evolution. *Genome Biol* **7:** 243.

Lipid Mediators in the Resolution of Inflammation

Charles N. Serhan[1], Nan Chiang[1], Jesmond Dalli[1], and Bruce D. Levy[2]

[1]Center for Experimental Therapeutics and Reperfusion Injury, Department of Anesthesiology, Perioperative and Pain Medicine, Harvard Institutes of Medicine, Brigham and Women's Hospital, and Harvard Medical School, Boston, Massachusetts 02115

[2]Pulmonary and Critical Care Medicine, Department of Internal Medicine, Harvard Institutes of Medicine, Brigham and Women's Hospital and Harvard Medical School, Boston, Massachusetts 02115

Correspondence: cnserhan@zeus.bwh.harvard.edu

Mounting of the acute inflammatory response is crucial for host defense and pivotal to the development of chronic inflammation, fibrosis, or abscess formation versus the protective response and the need of the host tissues to return to homeostasis. Within self-limited acute inflammatory exudates, novel families of lipid mediators are identified, named resolvins (Rv), protectins, and maresins, which actively stimulate cardinal signs of resolution, namely, cessation of leukocytic infiltration, counterregulation of proinflammatory mediators, and the uptake of apoptotic neutrophils and cellular debris. The biosynthesis of these resolution-phase mediators in sensu stricto is initiated during lipid-mediator class switching, in which the classic initiators of acute inflammation, prostaglandins and leukotrienes (LTs), switch to produce specialized proresolving mediators (SPMs). In this work, we review recent evidence on the structure and functional roles of these novel lipid mediators of resolution. Together, these show that leukocyte trafficking and temporal spatial signals govern the resolution of self-limited inflammation and stimulate homeostasis.

Resolution of an acute inflammatory response is the ideal outcome of this protective host response with return of the tissue to homeostasis (Majno and Joris 2004; Serhan et al. 2010). Lipid mediators are widely appreciated for their important roles in initiating the leukocyte traffic required in host defense (Cotran et al. 1999). These include the classic eicosanoids, prostaglandins (PGs) and leukotrienes (LTs) (Samuelsson et al. 1987; Samuelsson 2012), that stimulate blood flow changes, edema, and neutrophil influx to tissues (Flower 2006). Novel resolution-phase mediators that possess potent proresolving actions were identified and named resolvins, protectins, and maresins. Further studies established that these three families as well as lipoxins function together with their aspirin-triggered (AT) forms (collectively termed specialized proresolving mediators [SPMs]) and are biosynthesized during active resolution (Serhan 2004; Serhan and Chiang 2013). The complete stereochemistry of each of the main SPMs is established and their potent actions confirmed via total organic synthesis (Serhan and Petasis 2011). Given increased availability of certain SPMs, a body of

Cite this article as *Cold Spring Harb Perspect Biol* doi: 10.1101/cshperspect.a016311

literature emerged that expands their potent proresolving and anti-inflammatory actions and functions originally identified for the SPMs. In this work, we review and update the roles and actions of the SPMs, focusing on recent results with resolvins, protectins, and maresins, in active resolution mechanisms.

Professor Rod Flower of the William Harvey Research Institute, University of London once recited the quotation from Juvenal, a Roman poet, to introduce these new concepts and findings: *Quis custodiet ipsos custodes?* Who will guard the guards themselves? Hence, this quote is apropos to begin this article focusing on novel chemical mediators of resolution. The guards, the innate immune system phagocytes, certainly require direction (Serhan 2004; Perretti and D'Acquisto 2009) in the form of chemoattractants and chemical signals to appropriately control their function(s) and permit clearance of microbes and cellular debris without tissue injury; the cardinal signs of resolution.

THE ORCHESTRA AND PLAYERS OF THE RESOLUTION PHASE

In the acute inflammatory response, some chemical signals are from exogenous microbial origins, whereas others are biosynthesized by the host in response to tissue injury and invasion (Cotran et al. 1999; Lawrence et al. 2002; Serhan et al. 2010). Among the chemical signals at the site of an acute inflammatory response (Buckley et al. 2013), those that originate from host essential fatty acids are of particular interest because of their nutritional regulation and the potential to design synthetic mimetics of these naturally optimized molecules (Serhan 2004). Prostaglandins and leukotriene B_4 are involved in the initiating steps that permit leukocytes and specifically neutrophils to leave, via diapedesis, postcapillary venules (Malawista et al. 2008; Serhan et al. 2010). We focused on mechanisms involved in endogenous anti-inflammation and its resolution (Serhan et al. 2000, 2002; Levy et al. 2001; Serhan 2004). Using a systems approach with LC-MS-MS (liquid chromatography-tandem mass spectrometry)–based lipidomics, in vivo animal models, self-limited resolving inflammatory exudates, and functional assessment with isolated human leukocytes, we identified novel bioactive mediators produced in the resolution phase of acute sterile inflammation (Fig. 1) that activate new proresolving mechanisms (Serhan et al. 2000, 2002; Hong et al. 2003).

Focusing on self-limited resolving exudates also permitted a direct assessment of the host's responses that enables the return to homeostasis. For example, a key bioassay that proved critical in our initial studies focused on stopping human polymorphonuclear neutrophil (PMN) transmigration across vascular endothelial cells and mucosal epithelial cells (Serhan et al. 2000). We focused on neutrophils because they are among the first responders to injury, alarms, and microbial invasion. PMNs, given their high numbers, can amplify inflammation within tissues when inadvertently activated, causing collateral damage. Our hypothesis that endogenous chemical mediators are produced via cell–cell interactions within inflammatory exudates (i.e., pus) that control the size, magnitude, and duration of the inflammatory event proved to be the case and is relevant to human translation (Tabas and Glass 2013). Anti-PMN therapy (Takano et al. 1998) that limits tissue damage to control inflammation has increasing appeal. The milestones in resolution of inflammation from observation to active resolution, to new resolution therapeutics first in humans, are reviewed in Serhan (2011, 2014). In ancient medical texts of the 11th and 12th centuries, the notion of treating inflammation with resolvents to resolve disease is present (Avicenna, adapted by Laleh Bakhtiar, 1999). However, the concept was apparently lost until the structures and actions of endogenous resolution mediators were elucidated (Serhan et al. 2002). Within exudates resolving to homeostasis, the fundamental cellular processes impacted by SPMs, namely resolvins, protectins, and maresins, proved predictive of their actions in disease models in vivo, because cessation of PMN entry into tissue and the removal of dead PMNs are central to many disease pathologies in which uncontrolled inflammation is involved.

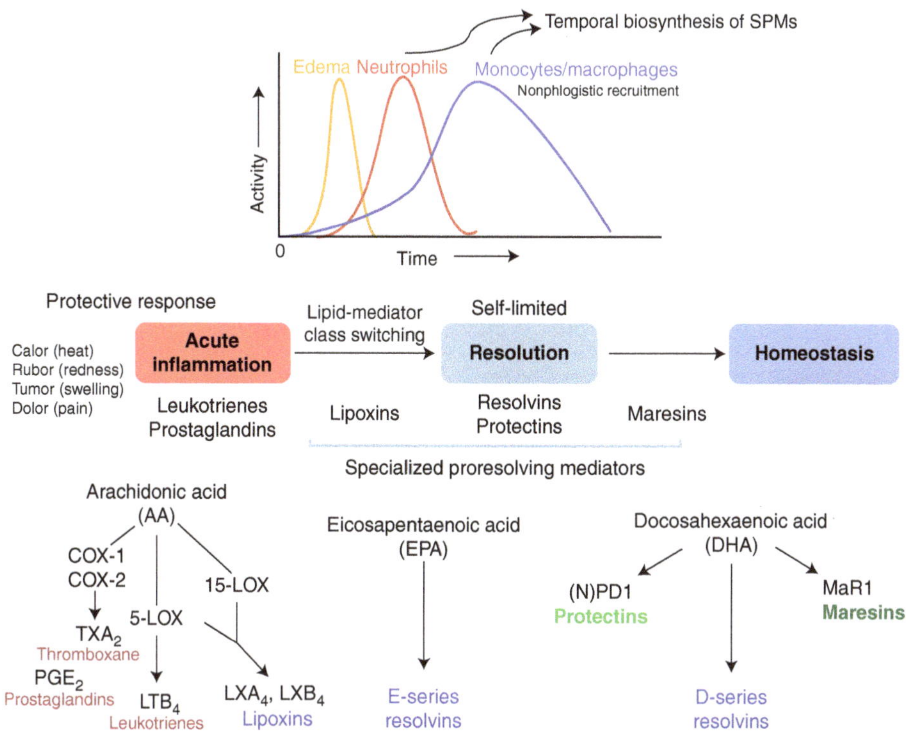

Figure 1. Lipid-mediator biosynthesis in exudate cell traffic in resolution of acute inflammation. Specialized proresolving mediators (SPMs) are generated during inflammation resolution and control the early events in acute inflammation such as edema formation, leukocyte trafficking, and functions (see text).

Metchnikoff observed more than 100 years ago that neutrophils are ingested by tissue macrophages ("big eaters") and that this clearance of neutrophils resolves tissue inflammation (Tauber and Chernyak 1991; Cotran et al. 1999). Subsequent decades of research identified the "go" signals (for example, complement components, cytokines, chemokines, and certain eicosanoids) that promote the recruitment of leukocytes from the blood to the inflamed tissue. Investigators believed that the removal of the inflammatory stimulus prevented the production of chemoattractants that promote leukocyte recruitment and that simple dilution of the chemoattractants prevented further leukocyte cell recruitment and that these passive events brought about the ending of inflammation.

Evidence that the resolution of inflammation is an active process came from our studies on acute self-limiting responses using a systems-based approach (Serhan et al. 2000, 2002; Levy et al. 2001; Hong et al. 2003). Results from these studies showed that, in resolving inflammatory exudates, cell–cell interactions lead to the biosynthesis of active signals that limit further neutrophil recruitment to the tissue (cessation of PMN influx) and enhance the engulfment of apoptotic neutrophils by macrophages, the two cardinal signs of resolution, promoting a return of the tissue to homeostasis. In active resolution, we uncovered a key process coined lipid-mediator class switching in exudates (see Fig. 1). That is, prostaglandins involved in the initiation phase of inflammation activate the translation of mRNAs encoding enzymes (Levy et al. 2001) that are needed for production of proresolvent mediators (lipoxins, resolvins, and protectins) during the resolution phase (Fig. 1). Low-dose aspirin jump-starts the resolution phase by triggering endogenous epimers of these SPMs (Serhan 2007), which is shown in animal disease (Chan and Moore 2010; Brancaleone et al. 2013) and in human skin blisters (Morris et al. 2009) and infantile eczema (Wu et al. 2013).

THE ENTRY OF OTHER SUBSTRATES IN RESOLUTION

We also learned that n-3 essential fatty acids are substrates within these inflammatory resolving exudates for the biosynthesis of potent anti-inflammatory and proresolving mediators (Serhan et al. 2002; Hong et al. 2003). Identification of eicosapentaenoic acid (EPA) and docosahexaenoic acid (DHA) as precursors of mediators that activate proresolving mechanisms opened new avenues to consider for appreciating mechanisms underlying uncontrolled inflammation. It is worth noting that a large body of literature addressing the anti-inflammatory impact of EPA and DHA is present (for a recent review, see De Caterina 2011; Calder 2013); yet, the molecular mechanism(s) by which these essential nutrients exert their anti-inflammatory actions remained the subject of debate. DHA and EPA have many known critical functions in mammalian biology. Neither EPA nor DHA is produced by humans to any great extent, requiring their dietary intake (Calder 2013). DHA is an ancient molecule that has functional roles in brain and eye optimized by evolutionary pressure for its physical properties in membranes (Crawford et al. 2013). Uncovering novel chemical mediators that are biosynthesized within self-limited inflammatory responses with exudates in murine systems with functions on individual mammalian and human leukocytes has far-reaching implications (Fig. 1). Along these lines, n-3-derived SPMs are documented in humans in health and disease, including plasma, milk, adipose tissue, and synovial fluid (see Table 1 for details). For recent detailed reviews on resolvins, protectins, and maresins, interested readers are directed to the following: SPM biosynthesis (Bannenberg and Serhan 2010), actions (Recchiuti and Serhan 2012; Serhan and Chiang 2013), and total organic synthesis (Serhan and Petasis 2011). At this point, we address the initial observations of the SPM biosynthesis and activities on phagocytes and in animal models of disease now confirmed in many laboratories (Tables 2 and 3).

Endogenous Anti-Inflammation and Proresolution Are *Not* Equivalent Processes

Each SPM possesses potent proresolving actions that are fundamental to resolution (Serhan 2004) including limiting or cessation of neutrophil tissue infiltration, counterregulation of chemokines and cytokines (Serhan et al. 2002; Hong et al. 2003), reduction in pain (Xu et al. 2010), and stimulation of macrophage-mediated actions, namely, phagocytosis of apoptotic PMNs and bacterial and debris clearance (Table 2) (Schwab et al. 2007; Chiang et al. 2012). The SPMs are multitarget agonists in that they each act on both PMNs and macrophages separately to stimulate resolution. Given this unique proresolving mechanism, resolvins and other SPMs each display potent actions in many animal disease models (Table 3). The actions of resolvins and all SPMs are stereochemically selective, reflecting their routes of biosynthesis and underlying their ability to activate receptors (G-protein-coupled receptor [GPCR]) that amplify and transduce their tissue response. Thus, establishing the complete stereochemical assignments for each resolvin, protectin, and maresin (SPM) family was key to confirming their novel leukocyte functions. Given their ability to stimulate resolution of inflammation without systemic immune suppression of host defense (Spite et al. 2009; Oh et al. 2011), we recognize that SPMs stimulate resolution of inflammation and bacterial infection (Serhan 2011; Chiang et al. 2012).

BIOSYNTHESIS, FUNCTION, AND STRUCTURAL ELUCIDATION

It was essential to confirm the proposed structures and novel potent actions for each of the resolvins and other SPMs (Figs. 2 and 3). To this end, a systematic approach was devised to match endogenous SPMs to those prepared by total organic synthesis (Serhan and Petasis

Table 1. Humans and SPMs

SPM	Disease/tissues	Formation
Lipoxins and aspirin-triggered lipoxins (ATLs)	Asthma	Higher urinary ATL levels in aspirin-tolerant asthma than in aspirin-intolerant asthma (Sanak et al. 2000; Levy et al. 2005; Yamaguchi et al. 2011) and regulate natural killer (NK) cell and innate lymphoid cell activation (Barnig et al. 2013; Peebles 2013)
	Alzheimer's disease (AD)	LXA_4 levels are reduced in AD brain and CSF (Wang et al. 2014)
	Colitis	Elevated mucosal LXA_4 promotes remission in individuals with ulcerative colitis (Vong et al. 2012)
	Type 2 diabetes	Increased plasma ATL with intake of pioglitazone (Gutierrez et al. 2012)
	Rheumatoid arthritis	LXA_4 in synovial fluid from rheumatoid arthritis patients (Giera et al. 2012)
	Localized aggressive periodontitis (LAP)	Less LXA_4 in LAP whole blood compared with healthy individuals (Fredman et al. 2011)
	Peripheral artery disease	Plasma levels of ATL are lower in patients with symptomatic peripheral artery disease (Ho et al. 2010)
	Adipose tissues	LXA_4 identified in human adipocytes from obese patients (Clària et al. 2012)
	Milk	Lipoxins and resolvins at very high levels in the first month of lactation (Weiss et al. 2013)
Resolvins	Synovial fluid	RvD5 present in synovial fluid from rheumatoid arthritis patients (Giera et al. 2012)
	Blood (healthy volunteers)	Plasma RvD1 and RvD2 identified with oral omega-3 supplementation (Mas et al. 2012; Colas et al. 2014)
	Adipose tissues	RvD1 and RvD2 identified in human adipocytes from obese patients (Clària et al. 2012)
	Human plasma and milk	RvE1 identified in human plasma (Psychogios et al. 2011) and milk (Weiss et al. 2013)
	Multiple sclerosis	RvD1 was detected and up-regulated in serum and cerebrospinal fluid in the highly active group (Pruss et al. 2013)
	Human IgA nephropathy	RvE1 identified in patients supplemented with fish oil n-3 (Zivkovic et al. 2012)
Protectin	Asthma	PD1 in exhaled breath condensates in asthma exacerbation (Levy et al. 2007); decreased PD1 in eosinophils from patients with severe asthma compared with healthy individuals (Miyata et al. 2013)
	Embryonic stem cells	PD1 produced in embryonic stem cells (Yanes et al. 2010)
	Multiple sclerosis	NPD1 was detected in serum and cerebrospinal fluid in the highly active group (Pruss et al. 2013)
Maresins	Synovial fluid	MaR1 identified in synovial fluid from arthritis patients (Giera et al. 2012)

2011). This approach was necessary because SPMs are isolated in pure form in only small quantities from exudates (picogram to nanogram range), are locally active, and are inactivated via further metabolism (Arita et al. 2006; Clària et al. 2012). These transient and small quantities preclude direct NMR analysis. The original identification of the D-series resolvins reported the structural elucidation of several distinct bioactive structures that stopped PMN influx and migration, denoted resolvin D1 through resolvin D6, from resolving self-limited murine exudates. Their biosynthetic pathway(s) were established with isolated human leukocytes (Figs. 2 and 3), and potent in vivo actions were determined in murine (Table 3) as well as human

Table 2. Host defense: Enhanced phagocytosis and the roles of SPMs

	Phagocytosis in vivo	PMN	Macrophages
LXA_4 ATL	STZ (Schwab et al. 2007) Apop PMN (El Kebir et al. 2009) *Escherichia coli* (El Kebir et al. 2009) Multimicrobial sepsis/CLP (Walker et al. 2011)		Apop PMN (Godson et al. 2000; Schwab et al. 2007) Serum-treated zymosan (Schwab et al. 2007) Latex beads (Schwab et al. 2007) *E. coli* (Prescott and McKay 2011)
RvE1	STZ (Schwab et al. 2007) HSV-1 (Rajasagi et al. 2011) *E. coli* (Seki et al. 2010; El Kebir et al. 2012)	*Candida albicans* (Haas-Stapleton et al. 2007)	Apop PMN (Schwab et al. 2007; Oh et al. 2011) Serum-treated zymosan (Schwab et al. 2007; Ohira et al. 2010; Oh et al. 2011) Latex beads (Schwab et al. 2007) *E. coli* (Oh et al. 2011)
18S-RvE1			Apop PMN (Oh et al. 2011) Serum-treated zymosan (Oh et al. 2011) *E. coli* (Oh et al. 2011)
RvE2			Serum-treated zymosan (Oh et al. 2011)
PD1	STZ (Schwab et al. 2007) Apop PMN (El Kebir et al. 2009) *E. coli* (Chiang et al. 2012)	*E. coli* (Chiang et al. 2012)	Serum-treated zymosan (Schwab et al. 2007) Apop PMN (Schwab et al. 2007) Latex beads (Schwab et al. 2007) *E. coli* (Chiang et al. 2012)
RvD1	*E. coli* (Chiang et al. 2012) Apop PMN (Hsiao et al. 2013)	*E. coli* (Chiang et al. 2012)	Serum-treated zymosan (Krishnamoorthy et al. 2010) Apop PMN (Krishnamoorthy et al. 2010) *E. coli* (Chiang et al. 2012)
AT-RvD1			*E. coli* (Palmer et al. 2011) IgG-OVA-coated beads (Rogerio et al. 2012)
RvD2	Multimicrobial sepsis/CLP (Spite et al. 2009)	*E. coli* (Spite et al. 2009)	Serum-treated zymosan (Spite et al. 2009)
RvD3			Serum-treated zymosan (Dalli et al. 2013a)
AT-RvD3			Apop PMN
RvD5	*E. coli* (Chiang et al. 2012)	*E. coli* (Chiang et al. 2012)	*E. coli* (Chiang et al. 2012)
MaR1			Serum-treated zymosan (Serhan et al. 2009)
MaR2			Apop PMN (Serhan et al. 2012; Deng et al. 2014)

ATL, aspirin-triggered lipoxins; STZ, serum-treated zymosan; CLP, common lymphoid progenitors.

inflammation (Serhan et al. 2002; Morris et al. 2009; Wu et al. 2009, 2013).

Recently, the stereochemical structures of resolvin D1 (RvD1; 7*S*,8*R*,17*S*-trihydroxy-4*Z*,9*E*,11*E*,13*Z*,15*E*,19*Z*-docosahexaenoic acid), its AT 17*R*-epimer (Sun et al. 2007), RvD2 (resolvin D2, 7*S*,16*R*,17*S*-trihydroxy-4*Z*, 8*E*,10*Z*,12*E*,14*E*,19*Z*-docosahexaenoic acid) (Spite et al. 2009), AT-protectin D1 (PD1; protectin D1/neuroprotectin D1, 10*R*,17*S*-dihydroxy-4*Z*,7*Z*,11*E*,13*E*,15*Z*,19*Z*-docosahexaenoic acid) (Serhan et al. 2011), and maresin 1 (MaR1; maresin 1, 7*R*,14*S*-dihydroxy-docosa-4*Z*,8*E*, 10*E*,12*Z*,16*Z*,19*Z*-hexaenoic acid) (Serhan et al. 2012) were each assigned, as well as their biosynthetic-related isomers, and several made commercially available.

Recently, we also establish the complete stereochemistry of RvD3 (resolvin D3, 4*S*,11*R*,17*S*-trihydroxydocosa-5*Z*,7*E*,9*E*,13*Z*,15*E*,19*Z*-hexaenoic acid) (Fig. 3) and its AT-RvD3 (4*S*,11*R*,17*R*-trihydroxydocosa-5*Z*,7*E*,9*E*,13*Z*,15*E*,19*Z*-hexaenoic acid) *natural* epimer (Dalli et al. 2013a). Using LC-MS-MS metabololipidomics,

Table 3. Update on SPM actions in disease models

Disease	SPM	Bioaction
Alzheimer's disease (AD)	RvD1	Stimulates phagocytosis of Aβ by AD macrophages (Mizwicki et al. 2013)
Burn wound	RvD2	Prevents secondary thrombosis and necrosis (Bohr et al. 2013)
Chronic pancreatitis	RvD1	Reverses allodynia (Feng et al. 2012)
Diabetic wounds	RvD1	Accelerates wound healing (Tang et al. 2013)
Dermatitis	RvE1	Ameliorates dermatitis (Kim et al. 2012)
Pulmonary inflammation	RvE1	Promotes apoptosis and accelerates airway resolution (Seki et al. 2010)
Peripheral nerve injury	RvE1	Inhibits neuropathic pain (Xu et al. 2013)
Obesity	RvD1, RvD2	Govern inflammatory tone (Clària et al. 2012)
Allergic airway response	RvD1, AT-RvD1, RvE1, PD1	Promote resolution (Levy et al. 2007; Haworth et al. 2011; Rogerio et al. 2012)
Amyotrophic lateral sclerosis	RvD1	Inhibits inflammation (Liu et al. 2012)
Acute lung injury	AT-RvD1	Reduces mucosal inflammation (Eickmeier et al. 2013)
Fibrosis	RvE1, RvD1	Inhibit kidney fibrosis (Qu et al. 2012)
Bacterial infection	RvD1, RvD5, PD1	Increase survival and lower antibiotic requirement (Chiang et al. 2012)
Peritonitis	RvD1	Limits PMN recruitment and accelerate resolution (Recchiuti et al. 2011; Norling et al. 2012)
Dry eye	RvE1 and analog	Protect from goblet cell loss (de Paiva et al. 2012); improves tear production (Li et al. 2010)
Tissue regeneration	RvE1, MaR1	Promote tissue regeneration in planaria (Serhan et al. 2012)
Pain	MaR1, RvD1, AT-RvD1, RvD2, RvE1	Control inflammatory pain (Bang et al. 2010, 2012; Xu et al. 2010; Park et al. 2011; Serhan et al. 2012)
Adipose tissue inflammation	RvD1	Elicits macrophage polarization and promote resolution (Titos et al. 2011)
Localized aggressive periodontitis	RvE1	Rescues impaired phagocytosis (Fredman et al. 2011)
Colitis	RvD1, RvD2, RvE1	Prevent colitis (Ishida et al. 2010; Bento et al. 2011)
Temporomandibular joint inflammation	AT-RvD1	Limits PMN infiltration to CFA-inflamed TMJ (Norling et al. 2012)
Arthritis	AT-RvD1	Antihyperalgesic (Lima-Garcia et al. 2011)
Postoperative pain	RvD1	Prevents and reduces pain (Huang et al. 2011)
Postsurgical cognitive decline	AT-RvD1	Improves postoperative decline and attenuates memory neuronal dysfunction (Terrando et al. 2013)
Endotoxin shock	RvD1	Suppresses septic mediators (Murakami et al. 2011)
HSV-keratitis	RvE1	Controls ocular inflammatory lesions (Rajasagi et al. 2011)
Allograft rejection	RvE1	Preserves organ function (Levy et al. 2011)
Heart ischemia	RvE1	Protects heart against reperfusion injury (Keyes et al. 2010)
Bacterial pneumonia	RvE1	Protects mice from pneumonia (Seki et al. 2010)
Cigarette smoke-induced lung inflammation	RvD1	Promotes M2 macrophages and efferocytosis as well as accelerates resolution of lung inflammation (Hsiao et al. 2013)
Vascular inflammation (arterial angioplasty)	RvD1	Attenuates cell proliferation, leukocyte recruitment, and neointimal hyperplasia (Miyahara et al. 2013)
Fibromyalgia	AT-RvD1, RvD2	Reduces mechanical allodynia and thermal sensitization and prevent depressive behavior (Klein et al. 2014)
Vagotomy	RvD1	Rescues hyperinflammation (Mirakaj et al. 2014)

TMJ, temporomandibular joint.

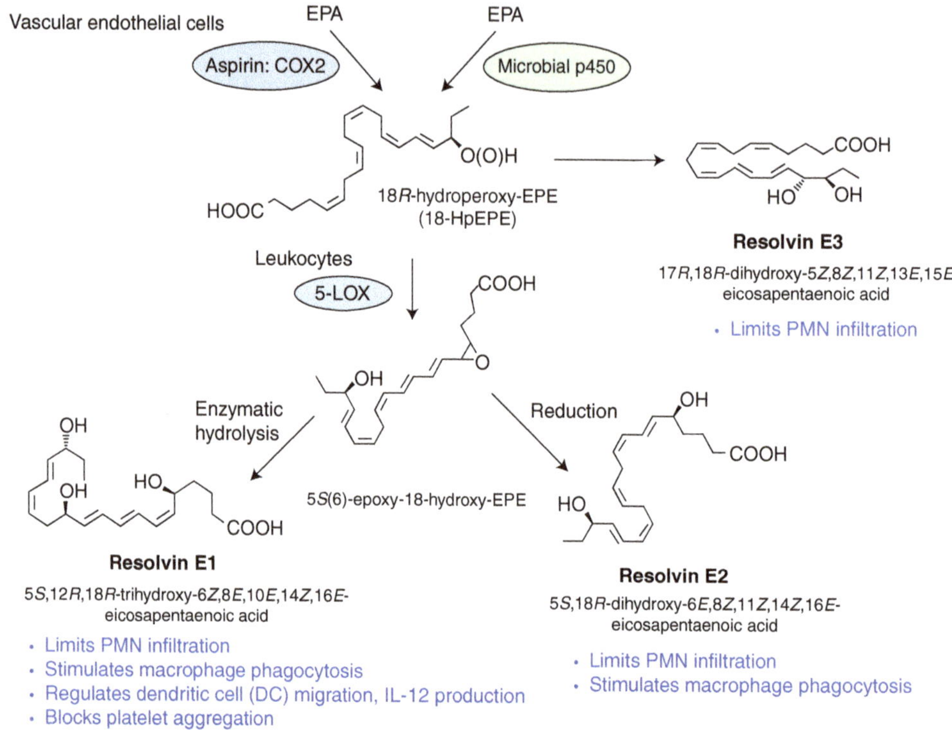

Figure 2. E-series resolvin biosynthesis and major function (see text for details).

we matched the physical properties of RvD3 with those of synthetic materials possessing the stereochemistry that proved to be 4*S*,11*R*,17*S*-trihydroxydocosa-5*Z*,7*E*,9*E*,13*Z*,15*E*,19*Z*-hexaenoic acid (Fig. 3) and the AT-RvD3, or aspirin-triggered form, matched synthetic 4*S*, 11*R*,17*R*-trihydroxydocosa-5*Z*,7*E*,9*E*,13*Z*, 15*E*, 19*Z*-hexaenoic acid (Dalli et al. 2013a). When administered in vivo, both of these synthetic epimers, at doses as low as 10 pg/mouse, gave potent reduction (40%–50%) of murine PMN recruitment to sites of inflammation. Both RvD3 and AT-RvD3 increased exudate IL-10 and reduced IL-6 and eicosanoids (Dalli et al. 2013a).

With human leukocytes, RvD3 and AT-RvD3 each potently regulate leukocyte functions enhancing peritoneal macrophage phagocytosis and efferocytosis in a dose-dependent manner while reducing human neutrophil transendothelial migration in response to tumor necrosis factor (TNF)-α (Dalli et al. 2013a). These results establish the complete stereochemistry and confirmed the potent anti-inflammatory and proresolving actions of RvD3 and its aspirin-triggered epimer denoted AT-RvD3 (see Fig. 3). Moreover, lipid-mediator metabololipidomic profiling of self-resolving exudates also placed RvD3 uniquely within the time course of inflammation resolution to vantage complete resolution, namely, in the later stages (Dalli et al. 2013a).

MARESINS: MACROPHAGE MEDIATORS IN RESOLVING INFLAMMATION

Recently, we also established the stereochemical assignments for both AT-PD1 (Serhan et al. 2011) and maresin 1 (MaR1) (Serhan et al. 2012). MaR1 produced by human macrophages (MΦ) from endogenous docosahexaenoic acid (DHA) matched the stereochemistry of synthetic 7*R*,14*S*-dihydroxydocosa-4*Z*,8*E*,10*E*,12*Z*, 16*Z*,19*Z*-hexaenoic acid. MaR1 alcohols groups

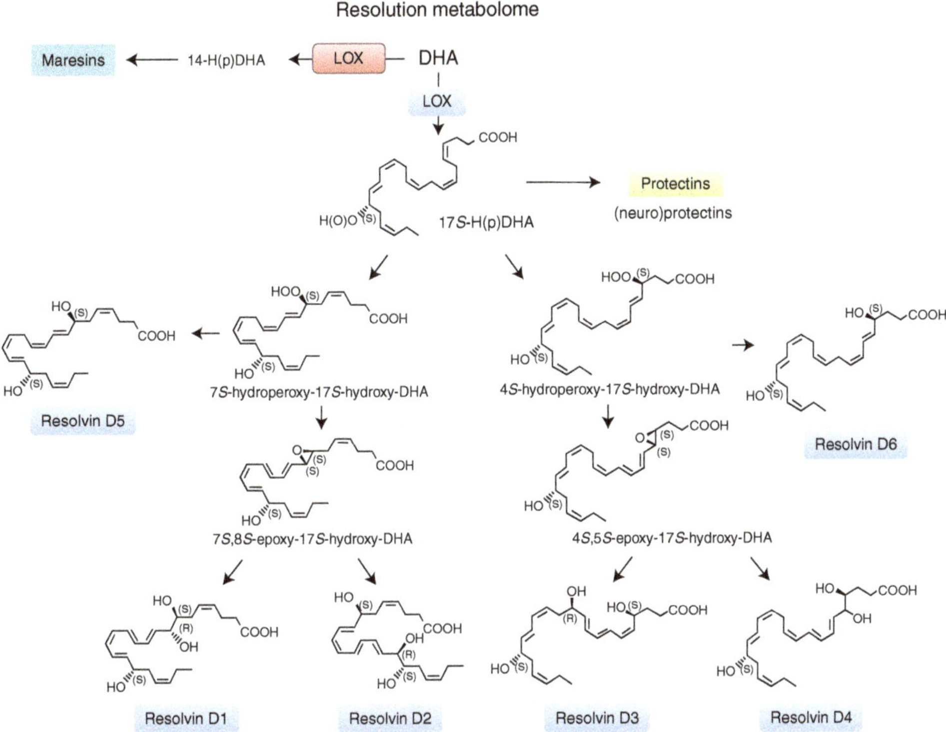

Figure 3. D-series resolvin biosynthesis. The complete stereochemistry of RvD1, RvD2, and RvD3 is established (see Dalli et al. 2013a and text for further details).

and *Z*/*E* geometry of conjugated double bonds were assigned using isomers prepared by total organic synthesis. MaR1's potent defining actions were confirmed with synthetic MaR1 (i.e., limiting neutrophil [PMN]) infiltration in murine peritonitis (ng/mouse range) as well as enhancing human macrophage uptake of apoptotic PMNs.

MaR1 is slightly more potent at 1 nM than Resolvin D1 (RvD1) in stimulating human MΦ efferocytosis, an action not shared by leukotriene B_4. Importantly, MaR1 also accelerates surgical regeneration in planaria, increasing the rate of head reappearance. On injury of the planaria (when cut in half), MaR1 is biosynthesized from deuterium-labeled (d_5)-DHA. MaR1 dose-dependently inhibited TRPV1 currents in neurons, blocked capsaicin-induced inward currents $IC_{50} \approx 0.5$ nM, and reduced both inflammatory and chemotherapy-induced neuropathic pain in mice (Serhan et al. 2012). Hence, MaR1 has potent actions in regulating inflammation resolution, tissue regeneration, and resolving pain. These findings also suggest that chemical signals are shared in resolution cellular trafficking that is key in tissue regeneration across phyla from worms to humans.

Of special interest, the total organic synthesis of MaR1 was also achieved by Rodriguez and Spur using Sonogashira coupling (2012c), who also reported resolvin D6 (2012a) and organic synthesis of resolvin E2 (2012b) (Figs. 2 and 3). Kobayashi et al. also reported stereoselective total organic synthesis of protectin D1 (Ogawa and Kobayashi 2011), resolvin E2 (Ogawa et al. 2009), and resolvin E1 (Ogawa and Kobayashi 2009). The total organic synthesis of the 18-HEPE, a precursor of E-series resolvins, was also reported (see Fig. 2) (Krishnamurthy et al. 2011). Importantly, the stereoselective actions of each SPM proved highly effective in regulating human PMNs and monocytes in microfluidic chambers (Kasuga et al. 2008; Jones et al. 2012), clearly establishing the direct actions on human cells and hence their potential in translational medicine.

MICROPARTICLES IN RESOLUTION AND LEUKOCYTE SUBPOPULATIONS

With complete stereochemistry of many of the main SPMs established (Figs. 2 and 3), it was possible to carry out lipid-mediator (LM) metabololipidomics profiling via targeted LC-MS-MS-based analyses with distinct phagocyte populations, namely, neutrophils (PMNs), apoptotic PMNs, and macrophages (Dalli and Serhan 2012). Efferocytosis increased SPM biosynthesis, including RvD1, RvD2, and RvE2 (resolving E2, 5*S*,18*R*-trihydroxy-6*E*,8*Z*,11*Z*,14*Z*, 16*E*-eicosapentaenoic acid) (Figs. 2 and 3), which are further elevated by PMN microparticles (Norling et al. 2011). Apoptotic PMNs produced prostaglandin E_2, lipoxin B_4, and RvE2, whereas zymosan-stimulated PMNs showed predominantly leukotriene B_4 and 20-hydroxy-leukotriene B_4, as well as lipoxin biosynthesis pathway marker 5,15-diHETE. Using deuterium-labeled precursors (d_8-arachidonic acid, d_5-eicosapentaenoic acid, and d_5-docosahexaenoic acid), apoptotic PMNs and microparticles each contribute to SPM biosynthesis during the process of efferocytosis. Also, M2 macrophage phenotype (Lawrence and Natoli 2011) produces SPMs including MaR1 and LXA_4 (lipoxin A_4, 5*S*,6*R*,15*S*-trihydroxy-7*E*,9*E*, 11*Z*,13*E*-eicosatetraenoic acid) with lower amounts of LTB_4 (leukotriene B_4, 5*S*,12*R*-dihydroxy-6*Z*,8*E*,10*E*,14*Z*-eicosatetraenoic acid) and prostaglandin (PG) than the macrophages of the M1 phenotype (Dalli and Serhan 2012). Of interest, the uptake of apoptotic PMNs by both macrophage subtypes led to modulation of their LM profiles and activation of transcellular SPM biosynthesis. These results establish LM signature profiles of human PMNs, apoptotic PMNs, and macrophage subpopulations (Dalli and Serhan 2012). Hence, microparticle regulation of specific endogenous LMs during defined stages of the acute inflammatory process and their dynamic changes in LM signatures are influenced by transcellular biosynthesis between apoptotic cells, microparticles, and macrophages.

RESOLUTION AND INFECTION: WHAT IS THEIR RELATIONSHIP?

How bacterial infections contribute to active resolution of inflammation is of wide interest. Hence, we focused on exudate leukocyte trafficking and mediator metabololipidomics with murine peritoneal *Escherichia coli* infections (Chiang et al. 2012) and documented the temporal identification of both proinflammatory (PG and LT) and SPMs. In self-resolving *E. coli* exudates (10^5 CFU), the dominant SPMs were RvD5 (resolvin D5, 7*S*,17*S*-dihydroxy-4*Z*,8*E*,10*Z*,13*Z*,15*E*,19*Z*-docosahexaenoic acid) and PD1, which at 12 h were greater than levels in exudates from higher titer *E. coli* (10^7 CFU) -infected mice. Of interest, germ-free mice produced endogenous RvD1 and PD1 levels higher than in conventional mice. RvD1 and RvD5 (ng/mouse) each reduced bacterial titers in blood and exudates, *E. coli*–induced hypothermia, and increased survival.

To translate these to humans, both human PMNs and macrophages were tested with RvD1, RvD5, and PD1, which each directly enhanced phagocytosis of *E. coli*, and both RvD1 and RvD5 counterregulate a panel of proinflammatory genes, including NF-κB and TNF-α. RvD5 activated the RvD1 receptor GPR32 to enhance phagocytosis. With self-limited *E. coli* infections, RvD1 and an antibiotic ciprofloxacin accelerated resolution, and each shortened resolution intervals (R_i). Host-directed RvD1 actions enhanced ciprofloxacin's therapeutic actions. In 10^7 CFU *E. coli* infections, SPMs (RvD1, RvD5, PD1) together with ciprofloxacin also heightened host antimicrobial responses, enhancing clearance of *E. coli* in blood and exudates. In skin infections, SPMs stimulated vancomycin clearance of *Staphylococcus aureus*. Hence, specific SPMs are temporally and differentially regulated during infections. They are antiphlogistic, enhance containment, and lower antibiotic requirements for bacterial clearance. These endogenous resolution mechanisms are of interest in host defense because initiation of the host response is controlled in part by PG and LT (von Moltke et al. 2012), which when uncontrolled can lead to increased mortality from infection (Chiang et al. 2012), as also observed in zebrafish infections (Tobin et al. 2012). This goes beyond bacteria to viral and fungal infections. Of special interest, PD1 (Fig. 3) produced by the host was identified as a novel antiviral that directly blocks viral replication and increases host survival to influenza viral infection (Baillie and Digard 2013; Morita et al. 2013), which also suggests treating the host as with bacterial infections rather than treating the microbes alone with antibiotics (Chiang et al. 2012). The EPA-derived RvE1 (resolvin E1, 5*S*,12*R*,18*R*-trihydroxy-6*Z*,8*E*,10*E*,14*Z*,16*E*-eicosapentaenoic acid) (Fig. 2) also controls fungal infections as observed with *Candida albicans* (Haas-Stapleton et al. 2007).

GPC RECEPTORS IN RESOLUTION

We identified two GPCRs for RvD1 on human phagocytes, namely, ALX and GPR32. ALX/FPR2 is the lipoxin A_4 receptor, and GPR32 was an orphan receptor. RvD1 displays specific binding and reduces actin polymerization and CD11b on PMNs, as well as stimulates macrophage phagocytosis, an action dependent on ALX and GPR32 (Krishnamoorthy et al. 2010). In addition to RvD1, its AT epimer 17R-RvD1 and stable analog 17-R/S-methyl-RvD1 each dose-dependently activates ALX/FPR2 and GPR32 in GPCR-overexpressing β-arrestin systems and electric cell-substrate impedance sensing (Krishnamoorthy et al. 2012). Of interest, we showed that RvD5 also activates human GPR32 in the GPR32-β-arrestin systems, and stimulates macrophage phagocytosis of *E. coli* in a GPR32-dependent manner (Chiang et al. 2012). In addition, RvD3 and AT-RvD3 each activates GPR32, contributing to their proresolving actions in stimulating macrophage uptake of microbial particles (Dalli et al. 2013a).

A specific receptor for RvE1, *ChemR23*, is closely related to LX and LT receptors in deduced amino acid sequences. ChemR23 displays specific RvE1 binding and RvE1-dependent signals to activate monocyte, and reduce dendritic cell migration and IL-12 production (Arita et al. 2005a). RvE1-ChemR23 interactions also stimulate macrophage phagocytosis via phosphorylation-signaling pathways including Ribosomal protein S6, a downstream target of the PI3K/Akt signaling pathway and the Raf/ERK pathway (Ohira et al. 2010). 18S-RvE1 also binds to ChemR23 with increased affinity and potency compared with the R-epimer, but was more rapidly inactivated than RvE1 (Oh et al. 2011). RvE2 is a partial agonist for ChemR23 (Oh et al. 2012). A leukotriene B_4 receptor, *BLT1*, also directly interacts with RvE1, which inhibits calcium mobilization, NF-κB activation in vitro, and PMN infiltration in vivo (Arita et al. 2007). 18S-RvE1 and RvE2 also bind to BLT1 (Oh et al. 2011, 2012). Hence, RvE1 gives cell-type-specific actions: It functions as an agonist for ChemR23 on mononuclear and dendritic cells as well as an antagonist for BLT1 signals on PMNs. Recently, ChemR23-dependent actions of RvE1 were confirmed in mouse renal fibrosis (Qu et al. 2012).

Genetically Engineered Mice

To prepare transgenic (TG) mice for human ALX/FPR2, hALX transgene was placed under control of CD11b promoter that directed receptor expression in myeloid cells (Levy et al. 2002; Devchand et al. 2003). In non-TG littermates, RvD1 as low as 10 ng given together with zymosan, reduced leukocyte numbers by ∼38% at 24 h. This action was further enhanced in ALX-TG mice giving 53% reduction of leukocytes. Also with RvD1 treatment, PMN numbers in TG mice was 50% lower than non-TG controls (Krishnamoorthy et al. 2012). We also prepared transgenic mice overexpressing human ChemR23, the RvE1 receptor, on myeloid cells. In these TG mice, RvE1 is 10-fold more potent in limiting PMN infiltration in zymosan-initiated peritonitis, compared with non-TG littermates. In addition, ligature-induced alveolar bone loss was diminished in ChemR23tg mice (Gao et al. 2013). Local RvE1 treatment of uniform craniotomy in the parietal bone significantly accelerated regeneration of the bone defect, indicating that RvE1 modulates osteoclast differentiation and bone remodeling by direct actions on bone.

In mice deficient in ALX/fpr2 (mouse ortholog of human ALX), the anti-inflammatory actions of RvD1 were dampened. Administration of RvD1 (1 ng/mouse) significantly reduces PMN infiltration in wild-type mice, but not in fpr2-deficient mice. Also in peritoneal exudates, RvD1 activates LX biosynthesis stimulating the production of the anti-inflammatory mediator LXB_4 (lipoxin B_4, 5*S*,14*R*,15*S*-trihydroxy-6*E*,8*Z*,10*E*,12*E*-eicosatetraenoic acid) and stimulated the biosynthesis of the cyclooxygenase-derived PGE_2 (prostaglandin E_2, 9-oxo-11*R*,15*S*-dihydroxy-5*Z*,13*E*-prostadienoic acid) while down-regulating production of the proinflammatory LTB_4. This regulation of lipid mediator by RvD1 is lost in the fpr2-deficient mice (Norling et al. 2012). These results indicate that RvD1 dampens acute inflammation in part via ALX receptor. Also, 15-epi-LXA_4 interacts with ALX/FPR in vivo controlled by aspirin (Brancaleone et al. 2013). In BLT1 knockout mice, in vivo anti-inflammatory actions of RvE1 were sharply reduced when given at low doses (100 ng i.v.) in peritonitis. In contrast, RvE1 at higher doses (1.0 µg i.v.) significantly reduced PMN infiltration in a BLT1-independent manner (Arita et al. 2007). Hence, RvE1 binds to BLT1 as a partial agonist, serving as a local damper of BLT1 signals on leukocytes along with other receptors (e.g., ChemR23 receptor-mediated counterregulatory actions) to mediate the resolution of inflammation.

MICRO RNAS OF RESOLUTION: SPM-RECEPTOR-microRNA CIRCUITS

RvD1 controls specific miRNA expression in vivo and in vitro (Recchiuti et al. 2011). This panel of miRs, including miR-146b, 208a, and 219, was temporally regulated during self-limited inflammation and regulated by RvD1 in vivo as well as in a RvD1-GPCR- (ALX and GPR32) dependent manner in human macrophages (Recchiuti et al. 2011). Macrophages overexpressing miR-219 significantly down-regulate 5-LOX and reduce LTB_4. Hence, 5-LOX is a target of miR-219 (Recchiuti et al. 2011). In addition, RvD1 at low dose (10 ng) significantly increases miR-219 in ALX-TG mice, whereas this dose of RvD1 was not effective in non-TG controls (Krishnamoorthy et al. 2012). Additionally, delayed resolution initiated by high-dose zymosan challenges decreases miR-219-5p expression along with higher LTB_4 and lower SPMs (Fredman et al. 2012). Therefore, RvD1 initiated a resolution circuit that involves activation of ALX and miR-219.

CAN RESOLVINS AND SPMs REVERSE ONGOING INFLAMMATION IN HUMANS?

Clinical Development

An RvE1 analog significantly improved signs and symptoms in a phase 2 clinical trial in patients with dry eye syndrome. This is the first demonstration of clinical efficacy for the novel class of resolvin therapeutics. The phase III clinical trial is now in progress (Safety and Efficacy Study of RX-10045 on the Signs and Symptoms of Dry Eye, identifier NCT00799552; www.clinical trials.gov). The AT 15-epi-LXA_4 analog, 15-*R*/*S*-methyl-LXA_4, reduced infantile eczema, showing no apparent toxicity or side effects (Wu et al. 2013).

New Uses of SPMs in Animal Disease Models

For an early review and detailed descriptions of initial animal models defining SPM proresolution action, see Serhan 2007. Recently, with conjunctiva goblet cells, both RvD1 and RvE1 reduced LTD_4- and histamine-stimulated conjunctival goblet cell secretion (Dartt et al. 2011; Li et al. 2013). RvE1 delivered as its methyl ester in a murine model of dry eye improves the outcome measures of corneal staining and goblet cell density, indicating the potential of resolvins in the treatment of dry eye (de Paiva et al. 2012). In HSV-induced ocular inflammation, RvE1 significantly reduces cornea lesions and angiogenesis as well as T cells and PMNs. These results indicate that RvE1 represents a novel approach to control virus-induced diseases (Rajasagi et al. 2011). A recent study showed a phenotype of delayed wound healing in cornea of female mice. Also in human corneal epithelial cells, estradiol reduced 15-LOX type-I and LXA_4. LXA_4 addition rescues the estradiol-abrogated wound healing, demonstrating gender-specific differences in the corneal repair mediated by the 15-LOX-LXA_4 circuit (Wang et al. 2012). In uveitis in rats, bolus intravenous injection of RvD1 (10–1000 ng/kg) significantly and dose-dependently reduced LPS-induced ocular derangement and PMNs, T-lymphocytes, as well as cytokines within the eye (Settimio et al. 2012).

In localized aggressive periodontitis (LAP) patients, macrophages show reduced phagocytosis. RvE1 rescues impaired phagocytic activity of LAP macrophages (Fredman et al. 2011). Humanized nanoparticles containing 17R-RvD1 or LXA_4 analog protect against inflammation in the temporomandibular joint, a model of temporomandibular joint disease (Norling et al. 2011). Exposure of salivary epithelium to TNF-α and/or interferon (IFN)-γ alters tight junction integrity, leading to secretory dysfunction. RvD1 (100 ng/mL) rescues TNF-α-induced tight junction and cytoskeletal disruption, and enhances cell migration and polarity in an ALX-dependent manner. Hence, RvD1 promotes tissue repair in salivary epithelium and restores salivary gland dysfunction associated with Sjögren's syndrome (Odusanwo et al. 2012). In rabbit arterial angioplasty, endogenous biosynthesis of proresolving lipid mediators, including RvD5 and LXB_4, was identified in the artery wall. Resolvins also reduce human smooth muscle cell proliferation and attenuate leukocyte recruitment and neointimal hyperplasia in rabbit balloon-injured arteries (Table 3) (Miyahara et al. 2013).

RvE1 promotes resolution in part via reducing IL-23 and IL-6 in allergic airways of mice as well as increasing IFN-γ (Haworth et al. 2008). Also, RvE1 regulates natural killer (NK) cell migration and cytotoxicity (Haworth et al. 2011), and LXA_4 regulates NK cells and type 2 innate lymphoid cell activation in asthma (Barnig et al. 2013). AT-RvD1 and RvD1 each markedly shorten resolution intervals for lung eosinophilia and reduce select inflammatory peptides and lipid mediators (Rogerio et al. 2012). In acute lung injury, AT-RvD1 improves epithelial and endothelial barrier integrity, decreases airway resistance, and increases epinephrine levels in bronchoalveolar lavage fluid (BALF) (Eickmeier et al. 2013). Of interest, Fat-1 transgenic mice that have increased endogenous lung n-3 PUFA (Hudert et al. 2006) also show higher PD1 and RvE1 levels after bronchoprovocative challenge (Bilal et al. 2011). These transgenic mice, which do not require dietary EPA and DHA to maintain high tissue levels, suggest a protective role for endogenous SPMs in allergic airway responses.

Human eosinophils biosynthesize PD1 as one of their main proresolving mediators, and PD1 production by eosinophils is impaired in patients with severe asthma. PD1, in nanomolar concentrations, reduces eosinophil chemotaxis and adhesion molecules (Miyata et al. 2013). In cigarette smoke-induced lung inflammation in the airways of mice, RvD1 protects and reduces PMN infiltration. RvD1 also promotes differentiation of M2 macrophages and efferocytosis in vivo, one of the cardinal signs of resolution. RvD1 also accelerated resolution of lung inflammation, demonstrating potential of SPMs to resolve lung injuries caused by toxicants such as cigarette smoke (Hsiao et al. 2013).

Resolvin D1 and Resolvin E1 potently regulate inflammatory pain (Xu et al. 2010), and intrathecal injections of RvD1 in rats reduces postoperative surgical pain (Huang et al. 2011). Along these lines, RvD1 (100 ng/kg) decreases TNBS-induced mechanical allodynia and blocked cytokine production in spinal dorsal horn (Quan-Xin et al. 2012), and RvD2 (0.01–1 ng) prevents formalin-induced pain. As part of the molecular mechanisms, RvD2, RvE1, and RvD1 differentially regulated transient receptor potential (TRP) channels (Park et al. 2011). AT-RvD1 significantly reverses the thermal hypersensitivity, and knockdown of epidermal TRPV3 blunts these antinociceptive actions (Bang et al. 2012). In arthritis, AT-RvD1 shows marked antihyperalgesia, decreases production of TNF-α and IL-1β in rat hind paw (Lima-Garcia et al. 2011), and RvD1 reduces neuroinflammation, stimulating phagocytosis of amyloid-β (Aβ) by Alzheimer's disease macrophages and inhibits fibrillar Aβ-induced apoptosis. These actions are dependent on GPR32 (Mizwicki et al. 2013).

In the original structure elucidation of resolvin E1, PMN infiltration to mouse skin (dorsal air pouch) was used as one of the biossay outcomes (Serhan et al. 2000). In burn models, RvD2 at 25 pg/g/animal given systemically post burn prevents thrombosis of the deep dermal vasculature, dermal necrosis, and PMN-mediated damage (Bohr et al. 2013). RvD2 restored PMN directionality in this system and increased survival after a second septic challenge (Kurihara et al. 2013). In DNFB-stimulated atopic dermatitis, RvE1 reduces skin lesions, lowers both IL-4 and IFN-γ, stimulates recruitment of $CD4^+$ T cells, and decreases serum IgE levels (Kim et al. 2012).

In murine models of colitis, systemic RvE1, AT-RvD1, RvD2, or 17R-HDHA (17*R*-hydroxy-4Z,7Z,10Z,13Z,15E,19Z-docosahexaenoic acid) in nanogram ranges improve disease severity, prevent body weight loss, colonic damage, and PMN infiltration as well as lower select colonic cytokines. The results suggest that some of the SPMs have potential in treating inflammatory bowel diseases (Arita et al. 2005b; Bento et al. 2011).

In adipose macrophages, RvD1 stimulates nonphlogistic phagocytosis and reduces macrophage reactive oxygen species production (Titos et al. 2011). In leptin receptor-deficient (db/db) mice, RvD1 (2 μg/kg) improves glucose tolerance, decreases fasting blood glucose, and increases adiponectin production and markers of alternatively activated M2 macrophages (Hellmann et al. 2011). Here, LXA_4 (1 nM) attenuates adipose inflammation and improves insulin sensitivity in a model of age-associated adipose inflammation (Borgeson et al. 2012). In diabetic wounds, local application of RvD1 accelerates wound closure and reduces accumulation of apoptotic cells in the wounds (Tang et al. 2013). It is noteworthy that in streptozotocin (STZ)-induced diabetes, fat-1 transgenic mice do not develop hyperglycemia and β-cell destruction compared with wild-type mice. RvE1 levels are highly increased in these fat-1 mice, emphasizing endogenous roles for RvE1 in diabetes (Bellenger et al. 2011). Thus, select SPMs and their mimetics may be novel approaches to reduce adipose inflammation and insulin resistance, key in type 2 diabetes.

Fibrosis

In a unilateral ureteric obstruction (UUO)-driven murine fibrosis, RvE1 at only 300 ng/day/mouse reduces accumulation of myofibroblasts, deposition of collagen IV, and myofibroblast proliferation. RvE1 (~1–30 nM) directly inhibits PDGF-BB-induced proliferation of fibroblasts, an action dependent on ChemR23 receptor expression (Qu et al. 2012).

CONCLUDING REMARKS

It is now clear from our efforts that pus, specifically contained resolving inflammatory exudates, produced from nutrient essential fatty acids, contains potent molecules that stimulate resolution and the return of tissues to homeostasis. The main families of molecules reviewed herein are the resolvins, protectins, and maresins, which together with the lipoxins and their AT epimeric versions (Serhan 2005) form a larger genus of molecules we denote as SPMs. Each SPM shows temporal and spatial formation dependent functions within contained inflammatory exudates. The key molecules from each of the families, structures, and actions have been confirmed via total organic synthesis, and their complete stereochemistries assigned. Availability of commercial resolvins and lipoxins has led to a recent surge in reports documenting their novel proresolving and anti-inflammatory actions in many disease models, some of which are reviewed here.

Although the resolvins and SPMs are locally produced and act as autacoids to terminate acute inflammatory responses, recent evidence indicates that these molecules can also reach circulating levels, for example, in human peripheral blood (Oh et al. 2011; Mas et al. 2012). These results demonstrating circulating levels of SPMs that are in the concentration ranges found for their anti-inflammatory and proresolving actions suggest that the SPMs, in addition to their local production and action, can also influence other organs. In this context, they might be able to signal anti-inflammation in second organs in addition to their target tissue exudates of origin. Hence, it is particularly relevant that resolvins and protectins can reduce neutrophilic infiltration and protect organs in the response to acute second organ ischemia reperfusion injury (Kasuga et al. 2008) and serve as immunoresolvents (Dalli et al. 2013a). Also, high levels of lipoxins and resolvins were recently identified in human breast milk (Weiss et al. 2013) as well as in animal placenta (Jones et al. 2013). These findings raise important new far-reaching implications for whether SPMs can play roles in cell traffic associated with organ development as well as regulate acute inflammatory responses around surgical events and tissue regeneration (Pillai et al. 2012; Dalli et al. 2013b). The presence of these proresolving lipid mediators can impact, for example, child development, and could impact diseases such as allergic asthma encountered later in life (Peebles 2013), as they are encountered during development and serve as determinants in later disease pathologies.

With the availability of commercial resolvins, we have also learned that they are effective in reducing inflammation in a wide range of inflammation-associated diseases (Tables 2 and 3), which taken together support the concept that the return of acute inflammatory responses to tissue homeostasis involves the active biosynthesis of proresolving mediators that function as local autacoids. Because the cell types that are responsible for the biosynthesis of SPMs travel within blood within vasculature that interact with the vascular endothelium and mucosal surfaces, their production is ubiquitous throughout the body as is the flow and traffic of leukocytes. Hence, in addition to their temporal and targeted actions within contained inflammatory exudates and their resolution, additional new biological functions of these mediators are likely to unfold in the years ahead, because they are also produced in vital nutrients in mammals such as in human milk, and, thus, SPMs may also impact both physiologic as well as resolving pathophysiologic processes in humans.

ACKNOWLEDGMENTS

The authors thank Mary Halm Small for expert assistance in manuscript preparation and support from National Institutes of Health Grants P01GM095467 and R01GM38765 (C.N.S.) and R01HL68669 (B.D.L.).

CONFLICT OF INTEREST DISCLOSURE

Disclosure: C.N.S. and B.D.L. are inventors on patents [resolvins] assigned to BWH and licensed to Resolvyx Pharmaceuticals. C.N.S. was scientific founder of Resolvyx Pharmaceuticals and owns equity in the company. The interests of C.N.S. and B.D.L. were reviewed and are managed by the Brigham and Women's Hospital and Partners HealthCare in accordance with their conflict of interest policies.

J.D. and N.C. have no financial conflicts of interest to disclose.

REFERENCES

Arita M, Bianchini F, Aliberti J, Sher A, Chiang N, Hong S, Yang R, Petasis NA, Serhan CN. 2005a. Stereochemical assignment, anti-inflammatory properties, and receptor for the omega-3 lipid mediator resolvin E1. *J Exp Med* **201:** 713–722.

Arita M, Yoshida M, Hong S, Tjonahen E, Glickman JN, Petasis NA, Blumberg RS, Serhan CN. 2005b. Resolvin E1, an endogenous lipid mediator derived from omega-3 eicosapentaenoic acid, protects against 2,4,6-trinitrobenzene sulfonic acid-induced colitis. *Proc Natl Acad Sci* **102:** 7671–7676.

Arita M, Oh S, Chonan T, Hong S, Elangovan S, Sun Y-P, Uddin J, Petasis NA, Serhan CN. 2006. Metabolic inactivation of resolvin E1 and stabilization of its anti-inflammatory actions. *J Biol Chem* **281:** 22847–22854.

Arita M, Ohira T, Sun YP, Elangovan S, Chiang N, Serhan CN. 2007. Resolvin E1 selectively interacts with leukotriene B_4 receptor BLT1 and ChemR23 to regulate inflammation. *J Immunol* **178:** 3912–3917.

Avicenna (Abu Ali Sina adapted by Lalech Bakhtiar). 1999. *The canon of medicine [al-Qanun fi'l-tibb]*. Great Books of the Islamic World, Chicago.

Baillie JK, Digard P. 2013. Influenza—Time to target the host? *N Engl J Med* **369:** 191–193.

Bang S, Yoo S, Yang TJ, Cho H, Kim YG, Hwang SW. 2010. Resolvin D1 attenuates activation of sensory transient receptor potential channels leading to multiple anti-nociception. *Br J Pharmacol* **161:** 707–720.

Bang S, Yoo S, Yang TJ, Cho H, Hwang SW. 2012. 17(R)-resolvin D1 specifically inhibits transient receptor potential ion channel vanilloid 3 leading to peripheral antinociception. *Br J Pharmacol* **165:** 683–692.

Bannenberg G, Serhan CN. 2010. Specialized pro-resolving lipid mediators in the inflammatory response: An update. *Biochim Biophys Acta* **1801:** 1260–1273.

Barnig C, Cernadas M, Dutile S, Liu X, Perrella MA, Kazani S, Wechsler ME, Israel E, Levy BD. 2013. Lipoxin A4 regulates natural killer cell and type 2 innate lymphoid cell activation in asthma. *Sci Transl Med* **5:** 174ra126.

Bellenger J, Bellenger S, Bataille A, Massey KA, Nicolaou A, Rialland M, Tessier C, Kang JX, Narce M. 2011. High pancreatic n-3 fatty acids prevent STZ-induced diabetes in fat-1 mice: Inflammatory pathway inhibition. *Diabetes* **60:** 1090–1099.

Bento AF, Claudino RF, Dutra RC, Marcon R, Calixto JB. 2011. Omega-3 fatty acid-derived mediators 17(R)-hydroxy docosahexaenoic acid, aspirin-triggered resolvin D1 and resolvin D2 prevent experimental colitis in mice. *J Immunol* **187:** 1957–1969.

Bilal S, Haworth O, Wu L, Weylandt KH, Levy BD, Kang JX. 2011. Fat-1 transgenic mice with elevated omega-3 fatty acids are protected from allergic airway responses. *Biochim Biophys Acta* **1812:** 1164–1169.

Bohr S, Patel SJ, Sarin D, Irimia D, Yarmush ML, Berthiaume F. 2013. Resolvin D2 prevents secondary thrombosis and necrosis in a mouse burn wound model. *Wound Repair Regen* **21:** 35–43.

Borgeson E, McGillicuddy FC, Harford KA, Corrigan N, Higgins DF, Maderna P, Roche HM, Godson C. 2012. Lipoxin A4 attenuates adipose inflammation. *FASEB J* **26:** 4287–4294.

Brancaleone V, Gobbetti T, Cenac N, le Faouder P, Colom B, Flower RJ, Vergnolle N, Nourshargh S, Perretti M. 2013. A vasculo-protective circuit centered on lipoxin A4 and aspirin-triggered 15-epi-lipoxin A4 operative in murine microcirculation. *Blood* **122:** 608–617.

Buckley CD, Gilroy DW, Serhan CN, Stockinger B, Tak PP. 2013. The resolution of inflammation. *Nat Rev Immunol* **13:** 59–66.

Calder PC. 2013. Omega-3 polyunsaturated fatty acids and inflammatory processes: Nutrition or pharmacology? *Br J Clin Pharmacol* **75:** 645–662.

Chan MM-Y, Moore AR. 2010. Resolution of inflammation in murine autoimmune arthritis is disrupted by cyclooxygenase-2 inhibition and restored by prostaglandin E2-mediated lipoxin A4 production. *J Immunol* **184:** 6418–6426.

Chiang N, Fredman G, Bäckhed F, Oh SF, Vickery TW, Schmidt BA, Serhan CN. 2012. Infection regulates pro-resolving mediators that lower antibiotic requirements. *Nature* **484:** 524–528.

Clària J, Dalli J, Yacoubian S, Gao F, Serhan CN. 2012. Resolvin D1 and resolvin D2 govern local inflammatory tone in obese fat. *J Immunol* **189:** 2597–2605.

Colas RA, Shinohara M, Dalli J, Chiang N, Serhan CN. 2014. Identification and signature profiles for pro-resolving and inflammatory lipid mediators in human tissue. *Am J Physiol Cell Physiol* **307:** C39–C54.

Cotran RS, Kumar V, Collins T, ed. 1999. *Robbins pathologic basis of disease*. W.B. Saunders, Philadelphia.

Crawford MA, Broadhurst CL, Guest M, Nagar A, Wang Y, Ghebremeskel K, Schmidt WF. 2013. A quantum theory for the irreplaceable role of docosahexaenoic acid in neural cell signalling throughout evolution. *Prostaglandins Leukot Essent Fatty Acids* **88:** 5–13.

Dalli J, Serhan CN. 2012. Specific lipid mediator signatures of human phagocytes: Microparticles stimulate macrophage efferocytosis and pro-resolving mediators. *Blood* **120:** e60–e72.

Dalli J, Winkler JW, Colas RA, Arnardottir H, Cheng CYC, Chiang N, Petasis NA, Serhan CN. 2013a. Resolvin D3 and aspirin-triggered resolvin D3 are potent immunoresolvents. *Chem Biol* **20:** 188–201.

Dalli J, Zhu M, Vlasenko NA, Deng B, Haeggstrom JZ, Petasis NA, Serhan CN. 2013b. The novel 13S,14S-epoxy-maresin is converted by human macrophages to maresin1 (MaR1), inhibits leukotriene A4 hydrolase (LTA4H), and shifts macrophage phenotype. *FASEB J* **27:** 2573–2583.

Dartt DA, Hodges RR, Li D, Shatos MA, Lashkari K, Serhan CN. 2011. Conjunctival goblet cell secretion stimulated by leukotrienes is reduced by resolvins D1 and E1 to promote resolution of inflammation. *J Immunol* **186:** 4455–4466.

De Caterina R. 2011. n-3 fatty acids in cardiovascular disease. *N Engl J Med* **364:** 2439–2450.

Deng B, Wang CW, Arnardottir HH, Li Y, Cheng CY, Dalli J, Serhan CN. 2014. Maresin biosynthesis and identification of maresin 2, a new anti-inflammatory and pro-resolving mediator from human macrophages. *PloS ONE* **9:** e102362.

de Paiva CS, Schwartz CE, Gjorstrup P, Pflugfelder SC. 2012. Resolvin E1 (RX-10001) Reduces corneal epithelial barrier disruption and protects against goblet cell loss in a murine model of dry eye. *Cornea* **31:** 1299–1303.

Devchand PR, Arita M, Hong S, Bannenberg G, Moussignac R-L, Gronert K, Serhan CN. 2003. Human ALX receptor regulates neutrophil recruitment in transgenic mice: Roles in inflammation and host-defense. *FASEB J* **17:** 652–659.

Eickmeier O, Seki H, Haworth O, Hilberath JN, Gao F, Uddin M, Croze RH, Carlo T, Pfeffer MA, Levy BD. 2013. Aspirin-triggered resolvin D1 reduces mucosal inflammation and promotes resolution in a murine model of acute lung injury. *Mucosal Immunol* **6:** 256–266.

El Kebir D, József L, Pan W, Wang L, Petasis NA, Serhan CN, Filep JG. 2009. 15-Epi-lipoxin A4 inhibits myeloperoxidase signaling and enhances resolution of acute lung injury. *Am J Respir Crit Care Med* **180:** 311–319.

El Kebir D, Gjorstrup P, Filep JG. 2012. Resolvin E1 promotes phagocytosis-induced neutrophil apoptosis and accelerates resolution of pulmonary inflammation. *Proc Natl Acad Sci* **109:** 14983–14988.

Feng Q, Feng F, Feng X, Li S, Wang S, Liu Z, Zhang X, Zhao Q, Wang W. 2012. Resolvin D1 reverses chronic pancreatitis-induced mechanical allodynia, phosphorylation of NMDA receptors, and cytokines expression in the thoracic spinal dorsal horn. *BMC Gastroenterol* **12:** 148.

Flower RJ. 2006. Prostaglandins, bioassay and inflammation. *Br J Pharmacol* **147:** S182–S192.

Fredman G, Oh SF, Ayilavarapu S, Hasturk H, Serhan CN, Van Dyke TE. 2011. Impaired phagocytosis in localized aggressive periodontitis: Rescue by resolvin E1. *PLoS ONE* **6:** e24422.

Fredman G, Li Y, Dalli J, Chiang N, Serhan CN. 2012. Self-limited versus delayed resolution of acute inflammation: Temporal regulation of pro-resolving mediators and microRNA. *Sci Rep* **2:** 639.

Gao L, Faibish D, Fredman G, Herrera BS, Chiang N, Serhan CN, Van Dyke TE, Gyurko R. 2013. Resolvin E1 and chemokine-like receptor 1 mediate bone preservation. *J Immunol* **190:** 689–694.

Giera M, Ioan-Facsinay A, Toes R, Gao F, Dalli J, Deelder AM, Serhan CN, Mayboroda OA. 2012. Lipid and lipid mediator profiling of human synovial fluid in rheumatoid arthritis patients by means of LC-MS/MS. *Biochim Biophys Acta* **1821:** 1415–1424.

Godson C, Mitchell S, Harvey K, Petasis NA, Hogg N, Brady HR. 2000. Cutting edge: Lipoxins rapidly stimulate nonphlogistic phagocytosis of apoptotic neutrophils by monocyte-derived macrophages. *J Immunol* **164:** 1663–1667.

Gutierrez AD, Sathyanarayana P, Konduru S, Ye Y, Birnbaum Y, Bajaj M. 2012. The effect of pioglitazone treatment on 15-epi-lipoxin A4 levels in patients with type 2 diabetes. *Atherosclerosis* **223:** 204–208.

Haas-Stapleton EH, Lu Y, Hong S, Arita M, Favoreto S, Nigam S, Serhan CN, Agabian N. 2007. *Candida albicans* modulates host defense by biosynthesizing the pro-resolving mediator resolvin E1. *PLoS ONE* **2:** e1316.

Haworth O, Cernadas M, Yang R, Serhan CN, Levy BD. 2008. Resolvin E1 regulates interleukin-23, interferon-gamma and lipoxin A_4 to promote resolution of allergic airway inflammation. *Nat Immunol* **9:** 873–879.

Haworth O, Cernadas M, Levy BD. 2011. NK cells are effectors for resolvin E1 in the timely resolution of allergic airway inflammation. *J Immunol* **186:** 6129–6135.

Hellmann J, Tang Y, Kosuri M, Bhatnagar A, Spite M. 2011. Resolvin D1 decreases adipose tissue macrophage accumulation and improves insulin sensitivity in obese-diabetic mice. *FASEB J* **25:** 2399–2407.

Ho KJ, Spite M, Owens CD, Lancero H, Kroemer AH, Pande R, Creager MA, Serhan CN, Conte MS. 2010. Aspirin-triggered lipoxin and resolvin E1 modulate vascular smooth muscle phenotype and correlate with peripheral atherosclerosis. *Am J Pathol* **177:** 2116–2123.

Hong S, Gronert K, Devchand P, Moussignac R-L, Serhan CN. 2003. Novel docosatrienes and 17S-resolvins generated from docosahexaenoic acid in murine brain, human blood and glial cells: Autacoids in anti-inflammation. *J Biol Chem* **278:** 14677–14687.

Hsiao HM, Sapinoro RE, Thatcher TH, Croasdell A, Levy EP, Fulton RA, Olsen KC, Pollock SJ, Serhan CN, Phipps RP, et al. 2013. A novel anti-inflammatory and pro-resolving role for resolvin D1 in acute cigarette smoke-induced lung inflammation. *PLoS ONE* **8:** e58258.

Huang L, Wang C-F, Serhan CN, Strichartz G. 2011. Enduring prevention and transient reduction of post-operative pain by intrathecal resolvin D1. *Pain* **152:** 557–565.

Hudert CA, Weylandt KH, Wang J, Lu Y, Hong S, Dignass A, Serhan CN, Kang JX. 2006. Transgenic mice rich in endogenous n-3 fatty acids are protected from colitis. *Proc Natl Acad Sci* **103:** 11276–11281.

Ishida T, Yoshida M, Arita M, Nishitani Y, Nishiumi S, Masuda A, Mizuno S, Takagawa T, Morita Y, Kutsumi H, et al. 2010. Resolvin E1, an endogenous lipid derived from eicosapentaenoic acid, prevents dextran sulfate sodium-induced colitis. *Inflamm Bowel Dis* **16:** 87–95.

Jones CN, Dalli J, Dimisko L, Wong E, Serhan CN, Irimia D. 2012. Microfluidic chambers for monitoring leukocyte trafficking and humanized nano-proresolving medicines interactions. *Proc Natl Acad Sci* **109:** 20560–20565.

Jones ML, Mark PJ, Keelan JA, Barden A, Mas E, Mori TA, Waddell BJ. 2013. Maternal dietary omega-3 fatty acid intake increases resolvin and protectin levels in the rat placenta. *J Lipid Res* **54:** 2247–2254.

Kasuga K, Yang R, Porter TF, Agrawal N, Petasis NA, Irimia D, Toner M, Serhan CN. 2008. Rapid appearance of resolvin precursors in inflammatory exudates: Novel mechanisms in resolution. *J Immunol* **181:** 8677–8687.

Keyes KT, Ye Y, Lin Y, Zhang C, Perez-Polo JR, Gjorstrup P, Birnbaum Y. 2010. Resolvin E1 protects the rat heart against reperfusion injury. *Am J Physiol Heart Circ Physiol* **299:** H153–H164.

Kim TH, Kim GD, Jin YH, Park YS, Park CS. 2012. Omega-3 fatty acid-derived mediator, Resolvin E1, ameliorates 2,4-dinitrofluorobenzene-induced atopic dermatitis in NC/Nga mice. *Int Immunopharmacol* **14:** 384–391.

Klein CP, Sperotto ND, Maciel IS, Leite CE, Souza AH, Campos MM. 2014. Effects of D-series resolvins on behavioral and neurochemical changes in a fibromyalgia-like model in mice. *Neuropharmacology* **86C:** 57–66.

Krishnamoorthy S, Recchiuti A, Chiang N, Yacoubian S, Lee C-H, Yang R, Petasis NA, Serhan CN. 2010. Resolvin D1 binds human phagocytes with evidence for pro-resolving receptors. *Proc Natl Acad Sci* **107:** 1660–1665.

Krishnamoorthy S, Recchiuti A, Chiang N, Fredman G, Serhan CN. 2012. Resolvin D1 receptor stereoselectivity and regulation of inflammation and pro-resolving microRNAs. *Am J Pathol* **180:** 2018–2027.

Krishnamurthy VR, Dougherty A, Haller CA, Chaikof EL. 2011. Total synthesis and bioactivity of 18R-hydroxy eicosapentaenoic acid. *J Org Chem* **76:** 5433–5437.

Kurihara T, Jones CN, Yu YM, Fischman AJ, Watada S, Tompkins RG, Fagan SP, Irimia D. 2013. Resolvin D2 restores neutrophil directionality and improves survival after burns. *FASEB J* **27:** 2270–2281.

Lawrence T, Natoli G. 2011. Transcriptional regulation of macrophage polarization: Enabling diversity with identity. *Nat Rev Immunol* **11:** 750–761.

Lawrence T, Willoughby DA, Gilroy DW. 2002. Anti-inflammatory lipid mediators and insights into the resolution of inflammation. *Nat Rev Immunol* **2:** 787–795.

Levy BD, Clish CB, Schmidt B, Gronert K, Serhan CN. 2001. Lipid mediator class switching during acute inflammation: signals in resolution. *Nat Immunol* **2:** 612–619.

Levy BD, De Sanctis GT, Devchand PR, Kim E, Ackerman K, Schmidt BA, Szczeklik W, Drazen JM, Serhan CN. 2002. Multi-pronged inhibition of airway hyper-responsiveness and inflammation by lipoxin A_4. *Nat Med* **8:** 1018–1023.

Levy BD, Bonnans C, Silverman ES, Palmer LJ, Marigowda G, Israel E, Severe Asthma Research Program National Heart Lung and Blood Institute. 2005. Diminished lipoxin biosynthesis in severe asthma. *Am J Respir Crit Care Med* **172:** 824–830.

Levy BD, Kohli P, Gotlinger K, Haworth O, Hong S, Kazani S, Israel E, Haley KJ, Serhan CN. 2007. Protectin D1 is generated in asthma and dampens airway inflammation and hyper-responsiveness. *J Immunol* **178:** 496–502.

Levy BD, Zhang QY, Bonnans C, Primo V, Reilly JJ, Perkins DL, Liang Y, Arnaout MA, Nikolic B, Serhan CN. 2011. The endogenous pro-resolving mediators lipoxin A_4 and resolvin E1 preserve organ function in allograft rejection. *Prostaglandins Leukot Essent Fatty Acids* **84:** 43–50.

Li N, He J, Schwartz CE, Gjorstrup P, Bazan HEP. 2010. Resolvin E1 improves tear production and decreases inflammation in a dry eye mouse model. *J Ocul Pharmacol Ther* **26:** 431–439.

Li D, Hodges RR, Jiao J, Carozza RB, Shatos MA, Chiang N, Serhan CN, Dartt DA. 2013. Resolvin D1 and aspirin-triggered resolvin D1 regulate histamine-stimulated conjunctival goblet cell secretion. *Mucosal Immunol* **6:** 1119–1130.

Lima-Garcia J, Dutra R, da Silva K, Motta E, Campos M, Calixto J. 2011. The precursor of resolvin D series and aspirin-triggered resolvin D1 display anti-hyperalgesic properties in adjuvant-induced arthritis in rats. *Br J Pharmacol* **164:** 278–293.

Liu G, Fiala M, Mizwicki MT, Sayre J, Magpantay L, Siani A, Mahanian M, Chattopadhyay M, La Cava A, Wiedau-Pazos M. 2012. Neuronal phagocytosis by inflammatory macrophages in ALS spinal cord: Inhibition of inflammation by resolvin D1. *Am J Neurodegener Dis* **1:** 60–74.

Majno G, Joris I. 2004. *Cells, tissues, and disease: Principles of general pathology.* Oxford University Press, New York.

Malawista SE, de Boisfleury Chevance A, van Damme J, Serhan CN. 2008. Tonic inhibition of chemotaxis in human plasma. *Proc Natl Acad Sci* **105:** 17949–17954.

Mas E, Croft KD, Zahra P, Barden A, Mori TA. 2012. Resolvins D1, D2, and other mediators of self-limited resolution of inflammation in human blood following n-3 fatty acid supplementation. *Clin Chem* **58:** 1476–1484.

Mirakaj V. Dalli J, Granja T, Rosenberger P, Serhan CN. 2014. Vagus nerve controls resolution and pro-resolving mediators of inflammation. *J Exp Med* **211:** 1037–1048.

Miyahara T, Runge S, Chatterjee A, Chen M, Mottola G, Fitzgerald JM, Serhan CN, Conte MS. 2013. D-series resolvins attenuate vascular smooth muscle cell activation and neointimal hyperplasia following vascular injury. *FASEB J* **27:** 2220–2232.

Miyata J, Fukunaga K, Iwamoto R, Isobe Y, Niimi K, Takamiya R, Takihara T, Tomomatsu K, Suzuki Y, Oguma T, et al. 2013. Dysregulated synthesis of protectin D1 in eosinophils from patients with severe asthma. *J Allergy Clin Immunol* **131:** 353–360.

Mizwicki MT, Liu G, Fiala M, Magpantay L, Sayre J, Siani A, Mahanian M, Weitzman R, Hayden E, Rosenthal MJ, et al. 2013. 1α,25-Dihydroxyvitamin D3 and resolvin D1 retune the balance between amyloid-β phagocytosis and inflammation in Alzheimer's disease patients. *J Alzheimers Dis* **34:** 155–170.

Morita M, Kuba K, Ichikawa A, Nakayama M, Katahira J, Iwamoto R, Watanebe T, Sakabe S, Daidoji T, Nakamura S, et al. 2013. The lipid mediator protectin D1 inhibits influenza virus replication and improves severe influenza. *Cell* **153:** 112–125.

Morris T, Stables M, Hobbs A, de Souza P, Colville-Nash P, Warner T, Newson J, Bellingan G, Gilroy DW. 2009. Effects of low-dose aspirin on acute inflammatory responses in humans. *J Immunol* **183:** 2089–2096.

Murakami T, Suzuki K, Tamura H, Nagaoka I. 2011. Suppressive action of resolvin D1 on the production and release of septic mediators in D-galactosamine-sensitized endotoxin shock mice. *Exp Ther Med* **2:** 57–61.

Norling LV, Spite M, Yang R, Flower RJ, Perretti M, Serhan CN. 2011. Cutting edge: Humanized nano-proresolving medicines mimic inflammation-resolution and enhance wound healing. *J Immunol* **186:** 5543–5547.

Norling LV, Dalli J, Flower RJ, Serhan CN, Perretti M. 2012. Resolvin D1 limits polymorphonuclear leukocytes recruitment to inflammatory loci: Receptor dependent actions. *Arterioscler Thromb Vasc Biol* **32:** 1970–1978.

Odusanwo O, Chinthamani S, McCall A, Duffey ME, Baker OJ. 2012. Resolvin D1 prevents TNF-α-mediated disruption of salivary epithelial formation. *Am J Physiol Cell Physiol* **302:** C1331–C1345.

Ogawa N, Kobayashi Y. 2009. Total synthesis of resolvin E1. *Tetrahedron Lett* **50:** 6079–6082.

Ogawa N, Kobayashi Y. 2011. Total synthesis of the antiinflammatory and proresolving protectin D1. *Tetrahedron Lett* **52:** 3001–3004.

Ogawa S, Urabe D, Yokokura Y, Arai H, Arita M, Inoue M. 2009. Total synthesis and bioactivity of resolvin E2. *Org Lett* **11:** 3602–3605.

Oh SF, Pillai PS, Recchiuti A, Yang R, Serhan CN. 2011. Pro-resolving actions and stereoselective biosynthesis of 18*S* E-series resolvins in human leukocytes and murine inflammation. *J Clin Invest* **121:** 569–581.

Oh SF, Dona M, Fredman G, Krishnamoorthy S, Irimia D, Serhan CN. 2012. Resolvin E2 formation and impact in inflammation resolution. *J Immunol* **188:** 4527–4534.

Ohira T, Arita M, Omori K, Recchiuti A, Van Dyke TE, Serhan CN. 2010. Resolvin E1 receptor activation signals phosphorylation and phagocytosis. *J Biol Chem* **285:** 3451–3461.

Palmer CD, Mancuso CJ, Weiss JP, Serhan CN, Guinan EC, Levy O. 2011. 17(R)-Resolvin D1 differentially regulates TLR4-mediated responses of primary human macrophages to purified LPS and live *E. coli*. *J Leukoc Biol* **90:** 459–470.

Park CK, Lü N, Xu ZZ, Liu T, Serhan CN, Ji RR. 2011. Resolving TRPV1 and TNF-α-mediated spinal cord synaptic plasticity and inflammatory pain with neuroprotectin D1. *J Neurosci* **31:** 15072–15085.

Peebles RS Jr. 2013. A new horizon in asthma: Inhibiting ILC function. *Sci Transl Med* **5:** 174fs7.

Perretti M, D'Acquisto F. 2009. Annexin A1 and glucocorticoids as effectors of the resolution of inflammation. *Nat Rev Immunol* **9:** 62–70.

Pillai PS, Leeson S, Porter TF, Owens CD, Kim JM, Conte MS, Serhan CN, Gelman S. 2012. Chemical mediators of inflammation and resolution in post-operative abdominal aortic aneurysm patients. *Inflammation* **35:** 98–113.

Prescott D, McKay DM. 2011. Aspirin-triggered lipoxin enhances macrophage phagocytosis of bacteria while inhibiting inflammatory cytokine production. *Am J Physiol Gastrointest Liver Physiol* **301:** G487–G497.

Pruss H, Rosche B, Sullivan AB, Brommer B, Wengert O, Gronert K, Schwab JM. 2013. Proresolution lipid mediators in multiple sclerosis—Differential, disease severity-dependent synthesis—A clinical pilot trial. *PLoS ONE* **8:** e55859.

Psychogios N, Hau DD, Peng J, Guo AC, Mandal R, Bouatra S, Sinelnikov I, Krishnamurthy R, Eisner R, Gautam B, et al. 2011. The human serum metabolome. *PLoS ONE* **6:** e16957.

Qu X, Zhang X, Yao J, Song J, Nikolic-Paterson DJ, Li J. 2012. Resolvins E1 and D1 inhibit interstitial fibrosis in the obstructed kidney via inhibition of local fibroblast proliferation. *J Pathol* **228:** 506–519.

Quan-Xin F, Fan F, Xiang-Ying F, Shu-Jun L, Shi-Qi W, Zhao-Xu L, Xu-Jie Z, Qing-Chuan Z, Wei W. 2012. Resolvin D1 reverses chronic pancreatitis-induced mechanical allodynia, phosphorylation of NMDA receptors, and cytokines expression in the thoracic spinal dorsal horn. *BMC Gastroenterol* **12:** 148.

Rajasagi NK, Reddy PBJ, Suryawanshi A, Mulik S, Gjorstrup P, Rouse BT. 2011. Controlling herpes simplex virus-induced ocular inflammatory lesions with the lipid-derived mediator resolvin E1. *J Immunol* **186:** 1735–1746.

Recchiuti A, Serhan CN. 2012. Pro-resolving lipid mediators (SPMs) and their actions in regulating miRNA in novel resolution circuits in inflammation. *Front Immun* **3:** 298.

Recchiuti A, Krishnamoorthy S, Fredman G, Chiang N, Serhan CN. 2011. MicroRNAs in resolution of acute inflammation: Identification of novel resolvin D1-miRNA circuits. *FASEB J* **25:** 544–560.

Rodriguez AR, Spur BW. 2012a. First total synthesis of the anti-inflammatory lipid mediator Resolvin D6. *Tetrahedron Lett* **53:** 86–89.

Rodriguez AR, Spur BW. 2012b. Total synthesis of the anti-inflammatory lipid mediator Resolvin E2. *Tetrahedron Lett* **53:** 1912–1915.

Rodriguez AR, Spur BW. 2012c. Total synthesis of the macrophage derived anti-inflammatory lipid mediator Maresin 1. *Tetrahedron Lett* **53:** 4169–4172.

Rogerio AP, Haworth O, Croze R, Oh SF, Uddin M, Carlo T, Pfeffer MA, Priluck R, Serhan CN, Levy BD. 2012. Resolvin D1 and aspirin-triggered resolvin D1 promote resolution of allergic airways responses. *J Immunol* **189:** 1983–1991.

Samuelsson B. 2012. Role of basic science in the development of new medicines: Examples from the eicosanoid field. *J Biol Chem* **287:** 10070–10080.

Samuelsson B, Dahlen SE, Lindgren JA, Rouzer CA, Serhan CN. 1987. Leukotrienes and lipoxins: Structures, biosynthesis, and biological effects. *Science* **237:** 1171–1176.

Sanak M, Levy BD, Clish CB, Chiang N, Gronert K, Mastalerz L, Serhan CN, Szczeklik A. 2000. Aspirin-tolerant asthmatics generate more lipoxins than aspirin-intolerant asthmatics. *Eur Respir J* **16:** 44–49.

Schwab JM, Chiang N, Arita M, Serhan CN. 2007. Resolvin E1 and protectin D1 activate inflammation-resolution programmes. *Nature* **447:** 869–874.

Seki H, Fukunaga K, Arita M, Arai H, Nakanishi H, Taguchi R, Miyasho T, Takamiya R, Asano K, Ishizaka A, et al. 2010. The anti-inflammatory and proresolving mediator resolvin E1 protects mice from bacterial pneumonia and acute lung injury. *J Immunol* **184:** 836–843.

Serhan CN. 2004. A search for endogenous mechanisms of anti-inflammation uncovers novel chemical mediators: Missing links to resolution. *Histochem Cell Biol* **122:** 305–321.

Serhan CN, guest ed. 2005. Special issue on lipoxins and aspirin-triggered lipoxins. *Prostaglandins Leukot Essent Fatty Acids* **73:** 139–321.

Serhan CN. 2007. Resolution phases of inflammation: Novel endogenous anti-inflammatory and pro-resolving lipid mediators and pathways. *Annu Rev Immunol* **25:** 101–137.

Serhan CN. 2011. The resolution of inflammation: The devil in the flask and in the details. *FASEB J* **25:** 1441–1448.

Serhan CN. 2014. Pro-resolving lipid mediators are leads for resolution physiology. *Nature* **510:** 92–101.

Serhan CN, Chiang N. 2013. Resolution phase lipid mediators of inflammation: Agonists of resolution. *Curr Opin Pharmacol* **13:** 632–640.

Serhan CN, Petasis NA. 2011. Resolvins and protectins in inflammation-resolution. *Chem Rev* **111:** 5922–5943.

Serhan CN, Clish CB, Brannon J, Colgan SP, Chiang N, Gronert K. 2000. Novel functional sets of lipid-derived mediators with antiinflammatory actions generated from omega-3 fatty acids via cyclooxygenase 2-nonsteroidal antiinflammatory drugs and transcellular processing. *J Exp Med* **192:** 1197–1204.

Serhan CN, Hong S, Gronert K, Colgan SP, Devchand PR, Mirick G, Moussignac R-L. 2002. Resolvins: A family of bioactive products of omega-3 fatty acid transformation circuits initiated by aspirin treatment that counter proinflammation signals. *J Exp Med* **196:** 1025–1037.

Serhan CN, Yang R, Martinod K, Kasuga K, Pillai PS, Porter TF, Oh SF, Spite M. 2009. Maresins: Novel macrophage mediators with potent anti-inflammatory and pro-resolving actions. *J Exp Med* **206:** 15–23.

Serhan CN, Ward PA, Gilroy DW, ed. 2010. *Fundamentals of inflammation.* Cambridge University Press, New York.

Serhan CN, Fredman G, Yang R, Karamnov S, Belayev LS, Bazan NG, Zhu M, Winkler JW, Petasis NA. 2011. Novel proresolving aspirin-triggered DHA pathway. *Chem Biol* **18:** 976–987.

Serhan CN, Dalli J, Karamnov S, Choi A, Park CK, Xu ZZ, Ji RR, Zhu M, Petasis NA. 2012. Macrophage pro-resolving mediator maresin 1 stimulates tissue regeneration and controls pain. *FASEB J* **26:** 1755–1765.

Settimio R, Clara DF, Franca F, Francesca S, Michele D. 2012. Resolvin D1 reduces the immunoinflammatory response of the rat eye following uveitis. *Mediators Inflamm* **2012:** 318621.

Spite M, Norling LV, Summers L, Yang R, Cooper D, Petasis NA, Flower RJ, Perretti M, Serhan CN. 2009. Resolvin D2 is a potent regulator of leukocytes and controls microbial sepsis. *Nature* **461:** 1287–1291.

Sun Y-P, Oh SF, Uddin J, Yang R, Gotlinger K, Campbell E, Colgan SP, Petasis NA, Serhan CN. 2007. Resolvin D1 and its aspirin-triggered 17*R* epimer: Stereochemical assignments, anti-inflammatory properties and enzymatic inactivation. *J Biol Chem* **282:** 9323–9334.

Tabas I, Glass CK. 2013. Anti-inflammatory therapy in chronic disease: Challenges and opportunities. *Science* **339:** 166–172.

Takano T, Clish CB, Gronert K, Petasis N, Serhan CN. 1998. Neutrophil-mediated changes in vascular permeability are inhibited by topical application of aspirin-triggered 15-epi-lipoxin A_4 and novel lipoxin B_4 stable analogues. *J Clin Invest* **101:** 819–826.

Tang Y, Zhang MJ, Hellmann J, Kosuri M, Bhatnagar A, Spite M. 2013. Proresolution therapy for the treatment of delayed healing of diabetic wounds. *Diabetes* **62:** 618–627.

Tauber AI, Chernyak L. 1991. *Metchnikoff and the origins of immunology: From metaphor to theory.* Oxford University Press, New York.

Terrando N, Gomez-Galan M, Yang T, Carlstrom M, Gustavsson D, Harding RE, Lindskog M, Eriksson LI. 2013. Aspirin-triggered resolvin D1 prevents surgery-induced cognitive decline. *FASEB J* **27:** 3564–3571.

Titos E, Rius B, González-Périz A, López-Vicario C, Morán-Salvador E, Martínez-Clemente M, Arroyo V, Clária J. 2011. Resolvin D1 and its precursor docosahexaenoic acid promote resolution of adipose tissue inflammation by eliciting macrophage polarization toward a pro-resolving phenotype. *J Immunol* **187:** 5408–5418.

Tobin DM, Roca FJ, Oh SF, McFarland R, Vickery TW, Ray JP, Ko DC, Zou Y, Bang ND, Chau TT, et al. 2012. Host genotype-specific therapies can optimize the inflammatory response to mycobacterial infections. *Cell* **148:** 434–446.

Vong L, Ferraz JG, Dufton N, Panaccione R, Beck PL, Sherman PM, Perretti M, Wallace JL. 2012. Up-regulation of Annexin-A1 and lipoxin A_4 in individuals with ulcerative colitis may promote mucosal homeostasis. *PLoS ONE* **7:** e39244.

von Moltke J, Trinidad NJ, Moayeri M, Kintzer AF, Wang SB, van Rooijen N, Brown CR, Krantz BA, Leppla SH, Gronert K, et al. 2012. Rapid induction of inflammatory lipid mediators by the inflammasome in vivo. *Nature* **490:** 107–111.

Walker J, Dichter E, Lacorte G, Kerner D, Spur B, Rodriguez A, Yin K. 2011. Lipoxin a4 increases survival by decreasing systemic inflammation and bacterial load in sepsis. *Shock* **36:** 410–416.

Wang SB, Hu KM, Seamon KJ, Mani V, Chen Y, Gronert K. 2012. Estrogen negatively regulates epithelial wound healing and protective lipid mediator circuits in the cornea. *FASEB J* **26:** 1506–1516.

Wang X, Zhu M, Hjorth E, Cortés-Toro V, Eyjolfsdottir H, Graff C, Nennesmo I, Palmblad J, Eriksdotter M, Sambamurti K, et al. 2014. Resolution of inflammation is altered in Alzheimer's disease. *Alzheimer's Dement* doi: 10.1016/j.jalz.2013.1012.1024.

Weiss GA, Troxler H, Klinke G, Rogler D, Braegger C, Hersberger M. 2013. High levels of anti-inflammatory and pro-resolving lipid mediators lipoxins and resolvins and declining docosahexaenoic acid levels in human milk during the first month of lactation. *Lipids Health Dis* **12:** 89.

Wu S-H, Liao P-Y, Yin P-L, Zhang Y-M, Dong L. 2009. Elevated expressions of 15-lipoxygenase and lipoxin A_4 in children with acute poststreptococcal glomerulonephritis. *Am J Pathol* **174:** 115–122.

Wu SH, Chen XQ, Liu B, Wu HJ, Dong L. 2013. Efficacy and safety of 15(R/S)-methyl-lipoxin A_4 in topical treatment of infantile eczema. *Br J Dermatol* **168:** 172–178.

Xu Z-Z, Zhang L, Liu T, Park J-Y, Berta T, Yang R, Serhan CN, Ji R-R. 2010. Resolvins RvE1 and RvD1 attenuate inflammatory pain via central and peripheral actions. *Nat Med* **16:** 592–597.

Xu ZZ, Berta T, Ji RR. 2013. Resolvin E1 inhibits neuropathic pain and spinal cord microglial activation following peripheral nerve injury. *J Neuroimmune Pharmacol* **8:** 37–41.

Yamaguchi H, Higashi N, Mita H, Ono E, Komase Y, Nakagawa T, Miyazawa T, Akiyama K, Taniguchi M. 2011. Urinary concentrations of 15-epimer of lipoxin A_4 are lower in patients with aspirin-intolerant compared with aspirin-tolerant asthma. *Clin Exp Allergy* **41:** 1711–1718.

Yanes O, Clark J, Wong DM, Patti GG, Sánchez-Ruiz A, Benton HP, Trauger SA, Desponts C, Ding S, Siuzdak G. 2010. Metabolic oxidation regulates embryonic stem cell differentiation. *Nat Chem Biol* **6:** 411–417.

Zivkovic AM, Yang J, Georgi K, Hegedus C, Nording ML, O'Sullivan A, German JB, Hogg RJ, Weiss RH, Bay C, et al. 2012. Serum oxylipin profiles in IgA nephropathy patients reflect kidney functional alterations. *Metabolomics* **8:** 1102–1113.

DNA Degradation and Its Defects

Kohki Kawane[1], Kou Motani[1], and Shigekazu Nagata[1,2]

[1]Department of Medical Chemistry, Kyoto University Graduate School of Medicine, Yoshida-Konoe, Kyoto 606-8501, Japan

[2]Core Research for Evolutional Science and Technology, Japan Science and Technology Corporation, Yoshida-Konoe, Kyoto 606-8501, Japan

Correspondence: snagata@mfour.med.kyoto-u.ac.jp

DNA is one of the most essential molecules in organisms, containing all the information necessary for organisms to live. It replicates and provides a mechanism for heredity and evolution. Various events cause the degradation of DNA into nucleotides. DNA also has a darker side that has only recently been recognized; DNA that is not properly degraded causes various diseases. In this review, we discuss four deoxyribonucleases that function in the nucleus, cytosol, and lysosomes and how undigested DNA causes such diseases as cancer, cataract, and autoinflammation. Studies on the biochemical and physiological functions of deoxyribonucleases should continue to increase our understanding of cellular functions and human diseases.

Chromosomal DNA replicates semiconservatively; it is constructed in growing cells and is not thereafter metabolized within the cell. Both animal and plant cells carry several DNA-degrading enzymes (called deoxyribonuclease, or DNase). DNases have primarily been regarded as enzymes that digest the DNA in food into nucleotides for use in rebuilding the organism's own DNA, just as proteases digest food proteins (from fish, meat, or vegetables) into amino acids. For many years, studies on DNases focused almost exclusively on their enzymatic activity, and not on their physiological or pathological roles. This changed with the discovery that chromosomal DNA is digested in apoptotic cells (Wyllie 1980). Since then, DNA degradation has been observed in the differentiation processes of red blood cells, skin, and optic lens (Bassnett 2002; McGrath et al. 2008; Eckhart et al. 2013). Reverse-transcribed DNA from endogenous retro elements is digested in the cytoplasm (Stetson et al. 2008), and in inflammation, extracellular DNA released from dead cells is actively degraded in the circulation (Rekvig and Mortensen 2012). Here, we discuss how DNA is digested in physiological and pathological settings, and what happens to the organism if DNA is not properly digested.

A DNase THAT CLEAVES DNA IN THE NUCLEUS

Apoptosis

Apoptosis destroys surplus cells generated during animal development, cells infected with a virus or bacteria, tumor cells, and senescent

Cite this article as *Cold Spring Harb Perspect Biol* doi: 10.1101/cshperspect.a016394

cells in a process involving cell shrinkage, chromatin condensation, and membrane blebbing (Jacobson et al. 1997; Strasser et al. 2009). Apoptosis occurs at a rate of a few million cells each second in the human body. It can be triggered by death factors, anticancer drugs, γ-ray irradiation, or deprivation of essential factors. Whatever the trigger may be, apoptosis is in most cases executed by caspases, a cysteine proteases family consisting of 14 members that are divided into initiator and effector caspases. Each caspase recognizes a specific sequence of four amino acids, and specifically cleaves proteins after an aspartic acid. Death factors such as Fas ligand (FasL), TNF (tumor necrosis factor), and TRAIL (TNF-related apoptosis-inducing factor) bind their receptors (Fas, TNF-R1, and DR4 and DR5, respectively) to activate caspase-8 via an adaptor, Fas-associated protein with death domain. Caspase-8 activates caspase-3 and caspase-7, alone or together, by cleaving their precursors. Anticancer drugs, γ-ray irradiation, and factor deprivation, trigger apoptosis by activating BH3-only proteins of the Bcl-2 family, such as Bim, Bad, or Noxa. These in turn activate Bax or Bak (also members of the Bcl-2 family), prompting mitochondria to release cytochrome *c*, which works with APAF-1 to activate caspase-9. Caspase-9 activates caspase-3 and caspase-7, individually or together. Caspase-3 and caspase-7 are effector caspases that cleave more than 400 substrates to affect apoptosis and kill the cell (Nagata 1997; Susin et al. 2000; Dix et al. 2008; Mahrus et al. 2008).

Caspase-Activated DNase

The DNase responsible for apoptotic DNA fragmentation is CAD (caspase-activated DNase), also called DFF-40 (DNA fragmentation factor 40) (Nagata 2005). In healthy cells, CAD is locked in a complex with its inhibitor, ICAD (inhibitor of CAD), also called DFF-45 (DNA fragmentation factor 45) (Liu et al. 1997; Enari et al. 1998; Sakahira et al. 1998). ICAD also acts as a molecular chaperone for CAD; it binds to the nascent CAD chain on the ribosomes, assists in folding the CAD protein, and remains with CAD when it is released from the ribosomes as a part of the CAD-ICAD complex (Sakahira et al. 2000; Sakahira and Nagata 2002). This sophisticated chaperone system for generating CAD, a potentially hazardous molecule with the ability to destroy DNA, guarantees fail-safe machinery for apoptotic DNA fragmentation. Because CAD cannot be produced without ICAD, apoptotic DNA fragmentation does not occur in *CAD*-null or *ICAD*-null cells (Zhang et al. 1998).

ICAD contains the caspase-3 or caspase-7 recognition sequence at two positions. When cells receive apoptotic stimuli, ICAD is cleaved by activated caspase-3 or caspase-7, releasing CAD as a homodimer with a scissor-like structure (Woo et al. 2004). The CAD dimer digests DNA at spacer regions between nucleosomes, whereas the chromatin structure is still intact, thus producing multimers of DNA fragments with nucleosomal units (Fig. 1). Apoptotic cells are detected by terminal deoxynucleotidyl transferase (TdT)-mediated dUTP nick-end labeling (TUNEL) staining (Gavrieli et al. 1992), in which the terminal transferase adds a nucleotide to a free hydroxyl group at the 3′ end of DNA fragments severed by CAD.

In apoptosis, chromosomal DNA is degraded in two steps: first into large (50–100 kb) units, and then into nucleosomal units. Regardless of the apoptotic stimuli or cell type, CAD is responsible for both steps because DNA degradation does not follow this pattern in *CAD*- or *ICAD*-null cells (Kawane et al. 2003; Nagase et al. 2003). Thus, other nucleases that have been reported to degrade DNA in apoptosis, such as endonuclease G, DNase γ, and AIF (Penninger and Kroemer 2003; Mizuta et al. 2009) have little, if any, role in this process. DNA fragmentation downstream from the caspase cascade has been thought to affect cell death. In fact, microinjecting activated CAD or CAD dimers into a healthy cell quickly kills it (Susin et al. 2000). However, apoptotic stimuli kill *CAD*- or *ICAD*-null cells as efficiently as *CAD*-competent cells (Sakahira et al. 1998). This indicates that there are many ways to kill a cell once caspases have been activated, and that DNA fragmentation itself is dispensable for apoptotic cell death.

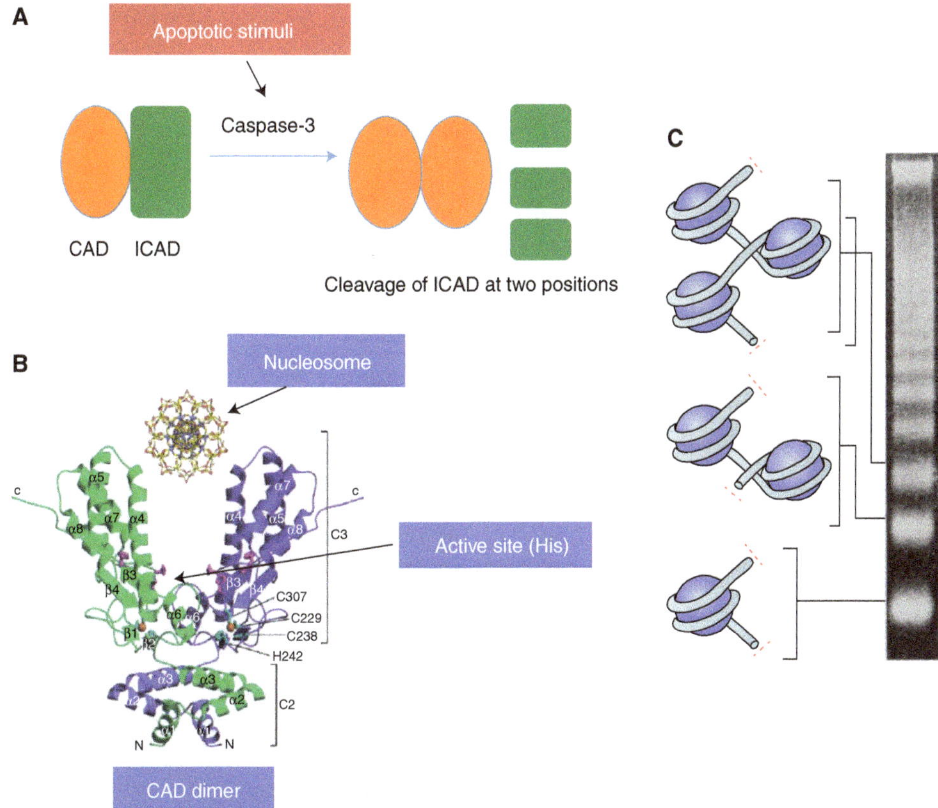

Figure 1. CAD-mediated DNA fragmentation in apoptosis. In healthy cells, CAD (caspase-activated DNase) is present as a complex with ICAD (inhibitor of CAD). When cells receive apoptotic stimuli, caspase-3 cleaves ICAD at two positions, and the cleaved ICAD loses the affinity to CAD (*A*). The activated CAD, or CAD freed from ICAD, forms a homodimer with a scissor-like structure, in which the active site carrying histidine residues is in the cleft. Because of the structure of the CAD dimer, DNA between nucleosomes, but not DNA on nucleosomes, has access to the active site of the enzyme (*B*). This explains why chromosomal DNA is cleaved into nucleosomal units during apoptotic cell death (*C*). (*B*, Adapted from data in Woo et al. 2004.)

Defective CAD Leads to Cancer and Autoimmune Disease

Despite CAD's indispensable role in fragmenting DNA during apoptosis, *CAD*-null mice have no apparent gross abnormality (Kawane et al. 2003). On the other hand, the human *CAD* and *ICAD* genes at chromosome 1p36.2 (Mukae et al. 1998) are aberrantly expressed in various human tumors, and are often mutated in neuroblastoma and germ cell tumors (Yang et al. 2001). *CAD*-deficient cells are sensitive to radiation-induced chromosome instability and are easily transformed into anchorage-independent tumor cells (Yan et al. 2006). Although CAD cleaves DNA randomly, its preferred targets largely overlap with sites where mutations are introduced in cancer cells (Fullwood et al. 2011), suggesting that CAD removes DNA mutated by DNA-damaging agents at an early stage of apoptosis. Tang et al. (2012) recently reported that apoptotic cells can be recovered to a viable state from even late stages of apoptosis. If CAD deficiency prevents the removal of damaged DNA, the damaged DNA sequences in cells rescued from apoptosis may cause cells to undergo transformation. CAD-mediated DNA fragmentation may also be involved in an amplification step of apoptosis (Boulares et al. 2001). In the apoptosis induced by specific agents, the CAD-fragmented DNA activates poly(ADP-ribose) polymerase, which quickly depletes the cellular NAD, efficiently executing the cells.

CAD also regulates autoimmunity. Apoptotic cells, or apoptotic bodies, induce autoantibody production in systemic lupus erythematosus (SLE)-type autoimmune diseases (Casciola-Rosen et al. 1994). Jog et al. (2012) recently showed that a *CAD*-null mutation in lupus-prone mice accelerates autoantigen production and aggravates renal disease, suggesting that CAD plays an important role in maintaining immunological tolerance against nuclear antigens. This interesting observation should lead to further investigations as to how this is accomplished.

DNases IN LYSOSOMES

Engulfment of Apoptotic Cells by Macrophages

Cell-autonomous DNA fragmentation does not occur in *CAD*-deficient cells, yet tissues in *CAD*-deficient mice do not accumulate undigested DNA. Close examination of *CAD*-null tissues revealed that the DNA from dead cells is degraded in the lysosomes of macrophages (McIlroy et al. 2000). The macrophages recognize an "eat me" phosphatidylserine (PtdSer) signal on apoptotic cells and swiftly engulf the dying cells (Fadok et al. 1992), preventing the release of noxious materials. PtdSer, one of the most abundant phospholipids in the plasma membrane, is kept tightly in the inner leaflet of the plasma membrane in healthy cells (Leventis and Grinstein 2010). We recently showed that when cells undergo apoptosis, caspase-3 or -7 cleaves off the carboxy-terminal tail of Xkr8, a plasma-membrane protein with six transmembrane regions (Suzuki et al. 2013). Xkr8 then mediates phospholipid scrambling between the inner and outer leaflets of the plasma membrane, exposing PtdSer on the cell surface as an "eat me" signal.

Macrophages express PtdSer receptors such as Tim4, BAI1, and stabillin2 (Miyanishi et al. 2007; Park et al. 2007, 2008), and Tim4 has been shown to function in the binding or tethering of apoptotic cells (Toda et al. 2012). The apoptotic cells tethered by Tim4 are transferred to $\alpha_v\beta_3$ integrin or MER receptor tyrosine kinases of the TAM family via PtdSer-binding opsonins, such as MFG-E8 and Gas6 that are secreted from macrophages (Scott et al. 2001; Hanayama et al. 2002; Xiong et al. 2008). Apoptotic cells are then internalized by macrophages via "phagocytic cups" assembled by Rac1-dependent actin polymerization (Kinchen and Ravichandran 2007; Nakaya et al. 2008), and are transported into lysosomes via early endosomes by the small GTPase Rab5 (Kitano et al. 2008). Lysosomes contain a variety of degradative enzymes (proteases, glycosidases, lipases, and nucleases) that digest components from dead cells into building units for reuse (amino acids, sugars, fatty acids, and nucleotides) (von Figura and Hasilik 1986).

DNase II and Its Degradation of the DNA of Apoptotic Cells

The enzyme that digests DNA in macrophage lysosomes is deoxyribonuclease II (DNase II), also called DNase IIα or acid DNase, which has the optimal pH at acid condition (Evans and Aguilera 2003). *DNase II*$^{-/-}$ mouse embryos carry numerous peculiar macrophages, filled with Feulgen-positive DNA in various tissues where apoptosis occurs extensively during development (Kawane et al. 2001, 2003; Krieser et al. 2002). Electron microscope analyses suggested that the *DNase II*$^{-/-}$ macrophages engulf apoptotic cells and digest all of the cell components except DNA. The DNA that accumulates in the lysosomes of *DNase II*$^{-/-}$ macrophages is fragmented and strongly TUNEL-positive. However, the DNA that accumulates in the macrophages of *CAD*$^{-/-}$*DNase II*$^{-/-}$ double knockout embryos appears intact, and is not stained by TUNEL. These results confirm that the DNA in the apoptotic cells is digested in two steps: first, it is fragmented into nucleosomal units by CAD in the apoptotic cells, and then it is completely digested into nucleotides by DNase II in the lysosomes of the engulfing macrophage (Kawane et al. 2003). DNase II in *C. elegans* is called Nuc-1, and both a *Nuc-1*-null mutation in *C. elegans* (Wu et al. 2000) and a *DNase II*–null mutation in *Drosophila* (Mukae et al. 2002) strongly enhance the TUNEL positivity. This suggests that the mechanism of two-step apoptotic DNA degradation is evolutionarily well conserved, although DNase II may also function cell-autonomously in the nonapoptotic cell death of nurse cells in late oogenesis in *Drosophila* (Bass et al. 2009).

Degradation of Pyrenocyte DNA by DNase II

Each day, our bodies generate about 200 billion red blood cells, at least 10 times number of cells undergoing apoptosis. In the *DNase II*$^{-/-}$ embryo, a large number of macrophages carrying undigested DNA in their lysosomes are found in the fetal liver (Kawane et al. 2001; Krieser et al. 2002), where definitive erythropoiesis takes place in late embryogenesis (Palis 2008). The macrophages carrying DNA are the central macrophages in erythroblastic islands, where the erythropoiesis takes place (Chasis and Mohandas 2008). Erythroblasts proliferate and differentiate on these macrophages, and finally divide asymmetrically into reticulocytes and pyrenocytes (nucleus covered by plasma membrane) (McGrath et al. 2008; Rhodes et al. 2008). The pyrenocytes quickly expose PtdSer on their surface in a caspase-independent manner, and the macrophages in the center of the islands recognize PtdSer as the "eat-nucleus" signal and engulf the pyrenocytes (Yoshida et al. 2005a). DNase II in these erythroblastic islands is responsible for digesting the DNA ($\sim$1.0 g) of 200 billion pyrenocytes per day.

Defective DNase II Leads to Autoinflammation Activation

DNase II–null mice die of severe anemia in late embryogenesis (Kawane et al. 2001). Because T-cell development is arrested at the CD4^{-}CD8^{-} stage, the *DNase II*–null fetal thymus is one-third the size of the wild-type one (Kawane et al. 2003). A set of interferon-(IFN) inducible genes is strongly expressed in these mutant embryos (Kawane et al. 2003; Yoshida et al. 2005b). Accordingly, in situ hybridization indicated that the *IFN-β* gene is expressed in macrophages carrying undigested DNA (Yoshida et al. 2005b). Type I IFNs, such as IFN-β, are cytokines that confer virus resistance to cells (Borden et al. 2007) and that are cytotoxic to various cancer cells, erythroid cells, and T-cell precursors (pro-T cells) (Su et al. 1997; Gómez-Benito et al. 2007; Jain and Zoellner 2010). The null mutation of *IFN-IR*, which encodes the type-I IFN receptor for IFN-α and IFN-β, rescues the lethality of *DNase II*–null embryos, and *DNase II*$^{-/-}$ *IFN-IR*$^{-/-}$ mice are born at nearly a Mendelian ratio (Yoshida et al. 2005b). The number of peripheral red blood cells and the thymus size return to normal, indicating that if the engulfed DNA from apoptotic cells or pyrenocytes is not properly degraded in the macrophages, the *IFN-β* gene is activated, causing lethal anemia and thymic atrophy.

The rescued *DNase II*$^{-/-}$*IFN-IR*$^{-/-}$ mice develop polyarthritis as they age (Kawane et al. 2006). At 5–7 months, all the joints show synovitis with well-established pannus, an inflammatory structure composed of synoviocytes, macrophages, fibroblasts, and infiltrated immune cells, and the cartilage and bone around the pannus are destroyed. If the *DNase II* gene is inducibly deleted, the mice develop arthritis with the same properties. The histology of the arthritis in *DNase II*–deficient mice resembles that of human rheumatoid arthritis (RA), a chronic inflammation characterized by a massive production of inflammatory cytokines in the joints (Scott et al. 2010). Similarly, inflammatory cytokines and chemokines are highly upregulated in the joints of *DNase II*–deficient mice, and high levels of rheumatoid factor, anti-cyclic citrullinated protein antibody, matrix metalloprotease-3 (MMP-3), and interleukin (IL)-18 are found in the serum. The *DNase II*–null mutation induces arthritis in various mice strains (C57BL6 and Balb/c) without gender preference. These properties of the arthritis in *DNase II*–null mice are similar to those observed in human systemic-onset juvenile idiopathic arthritis (soJIA, also called Still's disease) (Vastert et al. 2009). The arthritis in *DNase II*–deficient mice does not require acquired immunity, but depends strongly on inflammatory cytokines. That is, although the loss of *Rag2* (recombination activation gene), which is essential for the V(D)J gene rearrangement in lymphocytes, has no effect on the arthritis (Kawane et al. 2010), the arthritis is almost completely blocked by deleting the *IL-1* receptor, *IL-6*, or *TNF-α* genes (Kawane et al. 2010). Remarkably, a lack of one inflammatory cytokine blocks the expression of other cytokines in the joints; deleting the *TNF-α* gene blocks *IL-6* and *IL-1β* expression, while deleting *IL-6* gene blocks *TNF-α* and *IL-1β* expression, indicating that these inflammatory cytokines regulate the expression of other genes in the joint. Accordingly, administrating antibodies against TNF-α, IL-1 receptor (IL-1R), or IL-6 receptor (IL-6R) efficiently cures the disease. This is similar to the therapeutic effect of anti-TNF-α, anti-IL-6R, and anti-IL-1R antibodies in human soJIA patients.

In *DNase II*–deficient adult mice, macrophages carrying DNA in their lysosomes are present in the bone marrow and red splenic pulp, suggesting that cytokines produced in the macrophages induce arthritis at the joints. Because TNF-α-transgenic mice develop arthritis (Keffer et al. 1991), TNF-α is a strong candidate for the arthritis trigger in *DNase II*–null mice. In fact, macrophages carrying undigested DNA in their lysosomes constitutively produce TNF-α (Kawane et al. 2006). It is likely that the TNF-α in turn stimulates the synovial cells in joints to produce cytokines such as IL-1β, IL-6, granulocyte colony–stimulating factor (G-CSF), and chemokines. IL-1β, TNF-α, and IL-6 in the articular cavity then activate synovial cells and fibroblasts to produce other inflammatory cytokines. Inflammatory cytokines and chemokines accumulate in high concentrations in the articular cavity, where they stimulate the growth of synovial cells and fibroblasts, and recruit lymphocytes, neutrophils, and macrophages into the joints. Pannus, thus formed, destroys cartilage and bones, and arthritis develops (Fig. 2).

DLAD, a DNase in Lens Fiber Cells

DLAD (DNase II-like acid DNase, also called DNase IIβ) has a 38% amino acid sequence identity with DNase II (Shiokawa and Tanuma 1999) and is also localized to the lysosomes (Nakahara et al. 2007). Unlike DNase II, which is ubiquitously expressed in various tissues and cells, DLAD is specifically expressed in fiber cells of the optic lens (Fig. 3) (Nishimoto et al. 2003). This avascular tissue, which focuses light onto the retina, consists of packed fiber cells bounded at the anterior by a monolayer of epithelial cells. The epithelial cells continuously differentiate into fiber cells near the lens equator, and as there is no cell turnover, the lens continues to grow throughout life (McAvoy et al. 1999). As the epithelial cells differentiate into fiber cells, they synthesize lens-fiber-specific proteins such as crystallins. The DLAD gene is also activated during this differentiation process (Nakahara et al. 2007).

This differentiation of epithelial cells to fiber cells in the eye is accompanied by the loss of intracellular organelles such as mitochondria and nuclei, to form a light-transmittable organ (Bassnett 2009). Organelles are degraded cell-autonomously in the differentiating cells, and DLAD is responsible for degrading the DNA (Nishimoto et al. 2003). The lens fiber cells in $DLAD^{-/-}$ mice carry undigested DNA in their cytoplasm, although all the other organelles and nuclear components are degraded normally (Fig. 3) (Nishimoto et al. 2003). This cell-autonomous degradation of organelles in the lysosomes is similar to autophagy. However, no abnormalities have been observed in the lens fiber cells of mice carrying a mutation in the genes (*atg*) for autophagy. It is also possible that the lysosomal and nuclear membranes are broken when lens cells differentiate to allow DLAD to attack the chromosomal DNA. Elucidation of the mechanism behind organelle degradation during lens cell differentiation would be an interesting research challenge.

A DNase THAT CLEAVES DNA IN THE CYTOSOL

TREX1 and Its Defect

Trex1 (three prime repair exonuclease, also called DNase III), ubiquitously expressed in var-

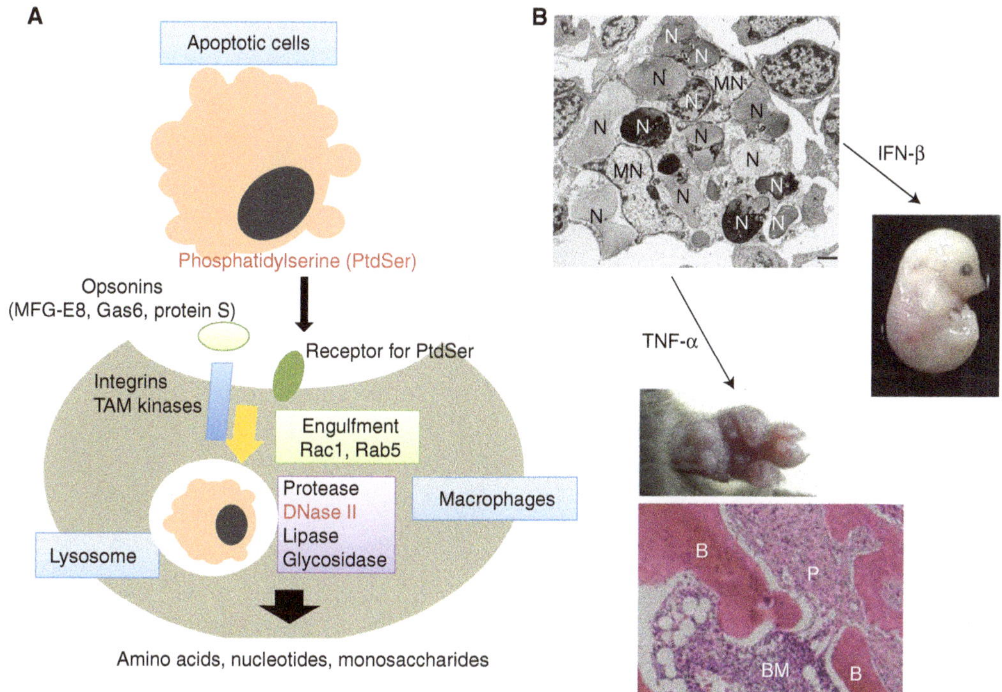

Figure 2. Engulfment of apoptotic cells by macrophages, and DNA degradation in lysosomes of macrophages. (*A*) Apoptotic cells expose phosphatidylserine (PtdSer) as an "eat me" signal, which is recognized by the PtdSer receptor (Tim4, Tim1, BAI1, and stabillin) expressed in macrophages. Apoptotic cells are engulfed by macrophages using the integrins and MER (TAM family kinases) system together with opsonins (MFG-E8, protein S, and Gas6) that bind PtdSer. The dead cells are then transported into lysosomes using the Rac1- and Rab5-mediated intracellular trafficking systems, and the dead cell components are degraded into amino acids, nucleotides, and monosaccharides. (*B*) The enzyme that digests DNA in macrophage lysosomes is DNase II. Macrophages in *DNase II*$^{-/-}$ embryos carry many undigested nuclei and produce cytokines such as IFN-β and TNF-α. IFN-β causes lethal anemia in mouse embryos, whereas TNF-α seems to be responsible for causing polyarthritis. (*B*, From Kawane et al. 2001; reprinted, with permission.)

ious cells, is the major 3′ exonuclease that cleaves mismatched and modified nucleotides. Because of its high homology with *Escherichia coli DnaQ*, Trex1 was originally identified as an enzyme that edits mismatched 3′ termini generated during DNA repair and DNA synthesis (Höss et al. 1999). However, Trex1 is expressed in both proliferating and nonproliferating cells, and its null mutation does not increase spontaneous mutation or tumorigenicity in mice; these observations suggest that Trex1 is either not active in base excision repair, or that it acts redundantly with other 3′ exonucleases (Morita et al. 2004). On the other hand, after studying the origin of DNA fragments accumulated in *TREX1*-null cells, Stetson et al. (2008) proposed that TREX1 cleaves the single-stranded DNA derived from retrotransposons. Retrotransposons are found in both humans and mice [L1 in humans, IAP (intracisternal particles), Etns (early transposons), and mammalian LTR-retrotransposons (MaLTRs) in mice], and comprise ~42% of the total mass of the human genome (Ostertag and Kazazian 2001). Although most retrotransposons are inactive, some intact retrotransposons carry reverse transcriptase, thus autonomously or nonautonomously transposing the elements (Sassaman et al. 1997; Prak and Kazazian 2000). The DNA is reverse-transcribed from RNA in the cytoplasm, and

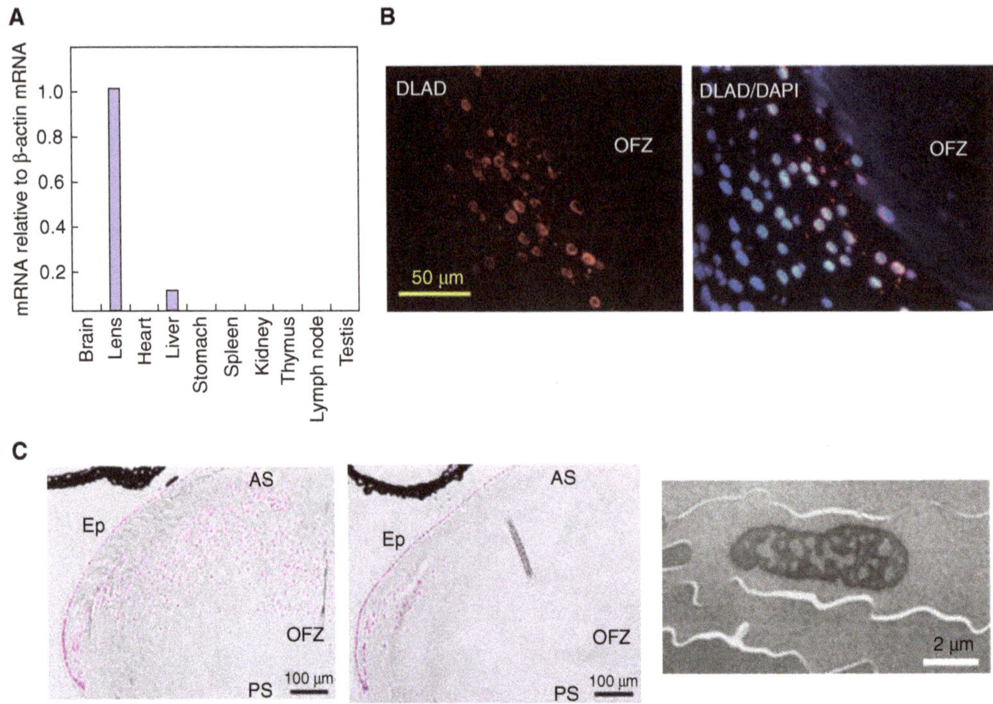

Figure 3. Degradation of DNA in lens fiber cells. DLAD (DNase IIβ) is exclusively expressed in eye lens (*A*). Its expression is induced during differentiation of lens epithelial cells into fiber cells. The DLAD protein is found at the cortical fiber cells that surround organelle-free zone (OFZ) (*B*), suggesting that it is expressed at the final stage of the lens cell differentiation. The lens fiber cells in *DLAD*$^{-/-}$ (*left* panel), but not wild-type (*middle* panel), mice have Feulgen-positive DNA in their cytoplasm (*right* panel; observed by an electron microscope) (*C*), indicating that DLAD is responsible for the cell-autonomous digestion of DNA in fiber cells. (*A,C,* From Nishimoto et al. 2003; reprinted, with permission; *B*, from Nakahara et al. 2007; reprinted, with permission.)

the TREX1 nuclease appears to be responsible for digesting this DNA (Fig. 4) (Stetson et al. 2008).

Aicardi-Goutières syndrome, a human disease characterized by cerebral atrophy, intracranial calcifications, and lymphocytosis with high IFN-α levels in the cerebrospinal fluid, is caused by a *Trex1* loss-of function (Crow and Rehwinkel 2009). Undigested single-stranded DNA that can activate IFN-β gene in *Trex1*$^{-/-}$ cells (Yan et al. 2010) is found in the cytoplasm of various cells in *Trex1*-deficient mice (Stetson et al. 2008). The *Trex1*$^{-/-}$ mice develop a lethal, IFN-α-dependent inflammatory myocarditis (Morita et al. 2004). These properties associated with the strong inflammation by IFN-β are similar to those in *DNase II*–null mice.

INTRACELLULAR SIGNALING, DNA-TO-GENE EXPRESSION

Signal Transduction

DNase II– or *Trex1*-null mice develop autoinflammation leading to anemia, arthritis, and myocarditis. In this process, undigested DNA induces the expression of cytokine genes such as type I IFN and TNF-α. Pathogenic viruses and bacteria are extracellularly or intracellularly recognized by pattern-recognition receptors (PRRs), leading to cytokine production (Medzhitov 2007; Ronald and Beutler 2010; Takeuchi and Akira 2010). Toll-like receptor (TLR) PRRs, a family of type I membrane proteins, recognize pathogenic bacteria or viruses and activate type I IFN and TNF-α genes through the transcrip-

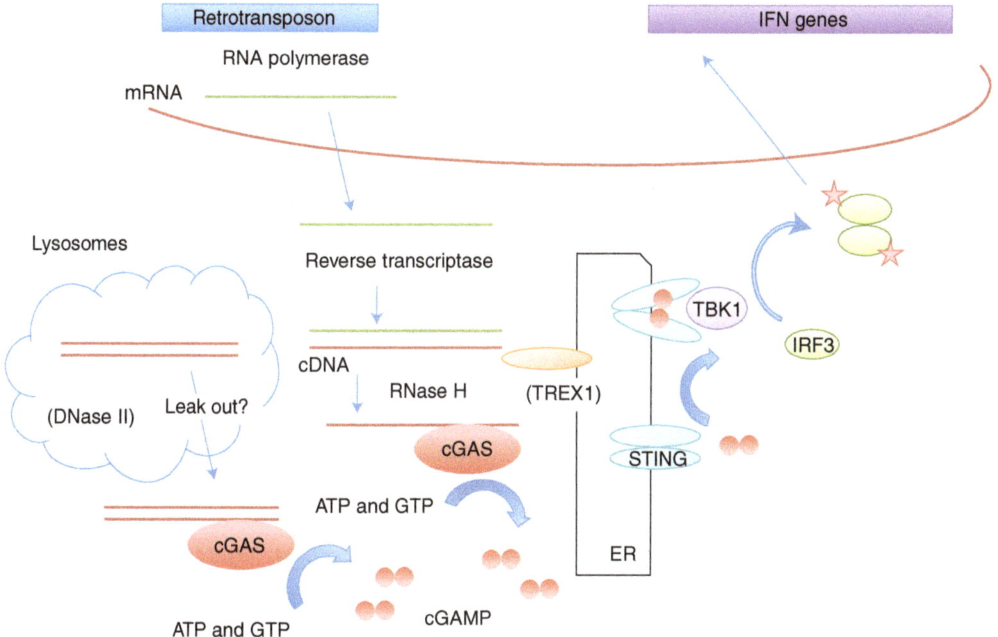

Figure 4. DNA-induced gene expression. Human and mouse cells contain in their genome many retrotransposons that are transcribed by RNA polymerase II into RNA. The RNA is transported into cytoplasm, and reverse-transcribed into cDNA. The complementary DNA (cDNA) is usually digested by TREX1 exonuclease at the endoplasmic reticulum (ER). When TREX1 cannot work, the cDNA activates cGAMP synthase (cGAS) to synthesize cGAMP from ATP and GTP. The cGAMP binds to STING at ER and induces its conformational change from inactive to active dimers, which provides a scaffold for TBK1 kinase to phosphorylate IRF3 transcription factor. The phosphorylated IRF3 dimerizes and enters the nucleus to activate type I IFN genes. The undigested DNA accumulated in the lysosomes of *DNase II*$^{-/-}$ macrophages may leak out from lysosomes and activate the type I IFN genes using the signal transduction described above.

tion factors IFN regulatory factor (IRF)3/IRF7 and NF-κB. The *IRF3/IRF7* double mutation rescues the lethality in *DNase II*–null embryos, indicating that the undegraded DNA in lysosomes leads to IRF3/IRF7 activation (Okabe et al. 2008). However, the double-null mutations of the TLR adaptors Myd88 and TRIF do not affect the lethality in *DNase II*$^{-/-}$ embryos, indicating that the TLR system is not involved in activating the IFN-β gene in macrophages carrying undigested DNA (Okabe et al. 2005). On the other hand, *STING* null mutations fully rescue the lethality in *DNase II*$^{-/-}$ embryos, although they do not reduce the number of macrophages carrying undigested DNA in the bone marrow (Ahn et al. 2012). Similarly, *STING*$^{-/-}$ *Trex1*$^{-/-}$ mice do not develop myocarditis (Gall et al. 2012). STING, also called TMEM173, is a transmembrane protein at the endoplasmic reticulum that is essential for the innate immune response against DNA and RNA viruses such as HSV, HIV, and RSV (Ishikawa and Barber 2008; Ishikawa et al. 2009). Once activated, STING forms a functional dimer and, together with the adaptor protein IPS-1, stimulates the protein kinases IKK and TBK1 to activate the transcription factors NF-κB and IRF3, respectively (Tanaka and Chen 2012). Eyes absent (EYA), a dual phosphatase, also binds IPS-1 and regulates DNA-induced innate immunity (Okabe et al. 2009). These results suggest that if DNA from retrotransposons, apoptotic cells, or pyrenocytes is not properly digested, it activates the innate immunity via the cytosolic pathogen-recognition system. The lysosomes carrying undigested

DNA in *DNase II*$^{-/-}$ macrophages are not apparently damaged, but the recently recognized labile nature of lysosomes (Terman et al. 2006) suggests that some DNA leaks from the lysosomes into the cytoplasm.

DNA Sensors

Cytosolic DNA must be recognized as a danger signal in the cytoplasm to activate the type I *IFN* and *TNF-α*. Many molecules have been proposed to be DNA sensors, including DAI, DDX41, DHX36, DHX9, IFI16, AIM2, Mre11, and HMGB1 (Desmet and Ishii 2012; Paludan and Bowie 2013). The ability of STING to directly bind DNA has also been reported (Abe et al. 2013). Among these sensors, Aim2 was convincingly shown to bind bacterial or viral DNA and to activate inflammasomes in macrophages to produce IL-1β and IL-18 (Fernandes-Alnemri et al. 2010; Rathinam et al. 2010), and is not for the production of type I IFN and TNF-α. Dr. Zhijian Chen's laboratory recently published a series of elegant papers demonstrating that cyclic GMP-AMP (cGAMP), a dicyclic nucleotide, is a second messenger for DNA-driven signal transduction (Gao et al. 2013; Li et al. 2013; Sun et al. 2013; Wu et al. 2013). They show that DNA from DNA viruses or reverse-transcribed from retrovirues binds to cGAMP synthase (cGAS), activating its enzymatic activity to catalyze the synthesis of cGAMP, which contains both $2'-5'$ and $3'-5'$ phosphodiester linkages (O'Neill 2013; Xiao and Fitzgerald 2013). Binding of cGAMP to STING induces its conformational changes to an active dimer, providing a scaffold for TBK1 kinase. Thus, the cGAMP-cGAS system is likely to function as a DNA sensor for the production of type I IFN and TNF-α in *DNase II*–null and *Trex1*-null mice (Fig. 4).

CONCLUDING REMARKS

Here, we reviewed four DNases, CAD, DNase II, DLAD, and Trex1, and diseases caused by their defects. CAD and DLAD defects cause cancer and cataracts, respectively. Considering the purposes of apoptosis and lens cell differentiation to remove damaged or dangerous cells, and to establish transparency, respectively, it is not surprising to find cancers and cataracts in animals that lack CAD or DLAD. On the other hand, the activation of inflammatory cytokine genes in *DNase II*– and *Trex1*-null mice appears to be a response to danger signals. The intracellular signal transduction system by which endogenous DNA activates the inflammatory cytokine genes seems to be identical to that used for bacterial or viral DNA. Bacterial or viral infection is often transient, and inflammation subsides when the infection is cleared. However, apoptotic cell death and definitive erythropoiesis continue throughout our lives. Retrotranspositions appear to be constitutive, as well. Thus, if DNase II or Trex1 does not function properly, cytokine genes are constitutively activated, resulting in chronic inflammation that can cause anemia, arthritis, or myocarditis. If the signaling pathway is identical in *DNase II*– and *Trex1*-null mice, it is not clear why one DNase defect causes arthritis, whereas the defect of another causes myocarditis; it may be because DNase II-mediated DNA degradation occurs mainly in the macrophages, whereas Trex1-mediated degradation occurs in all cell types. In this regard, it may be noteworthy that mice with a cardiac-specific *DNase II* deletion develop severe myocarditis and dilated cardiomyopathy when subjected to pressure overload (Oka et al. 2012).

There are many other DNases besides the four we have discussed. DNase I, a secreted DNase, is present in the blood serum. A loss-of-function *DNase I* mutation in mice (Napirei et al. 2004), and its heterozygous mutation in humans causes SLE-type autoimmune disease (Yasutomo et al. 2001), indicating that undigested extracellular DNA can activate acquired immunity. Endonuclease G, a DNase localized to the mitochondria, was suggested to be involved in apoptotic DNA degradation (Li et al. 2001). However, it was recently shown that endonuclease G stimulates DNA synthesis in mitochondria, and its defect impairs cardiac function (McDermott-Roe et al. 2011). Thus, various DNases are involved in different aspects of cell metabolism, and studies on the biochemical and physiological functions of DNases should contribute to our understanding of human diseases.

ACKNOWLEDGMENTS

We are grateful to members of our laboratory in the Graduate School of Medicine, Kyoto University. Work in our laboratory is supported in part by Grant-in-Aid from the Ministry of Education, Science, and Culture in Japan.

REFERENCES

Abe T, Harashima A, Xia T, Konno H, Konno K, Morales A, Ahn J, Gutman D, Barber GN. 2013. STING recognition of cytoplasmic DNA instigates cellular defense. *Mol Cell* **50:** 5–15.

Ahn J, Gutman D, Saijo S, Barber GN. 2012. STING manifests self DNA-dependent inflammatory disease. *Proc Natl Acad Sci* **109:** 19386–19391.

Bass BP, Tanner EA, Mateos San Martín D, Blute T, Kinser RD, Dolph PJ, McCall K. 2009. Cell-autonomous requirement for *DNaseII* in nonapoptotic cell death. *Cell Death Differ* **16:** 1362–1371.

Bassnett S. 2002. Lens organelle degradation. *Exp Eye Res* **74:** 1–6.

Bassnett S. 2009. On the mechanism of organelle degradation in the vertebrate lens. *Exp Eye Res* **88:** 133–139.

Borden EC, Sen GC, Uze G, Silverman RH, Ransohoff RM, Foster GR, Stark GR. 2007. Interferons at age 50: Past, current and future impact on biomedicine. *Nat Rev Drug Discov* **6:** 975–990.

Boulares AH, Zoltoski AJ, Yakovlev A, Xu M, Smulson ME. 2001. Roles of DNA fragmentation factor and poly(ADP-ribose) polymerase in an amplification phase of tumor necrosis factor-induced apoptosis. *J Biol Chem* **276:** 38185–38192.

Casciola-Rosen LA, Anhalt G, Rosen A. 1994. Autoantigens targeted in systemic lupus erythematosus are clustered in two populations of surface structures on apoptotic keratinocytes. *J Exp Med* **179:** 1317–1330.

Chasis JA, Mohandas N. 2008. Erythroblastic islands: Niches for erythropoiesis. *Blood* **112:** 470–478.

Crow Y, Rehwinkel J. 2009. Aicardi-Goutières syndrome and related phenotypes: Linking nucleic acid metabolism with autoimmunity. *Hum Mol Genet* **18:** R130–R136.

Desmet CJ, Ishii KJ. 2012. Nucleic acid sensing at the interface between innate and adaptive immunity in vaccination. *Nat Rev Immunol* **12:** 479–491.

Dix MM, Simon GM, Cravatt BF. 2008. Global mapping of the topography and magnitude of proteolytic events in apoptosis. *Cell* **134:** 679–691.

Eckhart L, Lippens S, Tschachler E, Declercq W. 2013. Cell death by cornification. *Biochim Biophys Acta* **1833:** 3471–3480.

Enari M, Sakahira H, Yokoyama H, Okawa K, Iwamatsu A, Nagata S. 1998. A caspase-activated DNase that degrades DNA during apoptosis, and its inhibitor ICAD. *Nature* **391:** 43–50.

Evans CJ, Aguilera RJ. 2003. DNase II: Genes, enzymes and function. *Gene* **322:** 1–15.

Fadok VA, Voelker DR, Campbell PA, Cohen JJ, Bratton DL, Henson PM. 1992. Exposure of phosphatidylserine on the surface of apoptotic lymphocytes triggers specific recognition and removal by macrophages. *J Immunol* **148:** 2207–2216.

Fernandes-Alnemri T, Yu J-W, Juliana C, Solorzano L, Kang S, Wu J, Datta P, McCormick M, Huang L, McDermott E, et al. 2010. The AIM2 inflammasome is critical for innate immunity to *Francisella tularensis*. *Nat Immunol* **11:** 385–393.

Fullwood MJ, Lee J, Lin L, Li G, Huss M, Ng P, Sung W-K, Shenolikar S. 2011. Next-generation sequencing of apoptotic DNA breakpoints reveals association with actively transcribed genes and gene translocations. *PLoS ONE* **6:** e26054.

Gall A, Treuting P, Elkon KB, Loo Y-M, Gale M, Barber GN, Stetson DB. 2012. Autoimmunity initiates in nonhematopoietic cells and progresses via lymphocytes in an interferon-dependent autoimmune disease. *Immunity* **36:** 120–131.

Gao D, Wu J, Wu Y-T, Du F, Aroh C, Yan N, Sun L, Chen ZJ. 2013. Cyclic GMP-AMP synthase is an innate immune sensor of HIV and other retroviruses. *Science* **341:** 903–906.

Gavrieli Y, Sherman Y, Ben-Sasson SA. 1992. Identification of programmed cell death in situ via specific labeling of nuclear DNA fragmentation. *J Cell Biol* **119:** 493–501.

Gómez-Benito M, Balsas P, Carvajal-Vergara X, Pandiella A, Anel A, Marzo I, Naval J. 2007. Mechanism of apoptosis induced by IFN-α in human myeloma cells: Role of Jak1 and Bim and potentiation by rapamycin. *Cell Signal* **19:** 844–854.

Hanayama R, Tanaka M, Miwa K, Shinohara A, Iwamatsu A, Nagata S. 2002. Identification of a factor that links apoptotic cells to phagocytes. *Nature* **417:** 182–187.

Höss M, Robins P, Naven TJ, Pappin DJ, Sgouros J, Lindahl T. 1999. A human DNA editing enzyme homologous to the *Escherichia coli* DnaQ/MutD protein. *EMBO J* **18:** 3868–3875.

Ishikawa H, Barber GN. 2008. STING is an endoplasmic reticulum adaptor that facilitates innate immune signalling. *Nature* **455:** 674–678.

Ishikawa H, Ma Z, Barber GN. 2009. STING regulates intracellular DNA-mediated, type I interferon-dependent innate immunity. *Nature* **461:** 788–792.

Jacobson MD, Weil M, Raff MC. 1997. Programmed cell death in animal development. *Cell* **88:** 347–354.

Jain MK, Zoellner C. 2010. Role of ribavirin in HCV treatment response: Now and in the future. *Expert Opin Pharmacother* **11:** 673–683.

Jog NR, Frisoni L, Shi Q, Monestier M, Hernandez S, Craft J, Prak ETL, Caricchio R. 2012. Caspase-activated DNase is required for maintenance of tolerance to lupus nuclear autoantigens. *Arthritis Rheum* **64:** 1247–1256.

Kawane K, Fukuyama H, Kondoh G, Takeda J, Ohsawa Y, Uchiyama Y, Nagata S. 2001. Requirement of DNase II for definitive erythropoiesis in the mouse fetal liver. *Science* **292:** 1546–1549.

Kawane K, Fukuyama H, Yoshida H, Nagase H, Ohsawa Y, Uchiyama Y, Okada K, Iida T, Nagata S. 2003. Impaired thymic development in mouse embryos deficient in apoptotic DNA degradation. *Nat Immunol* **4:** 138–144.

Kawane K, Ohtani M, Miwa K, Kizawa T, Kanbara Y, Yoshioka Y, Yoshikawa H, Nagata S. 2006. Chronic polyarthritis caused by mammalian DNA that escapes from degradation in macrophages. *Nature* **443:** 998–1002.

Kawane K, Tanaka H, Kitahara Y, Shimaoka S, Nagata S. 2010. Cytokine-dependent but acquired immunity-independent arthritis caused by DNA escaped from degradation. *Proc Natl Acad Sci* **107:** 19432–19437.

Keffer J, Probert L, Cazlaris H, Georgopoulos S, Kaslaris E, Kioussis D, Kollias G. 1991. Transgenic mice expressing human tumour necrosis factor: A predictive genetic model of arthritis. *EMBO J* **10:** 4025–4031.

Kinchen JM, Ravichandran KS. 2007. Journey to the grave: Signaling events regulating removal of apoptotic cells. *J Cell Sci* **120:** 2143–2149.

Kitano M, Nakaya M, Nakamura T, Nagata S, Matsuda M. 2008. Imaging of Rab5 activity identifies essential regulators for phagosome maturation. *Nature* **453:** 241–245.

Krieser R, MacLea K, Longnecker D, Fields J, Fiering S, Eastman A. 2002. Deoxyribonuclease IIα is required during the phagocytic phase of apoptosis and its loss causes perinatal lethality. *Cell Death Differ* **9:** 956–962.

Leventis PA, Grinstein S. 2010. The distribution and function of phosphatidylserine in cellular membranes. *Annu Rev Biophys* **39:** 407–427.

Li LY, Luo X, Wang X. 2001. Endonuclease G is an apoptotic DNase when released from mitochondria. *Nature* **412:** 95–99.

Li XD, Wu J, Gao D, Wang H, Sun L, Chen ZJ. 2013. Pivotal roles of cGAS-cGAMP signaling in antiviral defense and immune adjuvant effects. *Science* **341:** 1390–1394.

Liu X, Zou H, Slaughter C, Wang X. 1997. DFF, a heterodimeric protein that functions downstream of caspase-3 to trigger DNA fragmentation during apoptosis. *Cell* **89:** 175–184.

Mahrus S, Trinidad JC, Barkan DT, Sali A, Burlingame AL, Wells JA. 2008. Global sequencing of proteolytic cleavage sites in apoptosis by specific labeling of protein N termini. *Cell* **134:** 866–876.

McAvoy JW, Chamberlain CG, de Iongh RU, Hales AM, Lovicu FJ. 1999. Lens development. *Eye* **13:** 425–437.

McDermott-Roe C, Ye J, Ahmed R, Sun X-M, Serafín A, Ware J, Bottolo L, Muckett P, Cañas X, Zhang J, et al. 2011. Endonuclease G is a novel determinant of cardiac hypertrophy and mitochondrial function. *Nature* **478:** 114–118.

McGrath K, Kingsley P, Koniski A, Porter R, Bushnell T, Palis J. 2008. Enucleation of primitive erythroid cells generates a transient population of "pyrenocytes" in the mammalian fetus. *Blood* **111:** 2409–2417.

McIlroy D, Tanaka M, Sakahira H, Fukuyama H, Suzuki M, Yamamura K, Ohsawa Y, Uchiyama Y, Nagata S. 2000. An auxiliary mode of apoptotic DNA fragmentation provided by phagocytes. *Genes Dev* **14:** 549–558.

Medzhitov R. 2007. Recognition of microorganisms and activation of the immune response. *Nature* **449:** 819–826.

Miyanishi M, Tada K, Koike M, Uchiyama Y, Kitamura T, Nagata S. 2007. Identification of Tim4 as a phosphatidylserine receptor. *Nature* **450:** 435–439.

Mizuta R, Mizuta M, Araki S, Suzuki K, Ebara S, Furukawa Y, Shiokawa D, Tanuma S, Kitamura D. 2009. DNase γ-dependent and -independent apoptotic DNA fragmentations in Ramos Burkitt's lymphoma cell line. *Biomed Res* **30:** 165–170.

Morita M, Stamp G, Robins P, Dulic A, Rosewell I, Hrivnak G, Daly G, Lindahl T, Barnes DE. 2004. Gene-targeted mice lacking the Trex1 (DNase III) 3′→5′ DNA exonuclease develop inflammatory myocarditis. *Mol Cell Biol* **24:** 6719–6727.

Mukae N, Enari M, Sakahira H, Fukuda Y, Inazawa J, Toh H, Nagata S. 1998. Molecular cloning and characterization of human caspase-activated DNase. *Proc Natl Acad Sci* **95:** 9123–9128.

Mukae N, Yokoyama H, Yokokura T, Sakoyama Y, Nagata S. 2002. Activation of the innate immunity in *Drosophila* by endogenous chromosomal DNA that escaped apoptotic degradation. *Genes Dev* **16:** 2662–2671.

Nagase H, Fukuyama H, Tanaka M, Kawane K, Nagata S. 2003. Mutually regulated expression of caspase-activated DNase and its inhibitor for apoptotic DNA fragmentation. *Cell Death Differ* **10:** 142–143.

Nagata S. 1997. Apoptosis by death factor. *Cell* **88:** 355–365.

Nagata S. 2005. DNA degradation in development and programmed cell death. *Annu Rev Immunol* **23:** 853–875.

Nakahara M, Nagasaka A, Koike M, Uchida K, Kawane K, Uchiyama Y, Nagata S. 2007. Degradation of nuclear DNA by DNase II-like acid DNase in cortical fiber cells of mouse eye lens. *FEBS J* **274:** 3055–3064.

Nakaya M, Kitano M, Matsuda M, Nagata S. 2008. Spatiotemporal activation of Rac1 for engulfment of apoptotic cells. *Proc Natl Acad Sci* **105:** 9198–9203.

Napirei M, Wulf S, Mannherz H. 2004. Chromatin breakdown during necrosis by serum Dnase1 and the plasminogen system. *Arthritis Rheum* **50:** 1873–1883.

Nishimoto S, Kawane K, Watanabe-Fukunaga R, Fukuyama H, Ohsawa Y, Uchiyama Y, Hashida N, Ohguro N, Tano Y, Morimoto T, et al. 2003. Nuclear cataract caused by a lack of DNA degradation in the mouse eye lens. *Nature* **424:** 1071–1074.

Oka T, Hikoso S, Yamaguchi O, Taneike M, Takeda T, Tamai T, Oyabu J, Murakawa T, Nakayama H, Nishida K, et al. 2012. Mitochondrial DNA that escapes from autophagy causes inflammation and heart failure. *Nature* **485:** 251–255.

Okabe Y, Kawane K, Akira S, Taniguchi T, Nagata S. 2005. Toll-like receptor-independent gene induction program activated by mammalian DNA escaped from apoptotic DNA degradation. *J Exp Med* **202:** 1333–1339.

Okabe Y, Kawane K, Nagata S. 2008. IFN regulatory factor (IRF) 3/7-dependent and -independent gene induction by mammalian DNA that escapes degradation. *Eur J Immunol* **38:** 3150–3158.

Okabe Y, Sano T, Nagata S. 2009. Regulation of the innate immune response by threonine-phosphatase of eyes absent. *Nature* **460:** 520–524.

O'Neill LAJ. 2013. Immunology. Sensing the dark side of DNA. *Science* **339:** 763–764.

Ostertag EM, Kazazian HH. 2001. Biology of mammalian L1 retrotransposons. *Annu Rev Genet* **35:** 501–538.

Palis J. 2008. Ontogeny of erythropoiesis. *Curr Opin Hematol* **15:** 155–161.

Paludan SR, Bowie AG. 2013. Immune sensing of DNA. *Immunity* **38:** 870–880.

Park D, Tosello-Trampont AC, Elliott MR, Lu M, Haney LB, Ma Z, Klibanov AL, Mandell JW, Ravichandran KS. 2007. BAI1 is an engulfment receptor for apoptotic cells upstream of the ELMO/Dock180/Rac module. *Nature* **450:** 430–434.

Park SY, Jung MY, Kim HJ, Lee SJ, Kim SY, Lee BH, Kwon TH, Park RW, Kim IS. 2008. Rapid cell corpse clearance by stabilin-2, a membrane phosphatidylserine receptor. *Cell Death Differ* **15:** 192–201.

Penninger JM, Kroemer G. 2003. Mitochondria, AIF and caspases—Rivaling for cell death execution. *Nat Cell Biol* **5:** 97–99.

Prak ET, Kazazian HH. 2000. Mobile elements and the human genome. *Nat Rev Genet* **1:** 134–144.

Rathinam VAK, Jiang Z, Waggoner SN, Sharma S, Cole LE, Waggoner L, Vanaja SK, Monks BG, Ganesan S, Latz E, et al. 2010. The AIM2 inflammasome is essential for host defense against cytosolic bacteria and DNA viruses. *Nat Immunol* **11:** 395–402.

Rekvig OP, Mortensen ES. 2012. Immunity and autoimmunity to dsDNA and chromatin—The role of immunogenic DNA-binding proteins and nuclease deficiencies. *Autoimmunity* **45:** 588–592.

Rhodes MM, Kopsombut P, Bondurant MC, Price JO, Koury MJ. 2008. Adherence to macrophages in erythroblastic islands enhances erythroblast proliferation and increases erythrocyte production by a different mechanism than erythropoietin. *Blood* **111:** 1700–1708.

Ronald PC, Beutler B. 2010. Plant and animal sensors of conserved microbial signatures. *Science* **330:** 1061–1064.

Sakahira H, Nagata S. 2002. Co-translational folding of caspase-activated DNase with Hsp70, Hsp40 and inhibitor of caspase-activated DNase. *J Biol Chem* **277:** 3364–3370.

Sakahira H, Enari M, Nagata S. 1998. Cleavage of CAD inhibitor in CAD activation and DNA degradation during apoptosis. *Nature* **391:** 96–99.

Sakahira H, Iwamatsu A, Nagata S. 2000. Specific chaperone-like activity of inhibitor of caspase-activated DNase for caspase-activated DNase. *J Biol Chem* **275:** 8091–8096.

Sassaman DM, Dombroski BA, Moran JV, Kimberland ML, Naas TP, DeBerardinis RJ, Gabriel A, Swergold GD, Kazazian HH. 1997. Many human L1 elements are capable of retrotransposition. *Nat Genet* **16:** 37–43.

Scott RS, McMahon EJ, Pop SM, Reap EA, Caricchio R, Cohen PL, Earp HS, Matsushima GK. 2001. Phagocytosis and clearance of apoptotic cells is mediated by MER. *Nature* **411:** 207–211.

Scott D, Wolfe F, Huizinga T. 2010. Rheumatoid arthritis. *Lancet* **376:** 1094–1108.

Shiokawa D, Tanuma S. 1999. DLAD, a novel mammalian divalent cation-independent endonuclease with homology to DNase II. *Nucleic Acids Res* **27:** 4083–4089.

Stetson DB, Ko JS, Heidmann T, Medzhitov R. 2008. Trex1 prevents cell-intrinsic initiation of autoimmunity. *Cell* **134:** 587–598.

Strasser A, Jost PJ, Nagata S. 2009. The many roles of FAS receptor signaling in the immune system. *Immunity* **30:** 180–192.

Su DM, Wang J, Lin Q, Cooper MD, Watanabe T. 1997. Interferons α/β inhibit IL-7-induced proliferation of $CD4^- CD8^- CD3^- CD44^+ CD25^+$ thymocytes, but do not inhibit that of $CD4^- CD8^- CD3^- CD44^- CD25^-$ thymocytes. *Immunology* **90:** 543–549.

Sun L, Wu J, Du F, Chen X, Chen ZJ. 2013. Cyclic GMP-AMP synthase is a cytosolic DNA sensor that activates the type I interferon pathway. *Science* **339:** 786–791.

Susin SA, Daugas E, Ravagnan L, Samejima K, Zamzami N, Loeffler M, Costantini P, Ferri KF, Irinopoulou T, Prevost MC, et al. 2000. Two distinct pathways leading to nuclear apoptosis. *J Exp Med* **192:** 571–580.

Suzuki J, Denning DP, Imanishi E, Horvitz HR, Nagata S. 2013. Xk-related protein 8 and CED-8 promote phosphatidylserine exposure in apoptotic cells. *Science* **341:** 403–406.

Takeuchi O, Akira S. 2010. Pattern recognition receptors and inflammation. *Cell* **140:** 805–820.

Tanaka Y, Chen ZJ. 2012. STING specifies IRF3 phosphorylation by TBK1 in the cytosolic DNA signaling pathway. *Sci Signal* **5:** ra20–ra20.

Tang HL, Tang HM, Mak KH, Hu S, Wang SS, Wong KM, Wong CST, Wu HY, Law HT, Liu K, et al. 2012. Cell survival, DNA damage, and oncogenic transformation after a transient and reversible apoptotic response. *Mol Biol Cell* **23:** 2240–2252.

Terman A, Kurz T, Gustafsson B, Brunk UT. 2006. Lysosomal labilization. *IUBMB Life* **58:** 531–539.

Toda S, Hanayama R, Nagata S. 2012. Two-step engulfment of apoptotic cells. *Mol Cell Biol* **32:** 118–125.

Vastert SJ, Kuis W, Grom AA. 2009. Systemic JIA: New developments in the understanding of the pathophysiology and therapy. *Best Pract Res Clin Rheumatol* **23:** 655–664.

von Figura K, Hasilik A. 1986. Lysosomal enzymes and their receptors. *Annu Rev Biochem* **55:** 167–193.

Woo E-J, Kim Y-G, Kim M-S, Han W-D, Shin S, Robinson H, Park S-Y, Oh B-H. 2004. Structural mechanism for inactivation and activation of CAD/DFF40 in the apoptotic pathway. *Mol Cell* **14:** 531–539.

Wu YC, Stanfield GM, Horvitz HR. 2000. NUC-1, a *Caenorhabditis elegans* DNase II homolog, functions in an intermediate step of DNA degradation during apoptosis. *Genes Dev* **14:** 536–548.

Wu J, Sun L, Chen X, Du F, Shi H, Chen C, Chen ZJ. 2013. Cyclic GMP-AMP is an endogenous second messenger in innate immune signaling by cytosolic DNA. *Science* **339:** 826–830.

Wyllie AH. 1980. Glucocorticoid-induced thymocyte apoptosis is associated with endogenous endonuclease activation. *Nature* **284:** 555–556.

Xiao TS, Fitzgerald KA. 2013. The cGAS-STING pathway for DNA sensing. *Mol Cell* **51:** 135–139.

Xiong W, Chen Y, Wang H, Wu H, Lu Q, Han D. 2008. Gas6 and the Tyro 3 receptor tyrosine kinase subfamily regulate the phagocytic function of Sertoli cells. *Reproduction* **135:** 77–87.

Yan B, Wang H, Peng Y, Hu Y, Wang H, Zhang X, Chen Q, Bedford JS, Dewhirst MW, Li C-Y. 2006. A unique role of the DNA fragmentation factor in maintaining genomic stability. *Proc Natl Acad Sci* **103:** 1504–1509.

Yan N, Regalado-Magdos A, Stiggelbout B, Lee-Kirsch M, Lieberman J. 2010. The cytosolic exonuclease TREX1 inhibits the innate immune response to human immunodeficiency virus type 1. *Nat Immunol* **11:** 1005–1013.

Yang HW, Chen YZ, Piao HY, Takita J, Soeda E, Hayashi Y. 2001. *DNA fragmentation factor 45 (DFF45)* gene at 1p36.2 is homozygously deleted and encodes variant transcripts in neuroblastoma cell line. *Neoplasia* **3:** 165–169.

Yasutomo K, Horiuchi T, Kagami S, Tsukamoto H, Hashimura C, Urushihara M, Kuroda Y. 2001. Mutation of *DNASE1* in people with systemic lupus erythematosus. *Nat Genet* **28:** 313–314.

Yoshida H, Kawane K, Koike M, Mori Y, Uchiyama Y, Nagata S. 2005a. Phosphatidylserine-dependent engulfment by macrophages of nuclei from erythroid precursor cells. *Nature* **437:** 754–758.

Yoshida H, Okabe Y, Kawane K, Fukuyama H, Nagata S. 2005b. Lethal anemia caused by interferon-β produced in mouse embryos carrying undigested DNA. *Nat Immunol* **6:** 49–56.

Zhang J, Liu X, Scherer DC, van Kaer L, Wang X, Xu M. 1998. Resistance to DNA fragmentation and chromatin condensation in mice lacking the DNA fragmentation factor 45. *Proc Natl Acad Sci* **95:** 12480–12485.

Group 2 Innate Lymphoid Cells in Health and Disease

Brian S. Kim[1,2] and David Artis[3]

[1]Division of Dermatology, Department of Medicine, Washington University School of Medicine, St. Louis, Missouri 63110

[2]Center for the Study of Itch, Washington University School of Medicine, St. Louis, Missouri 63110

[3]Weill Cornell Medical College, Cornell University, New York, New York 10021

Correspondence: dartis@med.cornell.edu

Group 2 innate lymphoid cells (ILC2s) play critical roles in anti-helminth immunity, airway epithelial repair, and metabolic homeostasis. Recently, these cells have also emerged as key players in the development of allergic inflammation at multiple barrier surfaces. ILC2s arise from common lymphoid progenitors in the bone marrow, are dependent on the transcription factors RORα, GATA3, and TCF-1, and produce the type 2 cytokines interleukin (IL)-4, IL-5, IL-9, and/or IL-13. The epithelial cell–derived cytokines IL-25, IL-33, and TSLP regulate the activation and effector functions of ILC2s, and recent studies suggest that their responsiveness to these cytokines and other factors may depend on their tissue environment. In this review, we focus on recent advances in our understanding of the various factors that regulate ILC2 function in the context of immunity, inflammation, and tissue repair across multiple organ systems.

Innate lymphoid cells (ILCs) are part of a family of innate immune cells that are heterogeneous in their expression of transcription factors and production of effector cytokines (Bjorkstrom et al. 2013; Spits et al. 2013). ILCs do not express cell lineage (Lin) markers associated with T cells, B cells, dendritic cells (DCs), macrophages, and granulocytes, but do express CD90 (Thy1 antigen), CD25 (interleukin [IL]-2Rα), and CD127 (IL-7Rα) (Spits et al. 2013). These cells are derived from a common lymphoid progenitor, and their development is dependent on the common γ-chain (γc or CD132), IL-7, Notch, and the transcription factor inhibitor of DNA binding 2 (Id2) (Yokota et al. 1999; Satoh-Takayama et al. 2010; Monticelli et al. 2011; Wong et al. 2012). More recent studies indicate that the majority of ILCs are also dependent on the transcriptional repressor PLZF and that all ILC subsets arise from a Lin^- $Id2^+$ $CD127^+$ $CD25^-$ $\alpha_4\beta_7{}^+$ precursor (Constantinides et al. 2014; Klose et al. 2014). ILCs are currently categorized into three distinct populations based on their differential devel-

Cite this article as *Cold Spring Harb Perspect Biol* doi: 10.1101/cshperspect.a016337

opmental requirements, expression of defined transcription factors, and their expression of cell surface markers and effector cytokines (Spits and Cupedo 2012; Fuchs and Colonna 2013; Kim et al. 2013b; Spits et al. 2013, Walker et al. 2013): group 1 ILCs (ILC1s) include classical NK cells and T-bet-dependent, IFN-γ-producing ILCs; RORα- (Halim et al. 2012b; Wong et al. 2012), GATA3- (Hoyler et al. 2012; Klein et al. 2013), and TCF-1-dependent (Yang et al. 2013) group 2 ILCs (ILC2s) produce IL-4, IL-5, IL-9, IL-13, and/or amphiregulin (Monticelli et al. 2011); and RORγt-dependent group 3 ILCs (ILC3s) produce IL-17A and/or IL-22 (Fig. 1) (Sonnenberg and Artis 2012). These ILC populations are functionally analogous to the previously described T_H1, T_H2, and T_H17 $CD4^+$ T helper cell subsets, respectively. However, although ILCs exhibit shared functions with adaptive $CD4^+$ T cells, they are unique in that they respond to innate signals in the absence of antigen specificity, lack T-cell receptors, and have distinct phenotypic and functional profiles.

Different subsets of ILCs promote either tissue homeostasis or detrimental inflammatory processes at multiple epithelial barrier surfaces (Monticelli et al. 2011; Sonnenberg et al. 2012; Hepworth et al. 2013; Qiu et al. 2013). Further, these cells have been implicated in a variety of different disease states including allergy, autoimmunity, cancer, infection, and obesity (Sonnenberg and Artis 2012; Kim et al. 2013b; Molofsky et al. 2013; Nussbaum et al. 2013). The roles of the ILC1 and ILC3 subsets in various diseases have been covered elsewhere (Spits and Cupedo 2012; Fuchs and Colonna 2013; Sonnenberg 2013; Sonnenberg et al. 2013; Spits

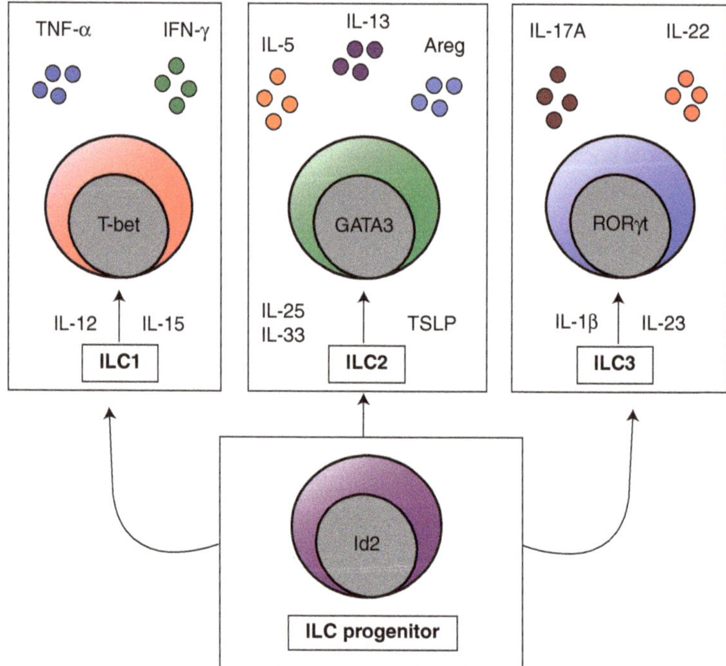

Figure 1. The innate lymphoid cell family. Innate lymphoid cells (ILCs) are a heterogeneous family of innate immune cells that arise from a common Id2-dependent lymphoid progenitor in the bone marrow. Group 1 ILCs (ILC1s) respond to IL-12 and IL-15, express the transcription factor T-bet, and produce tumor necrosis factor α (TNF-α) and interferon γ (IFN-γ). Group 2 ILCs (ILC2s) respond to IL-25, IL-33, and thymic stromal lymphopoietin (TSLP), express the transcription factor GATA3, and produce IL-4, IL-5, IL-9, IL-13, and amphiregulin (Areg). Group 3 ILCs (ILC3s) respond to IL-23 and IL-1β, express the transcription factor RORγt, and produce IL-17A and IL-22.

 Cite this article as *Cold Spring Harb Perspect Biol* doi: 10.1101/cshperspect.a016337

et al. 2013). Therefore, in this review, we will focus primarily on the emerging role of ILC2s in both health and disease across multiple organ systems. First, we will introduce ILC2s and the context in which these cells were originally identified. Second, we will give an overview of the factors that broadly regulate ILC2s and their known effector functions. Finally, we will discuss recent advances in our understanding of how ILC2s contribute to both homeostasis and inflammation in a tissue-specific and disease-oriented manner.

IDENTIFICATION OF ILC2s

ILC2s were originally identified as key contributors to the development of protective immunity to the parasite *Nippostrongylus brasiliensis* in the gut (Fort et al. 2001; Schmitz et al. 2005; Fallon et al. 2006; Moro et al. 2010; Neill et al. 2010; Price et al. 2010). In this context, ILC2s were found to be a critical source of IL-5 and IL-13, the latter of which promotes the induction of mucous secretion from goblet cells and smooth muscle contraction that contribute to anti-helminth immunity (Moro et al. 2010; Neill et al. 2010; Price et al. 2010). These studies were the first to uncover a previously unrecognized population of ILC2s that promotes type 2 cytokine-mediated immunity. Around the same time, an ILC2-like population, named multipotent progenitor type 2 (MPP^{type2}) cells, was also identified and shown to promote type 2 cytokine-mediated immunity to helminth infection (Saenz et al. 2010). Subsequently, MPP^{type2} cells have been distinguished from ILC2s by their preferential responsiveness to IL-25 rather than IL-33, their progenitor-like phenotype, and their ability to differentiate into multiple granulocyte populations (Saenz et al. 2013). Further, MPP^{type2} cells exhibit distinct developmental requirements based on their Id2-independence, altered genome-wide transcriptional profiles from ILC2s, and their capacity to undergo extramedullary hematopoiesis (Saenz et al. 2013). Given that MPP^{type2} cells and ILC2s are distinct populations, this review will focus specifically on the biology of ILC2s in the context of health and disease.

REGULATION AND EFFECTOR FUNCTIONS OF ILC2s

In the original studies that identified ILC2s, the epithelial cell-derived cytokines IL-25 and IL-33 were found to be potent activators of ILC2s, resulting in enhanced production of the key effector cytokines IL-5 and IL-13 (Fort et al. 2001; Schmitz et al. 2005; Fallon et al. 2006; Moro et al. 2010; Neill et al. 2010; Price et al. 2010). Subsequently, the predominantly epithelial cell–derived cytokine thymic stromal lymphopoietin (TSLP) has also been shown to be a key regulator of ILC2 function (Halim et al. 2012a; Kim et al. 2013a). Recent studies have shown that TSLP induces GATA3 expression in human ILC2s (Mjosberg et al. 2012) and promotes corticosteroid resistance to IL-33-mediated activation of ILC2s in the lung (Kabata et al. 2013). In addition, enforced expression of GATA3 in T cells and ILC2s results in elevated expression of IL-5 and IL-13 and enhanced susceptibility to allergic airway disease in mice (Kleinjan et al. 2014). These findings were consistent with prior studies by Mjosberg et al. showing that ectopic expression of GATA3 results in increased type 2 cytokine production in human ILC2s (Mjosberg et al. 2012). In addition to IL-5 and IL-13, in some circumstances, human and murine ILC2s can also produce the type 2 cytokines IL-4 and IL-9 (Wilhelm et al. 2011; Mjosberg et al. 2012; Doherty et al. 2013; Kleinjan et al. 2014). Recently, IL-9 has been shown to play a critical role in ILC2 survival in the context of lung infection with *N. brasiliensis* (Licona-Limon et al. 2013; Turner et al. 2013), suggesting that this ILC2 effector cytokine participates in a positive-feedback system. ILC2s are also a critical source of the epidermal growth factor receptor (EGFR) ligand Areg, which mediates lung epithelial repair following influenza infection (Monticelli et al. 2011). Although originally identified in association with lung ILC2s, Areg expression has also been shown in human skin ILC2s (Salimi et al. 2013). Taken together, these studies show that ILC2s are potently activated by the epithelial cell-derived cytokines IL-25, IL-33, and TSLP, and produce a variety of type 2 cytokines as well as the EGFR ligand Areg in mediating both immunity and epithelial repair (Fig. 2).

The regulation of ILC2 activation and acquisition of effector functions appears to involve a complex network of signals at the epithelial barrier surface. ILC2s are elicited by a variety of factors such as microbial pathogens, helminth parasites, and allergens (Monticelli et al. 2012; Sonnenberg and Artis 2012). In addition to the epithelial cell–derived cytokines IL-25, IL-33, and TSLP, ILC2s express the receptors for and are activated by IL-2 (CD25) and IL-7 (CD127), indicating that they also respond to other stromal and hematopoietic cell-derived cytokines (Roediger et al. 2013). Further, ILC2s have also recently been shown to respond directly to non-cytokine molecules, including eicosanoids. Specifically, ILC2s are activated by prostaglandin D_2 (PGD_2) (Barnig et al. 2013; Chang et al. 2013; Xue et al. 2013) and leukotriene D_4 (LTD_4) (Doherty et al. 2013) and are inhibited by lipoxin A_4 (LXA_4) (Barnig et al. 2013). In addition to activation, human ILC2s express the PGD_2 receptor chemoattractant receptor-homologous molecule expressed on T_H2 cells (CRTH2) and migrate in vitro in response to PGD_2 (Chang et al. 2013; Xue et al. 2013). Collectively, these studies show that ILC2s respond to a variety of signals that modulate both their cytokine production as well as chemotaxis (Fig. 2).

ILC2s IN THE GUT

ILC2s were originally identified in gastrointestinal tissue and fat-associated lymphoid clusters (FALCs), highlighting an important role for these cells in mediating protective immune responses and inflammation in the gut (Moro et al. 2010; Neill et al. 2010; Price et al. 2010). These original studies showed that IL-25- and IL-33-responsive ILC2s were critical for the development of type 2 cytokine–associated inflammation and goblet cell hyperplasia that facilitate expulsion of *N. brasiliensis* in the absence of adaptive immunity (Moro et al. 2010; Neill et al. 2010; Price et al. 2010). A more recent study has shown that IL-33 is critical for the induction of IL-13 production by ILC2s to mediate worm expulsion (Hung et al. 2013). In addition to their role in mediating protective immunity to helminth parasites, ILC2s also contribute to the

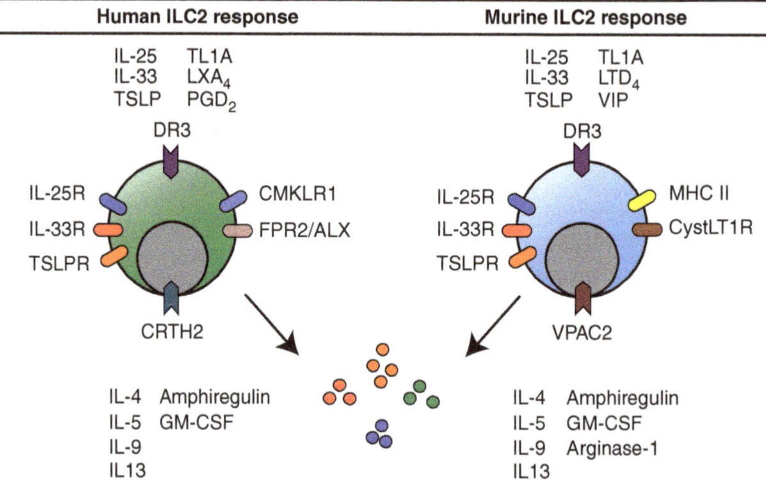

Figure 2. Regulation of human and murine ILC2 responses. Human ILC2s are activated by IL-25, IL-33, TSLP, TL1A, and prostaglandin D2 (PGD_2) and inhibited by lipoxin A_4 (LXA_4). Murine ILC2s are activated by IL-25, IL-33, TSLP, TL1A, leukotriene D4 (LTD_4), and vasoactive intestinal peptide (VIP). Both human and murine ILC2s express the receptors to respond directly to these mediators and produce IL-4, IL-5, IL-9, IL-13, granulocyte macrophage colony–stimulating factor (GM-CSF), and amphiregulin (Areg). Murine ILC2s express arginase-1.

Cite this article as *Cold Spring Harb Perspect Biol* doi: 10.1101/cshperspect.a016337

development of pathologic type 2 inflammation, as they promoted gut inflammation in an IL-25-dependent fashion in a murine model of oxazalone-induced colitis (Camelo et al. 2012). In the context of food allergy, elevated IL-25, IL-33, and TSLP responses have been observed in murine models and in patients (Blazquez et al. 2010; Herberth et al. 2010; Chu et al. 2013), provoking the hypothesis that these cytokines may promote ILC2 responses that contribute to the development of inflammation in the gut in response to food antigens through the expression of the type 2 cytokines IL-4, IL-5, and IL-13. However, the specific role of ILC2s in murine models of food allergy remains to be examined. Additionally, ILC2s have been characterized in human fetal gut and in the gut of healthy subjects as well as patients with inflammatory bowel disease (Mjosberg et al. 2011). However, accumulation of these cells in inflamed human intestinal tissue has not been shown. Further studies will be required to elucidate the role of ILC2s in promoting intestinal allergic inflammation in humans.

ILC2s IN THE RESPIRATORY TRACT

ILC2s and Influenza

Murine ILC2s in the lung were first described in murine models of influenza, where they were found to promote pathologic airway hyperreactivity (AHR) (Chang et al. 2011) as well as protective epithelial repair (Monticelli et al. 2011) in response to virus-induced lung inflammation. The latter study showed that ILC2s were a critical source of the EGFR ligand Areg in the lung following influenza virus infection, and treatment with recombinant Areg was sufficient to restore lung epithelial barrier integrity (Monticelli et al. 2011). A more recent study has shown that the interactions between ILC2s and NKT cells result in the production of IL-5 from ILC2s, which promotes the accumulation of eosinophils in the lung during the recovery phase of influenza infection (Gorski et al. 2013). However, the precise role of ILC2s in the pathogenesis of human influenza infection remains to be determined.

ILC2s and Allergic Airway Inflammation

Although ILC2s in the lung were first described in murine models of influenza (Chang et al. 2011; Monticelli et al. 2011), these cells have recently been shown to play a critical role in regulating the development of allergic airway inflammation. For example, IL-33-induced ILC2s that produce IL-13 contributed to the development of AHR in multiple murine asthma models in the absence of $CD4^+$ T cells (Bartemes et al. 2012; Beamer et al. 2012; Doherty et al. 2012; Kim et al. 2012; Salmond et al. 2012). Similarly, in models of allergen-induced airway inflammation, IL-25-, IL-33-, and TSLP-responsive ILC2s were critical for the development of allergic airway inflammation in lymphocyte-deficient mice (Halim et al. 2012a, 2012b; Klein et al. 2012). Although all of these studies showed that lung-resident ILC2s produce the effector cytokines IL-5 and IL-13, Wilhelm et al. further showed a critical role for ILC2-derived IL-9 in the context of papain-induced lung inflammation. Induction of allergic inflammation in IL-9 reporter mice revealed that ILC2s express IL-9 in an IL-2-dependent manner, which played a critical role in promoting the survival of ILC2s and in the induction of IL-5 and IL-13 expression (Wilhelm et al. 2011). A more recent study has also shown a critical role for IL-9 in mediating ILC2 survival in the lung during infection with *N. brasiliensis* (Turner et al. 2013).

Multiple reports indicate that IL-33 is a potent activator of IL-13-producing ILC2s in allergic airway inflammation (Bartemes et al. 2012; Beamer et al. 2012; Mjosberg et al. 2012; Salmond et al. 2012; Hung et al. 2013; Shaw et al. 2013), but other epithelial cell-derived cytokines, bioactive lipids, and inflammatory factors also contribute to the development of pathogenic ILC2 responses in the lung. A recent study has shown that TSLP can contribute to ILC2 activation that promotes corticosteroid resistance in the context of IL-33-mediated airway inflammation (Kabata et al. 2013). Further, recent reports suggest that other factors, such as eicosanoids, could also play a key role in promoting ILC2 responses in the inflamed lung

(Barnig et al. 2013; Doherty et al. 2013). For instance, lung ILC2s express the receptor for LTD_4 and ligation of this receptor rapidly induces IL-5 production by ILC2s. This process is abrogated by montelukast, a leukotriene receptor antagonist used in the treatment of asthma (Doherty et al. 2013). In the same study, LTD_4, but not IL-33, induced high levels of IL-4 production by ILC2s (Doherty et al. 2013). Furthermore, in vitro studies have shown that PGD_2 derived from mast cells activates ILC2s, up-regulates expression of receptors for IL-25 (IL-17RA) and IL-33 (ST2), and induces chemotaxis of human ILC2s (Barnig et al. 2013; Chang et al. 2013; Xue et al. 2013). Furthermore, LXA_4, a proresolving factor known to be decreased in severe asthma, directly inhibited PGD_2-mediated activation of human ILC2s from peripheral blood (Xue et al. 2013). LXA_4 is an endogenous ligand for the receptor FPR2/ALX and functions to limit inflammation in asthma. Importantly, LXA_4 in conjunction with other proresolving receptors such as CMKLR1 was found to be expressed on human ILC2s, suggesting that these anti-inflammatory pathways may be operative on ILC2s (Barnig et al. 2013). Finally, two groups independently identified that the TNF-family cytokine TL1A promotes allergic airway inflammation and pathology in response to papain (Yu et al. 2013; Meylan et al. 2014), and that human peripheral blood ILC2s are activated by TL1A in synergy with IL-25 and IL-33 (Yu et al. 2013). These studies indicate that various epithelial cell-derived cytokines might coordinately regulate ILC2 phenotype and function in inflamed tissue in conjunction with eicosinoids and TL1A. However, further studies will be required to fully characterize the factors that regulate ILC2 activation during allergic airway inflammation and the mechanisms by which these cells migrate into and out of lung tissue.

Beyond the factors that regulate pathogenic ILC2 responses during allergic airway inflammation, recent studies have identified ILC2s as novel sources of growth factors and enzymes that can regulate epithelial repair and inflammation in the lung. As mentioned above, Monticelli et al. (2011) identified that the EGFR ligand Areg critically regulates lung epithelial regeneration in the context of influenza infection (Monticelli et al. 2011). More recently, ILC2s have been shown to be a constitutive and dominant source of arginase-1 (Arg1) in healthy lung tissue (Bando et al. 2013). This was an unexpected finding, given that alternatively activated macrophages (AAMs) have traditionally been considered a significant source of Arg1, which regulates lung inflammation in asthma (Maarsingh et al. 2009; Pesce et al. 2009). However, the precise mechanisms and roles of these factors in regulating lung inflammation and repair remain poorly understood and is an active area of investigation (Fig. 3).

Although ILC2s have not been shown to accumulate in the lung of human asthmatic patients to date, numerous studies have suggested that human ILC2s in the lung could contribute to the development of allergic airway inflammation. Human lung ILC2s were first identified by flow cytometry as Lin^- $CD127^+$ $CRTH2^+$ and Lin^- $CD127^+$ $CD25^+$ $IL\text{-}33R^+$ cells in healthy fetal and adult lung tissue (Mjosberg et al. 2011; Monticelli et al. 2011), and were subsequently visualized by immunofluorescence as Lin^- c-Kit^+ $CD161^+$ cells (Barnig et al. 2013). Elevated expression of IL-25, IL-33, and TSLP has been shown in human asthmatic lung tissue (Prefontaine et al. 2010; Corrigan et al. 2011; Shikotra et al. 2012), and human peripheral blood ILC2s responded to asthma-associated PGD_2 by producing IL-13 (Barnig et al. 2013). In addition, pathogenic LTD_4-initiated lung ILC2 responses in mice were abrogated following treatment with montelukast (Doherty et al. 2013). Together, these studies suggest that ILC2s may be relevant targets in the treatment of asthma in patients. Further studies in human subjects will be required to determine whether ILC2s are directly pathogenic in human asthma and whether these cells could be targeted therapeutically.

ILC2s and Other Lung Diseases

Although the identification of ILC2s has generated significant interest in their role in allergic airway inflammation, ILC2s have been implicated in other pathologic pulmonary pro-

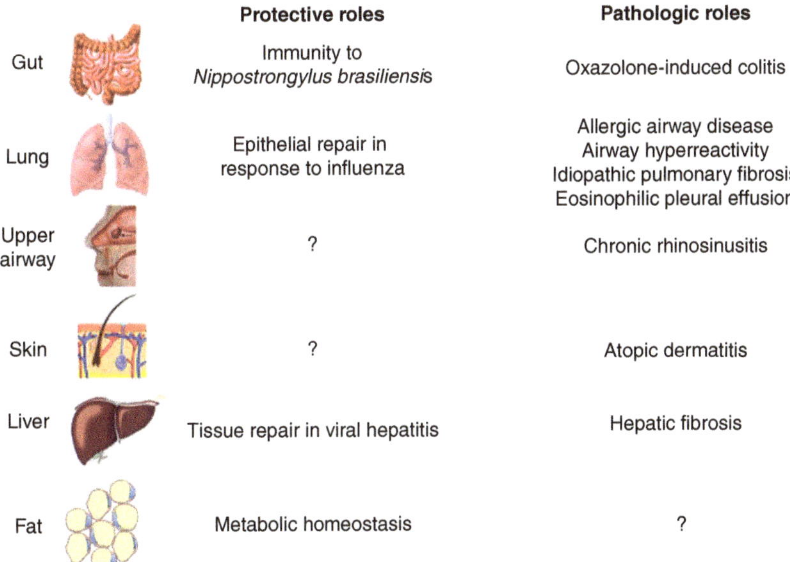

Figure 3. Protective versus pathogenic roles of ILC2s. Gut-associated ILC2s play an essential role in protective immunity to helminth parasites and can promote type 2 inflammatory colitis. Lung ILC2s have shown a protective role by mediating epithelial repair in response to influenza virus infection. Further, they have shown pathogenic roles by promoting allergic airway disease, airway hyperreactivity, and pulmonary fibrosis and have been implicated in eosinophilic pleural effusion. ILC2s in the upper airway have been implicated in chronic rhinosinusitis and found in the nasal poylps of patients. Skin ILC2s are highly enriched in atopic dermatitis (AD) lesions and promote AD-like disease in murine models. ILC2s in the liver have been shown to promote tissue repair in response to viral-induced injury as well as pathogenic fibrosis. ILC2s in white adipose tissue regulate metabolic homeostasis via cellular interactions with eosinophils and alternatively activated macrophages.

cesses in humans. A recent study in patients with eosinophilic pleural effusion (EPE) associated with primary spontaneous pneumothorax (PSP) reported elevated expression of TSLP and IL-33 in the pleural fluid along with elevated IL-5, eotaxin-3, and enhanced ILC2 responses (Kwon et al. 2013). This study provokes the hypothesis that ILC2s may be the key driver of eosinophilia in the context of EPE and PSP. Another recent study used a *Schistosoma mansoni* egg-induced model of pulmonary fibrosis to identify that IL-25 is a key mediator of this disease in mice (Hams et al. 2014). Further investigation identified that IL-13-expressing ILC2s were both necessary and sufficient to induce pulmonary fibrosis in mice and that Lin^- $CD127^+$ $CRTH2^+$ $IL\text{-}33R^+$ ILC2s were significantly enriched in the broncheoalveolar lavage fluid (BALF) of patients with idiopathic pulmonary fibrosis (IPF) (Hams et al. 2014). Collectively, these studies indicate that ILC2s may play different roles across multiple disease states in the lung including influenza, allergic airway disease, PSP, and IPF (Fig. 3).

Chronic Rhinosinusitis

Chronic rhinosinusitis (CRS) is a common complication arising from allergic rhinitis and, when associated with nasal polyps, is strongly associated with type 2 cytokine production in the nasal mucosa. CRS was the first human disease in which an accumulation of Lin^- $CRTH2^+$ $CD161^+$ ILC2s in inflamed tissue was clearly shown (Mjosberg et al. 2011). ILC2s in nasal polyps of CRS patients were originally identified as being responsive to IL-25 and IL-33, and were subsequently shown to respond to TSLP (Mjosberg et al. 2012). These studies also highlighted that human ILC2s produce IL-4, IL-5, IL-9, and IL-13, and express the TSLP receptor (TSLPR) (Mjosberg et al. 2012). A recent report confirmed that ILC2s are enriched in ethmoid sinus mucosa of patients with CRS and nasal polyps in comparison to control CRS patients without nasal polyps (Shaw et al. 2013). This study also showed that the ILC2s from CRS patients responded to IL-33-mediated stimulation by producing IL-13 (Shaw et al. 2013), further suggesting that ILC2s may contribute to the pathogenesis of allergic upper airway disease in humans. Consistent with previous findings (Mjosberg et al. 2012), TSLP has recently been shown to be highly expressed in the nasal polyps of CRS patients (Nagarkar et al. 2013). Although IL-33 appears to be more potent than TSLP in activating type 2 cytokine production from nasal polyp ILC2s (Mjosberg et al. 2012), further studies will be required to determine which cytokines optimally promote ILC2-mediated inflammation in CRS and whether these cells have a causal role in the development of nasal polyps or pathology in CRS patients.

ILC2s IN THE SKIN

Atopic Dermatitis

In addition to the role of ILC2s in promoting inflammation in the gut and lung, multiple studies have now shown that these cells also promote allergic inflammation in the skin. Lesional human AD skin has elevated expression of epithelial cell-derived cytokines that promote ILC2 responses including IL-25, IL-33, and TSLP (Soumelis et al. 2002; Hvid et al. 2011; Deleuran et al. 2012; Savinko et al. 2012), and ILC2s have been identified in both murine and human skin and are enriched in the lesional skin of human AD patients (Kim et al. 2013a; Roediger et al. 2013). Although IL-33 has emerged as the dominant cytokine in the activation of ILC2s from murine lung (Barlow et al. 2013), human blood, and nasal polyps (Mjosberg et al. 2012), the original studies identifying skin ILC2s found that murine skin ILC2s are IL-33- and IL-25-independent but dependent on TSLP for their activation during murine AD-like disease (Kim et al. 2013a). Further, skin-associated ILC2s could directly induce AD-like pathology and T_H2 cell responses in vivo (Kim et al. 2013a). A more recent study confirmed the presence of ILC2s in murine skin and used transgenic mice overexpressing IL-33 under a keratin 14 promoter to show that IL-33 expression can also drive AD-like inflammation and the expansion of ILC2s in the skin (Imai et al. 2013). Recently, Salimi et al. (2013) confirmed that skin ILC2s are dependent on TSLP in C57BL/6 mice and found that there is partial dependence on both IL-25 and IL-33 in the development of skin ILC2 responses in BALB/c (but not C57BL/6) mice during AD-like inflammation. They also showed the presence of ILC2s in human skin and their enrichment in lesional AD skin (Salimi et al. 2013). Further, they found that human skin ILC2s up-regulate expression of both type 2 cytokines and Areg in response to IL-33 (Salimi et al. 2013). Finally, recent work has also focused on the cellular interactions and migratory patterns of skin ILC2s. Roediger et al. used intravital multiphoton microscopy to directly visualize skin ILC2s and characterize their interactions with skin-resident mast cells (Roediger et al. 2013). In these studies, skin ILC2s were found to constitutively express IL-13, produce IL-5 in response to IL-2-mediated activation, and modulate cutaneous mast cell responses (Roediger et al. 2013). Collectively, these studies show that skin ILC2s promote type 2 cytokine–associated skin inflammation and coordinately interact with other innate and adaptive cells in the skin to influence their function (Fig. 4). However, further studies will be required to fully assess the factors that promote and regulate ILC2-mediated skin inflammation.

ILC2s IN METABOLIC TISSUES

ILC2s in the Liver

Although the importance of ILC2 responses at mucosal barriers such as the gut and lung has been appreciated, ILC2 responses have also been characterized in other organs that regulate metabolic homeostasis such as the liver and adipose tissue. The liver is a critical regulator of glucose metabolism, whereas the adipose tissue is a regulator of lipid homeostasis. In the context of liver disease, patients with cirrhosis and mice with CCL_4-induced hepatic fibrosis had significantly higher levels of IL-33 in the serum in comparison to controls (McHedlidze et al. 2013). Based on these findings, McHedlidze et al. (2013) showed that IL-33 mediates hepatic fibrogenesis in mice via ILC2s. They also identified that human cirrhotic livers have higher expression of IL-13 receptor components, suggesting that ILC2-derived IL-13 may be a key mediator of hepatic fibrosis in humans as well (McHedlidze et al. 2013). Another recent study identified that IL-33 attenuates liver damage in the context of adenovirus-induced hepatitis in mice, possibly through the ability of IL-33-dependent ILC2s to limit TNF-α production from hepatic T cells and macrophages (Liang et al. 2013). Adoptive transfer studies also suggested that ILC2s might mediate liver protection in vivo in response to viral hepatitis (Liang et al. 2013). However, the precise mechanisms by which ILC2s are protective in the liver remain to be determined. Collectively, these studies indicate that IL-33-dependent ILC2s can have either beneficial or detrimental effects on liver homeostasis depending on the context of liver injury (Fig. 3). Future studies will be required to determine the effector mechanisms by which ILC2s mediate these processes.

ILC2s in White Adipose Tissue

Recent work has highlighted a previously unappreciated role for ILC2s in mediating metabolic homeostasis in adipose tissue. Type 2 cytokine-associated eosinophil responses regulate AAMs to promote glucose and adipose tissue homeostasis (Wu et al. 2011) and ILC2s have recently been shown to regulate this process via production of IL-5 and its effect on eosinophil survival (Fig. 4) (Molofsky et al. 2013). In support of these findings, IL-33 was shown to limit obesity in mice (Miller et al. 2010), whereas another study showed that IL-25 elicits ILC2s to limit obesity (Hams et al. 2013). Similarly, depletion of ILC2s led to enhanced weight gain in lymphocyte-deficient *Rag1*$^{-/-}$ mice and loss of eosinophil and AAMs populations in the visceral adipose tissue (VAT) in response to a high fat diet (Hams et al. 2013; Molofsky et al. 2013). These findings were further explored in a study by Nussbaum et al., which reported that serum levels of IL-5 and blood eosinophils correlated with circadian variation (Nussbaum et al. 2013). In addition, this study showed that

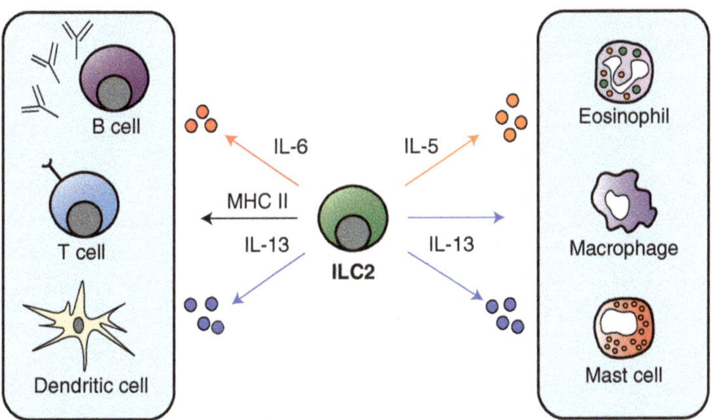

Figure 4. Effector functions of ILC2s. ILC2s in the lung have been proposed to promote adaptive CD4^{+} T-cell and B-cell responses via MHC class II–mediated antigen presentation and IL-6 production, respectively. Further, IL-13 has been shown to promote CD4^{+} T-cell responses indirectly via acting on dendritic cells. ILC2-derived IL-5 promotes eosinophil responses and IL-13 can regulate alternatively activated macrophage and mast cell functions.

ILC2 responses in the small intestine were enhanced in response to caloric input as determined by IL-13 expression (Nussbaum et al. 2013), suggesting that ILC2s may respond directly to signals that regulate feeding and circadian rhythms. In support of this hypothesis, ILC2s were shown to express VPAC2, the receptor for vasoactive intestinal peptide (VIP), a neuropeptide that is abundant in the intestine and tightly regulated by both feeding and circadian rhythms. Additionally, ILC2s were shown to produce IL-5 in response to both VIP and a VPAC2 agonist in vitro (Nussbaum et al. 2013). Collectively, these studies suggest that metabolic and circadian cues can influence ILC2 responses and their production of IL-5 and IL-13, which in turn regulate eosinophils and AAMs in the context of metabolic homeostasis (Fig. 4). Further studies will be required to fully dissect the mechanisms by which ILC2 responses regulate metabolic homeostasis and contribute to the development of various metabolic diseases.

CONCLUSIONS AND FUTURE DIRECTIONS

Originally described in the context of anti-helminth immunity, ILC2s appear to have diverse functions at multiple barrier surfaces including the upper and lower airways, skin, and gut (Fig. 3) (Tait Wojno and Artis 2012). Further, ILC2s have also been identified in multiple other tissues including the brain, heart, kidney, muscle, liver, and adipose tissue (Hams et al. 2013; Liang et al. 2013; McHedlidze et al. 2013; Molofsky et al. 2013; Nussbaum et al. 2013). However, there are a number of questions that remain regarding the function of ILC2s. For example, the potential tissue-specific factors that regulate ILC2s are complex and remain poorly understood. The current body of evidence suggests that IL-33 may be the dominant cytokine for the activation of lung and airway ILC2s (Mjosberg et al. 2012; Monticelli et al. 2012; Barlow et al. 2013), whereas IL-25 is critical for their role in gut inflammation (Camelo et al. 2012). However, IL-25 may also play a role in the regulation of pulmonary fibrosis (Hams et al. 2014). In the skin, ILC2s have been shown to be predominantly regulated by TSLP during AD-like disease (Kim et al. 2013a). However, recent studies showed that IL-25 and IL-33 may also have relevant roles in AD-like inflammation (Imai et al. 2013; Salimi et al. 2013). Collectively, whether IL-25, IL-33, or TSLP is the dominant cytokine for the activation and/or elicitation of ILC2s at different barrier surfaces remains to be determined.

Beyond epithelial cell–derived cytokine regulation, ILC2 activation and migration also appear to be regulated by other factors, such as TL1A (Yu et al. 2013; Meylan et al. 2014) and eicosanoids (Barnig et al. 2013; Chang et al. 2013; Doherty et al. 2013; Xue et al. 2013) in the intestine and lung, respectively. In the intestine, VIP, which is heavily influenced by caloric intake and circadian rhythms, has also been shown to activate cytokine production from ILC2s (Nussbaum et al. 2013). Furthermore, ILC2s are a novel source of growth factors (e.g., Areg) and enzymes (e.g., Arg1) that may regulate epithelial repair and inflammation at barrier surfaces (Monticelli et al. 2011; Bando et al. 2013). These newer studies show that the regulation and function of ILC2s may be much more complex than previously recognized.

Moreover, recent studies have shown that ILC2s interact with and/or regulate other innate cell populations such as mast cells (Roediger et al. 2013), eosinophils (Nussbaum et al. 2013), and macrophages (Molofsky et al. 2013), as well as adaptive T_H2 cell responses via type 2 cytokine–mediated activation of DCs (Fig. 4) (Halim et al. 2014). Additionally, MHC class II–mediated antigen presentation by ILC2s has been recently shown to induce $CD4^+$ T cell proliferation (Mirchandani et al. 2014). Thus, understanding how ILC2s regulate type 2 cytokine–associated inflammation through interactions with various innate and adaptive cell populations is an emerging field of great interest. Further, the role ILC2s play in food allergy and other allergic diseases that influence epithelial barriers, such as urticaria, eosinophilic gastrointestinal diseases, and anaphylaxis, remains to be explored. Future studies aimed at understanding the regulation and effector mechanisms of human ILC2s in different organ systems will be critical to developing therapeutics that target ILC2s to treat multiple inflammatory diseases.

REFERENCES

Bando JK, Nussbaum JC, Liang HE, Locksley RM. 2013. Type 2 innate lymphoid cells constitutively express arginase-I in the naïve and inflamed lung. *J Leukoc Biol* **94:** 877–884.

Barlow JL, Peel S, Fox J, Panova V, Hardman CS, Camelo A, Bucks C, Wu X, Kane CM, Neill DR, et al. 2013. IL-33 is more potent than IL-25 in provoking IL-13-producing nuocytes (type 2 innate lymphoid cells) and airway contraction. *J Allergy Clin Immunol* **132:** 933–941.

Barnig C, Cernadas M, Dutile S, Liu X, Perrella MA, Kazani S, Wechsler ME, Israel E, Levy BD. 2013. Lipoxin A_4 regulates natural killer cell and type 2 innate lymphoid cell activation in asthma. *Sci Transl Med* **5:** 174ra126.

Bartemes KR, Iijima K, Kobayashi T, Kephart GM, McKenzie AN, Kita H. 2012. IL-33–responsive lineage$^-$ CD25$^+$ CD44^{hi} lymphoid cells mediate innate type 2 immunity and allergic inflammation in the lungs. *J Immunol* **188:** 1503–1513.

Beamer CA, Girtsman TA, Seaver BP, Finsaas KJ, Migliaccio CT, Perry VK, Rottman JB, Smith DE, Holian A. 2012. IL-33 mediates multi-walled carbon nanotube (MWCNT)-induced airway hyper-reactivity via the mobilization of innate helper cells in the lung. *Nanotoxicology* **7:** 1070–1081.

Bjorkstrom NK, Kekalainen E, Mjosberg J. 2013. Tissue-specific effector functions of innate lymphoid cells. *Immunology* **139:** 416–427.

Blazquez AB, Mayer L, Berin MC. 2010. Thymic stromal lymphopoietin is required for gastrointestinal allergy but not oral tolerance. *Gastroenterology* **139:** 1301–1309.

Camelo A, Barlow JL, Drynan LF, Neill DR, Ballantyne SJ, Wong SH, Pannell R, Gao W, Wrigley K, Sprenkle J, et al. 2012. Blocking IL-25 signalling protects against gut inflammation in a type-2 model of colitis by suppressing nuocyte and NKT derived IL-13. *J Gastroenterol* **47:** 1198–1211.

Chang YJ, Kim HY, Albacker LA, Baumgarth N, McKenzie AN, Smith DE, Dekruyff RH, Umetsu DT. 2011. Innate lymphoid cells mediate influenza-induced airway hyper-reactivity independently of adaptive immunity. *Nat Immunol* **12:** 631–638.

Chang JE, Doherty TA, Baum R, Broide D. 2013. Prostaglandin D2 regulates human type 2 innate lymphoid cell chemotaxis. *J Allergy Clin Immunol* **133:** 899–901.

Chu DK, Llop-Guevara A, Walker TD, Flader K, Goncharova S, Boudreau JE, Moore CL, Seunghyun In T, Waserman S, Coyle AJ, et al. 2013. IL-33, but not thymic stromal lymphopoietin or IL-25, is central to mite and peanut allergic sensitization. *J Allergy Clin Immunol* **131:** e181–188.

Constantinides MG, McDonald BD, Verhoef PA, Bendelac A. 2014. A committed precursor to innate lymphoid cells. *Nature* **508:** 397–401.

Corrigan CJ, Wang W, Meng Q, Fang C, Eid G, Caballero MR, Lv Z, An Y, Wang YH, Liu YJ, et al. 2011. Allergen-induced expression of IL-25 and IL-25 receptor in atopic asthmatic airways and late-phase cutaneous responses. *J Allergy Clin Immunol* **128:** 116–124.

Deleuran M, Hvid M, Kemp K, Christensen GB, Deleuran B, Vestergaard C. 2012. IL-25 induces both inflammation and skin barrier dysfunction in atopic dermatitis. *Chem Immunol Allergy* **96:** 45–49.

Doherty TA, Khorram N, Chang JE, Kim HK, Rosenthal P, Croft M, Broide DH. 2012. STAT6 regulates natural helper cell proliferation during lung inflammation initiated by *Alternaria*. *Am J Physiol Lung Cell Mol Physiol* **303:** L577–588.

Doherty TA, Khorram N, Lund S, Mehta AK, Croft M, Broide DH. 2013. Lung type 2 innate lymphoid cells express cysteinyl leukotriene receptor 1, which regulates T2 cytokine production. *J Allergy Clin Immunol* **132:** 205–213.

Fallon PG, Ballantyne SJ, Mangan NE, Barlow JL, Dasvarma A, Hewett DR, McIlgorm A, Jolin HE, McKenzie AN. 2006. Identification of an interleukin (IL)-25-dependent cell population that provides IL-4, IL-5, and IL-13 at the onset of helminth expulsion. *J Exp Med* **203:** 1105–1116.

Fort MM, Cheung J, Yen D, Li J, Zurawski SM, Lo S, Menon S, Clifford T, Hunte B, Lesley R, et al. 2001. IL-25 induces IL-4, IL-5, and IL-13 and Th2-associated pathologies in vivo. *Immunity* **15:** 985–995.

Fuchs A, Colonna M. 2013. Innate lymphoid cells in homeostasis, infection, chronic inflammation and tumors of the gastrointestinal tract. *Curr Opin Gastroenterol* **29:** 581–587.

Gorski SA, Hahn YS, Braciale TJ. 2013. Group 2 innate lymphoid cell production of IL-5 is regulated by NKT cells during influenza virus infection. *PLoS Pathog* **9:** e1003615.

Halim TY, Krauss RH, Sun AC, Takei F. 2012a. Lung natural helper cells are a critical source of Th2 cell-type cytokines in protease allergen-induced airway inflammation. *Immunity* **36:** 451–463.

Halim TY, MacLaren A, Romanish MT, Gold MJ, McNagny KM, Takei F. 2012b. Retinoic-acid-receptor-related orphan nuclear receptor α is required for natural helper cell development and allergic inflammation. *Immunity* **37:** 463–474.

Halim TY, Steer CA, Matha L, Gold MJ, Martinez-Gonzalez I, McNagny KM, McKenzie AN, Takei F. 2014. Group 2 innate lymphoid cells are critical for the initiation of adaptive T helper 2 cell-mediated allergic lung inflammation. *Immunity* **40:** 425–435.

Hams E, Locksley RM, McKenzie AN, Fallon PG. 2013. Cutting edge: IL-25 elicits innate lymphoid type 2 and type II NKT cells that regulate obesity in mice. *J Immunol* **191:** 5349–5353.

Hams E, Armstrong ME, Barlow JL, Saunders SP, Schwartz C, Cooke G, Fahy RJ, Crotty TB, Hirani N, Flynn RJ, et al. 2014. IL-25 and type 2 innate lymphoid cells induce pulmonary fibrosis. *Proc Natl Acad Sci* **111:** 367–372.

Hepworth MR, Monticelli LA, Fung TC, Ziegler CG, Grunberg S, Sinha R, Mantegazza AR, Ma HL, Crawford A, Angelosanto JM, et al. 2013. Innate lymphoid cells regulate CD4$^+$ T-cell responses to intestinal commensal bacteria. *Nature* **498:** 113–117.

Herberth G, Daegelmann C, Roder S, Behrendt H, Kramer U, Borte M, Heinrich J, Herbarth O, Lehmann I. 2010. IL-17E but not IL-17A is associated with allergic sensitization: Results from the LISA study. *Pediatr Allergy Immunol* **21:** 1086–1090.

Hoyler T, Klose CS, Souabni A, Turqueti-Neves A, Pfeifer D, Rawlins EL, Voehringer D, Busslinger M, Diefenbach A. 2012. The transcription factor GATA-3 controls cell fate and maintenance of type 2 innate lymphoid cells. *Immunity* **37:** 634–648.

Hung LY, Lewkowich IP, Dawson LA, Downey J, Yang Y, Smith DE, Herbert DR. 2013. IL-33 drives biphasic IL-13 production for noncanonical type 2 immunity against hookworms. *Proc Natl Acad Sci* **110:** 282–287.

Hvid M, Vestergaard C, Kemp K, Christensen GB, Deleuran B, Deleuran M. 2011. IL-25 in atopic dermatitis: A possible link between inflammation and skin barrier dysfunction? *J Invest Dermatol* **131:** 150–157.

Imai Y, Yasuda K, Sakaguchi Y, Haneda T, Mizutani H, Yoshimoto T, Nakanishi K, Yamanishi K. 2013. Skin-specific expression of IL-33 activates group 2 innate lymphoid cells and elicits atopic dermatitis-like inflammation in mice. *Proc Natl Acad Sci* **110:** 13921–13926.

Kabata H, Moro K, Fukunaga K, Suzuki Y, Miyata J, Masaki K, Betsuyaku T, Koyasu S, Asano K. 2013. Thymic stromal lymphopoietin induces corticosteroid resistance in natural helper cells during airway inflammation. *Nat Commun* **4:** 2675.

Kim HY, Chang YJ, Subramanian S, Lee HH, Albacker LA, Matangkasombut P, Savage PB, McKenzie AN, Smith DE, Rottman JB, et al. 2012. Innate lymphoid cells responding to IL-33 mediate airway hyperreactivity independently of adaptive immunity. *J Allergy Clin Immunol* **129:** 216–227.

Kim BS, Siracusa MC, Saenz SA, Noti M, Monticelli LA, Sonnenberg GF, Hepworth MR, Van Voorhees AS, Comeau MR, Artis D. 2013a. TSLP elicits IL-33-independent innate lymphoid cell responses to promote skin inflammation. *Sci Transl Med* **5:** 170ra116.

Kim BS, Wojno ED, Artis D. 2013b. Innate lymphoid cells and allergic inflammation. *Curr Opin Immunol* **25:** 738–744.

Kleinjan A, Klein Wolterink RG, Levani Y, de Bruijn MJ, Hoogsteden HC, van Nimwegen M, Hendriks RW. 2014. Enforced expression of Gata3 in T cells and group 2 innate lymphoid cells increases susceptibility to allergic airway inflammation in mice. *J Immunol* **192:** 1385–1394.

Klein Wolterink RG, Kleinjan A, van Nimwegen M, Bergen I, de Bruijn M, Levani Y, Hendriks RW. 2012. Pulmonary innate lymphoid cells are major producers of IL-5 and IL-13 in murine models of allergic asthma. *Eur J Immunol* **42:** 1106–1116.

Klein Wolterink RG, Serafini N, van Nimwegen M, Vosshenrich CA, de Bruijn MJ, Fonseca Pereira D, Veiga Fernandes H, Hendriks RW, Di Santo JP. 2013. Essential, dose-dependent role for the transcription factor *Gata3* in the development of IL-5$^+$ and IL-13$^+$ type 2 innate lymphoid cells. *Proc Natl Acad Sci* **110:** 10240–10245.

Klose CS, Flach M, Mohle L, Rogell L, Hoyler T, Ebert K, Fabiunke C, Pfeifer D, Sexl V, Fonseca-Pereira D, et al. 2014. Differentiation of type 1 ILCs from a common progenitor to all helper-like innate lymphoid cell lineages. *Cell* **157:** 340–356.

Kwon BI, Hong S, Shin K, Choi EH, Hwang JJ, Lee SH. 2013. Innate type 2 immunity is associated with eosinophilic pleural effusion in primary spontaneous pneumothorax. *Am J Respir Crit Care Med* **188:** 577–585.

Liang Y, Jie Z, Hou L, Aguilar-Valenzuela R, Vu D, Soong L, Sun J. 2013. IL-33 induces nuocytes and modulates liver injury in viral hepatitis. *J Immunol* **190:** 5666–5675.

Licona-Limon P, Henao-Mejia J, Temann AU, Gagliani N, Licona-Limon I, Ishigame H, Hao L, Herbert DR, Flavell RA. 2013. Th9 cells drive host immunity against gastrointestinal worm infection. *Immunity* **39:** 744–757.

Maarsingh H, Zaagsma J, Meurs H. 2009. Arginase: A key enzyme in the pathophysiology of allergic asthma opening novel therapeutic perspectives. *Br J Pharmacol* **158:** 652–664.

McHedlidze T, Waldner M, Zopf S, Walker J, Rankin AL, Schuchmann M, Voehringer D, McKenzie AN, Neurath MF, Pflanz S, et al. 2013. Interleukin-33-dependent innate lymphoid cells mediate hepatic fibrosis. *Immunity* **39:** 357–371.

Meylan F, Hawley ET, Barron L, Barlow JL, Penumetcha P, Pelletier M, Sciume G, Richard AC, Hayes ET, Gomez-Rodriguez J, et al. 2014. The TNF-family cytokine TL1A promotes allergic immunopathology through group 2 innate lymphoid cells. *Mucosal Immunol* **7:** 958–968.

Miller AM, Asquith DL, Hueber AJ, Anderson LA, Holmes WM, McKenzie AN, Xu D, Sattar N, McInnes IB, Liew FY. 2010. Interleukin-33 induces protective effects in adipose tissue inflammation during obesity in mice. *Circ Res* **107:** 650–658.

Mirchandani AS, Besnard AG, Yip E, Scott C, Bain CC, Cerovic V, Salmond RJ, Liew FY. 2014. Type 2 innate lymphoid cells drive CD4$^+$ Th2 cell responses. *J Immunol* **192:** 2442–2448.

Mjosberg JM, Trifari S, Crellin NK, Peters CP, van Drunen CM, Piet B, Fokkens WJ, Cupedo T, Spits H. 2011. Human IL-25- and IL-33-responsive type 2 innate lymphoid cells are defined by expression of CRTH2 and CD161. *Nat Immunol* **12:** 1055–1062.

Mjosberg J, Bernink J, Golebski K, Karrich JJ, Peters CP, Blom B, te Velde AA, Fokkens WJ, van Drunen CM, Spits H. 2012. The transcription factor GATA3 is essential for the function of human type 2 innate lymphoid cells. *Immunity* **37:** 649–659.

Molofsky AB, Nussbaum JC, Liang HE, Van Dyken SJ, Cheng LE, Mohapatra A, Chawla A, Locksley RM. 2013. Innate lymphoid type 2 cells sustain visceral adipose tissue eosinophils and alternatively activated macrophages. *J Exp Med* **210:** 535–549.

Monticelli LA, Sonnenberg GF, Abt MC, Alenghat T, Ziegler CG, Doering TA, Angelosanto JM, Laidlaw BJ, Yang CY, Sathaliyawala T, et al. 2011. Innate lymphoid cells promote lung-tissue homeostasis after infection with influenza virus. *Nat Immunol* **12:** 1045–1054.

Monticelli LA, Sonnenberg GF, Artis D. 2012. Innate lymphoid cells: Critical regulators of allergic inflammation and tissue repair in the lung. *Curr Opin Immunol* **24:** 284–289.

Moro K, Yamada T, Tanabe M, Takeuchi T, Ikawa T, Kawamoto H, Furusawa J, Ohtani M, Fujii H, Koyasu S. 2010. Innate production of T_H2 cytokines by adipose tissue-associated c-Kit$^+$Sca-1$^+$ lymphoid cells. *Nature* **463:** 540–544.

Nagarkar DR, Poposki JA, Tan BK, Comeau MR, Peters AT, Hulse KE, Suh LA, Norton J, Harris KE, Grammer LC, et al. 2013. Thymic stromal lymphopoietin activity is increased in nasal polyps of patients with chronic rhinosinusitis. *J Allergy Clin Immunol* **132:** 593–600.

Neill DR, Wong SH, Bellosi A, Flynn RJ, Daly M, Langford TK, Bucks C, Kane CM, Fallon PG, Pannell R, et al. 2010. Nuocytes represent a new innate effector leukocyte that mediates type-2 immunity. *Nature* **464:** 1367–1370.

Nussbaum JC, Van Dyken SJ, von Moltke J, Cheng LE, Mohapatra A, Molofsky AB, Thornton EE, Krummel MF, Chawla A, Liang HE, et al. 2013. Type 2 innate lymphoid cells control eosinophil homeostasis. *Nature* **502:** 245–248.

Pesce JT, Ramalingam TR, Mentink-Kane MM, Wilson MS, El Kasmi KC, Smith AM, Thompson RW, Cheever AW, Murray PJ, Wynn TA. 2009. Arginase-1–expressing macrophages suppress Th2 cytokine-driven inflammation and fibrosis. *PLoS Pathog* **5:** e1000371.

Prefontaine D, Nadigel J, Chouiali F, Audusseau S, Semlali A, Chakir J, Martin JG, Hamid Q. 2010. Increased IL-33 expression by epithelial cells in bronchial asthma. *J Allergy Clin Immunol* **125:** 752–754.

Price AE, Liang HE, Sullivan BM, Reinhardt RL, Eisley CJ, Erle DJ, Locksley RM. 2010. Systemically dispersed innate IL-13-expressing cells in type 2 immunity. *Proc Natl Acad Sci* **107:** 11489–11494.

Qiu J, Guo X, Chen ZM, He L, Sonnenberg GF, Artis D, Fu YX, Zhou L. 2013. Group 3 innate lymphoid cells inhibit T-cell-mediated intestinal inflammation through aryl hydrocarbon receptor signaling and regulation of microflora. *Immunity* **39:** 386–399.

Roediger B, Kyle R, Yip KH, Sumaria N, Guy TV, Kim BS, Mitchell AJ, Tay SS, Jain R, Forbes-Blom E, et al. 2013. Cutaneous immunosurveillance and regulation of inflammation by group 2 innate lymphoid cells. *Nat Immunol* **14:** 564–573.

Saenz SA, Siracusa MC, Perrigoue JG, Spencer SP, Urban JF Jr, Tocker JE, Budelsky AL, Kleinschek MA, Kastelein RA, Kambayashi T, et al. 2010. IL25 elicits a multipotent progenitor cell population that promotes T_H2 cytokine responses. *Nature* **464:** 1362–1366.

Saenz SA, Siracusa MC, Monticelli LA, Ziegler CG, Kim BS, Brestoff JR, Peterson LW, Wherry EJ, Goldrath AW, Bhandoola A, et al. 2013. IL-25 simultaneously elicits distinct populations of innate lymphoid cells and multipotent progenitor type 2 (MPP^{type2}) cells. *J Exp Med* **210:** 1823–1837.

Salimi M, Barlow JL, Saunders SP, Xue L, Gutowska-Owsiak D, Wang X, Huang LC, Johnson D, Scanlon ST, McKenzie AN, et al. 2013. A role for IL-25 and IL-33-driven type-2 innate lymphoid cells in atopic dermatitis. *J Exp Med* **210:** 2939–2950.

Salmond RJ, Mirchandani AS, Besnard AG, Bain CC, Thomson NC, Liew FY. 2012. IL-33 induces innate lymphoid cell-mediated airway inflammation by activating mammalian target of rapamycin. *J Allergy Clin Immunol* **130:** e1156.

Satoh-Takayama N, Lesjean-Pottier S, Vieira P, Sawa S, Eberl G, Vosshenrich CA, Di Santo JP. 2010. IL-7 and IL-15 independently program the differentiation of intestinal CD3$^-$NKp46$^+$ cell subsets from Id2-dependent precursors. *J Exp Med* **207:** 273–280.

Savinko T, Matikainen S, Saarialho-Kere U, Lehto M, Wang G, Lehtimaki S, Karisola P, Reunala T, Wolff H, Lauerma A, et al. 2012. IL-33 and ST2 in atopic dermatitis: Expression profiles and modulation by triggering factors. *J Invest Dermatol* **132:** 1392–1400.

Schmitz J, Owyang A, Oldham E, Song Y, Murphy E, McClanahan TK, Zurawski G, Moshrefi M, Qin J, Li X, et al. 2005. IL-33, an interleukin-1-like cytokine that signals via the IL-1 receptor-related protein ST2 and induces T helper type 2-associated cytokines. *Immunity* **23:** 479–490.

Shaw JL, Fakhri S, Citardi MJ, Porter PC, Corry DB, Kheradmand F, Liu YJ, Luong A. 2013. IL-33-responsive innate lymphoid cells are an important source of IL-13 in chronic rhinosinusitis with nasal polyps. *Am J Resp Crit Care Med* **188:** 432–439.

Shikotra A, Choy DF, Ohri CM, Doran E, Butler C, Hargadon B, Shelley M, Abbas AR, Austin CD, Jackman J, et al. 2012. Increased expression of immunoreactive thymic stromal lymphopoietin in patients with severe asthma. *J Allergy Clin Immunol* **129:** e101–109.

Sonnenberg GF. 2013. Editorial: New tricks for innate lymphoid cells. *J Leukocyte Biol* **94:** 862–864.

Sonnenberg GF, Artis D. 2012. Innate lymphoid cell interactions with microbiota: Implications for intestinal health and disease. *Immunity* **37:** 601–610.

Sonnenberg GF, Monticelli LA, Alenghat T, Fung TC, Hutnick NA, Kunisawa J, Shibata N, Grunberg S, Sinha R, Zahm AM, et al. 2012. Innate lymphoid cells promote anatomical containment of lymphoid-resident commensal bacteria. *Science* **336:** 1321–1325.

Sonnenberg GF, Mjosberg J, Spits H, Artis D. 2013. SnapShot: Innate lymphoid cells. *Immunity* **39:** 622–622.

Soumelis V, Reche PA, Kanzler H, Yuan W, Edward G, Homey B, Gilliet M, Ho S, Antonenko S, Lauerma A, et al. 2002. Human epithelial cells trigger dendritic cell mediated allergic inflammation by producing TSLP. *Nat Immunol* **3:** 673–680.

Spits H, Cupedo T. 2012. Innate lymphoid cells: Emerging insights in development, lineage relationships, and function. *Annu Rev Immunol* **30:** 647–675.

Spits H, Artis D, Colonna M, Diefenbach A, Di Santo JP, Eberl G, Koyasu S, Locksley RM, McKenzie AN, Mebius RE, et al. 2013. Innate lymphoid cells—A proposal for uniform nomenclature. *Nat Rev Immunol* **13:** 145–149.

Tait Wojno ED, Artis D. 2012. Innate lymphoid cells: Balancing immunity, inflammation, and tissue repair in the intestine. *Cell Host Microbe* **12:** 445–457.

Turner JE, Morrison PJ, Wilhelm C, Wilson M, Ahlfors H, Renauld JC, Panzer U, Helmby H, Stockinger B. 2013. IL-9-mediated survival of type 2 innate lymphoid cells promotes damage control in helminth-induced lung inflammation. *J Exp Med* **210:** 2951–2965.

Walker JA, Barlow JL, McKenzie AN. 2013. Innate lymphoid cells—How did we miss them? *Nat Rev Immunol* **13:** 75–87.

Wilhelm C, Hirota K, Stieglitz B, Van Snick J, Tolaini M, Lahl K, Sparwasser T, Helmby H, Stockinger B. 2011. An IL-9 fate reporter demonstrates the induction of an innate IL-9 response in lung inflammation. *Nat Immunol* **12:** 1071–1077.

Wong SH, Walker JA, Jolin HE, Drynan LF, Hams E, Camelo A, Barlow JL, Neill DR, Panova V, Koch U, et al. 2012. Transcription factor RORα is critical for nuocyte development. *Nat Immunol* **13:** 229–236.

Wu D, Molofsky AB, Liang HE, Ricardo-Gonzalez RR, Jouihan HA, Bando JK, Chawla A, Locksley RM. 2011. Eosinophils sustain adipose alternatively activated macrophages associated with glucose homeostasis. *Science* **332:** 243–247.

Xue L, Salimi M, Panse I, Mjosberg JM, McKenzie AN, Spits H, Klenerman P, Ogg G. 2013. Prostaglandin D activates group 2 innate lymphoid cells through chemoattractant receptor-homologous molecule expressed on T2 cells. *J Allergy Clin Immunol* **133:** 1184–1194.

Yang Q, Monticelli LA, Saenz SA, Chi AW, Sonnenberg GF, Tang J, De Obaldia ME, Bailis W, Bryson JL, Toscano K, et al. 2013. T cell factor 1 is required for group 2 innate lymphoid cell generation. *Immunity* **38:** 694–704.

Yokota Y, Mansouri A, Mori S, Sugawara S, Adachi S, Nishikawa S, Gruss P. 1999. Development of peripheral lymphoid organs and natural killer cells depends on the helix-loop-helix inhibitor Id2. *Nature* **397:** 702–706.

Yu X, Pappu R, Ramirez-Carrozzi V, Ota N, Caplazi P, Zhang J, Yan D, Xu M, Lee WP, Grogan JL. 2013. TNF superfamily member TL1A elicits type 2 innate lymphoid cells at mucosal barriers. *Mucosal Immunol* **7:** 730–740.

Allergic Inflammation—Innately Homeostatic

Laurence E. Cheng[1] and Richard M. Locksley[2,3,4]

[1]Department of Pediatrics, University of California, San Francisco, San Francisco, California 94143

[2]Department of Medicine, University of California, San Francisco, San Francisco, California 94143

[3]Department of Microbiology and Immunology, University of California, San Francisco, San Francisco, California 94143

[4]Howard Hughes Medical Institute, University of California, San Francisco, San Francisco, California 94143

Correspondence: locksley@medicine.ucsf.edu

Allergic inflammation is associated closely with parasite infection but also asthma and other common allergic diseases. Despite the engagement of similar immunologic pathways, parasitized individuals often show no outward manifestations of allergic disease. In this perspective, we present the thesis that allergic inflammatory responses play a primary role in regulating circadian and environmental inputs involved with tissue homeostasis and metabolic needs. Parasites feed into these pathways and thus engage allergic inflammation to sustain aspects of the parasitic life cycle. In response to parasite infection, an adaptive and regulated immune response is layered on the host effector response, but in the setting of allergy, the effector response remains unregulated, thus leading to the cardinal features of disease. Further understanding of the homeostatic pressures driving allergic inflammation holds promise to further our understanding of human health and the treatment of these common afflictions.

Buoyed by the successes of prophylactic immunization against toxins at the turn of the 20th century, Portier and Richet began studies with hypnotoxin from the cnidarian, *Physalia physalis*, commonly known as the Portuguese man o' war, and a related toxin from the sea anemones, *Actinia equina* and *Anemonia sulcata*. These investigations led to the paradoxical discovery of immediate hypersensitivity reactions and even death among some immunized animals by a process termed "anaphylaxis" (Richet 1913). Recognized by a Nobel Prize, these seminal findings underpinned the modern field of allergy and led to the eventual identification of the transferable nature of the activating agent in serum, first noted by Richet, as immunoglobulin E (IgE) (Ishizaka et al. 1966). The rapid advances in molecular biology, genetics, and genomics from 1970 to 2000 elucidated the central role for cytokines, particularly the duplicated genes for interleukin (IL)-4, IL-13, IL-5, and IL-9, in mediating the effector functions of allergic immunity. Although initial studies fueled by the discoveries of helper T-cell subsets focused on T cells, designated Th2 cells, as sources of these cytokines, recent findings have increasingly highlighted the role of innate cells in allergic immunity. These discoveries have

Cite this article as *Cold Spring Harb Perspect Biol* doi: 10.1101/cshperspect.a016352

raised hopes that insights regarding the initiation and/or maintenance of tissue pathology mediated by interactions between innate and adaptive cells might translate to new therapeutic modalities for diseases underpinned by type 2 immunity.

PREVALENCE OF TYPE 2 IMMUNE MANIFESTATIONS AND THE PARADOX OF ALLERGY

Clinical manifestations of type 2 immune responses are commonplace worldwide in association with parasite infections and allergic diseases. It is estimated that 2–4 billion people worldwide harbor parasitic infections with the vast majority concentrated in developing nations (Chan 1997). Despite the paucity of parasitic infections in developed countries, type 2 immunity significantly impacts human health in the form of allergic diseases, including IgE-mediated anaphylaxis, allergic rhinitis, asthma, atopic dermatitis, eosinophilic gastrointestinal diseases, and food allergies. Worldwide, it is estimated that 300 million people have asthma and 400 million have allergic rhinitis (WHO 2007). In the United States, annual asthma costs approximate 56 billion dollars (CDC 2011).

In humans, "normal" values for IgE and eosinophils are defined using populations from developed countries where elevated levels are associated with pathologic states. In less developed countries, however, parasitic infestation is more widespread, and IgE levels and eosinophils are high. Indeed, hypereosinophilia was the commonest criteria underlying exclusion of healthy Uganda volunteers for vaccine trials (Eller et al. 2008). Although data collection is imperfect, the consensus view is that allergic diseases such as asthma are less prevalent in underdeveloped countries (Godfrey 1975; ISAAC 1998; Eller et al. 2008). In considering nonhuman vertebrates, domestic dogs have IgE levels 100 times greater than that in humans, and populations of Scandinavian wolves and a variety of horses show even higher levels (Ledin et al. 2006, 2008; Wagner 2009). As in humans from less developed countries, elevated IgE and eosinophils in feral vertebrates are associated with widespread parasitism, particularly intestinal helminths (usually multiple species) and ectoparasites, such as mites and ticks. Crocodiles, Antarctic petrels, Icelandic minke whales, penguins, and arctic mammals including bears, wolves, and cervids all show evidence of ecto- and endoparasitism (Jones 1988; Frenot et al. 2001; Lavikainen et al. 2011; La Grange et al. 2013; Olafsdottir and Shinn 2013). Taken together, these data suggest that manifestations of allergic inflammation are universal in nonhuman vertebrate populations in association with high levels of parasitism but there is little evidence for pathology associated with human allergic disease.

By extrapolation, it is likely that human evolution was marked by a higher "set point" for the cells and effector molecules, such as eosinophils and IgE, which are now associated with allergic manifestations that remain unusual or infrequent in wild and indigenous vertebrates. A number of possibilities have been considered to explain this apparent paradox. First, as a variant of the hygiene hypothesis, exposure to pathogens during critical developmental periods may be necessary to entrain the immune system to focus on exogenous organisms rather than innocuous allergens. Mechanisms proposed to underlie such "training" include induction of regulatory T cells or blocking antibodies that function to establish tolerance to antigens acquired later through food or inhalation. The data underlying such explanations have been reviewed elsewhere (Soyer et al. 2013). Such a mechanism may also underlie the dysregulated inflammatory responses that accompany many diseases of developed countries, including atherosclerosis, dementia, and obesity. A variant of this possibility, based on increasing information regarding immune cells that develop during fetal but not adult hematopoiesis (e.g., Langerhans cells and microglia) (Ginhoux et al. 2010; Mold et al. 2010; Hoeffel et al. 2012), is that certain cells in tissues may function in "anticipatory" roles, awaiting terminal differentiation by developmental or environmental signals, such as microbes and food, that the organism encounters postbirth. Alterations in these environmental signals during early developmental periods may bypass windows of differentiation that leave the organism more prone to inflammatory states in later life, whether or not accompanied by excesses of Th1- or Th2-associated pathology (Mold et al. 2010). A final possibility considered here is that intestinal helminths and ectoparasites elicit immune responses that mimic homeostatic responses used by the vertebrate host, but to facilitate aspects of the differentiation or development of the parasite. By this scenario, parasites have evolved to elicit tissue reactions that promote their own parasitism. Although the advantages of such evolution are not always readily apparent, the consideration is warranted owing to the extreme penetrance of parasitic infestation on the vertebrate immune system and the relatively small impact of these infections on survival through the reproductive age.

COMPONENTS OF THE ALLERGIC MODULE IN HOST DEFENSE AND ALLERGIC DISEASE

Studies in humans as well as model organisms have detailed the immunologic constituents of allergic inflammation. Here, we summarize recent insights regarding the functions of these various components (Fig. 1).

IgE, Mast Cells, and Basophils

IgE is the least represented serum immunoglobulin, consistent with a short serum half-life and distribution within peripheral tissues (Gould and Sutton 2008). Isotype switching of B cells to IgE requires IL-4-producing T follicular helper (T_{FH}) cells (Reinhardt et al. 2009; Crotty 2011). T-cell–B-cell collaboration is likely short-lived because IgE-switched B cells egress rapidly from germinal centers to become plas-

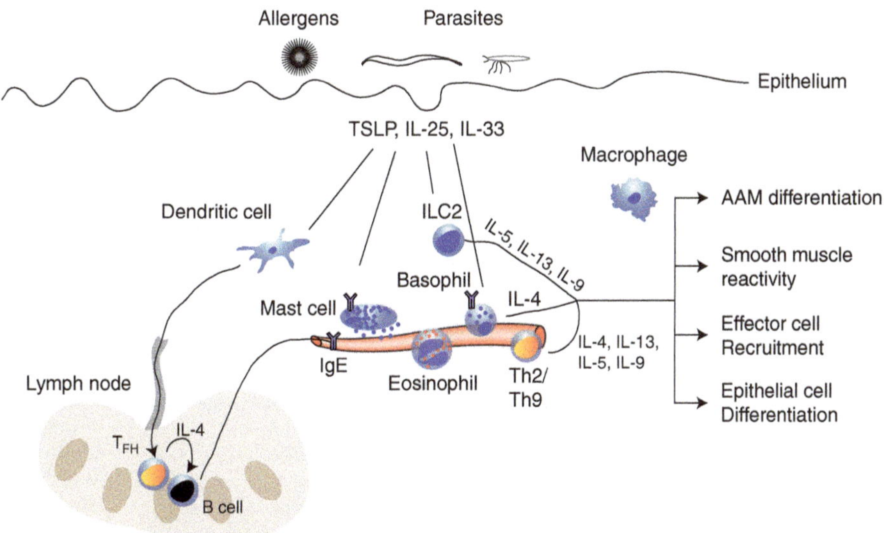

Figure 1. Constituents of allergic inflammation. Allergic inflammatory responses involve the coordinated response of (1) tissue-resident cells (dendritic cells, innate lymphoid cell [ILC]2, mast cells, and macrophages), (2) expansion of antigen-specific T cells and B cells in the draining lymph node, and (3) recruited cells from the blood (Th2/Th9 cells, basophils, and eosinophils). On allergen exposure or parasite infection, epithelial sensors evoke the release/production of epithelial cytokines (TSLP, IL-25, and IL-33), which in turn drive dendritic cells to migrate to lymph nodes and facilitate IgE production by B cells. Mast cells can be activated through IgE–antigen interactions as well as epithelial cytokines and modulate vascular tone through the release of granules. ILC2s, recruited basophils, and Th2/Th9 are activated by antigen and/or epithelial cytokines, and all three are significant producers of various type 2 cytokines. The sum total of this response includes alternatively activated macrophage (AAM) differentiation, modulation of smooth muscle and epithelial cell function, and further recruitment of effector cells.

ma cells resident in extrafollicular regions of the lymph node (Talay et al. 2012; Yang et al. 2012). Clarification is needed to explain how allergen-specific IgE antibodies develop extensive hypermutation consistent with their high-affinity states (Davies et al. 2013), whether arising from previously mutated IgG1 germinal center B cells or through some unknown process driving extensively hypermutated antibodies during chronic antigen exposure, as revealed by studies of neutralizing HIV antibodies (Zhou et al. 2010). IgE disseminates through the bloodstream, where free IgE is stabilized on the surface of cells bearing the high-affinity IgE receptor, FcεRI. Unlike other immunoglobulin isotypes, IgE effector function is related almost entirely to its capacity to bind to FcεRI before antigen recognition. In mice, mast cells and basophils constitutively express FcεRI, and is inducible on some dendritic cell populations (Grayson et al. 2007). In humans, FcεRI is also constitutively expressed on mast cells and basophils, but is more widely expressed, including on dendritic cell populations (Gould and Sutton 2008).

Mast cells are tissue-resident cells found near blood vessels in a perivascular distribution. This positioning allows mast cells to extend cellular processes across the endothelial barrier to acquire circulating IgE as well as to control vascular tone and permeability (Galli and Tsai 2010; Cheng et al. 2013). On activation, mast cells release preformed and synthesized mediators, including histamine, lipid mediators, and cytokines. In the skin, the clinical manifestations of this reaction include the "wheal and flare" seen in urticaria. The evolutionary benefit of these rapidly activated responses may be to mitigate threats from insects or noxious substances by promoting resolution of the threat or alerting the host to establish avoidance behavior (Palm et al. 2012). In the gut, these mediators also lead to neural stimulation and intestinal smooth muscle hypermotility, which might function to clear ingested toxins via emesis and diarrhea.

Although commonly associated with allergic conditions, mast cells also participate in host defense to bacterial pathogens (Malaviya et al. 1996; McLachlan et al. 2003). As during allergy, these nonallergic stimulants promote changes in vascular permeability, activate antigen-presenting cells to mature and migrate to draining lymph nodes, and engage components of adaptive immunity (Shelburne et al. 2009). Because mast cells are tissue-resident cells capable of sensing a variety of inputs associated with disparate inflammatory modules, these cells may have evolved the capacity to acquire IgE to broaden functional capacity and respond to environmental insults.

Basophils are closely related to mast cells as revealed by developmental similarities driven by common transcription factor modules (Qi et al. 2013). Basophils also bind IgE but, unlike mast cells, circulate in blood with a half-life similar to other circulating myeloid cells (Voehringer 2013). The scarcest of all circulating leukocytes, basophils degranulate after FcεRI-mediated activation, but unlike mast cells, contribute little to anaphylaxis (Ohnmacht et al. 2010). During migratory helminth infection, basophils enter involved tissues and interact closely with CD4 T cells, which mediate contact-dependent and IL-3-dependent IL-4 release from basophils; antigen-specific IgE enhances basophil IL-4 production (Sullivan et al. 2011). Basophils also promote eosinophil entry into skin models of chronic allergic inflammation and induce the differentiation of recruited monocytes to alternatively activated macrophages in tissues (see below).

CD4 T Cells: Th2 Cells, Th9 Cells, and T Regulatory Cells (Tregs)

Activation of helper T cells in allergy is dependent on dendritic cells (DCs), although the precise signals and surface phenotype of type 2–inducing DCs may differ in response to different stimuli or within different tissues (Pulendran et al. 2010). A gradient defined by integration of T-cell receptor (TCR) signal strength, T-cell precursor frequency, and antigen abundance underlies the proclivity of some cells to enter follicles and become IL-4-expressing T_{FH} cells (typically higher-affinity TCRs) or activate more extensive patterns of type 2 cytokine expression and leave lymph nodes to enter peripheral tissues as Th2 cells (Tubo et al. 2013). Tissue

Th2 cells show a range of cytokine patterns, including single and multiple combinations of IL-4, IL-5, IL-9, and IL-13. As shown most convincingly using knockout mice in models of helminth and allergic immune challenges, these cytokines have both unique and redundant functions in type 2 immunity (Fallon et al. 2002; Nath et al. 2007); importantly, absence of all of these attenuates most of the manifestations of allergy, in part owing to the inevitable outgrowth of effector T cells expressing inflammatory cytokines. Further information is needed to understand fully the distribution and life span of allergic memory cells to ascertain whether long-term tissue reservoirs exist for these cells as shown for other memory-effector T cells (Shin and Iwasaki 2013).

A second area of much importance and in need of clarification is the relationship of allergic immunity with Treg induction. Infants born with *FOXP3* mutations, and thus lacking Treg, suffer from a multisystem inflammatory disorder termed immunodysregulation, polyendocrinopathy, enteropathy, X-linked (IPEX). Although patients suffer from a wide range of inflammatory disease, allergic manifestations with severe eczema, elevated IgE, and eosinophilia are prominent (Ozcan et al. 2008). Other T-cell immunodeficiencies, including Wiskott-Aldrich and Omenn's syndromes, show inflammatory disorders that are characterized by unrestrained allergic inflammation with clinical features including increased IgE and eosinophils and eczematous skin conditions. Both disorders are accompanied by Treg deficiency and/or dysfunction to which the allergic manifestations have been attributed (Ozcan et al. 2008). Mice with defects in extrathymically derived Treg also show systemic allergic inflammation (Josefowicz et al. 2012). Ablation of GATA3, a transcription factor required for Th2 function, in Foxp3-expressing cells leads to uncontrolled allergic inflammation mimicking total Foxp3 gene ablation, thus emphasizing a primary role for these cells in the control of allergic inflammation to environmental antigens (Rudra et al. 2012). Helminth infections have been linked with Treg induction, consistent with resistance to allergy induction in infected mice (Maizels and Smith 2011), but why environmental allergens fail to induce similarly restraining Tregs in susceptible individuals remains unknown.

Innate Lymphoid Cell (ILC)2

The identification of innate lymphoid cells, now designated ILC2, as the major source of innate type 2 cytokines has kindled much interest in these cells in studies of allergic immunity (Spits and Cupedo 2012). ILC2 are dispersed throughout the body in most organs, and particularly in barrier tissues like the lung, gastrointestinal tract, and skin (Moro et al. 2010; Neill et al. 2010; Price et al. 2010; Nussbaum et al. 2013). Recent studies show that these cells accumulate in organs in the perinatal period and constitute a relatively long-lived subset of tissue lymphoid cells. As shown using cytokine reporter mice, small numbers of ILC2 in tissues constitutively produce IL-5, thus accounting for basal eosinophilopoiesis (Nussbaum et al. 2013). In response to migratory helminths, ILC2 respond to tissue alarmins, such as IL-33, IL-25, and thymic stromal lymphopoietin (TSLP) (Walker and McKenzie 2013), to express increased amounts and range of type 2 cytokines, including IL-13 and IL-9 but also growth factors, such as amphiregulin, that may contribute to local epithelial repair (Monticelli et al. 2011). As such, activated ILC2 are required to mediate early recruitment of tissue eosinophils from blood and the differentiation of tissue macrophages and recruited monocytes to an alternatively activated macrophage (AAM) phenotype (Molofsky et al. 2013). Although less well studied, activated ILC2 may also play a role in the early attraction of Th2 effectors to tissues, where these cells amplify the cytokine milieu, but also provide growth signals that stimulate the proliferation and/or recruitment of the tissue ILC2 population; some studies suggest a role for IL-9, both from Th2 cells and, in an autocrine fashion, from ILC2 themselves (Wilhelm et al. 2011). Although Th2 cells migrate to involved sites deeper in tissues to mediate local immunity, it remains unclear whether ILC2 migrate from their position near vascular tissues; further study is needed. The ultimate fate of expanded ILC2 populations in tissue and their ultimate interactions with Th2 memory cells and Treg remain important areas for investigation.

AAMs

Macrophages are constitutive in all tissues and increase in number and change their phenotype during inflammation (Galli et al. 2011; Van Dyken and Locksley 2013). In the setting of type 2 immunity, IL-4 and/or IL-13 promote the coordinate expression of a set of genes that characterize alternatively activated (or M2) macrophages. Under some conditions, AAM differentiation is driven by in situ proliferation of resident tissue macrophages (Jenkins et al. 2011), whereas differentiation of both resident and recruited blood monocytes into AAMs occurs in other situations (Reese et al. 2007). Recent findings that resident tissue macrophages, such as Langerhans cells and microglia, accumulate in tissues from fetal blood precursors emphasize the potential that certain populations of hematopoietic cells may be intimately involved during discrete developmental windows, with implications for tissue integrity based on the capacity to fully renew these populations in response to injury or senescence (Schulz et al. 2012). Along these lines, AAM-like cells are found dispersed throughout the body during embryonic development and peak during periods of growth, remodeling, and organization of developing tissues, such as the kidney (Rae et al. 2007).

The role of AAMs in type 2 immunity has been explored in various systems involving deletion of key elements involved in their differentiation, such as the IL-4Rα component of the type 1 and 2 IL-4 receptors. After infection with *Schistosoma mansoni*, mice lacking the capacity to generate IL-4/IL-13-mediated AAM differentiation fail to control intestinal epithelial integrity around trapped eggs and die from sepsis owing to enhanced translocation of intestinal bacteria and their products into the systemic circulation (Herbert et al. 2004). Thus, part of the function of AAMs may involve control of barrier integrity in response to antigens that elicit granulomatous type 2 immunity.

Eosinophils

Elevations of blood and tissue eosinophils are hallmarks of virtually all disorders of type 2 immunity (Rosenberg et al. 2013). Although decades of work highlighted the capacity of eosinophils to mediate parasite damage in various in vitro systems, data supporting a direct role for eosinophils in limiting the initiation or duration of adapted intestinal helminth infection in vivo is modest at best. The widespread prevalence of helminths in feral vertebrates, together with the relatively long life span of adult worms, despite prolonged eosinophilia, is difficult to reconcile with a primary role for these cells in limiting established parasitism. During secondary infections, however, the combination of memory-effector Th2 cells and expanded tissue ILC2 may accelerate and focus eosinophils at sites of larval migration, thus limiting further infection by a process termed “concomitant immunity” (a situation in which immunity against larval forms limits infections despite the presence of living adult parasites in the body that cannot be rejected). An emerging concept that requires further exploration is the inevitable co-accumulation of eosinophils with AAMs, reflecting the stereotyped response of tissues to the presence of activated ILC2 and Th2 that produce type 2 cytokines.

As discussed above, another possibility is that helminths evoke eosinophilia to elicit a tissue response that favors parasitism. After infection with *Trichinella spiralis*, newborn larvae migrate from the intestines to skeletal muscle to complete their development. Larval maturation involves the dedifferentiation of muscle cells to nurse cells, which provide the necessary environment. Although not entirely understood, eosinophils accumulate around nurse cells and are themselves required for continued larval development, presumably owing to their role in sustaining the nurse cell (Gebreselassie et al. 2012). As noted below, aspects of this response resemble that seen during muscle injury, suggesting that *Trichinella* elicits a gene program used by the host for one purpose but co-opts it to establish a developmental niche. Evidence that bacteria can also subvert cellular

 Cite this article as *Cold Spring Harb Perspect Biol* doi: 10.1101/cshperspect.a016352

gene programs to redirect fundamental pathways of differentiation suggests that further understanding of such highly involved interactions between microbes and hosts may yield fundamental insights into cellular reprogramming that might be applicable to many disease states (Masaki et al. 2013). Such examples offer the possibility that uncovering the basic roles for type 2 immunity in vertebrate homeostasis may lead to enhanced understanding of its stereotyped elicitation by helminths and allergens.

HOMEOSTATIC ROLES FOR COMPONENTS OF ALLERGIC MODULE

As noted above, the majority of studies of type 2 immunity have focused on pathologic associations with parasite infection and allergic diseases. Recent investigations have called attention to potential roles for type 2 immunity in metabolism and tissue homeostasis in response to injury that might suggest pathways by which these responses come to be linked with these two pathologic states (Fig. 2).

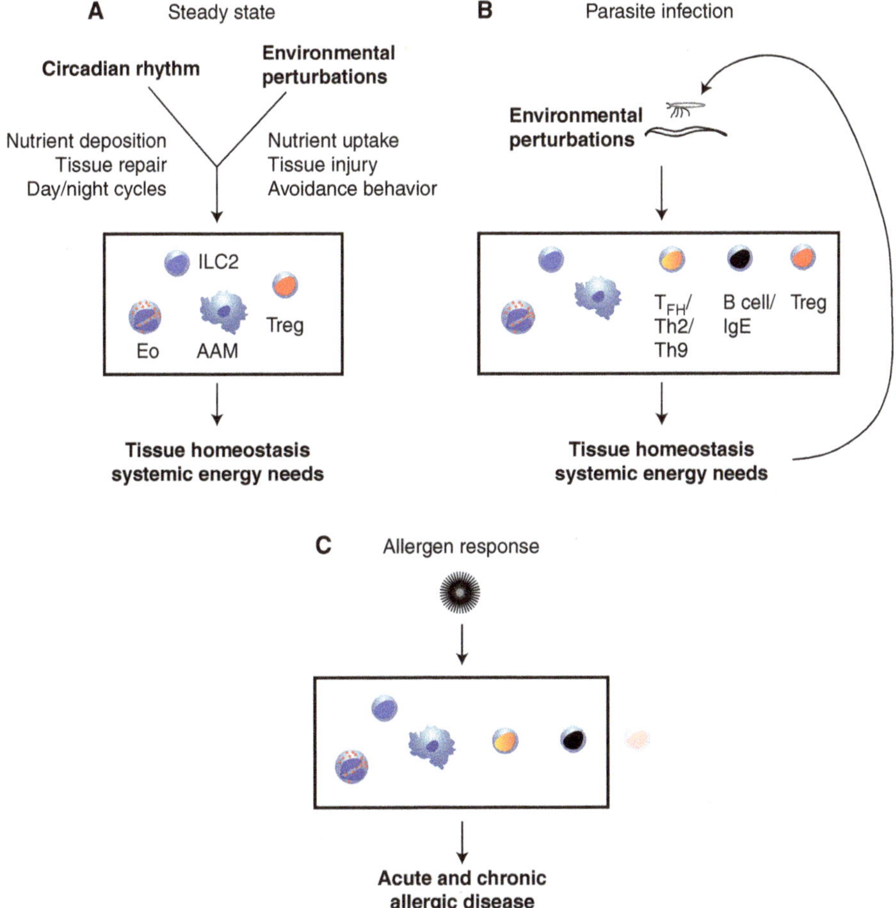

Figure 2. An integrated view of allergic inflammation in infection and disease. (*A*) Innate components of allergic inflammation (ILC2, eosinophils, and AAMs) along with Tregs integrate broad inputs related to circadian patterns and environmental perturbations to drive tissue homeostasis and meet systemic energy needs. (*B*) During parasite infection, the additional stress on tissue homeostasis and energy needs engages a broader but balanced adaptive immune response, which promotes both host and parasite fitness, and is characterized by a regulatory component. (*C*) Allergen-specific responses fail to establish this balance, with a lack of Treg activation, resulting in tissue injury and disease.

Metabolism

A number of studies over the last decade have provided evidence that obesity is accompanied by an inflammatory state, marked by infiltration of activated macrophages and other immune cells into visceral adipose with adverse effects on systemic insulin resistance and lipid metabolism (Mathis 2013). Conversely, lean animals and perhaps humans contain AAMs in white adipose tissue, and deleting these cells in animals results in proclivity to obesity and insulin resistance (Odegaard et al. 2007). AAMs may also play a role in brown adipose tissue by facilitating early production of norepinephrine and the transfer of fat from reservoirs in white to metabolizing brown fat tissue during adaptive thermogenesis (Nguyen et al. 2011). As alluded to above, AAMs in tissue are accompanied by eosinophils and ILC2s, and, indeed, both of these cells are not only present in resting visceral adipose, but their presence is necessary to restrain metabolic abnormalities incurred by challenge with a high-fat diet (Wu et al. 2011; Molofsky et al. 2013). Further evidence for the linkage between these core elements of innate type 2 immunity and Treg cells is the finding that the latter cells are also present in fat and required for the maintenance of metabolic homeostasis (Feuerer et al. 2009). Like ILC2, adipose Treg express GATA-3 and the IL-33 receptor; conversely, ILC2 express CD25, the high-affinity IL-2 receptor. Thus, ILC2 and Treg in adipose tissues share these and additional surface receptors and transcription factors, a finding that may underlie the capacity of these cells to respond to common environmental signals in specialized tissues undergoing cyclic nutrient exchange.

Eosinophils display prominent circadian cycling that was shown to be entrained by food intake more than 35 years ago (Pauly et al. 1975). A small percentage of tissue-dispersed ILC2s constitutively secrete IL-5 necessary for reactive eosinophilopoiesis (Nussbaum et al. 2013). Unexpectedly, IL-5 secretion was circadian, thus accounting for the cyclic nature of blood eosinophils, but secretion was enhanced by feeding and depressed by fasting (Nussbaum et al. 2013). Eosinophils, like Treg, are present constitutively in the intestines, particularly the small bowel, and ILC2 in the intestine activated IL-13 secretion in response to nutrient intake. ILC2 also express vasoactive intestinal polypeptide receptor (VPAC)2, the type 2 receptor for the neuronal peptide, vasoactive intestinal peptide (VIP), and activate IL-5 secretion rapidly after activation by VIP or VPAC2 analogs in vitro (Nussbaum et al. 2013). VIP is itself secreted from intestinal neurons in response to feeding, and helps to coordinate pancreatic secretions, smooth muscle contractility, and hepatic metabolic cycles linked with nutrient absorption and disposition (Lelievre et al. 2007). Mice deficient in VIP or VPAC2, both of which are highly expressed in neurons of the suprachiasmatic nucleus (Maywood et al. 2013), lose the ability to coordinate environmental signals with the circadian cycle, and show metabolic dysregulation (Harmar et al. 2002; Colwell et al. 2003). Thus, ILC2 respond to VIP, integrating their activation with central circadian rhythms, although precisely what role recruited eosinophils might play in nutrient acquisition or other aspects of metabolism remains unknown. Type 2 cytokines and Stat6-mediated signaling have been shown to affect primary hepatic metabolism directly as well (Sajic et al. 2013; Stanya et al. 2013).

An additional layer of complexity in metabolic homeostasis is the relationship of intestinal commensal organisms with epithelial and immune cell populations. In mice, key innate lymphoid cells are established after birth during exposure to dietary nutrients via signals sustained through the aryl hydrocarbon receptor (Kiss et al. 2011; Li et al. 2011; Lee et al. 2012). Development of these immune cell populations is constrained to a defined window of time postbirth, suggesting that alterations of this window might have adverse effects on gut homeostasis. Intriguingly, antibiotics, which have long been used to increase body size and fat content in animals farmed for food, have been shown to affect body size and intestinal microbial communities in adult mice, even when given in subtherapeutic levels during the perinatal period (Cho et al. 2012). Antibiotic use in early life has been linked to patterns of obesity at the population level (Trasande et al. 2013). Although causality has

 Cite this article as *Cold Spring Harb Perspect Biol* doi: 10.1101/cshperspect.a016352

yet to be established, the similarities in a number of the systemic consequences of dysbiotic microbial communities and the absence of AAMs, eosinophils, ILC2s, and/or Tregs raise the possibility that some or each of the components of this type 2 immune module are linked deeply with basal metabolic homeostasis (Fig. 2A).

Wound Healing

Maintenance of barrier integrity is critical for vertebrate survival, particularly at mucosal sites involved with gas exchange and nutrient acquisition in the lungs and bowel, respectively. Animals rendered unable to respond to IL-4/IL-13 signals from immune cells have been shown to suffer adverse epithelial injury against migratory and intestinal helminths in model systems (Gause et al. 2013). Recovery after toxin-mediated muscle injury was also enhanced by eosinophils and an intact type 2 cytokine system by a process linked to proliferation and differentiation of resident pluripotent fibro-/adipogenic precursor cells (FAPs) at the injury site (Heredia et al. 2013). FAPs were implicated in necrotic cell clearance and muscle regeneration; in the absence of IL-4/-13, the FAPs differentiated into adipocytes resulting in fatty degeneration. A similar role for eosinophils and IL-4/-13 was found in liver regeneration, but, in this case, IL-4 promoted hepatocyte proliferation and liver regeneration (Goh et al. 2013). In various skin models involving epicutaneous sensitization, a variety of type 2 immune cells, including basophils, eosinophils, ILC2, and AAMs, have been implicated along with IgE in mediating the manifestations of allergy (Mukai et al. 2005; Egawa et al. 2013). Thus, multiple components of the type 2 immune response have been implicated in barrier injury responses, although the precise molecular mechanisms by which these responses are mediated remain incompletely defined.

CONCLUDING REMARKS: SPECULATIONS AND FUTURE NEEDS

The deep penetrance of allergy prevalence into humans living in developed countries suggests that environmental alterations have impacted an evolved pathway that is deeply embedded in normal biology. Here, we have summarized recent support for involvement of type 2 immune cells in metabolism and tissue integrity, particularly at barriers. Although much more work is needed to flesh out the molecular details and mechanisms, the finding that multiple cell types, including Treg associated with type 2 immunity, localize to metabolically active tissues like adipose and small intestines is consistent with a role for these cells in more fundamental homeostatic processes of importance to vertebrate biology, such as tolerance to food or self-neoantigens exposed during normal tissue turnover. Chronic intestinal parasitism induces similar activation of type 2 immune cells, but these responses are not associated with pathologic responses associated with allergy, perhaps related to the capacity to induce a more balanced immune response, which includes Treg or additional suppressive mechanisms (Fig. 2B). This balanced response contrasts with allergen-specific responses in which suppressive mechanisms are lacking, and pathologic consequences emerge (Fig. 2C). Alternatively, the age of acquisition of these highly adapted parasites may have effects on immune system development or intestinal microbiota that protect against the subsequent dysregulated type 2 immune responses to innocuous environmental antigens. We propose that parasitic helminths, rather than being attacked by the components of type 2 immunity, have evolved to elicit these localized tissue responses to facilitate their own differentiation and maintenance within the vertebrate niche. Indeed, some of the properties of type 2 immunity we have outlined, including mobilization of nutrients and preservation of tissue integrity, may be appropriated by parasites for their own purposes. Numerous examples of bacterial commensals eliciting immune reactivity from the host that facilitates their colonization have been shown (Nussbaum and Locksley 2012), and the possibility that more complex parasites do the same thing would not be unexpected.

Despite evidence for this thesis, the mechanisms underlying these processes remain woefully undefined. In this way, study of parasite infections may yield information regarding the precise pathways that elicit the type 2 immune response, whether through induction of canonical alarmins or other mediators (Doherty et al. 2013; Gause et al. 2013). Conversely, more information regarding the mechanisms that control the tissue localization, organization, and activation of type 2 immune cells under homeostatic conditions will reveal the terrain that establishes this network within the greater context of vertebrate development and life span. Taken together, the study of type 2 immunity has evolved greatly in recent years and will be sure to contain additional surprises in the near future.

ACKNOWLEDGMENTS

We thank members of the Locksley Laboratory for insights, comments, and discussions, and we regret being unable to cite all of the relevant literature. Research support was provided from the National Institutes of Health, Howard Hughes Medical Institute, and the Sandler Asthma Basic Research Center at the University of California, San Francisco.

REFERENCES

CDC. 2011. *Vital Signs*, May 2011.

Chan MS. 1997. The global burden of intestinal nematode infections—Fifty years on. *Parasitol Today* **13:** 438–443.

Cheng LE, Hartmann K, Roers A, Krummel MF, Locksley RM. 2013. Perivascular mast cells dynamically probe cutaneous blood vessels to capture immunoglobulin E. *Immunity* **38:** 166–175.

Cho I, Yamanishi S, Cox L, Methe BA, Zavadil J, Li K, Gao Z, Mahana D, Raju K, Teitler I, et al. 2012. Antibiotics in early life alter the murine colonic microbiome and adiposity. *Nature* **488:** 621–626.

Colwell CS, Michel S, Itri J, Rodriguez W, Tam J, Lelievre V, Hu Z, Liu X, Waschek JA. 2003. Disrupted circadian rhythms in VIP- and PHI-deficient mice. *Am J Physiol Regul Integr Comp Physiol* **285:** R939–R949.

Crotty S. 2011. Follicular helper CD4 T cells (TFH). *Annu Rev Immunol* **29:** 621–663.

Davies JM, Platts-Mills TA, Aalberse RC. 2013. The enigma of IgE^{+} B-cell memory in human subjects. *J Allergy Clin Immunol* **131:** 972–976.

Doherty TA, Khorram N, Lund S, Mehta AK, Croft M, Broide DH. 2013. Lung type 2 innate lymphoid cells express cysteinyl leukotriene receptor 1, which regulates TH2 cytokine production. *J Allergy Clin Immunol* **132:** 205–213.

Egawa M, Mukai K, Yoshikawa S, Iki M, Mukaida N, Kawano Y, Minegishi Y, Karasuyama H. 2013. Inflammatory monocytes recruited to allergic skin acquire an anti-inflammatory M2 phenotype via basophil-derived interleukin-4. *Immunity* **38:** 570–580.

Eller LA, Eller MA, Ouma B, Kataaha P, Kyabaggu D, Tumusiime R, Wandege J, Sanya R, Sateren WB, Wabwire-Mangen F, et al. 2008. Reference intervals in healthy adult Ugandan blood donors and their impact on conducting international vaccine trials. *PLoS ONE* **3:** e3919.

Fallon PG, Jolin HE, Smith P, Emson CL, Townsend MJ, Fallon R, McKenzie AN. 2002. IL-4 induces characteristic Th2 responses even in the combined absence of IL-5, IL-9, and IL-13. *Immunity* **17:** 7–17.

Feuerer M, Herrero L, Cipolletta D, Naaz A, Wong J, Nayer A, Lee J, Goldfine AB, Benoist C, Shoelson S, et al. 2009. Lean, but not obese, fat is enriched for a unique population of regulatory T cells that affect metabolic parameters. *Nat Med* **15:** 930–939.

Frenot Y, de Oliveira E, Gauthier-Clerc M, Deunff J, Bellido A, Vernon P. 2001. Life cycle of the tick *Ixodesuriae* in penguin colonies: Relationships with host breeding activity. *Int J Parasitol* **31:** 1040–1047.

Galli SJ, Tsai M. 2010. Mast cells in allergy and infection: Versatile effector and regulatory cells in innate and adaptive immunity. *Eur J Immunol* **40:** 1843–1851.

Galli SJ, Borregaard N, Wynn TA. 2011. Phenotypic and functional plasticity of cells of innate immunity: Macrophages, mast cells and neutrophils. *Nat Immunol* **12:** 1035–1044.

Gause WC, Wynn TA, Allen JE. 2013. Type 2 immunity and wound healing: Evolutionary refinement of adaptive immunity by helminths. *Nat Rev Immunol* **13:** 607–614.

Gebreselassie NG, Moorhead AR, Fabre V, Gagliardo LF, Lee NA, Lee JJ, Appleton JA. 2012. Eosinophils preserve parasitic nematode larvae by regulating local immunity. *J Immunol* **188:** 417–425.

Ginhoux F, Greter M, Leboeuf M, Nandi S, See P, Gokhan S, Mehler MF, Conway SJ, Ng LG, Stanley ER, et al. 2010. Fate mapping analysis reveals that adult microglia derive from primitive macrophages. *Science* **330:** 841–845.

Godfrey RC. 1975. Asthma and IgE levels in rural and urban communities of The Gambia. *Clin Allergy* **5:** 201–207.

Goh YP, Henderson NC, Heredia JE, Red Eagle A, Odegaard JI, Lehwald N, Nguyen KD, Sheppard D, Mukundan L, Locksley RM, et al. 2013. Eosinophils secrete IL-4 to facilitate liver regeneration. *Proc Natl Acad Sci* **110:** 9914–9919.

Gould HJ, Sutton BJ. 2008. IgE in allergy and asthma today. *Nat Rev Immunol* **8:** 205–217.

Grayson MH, Cheung D, Rohlfing MM, Kitchens R, Spiegel DE, Tucker J, Battaile JT, Alevy Y, Yan L, Agapov E, et al. 2007. Induction of high-affinity IgE receptor on lung dendritic cells during viral infection leads to mucous cell metaplasia. *J Exp Med* **204:** 2759–2769.

Harmar AJ, Marston HM, Shen S, Spratt C, West KM, Sheward WJ, Morrison CF, Dorin JR, Piggins HD, Reubi JC, et al. 2002. The VPAC(2) receptor is essential for circadian function in the mouse suprachiasmatic nuclei. *Cell* **109:** 497–508.

Herbert DR, Holscher C, Mohrs M, Arendse B, Schwegmann A, Radwanska M, Leeto M, Kirsch R, Hall P, Mossmann H, et al. 2004. Alternative macrophage activation is essential for survival during schistosomiasis and downmodulates T helper 1 responses and immunopathology. *Immunity* **20:** 623–635.

Heredia JE, Mukundan L, Chen FM, Mueller AA, Deo RC, Locksley RM, Rando TA, Chawla A. 2013. Type 2 innate signals stimulate fibro/adipogenic progenitors to facilitate muscle regeneration. *Cell* **153:** 376–388.

Hoeffel G, Wang Y, Greter M, See P, Teo P, Malleret B, Leboeuf M, Low D, Oller G, Almeida F, et al. 2012. Adult Langerhans cells derive predominantly from embryonic fetal liver monocytes with a minor contribution of yolk sac-derived macrophages. *J Exp Med* **209:** 1167–1181.

ISAAC. 1998. Worldwide variation in prevalence of symptoms of asthma, allergic rhinoconjunctivitis, and atopic eczema: ISAAC. The International Study of Asthma and Allergies in Childhood (ISAAC) Steering Committee. *Lancet* **351:** 1225–1232.

Ishizaka K, Ishizaka T, Hornbrook MM. 1966. Physicochemical properties of human reaginic antibody: IV. Presence of a unique immunoglobulin as a carrier of reaginic activity. *J Immunol* **97:** 75–85.

Jenkins SJ, Ruckerl D, Cook PC, Jones LH, Finkelman FD, van Rooijen N, MacDonald AS, Allen JE. 2011. Local macrophage proliferation, rather than recruitment from the blood, is a signature of TH2 inflammation. *Science* **332:** 1284–1288.

Jones HI. 1988. Notes on parasites in penguins (Spheniscidae) and petrels (Procellariidae) in the Antarctic and Sub-Antarctic. *J Wildl Dis* **24:** 166–167.

Josefowicz SZ, Niec RE, Kim HY, Treuting P, Chinen T, Zheng Y, Umetsu DT, Rudensky AY. 2012. Extrathymically generated regulatory T cells control mucosal TH2 inflammation. *Nature* **482:** 395–399.

Kiss EA, Vonarbourg C, Kopfmann S, Hobeika E, Finke D, Esser C, Diefenbach A. 2011. Natural aryl hydrocarbon receptor ligands control organogenesis of intestinal lymphoid follicles. *Science* **334:** 1561–1565.

La Grange LJ, Govender D, Mukaratirwa S. 2013. The occurrence of *Trichinella zimbabwensis* in naturally infected wild crocodiles (*Crocodylus niloticus*) from the Kruger National Park, South Africa. *J Helminthol* **87:** 91–96.

Lavikainen A, Laaksonen S, Beckmen K, Oksanen A, Isomursu M, Meri S. 2011. Molecular identification of *Taenia* spp. in wolves (*Canis lupus*), brown bears (Ursus arctos) and cervids from North Europe and Alaska. *Parasitol Int* **60:** 289–295.

Ledin A, Bergvall K, Hillbertz NS, Hansson H, Andersson G, Hedhammar A, Hellman L. 2006. Generation of therapeutic antibody responses against IgE in dogs, an animal species with exceptionally high plasma IgE levels. *Vaccine* **24:** 66–74.

Ledin A, Arnemo JM, Liberg O, Hellman L. 2008. High plasma IgE levels within the Scandinavian wolf population, and its implications for mammalian IgE homeostasis. *Mol Immunol* **45:** 1976–1980.

Lee JS, Cella M, McDonald KG, Garlanda C, Kennedy GD, Nukaya M, Mantovani A, Kopan R, Bradfield CA, Newberry RD, et al. 2012. AHR drives the development of gut ILC22 cells and postnatal lymphoid tissues via pathways dependent on and independent of Notch. *Nat Immunol* **13:** 144–151.

Lelievre V, Favrais G, Abad C, Adle-Biassette H, Lu Y, Germano PM, Cheung-Lau G, Pisegna JR, Gressens P, Lawson G, et al. 2007. Gastrointestinal dysfunction in mice with a targeted mutation in the gene encoding vasoactive intestinal polypeptide: A model for the study of intestinal ileus and Hirschsprung's disease. *Peptides* **28:** 1688–1699.

Li Y, Innocentin S, Withers DR, Roberts NA, Gallagher AR, Grigorieva EF, Wilhelm C, Veldhoen M. 2011. Exogenous stimuli maintain intraepithelial lymphocytes via aryl hydrocarbon receptor activation. *Cell* **147:** 629–640.

Maizels RM, Smith KA. 2011. Regulatory T cells in infection. *Adv Immunol* **112:** 73–136.

Malaviya R, Ikeda T, Ross E, Abraham SN. 1996. Mast cell modulation of neutrophil influx and bacterial clearance at sites of infection through TNF-α. *Nature* **381:** 77–80.

Masaki T, Qu J, Cholewa-Waclaw J, Burr K, Raaum R, Rambukkana A. 2013. Reprogramming adult Schwann cells to stem cell-like cells by leprosy bacilli promotes dissemination of infection. *Cell* **152:** 51–67.

Mathis D. 2013. Immunological goings-on in visceral adipose tissue. *Cell Metab* **17:** 851–859.

Maywood ES, Drynan L, Chesham JE, Edwards MD, Dardente H, Fustin JM, Hazlerigg DG, O'Neill JS, Codner GF, Smyllie NJ, et al. 2013. Analysis of core circadian feedback loop in suprachiasmatic nucleus of mCry1-luc transgenic reporter mouse. *Proc Natl Acad Sci* **110:** 9547–9552.

McLachlan JB, Hart JP, Pizzo SV, Shelburne CP, Staats HF, Gunn MD, Abraham SN. 2003. Mast cell-derived tumor necrosis factor induces hypertrophy of draining lymph nodes during infection. *Nat Immunol* **4:** 1199–1205.

Mold JE, Venkatasubrahmanyam S, Burt TD, Michaelsson J, Rivera JM, Galkina SA, Weinberg K, Stoddart CA, McCune JM. 2010. Fetal and adult hematopoietic stem cells give rise to distinct T cell lineages in humans. *Science* **330:** 1695–1699.

Molofsky AB, Nussbaum JC, Liang HE, Van Dyken SJ, Cheng LE, Mohapatra A, Chawla A, Locksley RM. 2013. Innate lymphoid type 2 cells sustain visceral adipose tissue eosinophils and alternatively activated macrophages. *J Exp Med* **210:** 535–549.

Monticelli LA, Sonnenberg GF, Abt MC, Alenghat T, Ziegler CG, Doering TA, Angelosanto JM, Laidlaw BJ, Yang CY, Sathaliyawala T, et al. 2011. Innate lymphoid cells promote lung-tissue homeostasis after infection with influenza virus. *Nat Immunol* **12:** 1045–1054.

Moro K, Yamada T, Tanabe M, Takeuchi T, Ikawa T, Kawamoto H, Furusawa J, Ohtani M, Fujii H, Koyasu S. 2010. Innate production of T_H2 cytokines by adipose tissue-associated c-Kit$^+$Sca-1$^+$ lymphoid cells. *Nature* **463:** 540–544.

Mukai K, Matsuoka K, Taya C, Suzuki H, Yokozeki H, Nishioka K, Hirokawa K, Etori M, Yamashita M, Kubota T, et al. 2005. Basophils play a critical role in the development of IgE-mediated chronic allergic inflammation independently of T cells and mast cells. *Immunity* **23:** 191–202.

Nath P, Leung SY, Williams AS, Noble A, Xie S, McKenzie AN, Chung KF. 2007. Complete inhibition of allergic airway inflammation and remodelling in quadruple IL-4/5/9/13$^{-/-}$ mice. *Clin Exp Allergy* **37:** 1427–1435.

Neill DR, Wong SH, Bellosi A, Flynn RJ, Daly M, Langford TK, Bucks C, Kane CM, Fallon PG, Pannell R, et al. 2010. Nuocytes represent a new innate effector leukocyte that mediates type-2 immunity. *Nature* **464:** 1367–1370.

Nguyen KD, Qiu Y, Cui X, Goh YP, Mwangi J, David T, Mukundan L, Brombacher F, Locksley RM, Chawla A. 2011. Alternatively activated macrophages produce catecholamines to sustain adaptive thermogenesis. *Nature* **480:** 104–108.

Nussbaum JC, Locksley RM. 2012. Infectious (non)tolerance—Frustrated commensalism gone awry? *Cold Spring Harb Perspect Biol* **4:** a007328.

Nussbaum JC, Van Dyken SJ, von Moltke J, Cheng LE, Mohapatra A, Molofsky AB, Thornton EE, Krummel MF, Chawla A, Liang HE, et al. 2013. Type 2 innate lymphoid cells control eosinophil homeostasis. *Nature* **502:** 248–248.

Odegaard JI, Ricardo-Gonzalez RR, Goforth MH, Morel CR, Subramanian V, Mukundan L, Red Eagle A, Vats D, Brombacher F, Ferrante AW, et al. 2007. Macrophage-specific PPARγ controls alternative activation and improves insulin resistance. *Nature* **447:** 1116–1120.

Ohnmacht C, Schwartz C, Panzer M, Schiedewitz I, Naumann R, Voehringer D. 2010. Basophils orchestrate chronic allergic dermatitis and protective immunity against helminths. *Immunity* **33:** 364–374.

Olafsdottir D, Shinn AP. 2013. Epibiotic macrofauna on common minke whales, *Balaenoptera acutorostrata* Lacépède, 1804, in Icelandic waters. *Parasit Vectors* **6:** 105.

Ozcan E, Notarangelo LD, Geha RS. 2008. Primary immune deficiencies with aberrant IgE production. *J Allergy Clin Immunol* **122:** 1063–1054.

Palm NW, Rosenstein RK, Medzhitov R. 2012. Allergic host defences. *Nature* **484:** 465–472.

Pauly JE, Burns ER, Halberg F, Tsai S, Betterton HO, Scheving LE. 1975. Meal timing dominates the lighting regimen as a synchronizer of the eosinophil rhythm in mice. *Acta Anat (Basel)* **93:** 60–68.

Price AE, Liang HE, Sullivan BM, Reinhardt RL, Eisley CJ, Erle DJ, Locksley RM. 2010. Systemically dispersed innate IL-13-expressing cells in type 2 immunity. *Proc Natl Acad Sci* **107:** 11489–11494.

Pulendran B, Tang H, Manicassamy S. 2010. Programming dendritic cells to induce T_H2 and tolerogenic responses. *Nat Immunol* **11:** 647–655.

Qi X, Hong J, Chaves L, Zhuang Y, Chen Y, Wang D, Chabon J, Graham B, Ohmori K, Li Y, et al. 2013. Antagonistic regulation by the transcription factors C/EBPα and MITF specifies basophil and mast cell fates. *Immunity* **39:** 97–110.

Rae F, Woods K, Sasmono T, Campanale N, Taylor D, Ovchinnikov DA, Grimmond SM, Hume DA, Ricardo SD, Little MH. 2007. Characterisation and trophic functions of murine embryonic macrophages based upon the use of a Csf1r-EGFP transgene reporter. *Dev Biol* **308:** 232–246.

Reese TA, Liang HE, Tager AM, Luster AD, Van Rooijen N, Voehringer D, Locksley RM. 2007. Chitin induces accumulation in tissue of innate immune cells associated with allergy. *Nature* **447:** 92–96.

Reinhardt RL, Liang HE, Locksley RM. 2009. Cytokine-secreting follicular T cells shape the antibody repertoire. *Nat Immunol* **10:** 385–393.

Richet C. 1967. Anaphylaxis. In *Nobel lectures, physiology or medicine 1901–1921*. Elsevier, Amsterdam.

Rosenberg HF, Dyer KD, Foster PS. 2013. Eosinophils: Changing perspectives in health and disease. *Nat Rev Immunol* **13:** 9–22.

Rudra D, deRoos P, Chaudhry A, Niec RE, Arvey A, Samstein RM, Leslie C, Shaffer SA, Goodlett DR, Rudensky AY. 2012. Transcription factor Foxp3 and its protein partners form a complex regulatory network. *Nat Immunol* **13:** 1010–1019.

Sajic T, Hainard A, Scherl A, Wohlwend A, Negro F, Sanchez JC, Szanto I. 2013. STAT6 promotes bi-directional modulation of PKM2 in liver and adipose inflammatory cells in Rosiglitazone-treated mice. *Sci Rep* **3:** 2350.

Schulz C, Gomez Perdiguero E, Chorro L, Szabo-Rogers H, Cagnard N, Kierdorf K, Prinz M, Wu B, Jacobsen SE, Pollard JW, et al. 2012. A lineage of myeloid cells independent of Myb and hematopoietic stem cells. *Science* **336:** 86–90.

Shelburne CP, Nakano H, St John AL, Chan C, McLachlan JB, Gunn MD, Staats HF, Abraham SN. 2009. Mast cells augment adaptive immunity by orchestrating dendritic cell trafficking through infected tissues. *Cell Host Microbe* **6:** 331–342.

Shin H, Iwasaki A. 2013. Tissue-resident memory T cells. *Immunol Rev* **255:** 165–181.

Soyer OU, Akdis M, Ring J, Behrendt H, Crameri R, Lauener R, Akdis CA. 2013. Mechanisms of peripheral tolerance to allergens. *Allergy* **68:** 161–170.

Spits H, Cupedo T. 2012. Innate lymphoid cells: Emerging insights in development, lineage relationships, and function. *Annu Rev Immunol* **30:** 647–675.

Stanya KJ, Jacobi D, Liu S, Bhargava P, Dai L, Gangl MR, Inouye K, Barlow JL, Ji Y, Mizgerd JP, et al. 2013. Direct control of hepatic glucose production by interleukin-13 in mice. *J Clin Invest* **123:** 261–271.

Sullivan BM, Liang HE, Bando JK, Wu D, Cheng LE, McKerrow JK, Allen CDC, Locksley RM. 2011. Genetic analysis of basophil function in vivo. *Nat Immunol* **12:** 527–535.

Talay O, Yan D, Brightbill HD, Straney EE, Zhou M, Ladi E, Lee WP, Egen JG, Austin CD, Xu M, et al. 2012. IgE memory B cells and plasma cells generated through a germinal-center pathway. *Nat Immunol* **13:** 396–404.

Trasande L, Blustein J, Liu M, Corwin E, Cox LM, Blaser MJ. 2013. Infant antibiotic exposures and early-life body mass. *Int J Obesity* **37:** 16–23.

Tubo NJ, Pagan AJ, Taylor JJ, Nelson RW, Linehan JL, Ertelt JM, Huseby ES, Way SS, Jenkins MK. 2013. Single naive $CD4^+$ T cells from a diverse repertoire produce different effector cell types during infection. *Cell* **153:** 785–796.

Van Dyken SJ, Locksley RM. 2013. Interleukin-4- and interleukin-13-mediated alternatively activated macrophages: Roles in homeostasis and disease. *Annu Rev Immunol* **31:** 317–343.

Voehringer D. 2013. Protective and pathological roles of mast cells and basophils. *Nat Rev Immunol* **13:** 362–375.

Wagner B. 2009. IgE in horses: Occurrence in health and disease. *Vet Immunol Immunopathol* **132:** 21–30.

Walker JA, McKenzie AN. 2013. Development and function of group 2 innate lymphoid cells. *Curr Opin Immunol* **25:** 148–155.

WHO. 2007. *Global surveillance, prevention and control of chronic respiratory diseases.* WHO, Switzerland.

Wilhelm C, Hirota K, Stieglitz B, Van Snick J, Tolaini M, Lahl K, Sparwasser T, Helmby H, Stockinger B. 2011. An IL-9 fate reporter demonstrates the induction of an innate IL-9 response in lung inflammation. *Nat Immunol* **12:** 1071–1077.

Wu D, Molofsky AB, Liang HE, Ricardo-Gonzalez RR, Jouihan HA, Bando JK, Chawla A, Locksley RM. 2011. Eosinophils sustain adipose alternatively activated macrophages associated with glucose homeostasis. *Science* **332:** 243–247.

Yang Z, Sullivan BM, Allen CD. 2012. Fluorescent in vivo detection reveals that IgE^{+} B cells are restrained by an intrinsic cell fate predisposition. *Immunity* **36:** 857–872.

Zhou T, Georgiev I, Wu X, Yang ZY, Dai K, Finzi A, Kwon YD, Scheid JF, Shi W, Xu L, et al. 2010. Structural basis for broad and potent neutralization of HIV-1 by antibody VRC01. *Science* **329:** 811–817.

Inflammation and the Blood Microvascular System

Jordan S. Pober[1] and William C. Sessa[2]

[1]Department of Immunobiology, Yale University School of Medicine, New Haven, Connecticut 06520-8089

[2]Department of Pharmacology, Yale University School of Medicine, New Haven, Connecticut 06520-8089

Correspondence: jordan.pober@yale.edu

Acute and chronic inflammation is associated with changes in microvascular form and function. At rest, endothelial cells maintain a nonthrombogenic, nonreactive surface at the interface between blood and tissue. However, on activation by proinflammatory mediators, the endothelium becomes a major participant in the generation of the inflammatory response. These functions of endothelium are modified by the other cell populations of the microvessel wall, namely pericytes, and smooth muscle cells. This article reviews recent advances in understanding the roles played by microvessels in inflammation.

Inflammation, a major component of the innate immune response, is typically a local process historically characterized by cardinal features, such as rubor (redness) and calor (warmth), both caused by increased blood flow to the inflamed site, and tumor (swelling), caused by extravasation of fluid, plasma proteins, and leukocytes (changes attributable to actions of the local microvasculature). In this article, we will review how blood microvessels and the cells from which they are formed respond to innate inflammatory signals, leading to inflammation. We will not discuss recently reviewed topics, such as the lymphatic circulation (Alitalo 2011) or role of blood vessels in adaptive immunity (Pober and Tellides 2012). We begin with an overview of the blood microvascular system.

ORGANIZATION OF THE BLOOD MICROVASCULAR SYSTEM

The blood vascular system consists of two (systemic and pulmonary) closed loops organized into distinct segments. Large elastic arteries arising from the heart give rise to smaller muscular arteries that convey blood to specific organs. Within the organs, arteries further arborize, ending as arterioles having diameters in tens of micrometers. Arterioles, the most proximal segments of the microvasculature, are internally lined by a monolayer of endothelial cells (ECs), connected to each other by intermixed tight and adherens junctions, that is invested by circumferentially arranged layers (lamellae) of vascular smooth muscle cells (SMCs). ECs are separated from the SMC layers by both a condensed layer

Cite this article as *Cold Spring Harb Perspect Biol* doi: 10.1101/cshperspect.a016345

of extracellular matrix enriched in type IV collagen and laminin, known as basement membrane (BM), and deep to the BM, by a thick connective tissue band enriched in elastin fibers called the internal elastic lamina (IEL). The IEL marks the boundary between the arterial intima and the media. The intima between the BM and IEL is formed by a small (in arterioles) zone of loose connective tissue that may contain both SMCs and leukocytes. Focal discontinuities in both BM and the IEL allow ECs, which are electrically coupled to each other through gap junctions, to form gap junctions with underlying SMCs (Behringer and Segal 2012). A less well-defined outer elastic lamina marks the outer boundary of the media, separating it from the parenchyma of the surrounding organ.

At their distal end, arterioles arborize further to give rise to capillaries. This portion of the microvasculature varies widely in structure from tissue to tissue but has the general feature of being formed by a series of single ECs, each of which curves around to form a lumenized tube with diameters of <10 μm. Capillaries are very numerous and comprise the major surface for exchange of gases, fluid, and nutrients between blood and tissue. Heterogeneity of capillary ECs affects the degree to which they limit or permit such exchanges (Aird 2007a,b). Proteins may pass either through the junctions between capillary ECs (paracellular transit) or through the ECs themselves (transcellular transit via vesicles, channels, or fenestrae). Highly impermeant capillaries, such as those in the central nervous system (CNS), form many more tight junctions between adjacent ECs than other capillaries, limiting paracellular passage of proteins, and lack fenestrae and channels, forcing all exchanges through a limited vesicular transport system. At the other extreme, sinusoidal capillaries in liver, spleen, or bone marrow have open gaps between adjacent ECs. Some capillary ECs form fenestrae (holes) that allow transcellular passage of proteins. Fenestrae may be filled in by organized protein structures called diaphragms, partially limiting protein transit. Capillaries are invested and supported by a single discontinuous layer of contractile cells known as pericytes (PCs) (Dore-Duffy and Cleary 2011). PCs are arrayed longitudinally along the capillary and each PC may contact multiple ECs. The ratio of PCs to ECs varies among different tissues, being highest at 1:1 in the CNS where permeability is the lowest. Unlike SMCs of arterioles that form their own layers of extracellular matrix typically enriched for type I collagen and elastin, PCs reside within the collagen IV- and laminin-rich BM of the ECs. PCs lack tight junctions but may limit extravasation of plasma proteins indirectly through influence on EC junctions and/or BM composition (Goddard and Iruela-Arispe 2013).

At their termini, capillaries converge and drain their contents into larger caliber vessels known as venules. Postcapillary venules further converge into larger collecting vessels, ultimately connecting the microvasculature to veins that drain blood from the organ and return it to the heart. The EC lining of larger venules and veins is invested by layers of vascular SMCs, although these are typically less well organized into distinct lamellae and are fewer in number than those found in arterioles and the IEL is less well formed. In contrast, postcapillary venules, like capillaries, are primarily invested by PCs rather than SMCs. The ECs of the postcapillary venules are connected to each other by adherens junctions and lack tight junctions (Goddard and Iruela-Arispe 2013), making them more intrinsically leaky than the continuous capillaries that empty into them (Rous and Smith 1931). The capillaries themselves may be leakier at their venular end than they are at the arteriolar origin. In the skin, the BM of postcapillary venules has a distinct appearance by transmission electron microscopic, forming lamellae as opposed to the more homogeneous BM of arterioles and capillaries (Braverman 1989), implying a difference in composition and/or organization that may influence EC functions.

The embryological origin of the three principal cell types of the microvasculature are distinct. Almost all ECs in the adult organism can trace their lineage back to angioblasts arising in the blood islands of the aorto-gonadal-mesonephric region that migrate and differentiate into endothelial progenitor cells (EPCs) and/or fully differentiated ECs. Consequently, the

Cite this article as *Cold Spring Harb Perspect Biol* doi: 10.1101/cshperspect.a016345

tissue-specific features of capillary ECs develop in response to local environmental cues rather than to distinct embryological origins (Aird 2007a,b). The formation of new blood vessels in settings of chronic inflammation begins with outgrowth and replication of differentiated EC lining preexisting microvessels, although EPCs, probably residing within the vessel wall, may also contribute. Vascular SMCs of large vessels are more heterogeneous in their origin. Some, as in the proximal aorta, arise from the second heart field or neural crest, whereas others develop from local mesenchyme, often maintaining embryological patterns of Hox gene expression (Majesky 2007). Both SMCs and PCs of the microvasculature are probably derived from local mesenchyme but SMCs and PCs within the CNS may derive from neural crest and, at some sites, PCs may derive from ECs undergoing "endothelial to mesenchymal cell transition" (Armulik et al. 2011). Little is known about tissue-specific differences among PCs and how these may affect the contributions of these mural cells to inflammatory processes. Mesenchymal stem cells (MSCs, also called mesenchymal stromal cells) localize to the PC layer of microvessels (Crisan et al. 2008), but it is not clear whether all PCs have multipotency.

HOMEOSTATIC FUNCTIONS OF THE VASCULATURE

The vascular system under normal circumstances performs four major homeostatic functions: (1) it keeps blood fluid; (2) it regulates perfusion of different organs; (3) it prevents inappropriate activation of leukocytes; and (4) it regulates permselective exchange of macromolecules between blood and tissues. The prevention of coagulation is a general feature of the whole vascular system and will be discussed here. The other homeostatic properties of the vasculature are typically assigned to specific microvascular segments and will be discussed in subsequent sections.

ECs prevent intravascular coagulation by several mechanisms. First, ECs sequester phosphatidylserine (PS) to the inner leaflet of their plasma membrane, depriving circulating clotting factors of a surface required for assembly into functional complexes. Second, resting ECs express proteins that inhibit coagulation, notably tissue factor pathway inhibitor (TFPI) and thrombomodulin. TFPI prevents tissue factors from capturing and accelerating the catalytic activity of factor VIIa, the initiator of the intrinsic coagulation cascade. Thrombomodulin captures active thrombin and alters its substrate specificity from a procoagulant protease that converts fibrinogen to fibrin to an anticoagulant protease that cleaves and activates protein C. Activated protein C, when bound to its receptor on ECs and in combination with protein S (made by ECs among other cell types), cleaves and inactivates various coagulation factors, such as factor V (Bouwens et al. 2013). Third, ECs also express proteoglycans bearing heparin sulfate glycosaminoglycans (GAGs) that capture and activate antithrombin III, creating a substrate trap for active thrombin. Fourth, resting ECs synthesize plasminogen activators (both tissue type and urokinase type) as well as the receptor for urokinase-type plasminogen activator, converting circulating plasminogen to plasmin, a protease that cleaves fibrin and lyses incipient thrombi. Fifth, ECs prevent platelet activation by inhibiting thrombin, by preventing contact with BM or interstitial collagens, and by degrading extracellular adenosine-5′-triphosphate (ATP). Resting ECs synthesize von Willebrand factor (vWf), a platelet adhesive molecule important for platelet adhesion in settings of high shear stress, but sequester it internally within storage granules known as Weibel–Palade bodies (WPB) where it is inaccessible to platelets. Finally, resting ECs synthesize and release small quantities of PGI2 and nitric oxide (NO) sufficient to inhibit platelet activation by raising intracellular cAMP and cGMP, respectively. Inhibition of platelet activation by these mediators may act synergistically.

ARTERIOLES AND CONTROL OF BLOOD FLOW

The major roles of arterioles are to control pressure, flow, and nutrient delivery to the capillary beds. Arborizing branches of arterioles provide a physical buffer or resistance to normalize pressure gradients generated throughout the cardiac cycle. As pressure increases into proximal arteries during cardiac contraction, distal arterioles will vasoconstrict to limit pressure and flow into the microcirculation through a mechanism called myogenic vasoconstriction. This arteriole, smooth muscle intrinsic response is critical for maintaining constant flow to vital organs, such as the brain, kidney, and heart and serves as a mechanism for reducing pressure into more structurally fragile capillary beds. In addition to myogenic constriction in arterioles, ECs in arteries and arterioles can sense changes in flow and release local autacoids, such as NO, lipid metabolites, and other mediators to increase or decrease vessel diameter. During acute inflammatory responses, leukocyte-derived mediators such histamine and bradykinin will cause arteriolar dilation thereby increasing blood flow leading to rubor. Histamine-induced arteriolar dilation is abrogated in mice lacking the endothelial nitric oxide synthase gene (eNOS) (Payne et al. 2003); however, other vasodilatory mediators, such as PGI2 and endothelium-derived hyperpolarization factors may contribute to enhanced blood flow in response to proinflammatory molecules. The increase in flow and pressure will increase intravascular hydrostatic pressure providing a gradient for the extravasation of fluid and protein through postcapillary venules into tissue. This occurs simultaneously with retraction of EC lining the venules as highlighted below.

POSTCAPILLARY VENULES, PLASMA PROTEIN EXTRAVASATION, AND LEUKOCYTE RECRUITMENT

In their basal state, venular ECs form a barrier sufficient to retain most proteins, including albumin (Dejana and Giampietro 2012). The oncotic effect of these plasma proteins serves to limit the extravasation of fluid. Resting ECs also fail to recruit leukocytes, largely attributed to the absence of luminal molecules capable of capturing these cells from the circulation. Similarly, leukocyte-activating chemokines are also absent on resting ECs. However, certain leukocyte adhesion molecules (e.g., P-selectin [CD62P]) and some chemokines (e.g., interleukin [IL]-8 [CXCL8], MCP-1 [CCL2], and eotaxin 3 [CCL26]), are expressed in resting ECs; they are sequestered within WPB and, thus, unavailable to circulating leukocytes (Rondaij et al. 2006). Furthermore, basal NO production in ECs can potentially exert anti-inflammatory effects by inhibiting activation of leukocytes.

Changes in venular ECs alter the behavior of these cells to promote inflammation; such changes have been denoted as type I and/or type II activation (Pober and Sessa 2007). Many mediators of type I activation are vasoactive autacoids, such as histamine, that signal through G-protein-coupled receptors (GPCRs). Histamine receptors are more highly expressed on venular ECs than they are on arteriolar or capillary ECs (Heltianu et al. 1982), although histamine can stimulate vasodilator production from arteriolar ECs. The EC signaling pathways activated by these mediators have been reviewed elsewhere (Pober and Sessa 2007). Venular EC responses to histamine include: transient contraction of ECs, creating intercellular gaps and leading to parcellular escape of plasma proteins (Majno et al. 1961); regulated exocytosis of the contents of WPB, bringing molecules, such as vWf, P-selectin, and certain stored chemokines (IL-8, MCP-1, eotaxin-3) to the luminal cell surface (Lorant et al. 1991); and synthesis of lipid mediators, such as platelet-activating factor (PAF), a potent activator of leukocytes as well as platelets. Coexpression of leukocyte-binding proteins (P-selectin) and leukocyte-activating molecules (PAF, chemokines) on the EC plasma membrane has been described as a "juxtacrine" signaling that, in vitro, can capture and stimulate the transendothelial extravasation of neutrophils (Lorant et al. 1991). However, GPCR-induced signals are self-limited in duration to ~15 min and the same signals induce synthesis of other mediators that inhibit leukocyte activation, such as NO. Consequently, the result of histamine injection is simply a transient vasodilation and vascular leak ("wheal and flare") that quickly resolves without significant leukocyte recruitment.

 Cite this article as *Cold Spring Harb Perspect Biol* doi: 10.1101/cshperspect.a016345

Sustained tissue swelling and significant leukocyte recruitment is dependent on type II activation of venular EC, mediated by inflammatory cytokines. The prototypical examples of such cytokines are IL-1 (both IL-1α and IL-1β) and tumor necrosis factor (TNF, often designated as TNF-α). ECs express both the signaling IL-1 receptor (IL-1R1, designated CD121a) and, in the resting state, TNF receptor 1 (TNFR1 or CD120a) but not TNFR2 (CD120b) (Al-Lamki et al. 2005). The binding of IL-1 or TNF to these receptors activates transcription factors, thereby increasing expression of mRNAs encoding proteins that capture and activate leukocytes. For example, TNF or IL-1 induce de novo expression of leukocyte adhesion molecules E-selectin (CD62E) and vascular cell adhesion molecule (VCAM)-1 (CD104) and increase expression of intercellular adhesion molecule (ICAM)-1 (CD54) from a low but detectable basal state (Pober and Sessa 2007). In mice, but not humans, P-selectin transcription is induced as well (Pan et al. 1998). TNF also induces miRs that feedback and limit E-selectin and ICAM-1 expression (Suarez et al. 2010). TNF and IL-1 also induce synthesis of chemokines, notably of IL-8 and MCP-1, the principal human chemokines that activate neutrophils and monocytes, respectively. The de novo synthesis of these chemokines is independent of the release of prestored chemokines from WPB, a type I activation responses, as inflammatory cytokines do not cause WPB exocytosis (Zavoico et al. 1989). Chemokines released from EC or from leukocytes within the perivascular space bind to GAG expressed on the EC luminal surface (Mortier et al. 2012), which, along with leukocyte adhesion molecules, provide sustained juxtacrine signaling that leads to leukocyte infiltration in vivo. Fractalkine (CX3CL-1) represents a variation on this theme in that the chemokine moiety that interacts with its receptor, predominantly expressed on a subset of monocytes, is synthesized and expressed as the amino-terminal domain of an EC proteoglycan (Imaizumi et al. 2004). Finally, it should be noted that, because the signaling pathways activated by IL-1 binding to its receptor are also activated by Toll-like receptors (TLRs), with the exception of TLR3, and because human vascular EC express TLR1, TLR2, TLR4, TLR6, and TLR9, engagement of these receptors by appropriate ligands, typically one or more pathogen-associated molecular patterns or damage-associated molecular patterns, can produce type II activation responses (Opitz et al. 2007).

The general model for leukocyte recruitment by type II–activated venular ECs involves a multistep cascade (Ley et al. 2007). In humans, E-selectin typically mediates the initial tethering of the circulating neutrophil or monocyte by recognizing specific carbohydrate determinants, which in the case of neutrophils are attached to L-selectin expressed on the tips of microvillous projections. These adhesive interactions are of low affinity but are rapidly formed. Flowing blood pushes the tethered leukocyte, breaking selectin-mediated attachments that subsequently rapidly reform with new E-selectin molecules as the leukocyte is displaced in the direction of flow, resulting in leukocyte rolling. Monocytes and some T cells may instead (or additionally) be tethered and roll using leukocyte integrin VLA-4 (CD49d/CD29), that in its low affinity state, can also rapidly form weak attachments to endothelial VCAM-1. Some birectional signals may be transmitted during these adhesive interactions, but these are generally thought to provide insufficient activation of the leukocyte to progress to the next step in the cascade, known as firm adhesion. Instead, specific GPCRs on the rolling leukocyte must encounter a cognate ligand, typically an EC-bound chemokine, resulting in signaling that causes the leukocyte to increase affinity of its integrins for EC ligands, specifically LFA-1 (CD11a/CD18) and Mac-1 (CD11b/CD18) for ICAM-1 or VLA-4 (CD49d/CD29) for VCAM-1, changing from a round cell to one spread out on the surface of the ECs. These "firmly adherent" leukocytes acquire motility and use these same integrins and EC ligands to crawl toward EC junctions. At or near the junctions, the EC may project its luminal plasma membrane upward, forming an "adhesion cup" or "docking structure" that partially engulfs the bound leukocyte and within which EC adhesion molecules, specifically ICAM-1 and VCAM-1, cluster (Barreiro et al. 2004). The clustering of ICAM-1 and possibly VCAM-1 induces the membrane remodeling to occur at the EC junction, facilitating a path for transendothelial migration. At the same time, the leukocyte projects cytosolic processes ("invadosomes") that push between or through the ECs, initiating the process of extravasation (Carman et al. 2007).

Transit across the EC lining of the venule involves sequential molecular interactions with platelet/endothelial cell adhesion molecule PECAM-1 (CD31) and CD99, each of which forms homophilic adhesions with the same molecules expressed on the leukocyte (Muller 2011). VE-cadherin and certain tight junction proteins (e.g., junctional adhesion molecule [JAM] A or C) on the EC may also play a role at this stage. Unlike ICAM-1 and VCAM-1, the total levels of EC expression of PECAM-1 and CD99 are not increased by inflammatory cytokines. However, ECs sequester these molecules in perijunctional vesicles denoted as the lateral border recycling compartment. Attachment of leukocytes at or near the junctions induces the EC to bring PECAM-1 and CD99 to the surface of the junction or to the membrane of transcellular channels that form near the junction, lining the path through which the leukocyte traverses the EC monolayer. The signal transmitted to the ECs that leads to increased surface expression of PECAM-1 and CD99 may be clustering of molecules, such as ICAM-1.

Once through the EC layer, the leukocyte still must traverse the PC layer and the BM in which the PCs are embedded. PCs also respond to inflammatory cytokines, but their levels of induced expression of adhesion molecules, largely restricted to ICAM-1 and VCAM-1, are less than on ECs (Ayres-Sander et al. 2013). Consequently, cultured PC monolayers support only limited transit of neutrophils, although neutrophils that traverse an EC monolayer become altered in an unspecified manner that increases their ability to traverse PC monolayers. PCs do express CD99, but not PECAM-1, and CD99 may also be involved in traversing the PC layer of the venular wall. In addition, PCs are contractile cells that anchor to BM proteins and can manipulate the organization of the BM; PCs also steer extravasating leukocytes to regions in which the BM is more attenuated, known as "low expression regions" (Nourshargh et al. 2010). However, the detailed functions of PCs in the process of leukocyte extravasation are less well understood than those of ECs.

During inflammation, plasma protein and leukocyte extravasation both occur at the same sites. The extravasation of leukocytes through an EC monolayer in vitro can occur without inducing inter-EC gaps that result in increased paracellular leak of macromolecules (Huang et al. 1988). The increased paracellular leak in vivo probably results directly from inflammatory cytokine exposure rather than through gaps opened by leukocytes. The extravasated plasma proteins form a provisional matrix within the tissue that can support attachment, survival, and migration of subsequently extravasated leukocytes. In cell culture models, EC leaks in response to IL-1 or TNF occur in two stages: an early but transient leak that may be mediated by cytoskeletal contraction induced by activation of a small G protein, likely Arf6 (Zhu et al. 2012), and a later, more pronounced and sustained leak dependent on new protein synthesis (Clark et al. 2007). The newly synthesized protein(s) responsible for increased leakiness have not been defined.

Entry of leukocytes into the CNS has unique features that differ from extravasation elsewhere in the body (Man et al. 2007). Within the CNS, ECs form numerous intercellular tight junctions, the PC to EC ratio reaches 1:1, the highest in the body, and astrocyte and glial foot processes abut on the EC/PC BM, creating an additional barrier referred to as the glia limitans. As a result, resting CNS microvessels effectively prevent all protein transit, creating the "blood–brain barrier." It is unknown to what extent these features extend to the postcapillary venules of the CNS, but basal trafficking of leukocytes is very low. Inflammation may disrupt the blood–brain barrier and passage of leukocytes through the vessel wall seems to be particularly dependent on leukocyte integrin VLA-4 (capable of recognizing cellular fibronectin as well as VCAM-1). Low-affinity VLA-4 interactions with VCAM-1 can support leukocyte tethering and rolling, possibly explaining why knockout of selectins has little effect on entry of leukocytes into the CNS in mouse models of neuroinflammation. LFA-1 is also important for entry into the CNS, and it may engage ICAM-2, a constitutively expressed EC molecule, as well as ICAM-1 during crawling to the EC junctions. LFA-1 may also engage JAM-A, a tight junction protein during transendothelial migration. Those cells that do get through the EC/PC/BM barrier may still be prevented from entering the CNS because of the glia limitans, accumulating in the so-called Virchow–Robin space adjacent to the blood vessels. Leukocytes may also bypass the CNS vasculature to enter into the brain by crossing the more permeable choroid plexus where blood is filtered to produce cerebrospinal fluid.

In skin as in most other tissues, inflammatory cytokines induce E-selectin and VCAM-1 on venular ECs but not adjacent capillary ECs; capillary ECs are responsive to these same cytokines as shown by selective up-regulation ICAM-1 expression (Enis et al. 2005). This difference accounts for why leukocytes extravasate through venules and generally not capillaries, but the basis of this restriction of the response to cytokines is unknown. Possible contributors are differences in the shear stress detected by the ECs (high in capillaries, low in venules) and differences in BM composition. Application of shear stress to cultured microvascular ECs induces expression of the transcription factor KLF-4, which limits cytokine-induced adhesion molecule expression (Clark et al. 2011). As noted previously, the venular BM in skin microvessels has a distinct appearance by transmission electron microscopy from that of capillaries and in psoriatic lesions, when capillary loops in the dermal papillae are remodeled to become venules, these microvessels change the appearance of their BM concomitantly with the capacity of their ECs to express E-selectin and VCAM-1 (Petzelbauer et al. 1994). These data are consistent with the hypothesis that attachment to BM of different composition or organization can influence EC responses to cytokines. It should be noted that there are exceptions to the primary role of venules for extravasation. For example, in the lung, leukocytes exit into the bronchial wall via venules of the bronchial circulation but enter into the alveolus via alveolar capillaries of the pulmonary circulation. The capillary tufts within the renal glomeruli and the sinusoidal capillaries of the liver can also express adhesion molecules and support leukocyte extravasation.

The various stimuli that evoke inflammation may elicit inflammatory infiltrates that are enriched for particular types of leukocytes and these may change over time. Such selectivity and its evolution can often be explained by changes in the adhesion molecules and chemokines displayed on the EC surface. For example, E-selectin is more rapidly synthesized and displayed than is VCAM-1 or ICAM-1, but its expression on the cell surface is generally more transient, peaking at 4–6 h and falling to low levels by 24 h, and better correlates with neutrophil than mononuclear cell capture and recruitment, perhaps because mononuclear cells are more adept at tethering and rolling on VCAM-1. IL-4 augments VCAM-1 expression by EC and also causes ECs to synthesize eotaxin-3, a chemokine that favors recruitment of eosinophils. Thus, the combination of TNF plus IL-4 can lead to inflammatory infiltrates that are enriched for this cell type (Briscoe et al. 1992). This kind of "specialized" inflammation has historically been associated with adaptive immunity in which $CD4^+$, Th-1, Th-2, and, more recently, Th-17 cells favor their own recruitment and specific types of effector cells and, hence, are outside the scope of this review. However, it has recently been appreciated that other cell types, such as mast cells, basophils, and innate lymphocytes, also may display polarized cytokine profiles and elicit particular inflammatory patterns that are independent of antigen (Bochner and Schleimer 2001). It is likely that ECs will play a role in shaping the nature of these types of inflammatory reactions, although little is known to date what signals are involved. IL-17, for example, has little effect on ECs by itself but may modulate other cytokine responses (Bernink et al. 2013). Interestingly, IL-17 does have direct effects on PC production of chemokines, similar to previous observations using cultured

SMCs (Eid et al. 2009), and a more integrated view of venular activation rather than EC activation may be required to understand how inflammatory reactions evolve and differentiate.

VASCULAR DYSFUNCTION: FLOW DYSREGULATION, THROMBOSIS, AND CAPILLARY LEAK

As we noted earlier in this article, changes in ECs induced by vasoactive autacoids (type I activation) and inflammatory cytokines (type II activation) contribute to inflammation. Because resting ECs actively resist the development of inflammation, these responses may be seen as interfering with normal (basal) EC function. Alternatively, such changes can be viewed as being adaptive because local inflammation is an important mechanism of host defense and homeostasis. However, there are changes that occur in ECs that disrupt homeostasis without beneficial effect to the host. Some of these are simply exaggerated or inappropriate forms of activation, whereas others result from injury and have been linked together under the category of "endothelial dysfunction" (Pober et al. 2009). As discussed above, tissue perfusion is controlled by SMC tone in the terminal arterioles. SMCs normally respond to EC-derived signals, neural signals, and humoral (hormonal) signals. Perfusion to a specific tissue can be inappropriately diminished when arteriolar SMCs in the microvasculature of a particular tissue become refractory to vasodilatory signals and/or because EC-derived vasodilatory signals are decreased. A common pattern of EC dysfunction occurs when EC lose the capacity to synthesize NO because of diminished expression of eNOS owing to TNF-mediated destabilization of its mRNA mediated by TNF-induced miR155 (Sun et al. 2012). Additionally, oxidative stress associated with inflammation can turn eNOS into a generator of superoxide anion, a change called "eNOS uncoupling" thereby reducing NO bioactivity independent of changes in eNOS mRNA or protein levels (Forstermann and Sessa 2012). TNF may also cause ECs to increase synthesis of the vasoconstrictor peptide endothelin-1 (Marsden and Brenner 1992). Paradoxically, because flow is regulated by blood pressure in addition to vascular resistance, a global decrease in blood pressure caused by widespread vasodilation throughout the circulatory system, as occurs in septic shock, may also lead to hypoperfusion. A possible explanation is that TNF increases production of PGI2 by SMCs, as well as ECs, possibly through induction of cyclo-oxygenase 2 (Jimenez et al. 2005).

Thrombosis is another potential example of EC dysfunction (Pober et al. 2009). Local intravascular coagulation may be a means of preventing hematogenous dissemination of an infection and thus part of innate immunity. However, thrombosis can also produce tissue infarcts and, when widespread, paradoxical bleeding caused by consumption of clotting factors (Esmon 2004). ECs may contribute to this by shedding microparticles, for instance plasma membrane–derived vesicles with surface-exposed PS and thus can serve as a platform for assembly of coagulation factors. TNF-activated ECs may also lose expression of thrombomodulin through mRNA destabilization and may lose their anticoagulant heparin sulfates through cytokine-induced enzymatic degradation. Tissue factor pathway inhibitor may also be down-regulated at the same time that tissue factor is synthesized and "de-encrypted," the latter process describing transfer to a location, such as to shed microparticles, where it may encounter and catalytically enhance factor VIIa. The net effect of these changes is to enhance thrombin generation and to allow thrombin to cleave fibrinogen to fibrin. At the same time, ECs or other vascular cells may enhance production of plasminogen activator inhibitors, reducing the local capacity of ECs to activate plasminogen to plasmin, and lyse fibrin thrombi as they are formed.

We noted earlier that venular leak of plasma proteins plays an important role in the inflammatory process, providing the components of a provisional matrix to support extravasating leukocytes. We also noted that, in most tissues, capillaries form a much larger surface for exchange of nutrients and wastes between tissues and blood. Under normal circumstances, capillaries (and arterioles) do not leak as a part of the inflammatory process. This difference may relate to the observation that capillaries (and arterioles) form tight junctions between adjacent ECs, whereas venules do not (Simionescu et al. 1975). The process of disassembly of tight junctions may be viewed as pathological and as an example of dysfunction. Capillary leak is a characteristic feature of sepsis and its development is one of the causes of organ failure in that syndrome (Gustot 2011). The mechanisms by which mediators of sepsis open tight junctions is unknown and may be distinct from the process by which venular adherens junctions are disassembled. Capillary leak may also result from EC injury and death (Joris et al. 1990).

ANGIOGENESIS AND THE VASCULAR SYSTEM IN CHRONIC INFLAMMATION

Mononuclear leukocytes (monocytes) are recruited to sites of inflammation where they may differentiate into macrophages and promote angiogenesis. Angiogenesis is defined as the migration and proliferation of ECs lining venules into surrounding tissue resulting in the formation of a capillary plexus. The recruitment of macrophages to sites of inflammation is critical for the resolution of inflammation. However, if the signal inducing the acute inflammatory response is not eradicated, chronic inflammation may ensue, and there is evidence that the transition from acute to chronic inflammation relies on an angiogenic response as a means to provide blood supply to inflamed neotissue. In this context, sustained angiogenesis may amplify the extent of macrophage infiltration and edema, and worsen tissue damage. Indeed, strategies aimed at inhibiting macrophage subsets and/or angiogenesis can reduce the degree of inflammation in several preclinical model systems. Vascular endothelial cell growth factor-A (VEGFA) and TNF derived from macrophages are potent angiogenic factors and inducers of inflammation-driven vascular remodeling in a variety of inflammatory diseases, such as psoriasis, rheumatoid arthritis, inflammatory bowel disease, and asthma. Both VEGF and TNF are produced by activated macrophages and bind to their cognate receptors on endothelium to stimulate angiogenesis. VEGF binds to VEGF receptor 2 (VEGFR2) promoting angiogenesis (Baer et al. 2013), whereas TNF binds to TNFR2, which is induced on ECs during the early stages of inflammation and mediates its proangiogenic function (Luo et al. 2006). Moreover, TNF can prime endothelium by up-regulating VEGF receptor 2 (Sainson et al. 2008). Additional inflammatory mediators can also promote angiogenesis include IL-1, IL-6, IL-17, and IL-18, and there is substantial cross talk between these cytokines and VEGF/TNF signaling pathways (Huggenberger and Detmar 2011). Chemokines play an important role both as agents that recruit macrophages and as agents produced by macrophages (or by other cells, including ECs and PCs) that may then promote or inhibit angiogenesis (Owen and Mohamadzadeh 2013).

CONCLUDING REMARKS

The blood vascular system has evolved to control tissue homeostasis. Under normal conditions, the vasculature maintains a quiescent interface between blood and tissue. On encountering an inflammatory insult, the endothelium actively participates in controlling blood flow, permeability, leukocyte infiltration, and tissue edema, changes that serve to eradicate the initial stimulus. PCs and SMCs also participate in these processes, but their roles are less well understood than those of ECs. If the stimulus persists, inflammation evolves into a chronic phase. Sustained activation of the endothelium caused by persistent inflammation can promote macrophage recruitment and neotissue angiogenesis to sustain blood flow. Understanding the vascular changes that occur during the transition from acute to chronic inflammation is crucial for our development of novel therapeutic approaches associated with debilitating inflammatory diseases.

REFERENCES

Aird WC. 2007a. Phenotypic heterogeneity of the endothelium: I. Structure, function, and mechanisms. *Circ Res* **100:** 158–173.

Aird WC. 2007b. Phenotypic heterogeneity of the endothelium: II. Representative vascular beds. *Circ Res* **100:** 174–190.

Alitalo K. 2011. The lymphatic vasculature in disease. *Nat Med* **17:** 1371–1380.

Al-Lamki RS, Wang J, Vandenabeele P, Bradley JA, Thiru S, Luo D, Min W, Pober JS, Bradley JR. 2005. TNFR1- and TNFR2-mediated signaling pathways in human kidney are cell type-specific and differentially contribute to renal injury. *FASEB J* **19:** 1637–1645.

Armulik A, Genove G, Betsholtz C. 2011. Pericytes: Developmental, physiological, and pathological perspectives, problems, and promises. *Dev Cell* **21:** 193–215.

Ayres-Sander CE, Lauridsen H, Maier CL, Sava P, Pober JS, Gonzalez AL. 2013. Transendothelial migration enables subsequent transmigration of neutrophils through underlying pericytes. *PLoS ONE* **8:** e60025.

Baer C, Squadrito ML, Iruela-Arispe ML, De Palma M. 2013. Reciprocal interactions between endothelial cells and macrophages in angiogenic vascular niches. *Exp Cell Res* **319:** 1626–1634.

Barreiro O, Vicente-Manzanares M, Urzainqui A, Yanez-Mo M, Sanchez-Madrid F. 2004. Interactive protrusive structures during leukocyte adhesion and transendothelial migration. *Front Biosci* **9:** 1849–1863.

Behringer EJ, Segal SS. 2012. Spreading the signal for vasodilatation: Implications for skeletal muscle blood flow control and the effects of ageing. *J Physiol* **590:** 6277–6284.

Bernink J, Mjosberg J, Spits H. 2013. Th1- and Th2-like subsets of innate lymphoid cells. *Immunol Rev* **252:** 133–138.

Bochner BS, Schleimer RP. 2001. Mast cells, basophils, and eosinophils: Distinct but overlapping pathways for recruitment. *Immunol Rev* **179:** 5–15.

Bouwens EA, Stavenuiter F, Mosnier LO. 2013. Mechanisms of anticoagulant and cytoprotective actions of the protein C pathway. *J Thromb Haemost* **11:** 242–253.

Braverman IM. 1989. Ultrastructure and organization of the cutaneous microvasculature in normal and pathologic states. *J Invest Dermatol* **93:** 2S–9S.

Briscoe DM, Cotran RS, Pober JS. 1992. Effects of tumor necrosis factor, lipopolysaccharide, and IL-4 on the expression of vascular cell adhesion molecule-1 in vivo. Correlation with $CD3^{+}$ T cell infiltration. *J Immunol* **149:** 2954–2960.

Carman CV, Sage PT, Sciuto TE, de la Fuente MA, Geha RS, Ochs HD, Dvorak HF, Dvorak AM, Springer TA. 2007. Transcellular diapedesis is initiated by invasive podosomes. *Immunity* **26:** 784–797.

Clark PR, Manes TD, Pober JS, Kluger MS. 2007. Increased ICAM-1 expression causes endothelial cell leakiness, cytoskeletal reorganization and junctional alterations. *J Invest Dermatol* **127:** 762–774.

Clark PR, Jensen TJ, Kluger MS, Morelock M, Hanidu A, Qi Z, Tatake RJ, Pober JS. 2011. MEK5 is activated by shear stress, activates ERK5 and induces KLF4 to modulate TNF responses in human dermal microvascular endothelial cells. *Microcirculation* **18:** 102–117.

Crisan M, Yap S, Casteilla L, Chen CW, Corselli M, Park TS, Andriolo G, Sun B, Zheng B, Zhang L, et al. 2008. A perivascular origin for mesenchymal stem cells in multiple human organs. *Cell Stem Cell* **3:** 301–313.

Dejana E, Giampietro C. 2012. Vascular endothelial-cadherin and vascular stability. *Curr Opin Hematol* **19:** 218–223.

Dore-Duffy P, Cleary K. 2011. Morphology and properties of pericytes. *Methods Mol Biol* **686:** 49–68.

Eid RE, Rao DA, Zhou J, Lo SF, Ranjbaran H, Gallo A, Sokol SI, Pfau S, Pober JS, Tellides G. 2009. Interleukin-17 and interferon-γ are produced concomitantly by human coronary artery-infiltrating T cells and act synergistically on vascular smooth muscle cells. *Circulation* **119:** 1424–1432.

Enis DR, Shepherd BR, Wang Y, Qasim A, Shanahan CM, Weissberg PL, Kashgarian M, Pober JS, Schechner JS. 2005. Induction, differentiation, and remodeling of blood vessels after transplantation of Bcl-2-transduced endothelial cells. *Proc Natl Acad Sci* **102:** 425–430.

Esmon CT. 2004. The impact of the inflammatory response on coagulation. *Thromb Res* **114:** 321–327.

Forstermann U, Sessa WC. 2012. Nitric oxide synthases: Regulation and function. *Eur Heart J* **33:** 829–837, 837a–837d.

Goddard LM, Iruela-Arispe ML. 2013. Cellular and molecular regulation of vascular permeability. *Thromb Haemost* **109:** 407–415.

Gustot T. 2011. Multiple organ failure in sepsis: Prognosis and role of systemic inflammatory response. *Curr Opin Crit Care* **17:** 153–159.

Heltianu C, Simionescu M, Simionescu N. 1982. Histamine receptors of the microvascular endothelium revealed in situ with a histamine-ferritin conjugate: Characteristic high-affinity binding sites in venules. *J Cell Biol* **93:** 357–364.

Huang AJ, Furie MB, Nicholson SC, Fischbarg J, Liebovitch LS, Silverstein SC. 1988. Effects of human neutrophil chemotaxis across human endothelial cell monolayers on the permeability of these monolayers to ions and macromolecules. *J Cell Physiol* **135:** 355–366.

Huggenberger R, Detmar M. 2011. The cutaneous vascular system in chronic skin inflammation. *J Investig Dermatol Symp Proc* **15:** 24–32.

Imaizumi T, Yoshida H, Satoh K. 2004. Regulation of CX3CL1/fractalkine expression in endothelial cells. *J Atheroscler Thromb* **11:** 15–21.

Jimenez R, Belcher E, Sriskandan S, Lucas R, McMaster S, Vojnovic I, Warner TD, Mitchell JA. 2005. Role of Toll-like receptors 2 and 4 in the induction of cyclooxygenase-2 in vascular smooth muscle. *Proc Natl Acad Sci* **102:** 4637–4642.

Joris I, Cuenoud HF, Doern GV, Underwood JM, Majno G. 1990. Capillary leakage in inflammation. A study by vascular labeling. *Am J Pathol* **137:** 1353–1363.

Ley K, Laudanna C, Cybulsky MI, Nourshargh S. 2007. Getting to the site of inflammation: The leukocyte adhesion cascade updated. *Nat Rev Immunol* **7:** 678–689.

Lorant DE, Patel KD, McIntyre TM, McEver RP, Prescott SM, Zimmerman GA. 1991. Coexpression of GMP-140 and PAF by endothelium stimulated by histamine or thrombin: A juxtacrine system for adhesion and activation of neutrophils. *J Cell Biol* **115:** 223–234.

Luo D, Luo Y, He Y, Zhang H, Zhang R, Li X, Dobrucki WL, Sinusas AJ, Sessa WC, Min W. 2006. Differential functions of tumor necrosis factor receptor 1 and 2 signaling in ischemia-mediated arteriogenesis and angiogenesis. *Am J Pathol* **169:** 1886–1898.

Majesky MW. 2007. Developmental basis of vascular smooth muscle diversity. *Arterioscler Thromb Vasc Biol* **27:** 1248–1258.

Majno G, Palade GE, Schoefl GI. 1961. Studies on inflammation: II. The site of action of histamine and serotonin along the vascular tree: A topographic study. *J Biophys Biochem Cytol* **11:** 607–626.

Man S, Ubogu EE, Ransohoff RM. 2007. Inflammatory cell migration into the central nervous system: A few new twists on an old tale. *Brain Pathol* **17:** 243–250.

Marsden PA, Brenner BM. 1992. Transcriptional regulation of the endothelin-1 gene by TNF-α. *Am J Physiol* **262:** C854–C861.

Mortier A, Van Damme J, Proost P. 2012. Overview of the mechanisms regulating chemokine activity and availability. *Immunol Lett* **145:** 2–9.

Muller WA. 2011. Mechanisms of leukocyte transendothelial migration. *Annu Rev Pathol* **6:** 323–344.

Nourshargh S, Hordijk PL, Sixt M. 2010. Breaching multiple barriers: Leukocyte motility through venular walls and the interstitium. *Nat Rev Mol Cell Biol* **11:** 366–378.

Opitz B, Hippenstiel S, Eitel J, Suttorp N. 2007. Extra- and intracellular innate immune recognition in endothelial cells. *Thromb Haemost* **98:** 319–326.

Owen JL, Mohamadzadeh M. 2013. Macrophages and chemokines as mediators of angiogenesis. *Front Physiol* **4:** 159.

Pan J, Xia L, McEver RP. 1998. Comparison of promoters for the murine and human P-selectin genes suggests species-specific and conserved mechanisms for transcriptional regulation in endothelial cells. *J Biol Chem* **273:** 10058–10067.

Payne GW, Madri JA, Sessa WC, Segal SS. 2003. Abolition of arteriolar dilation but not constriction to histamine in cremaster muscle of eNOS$^{-/-}$ mice. *Am J Physiol Heart Circ Physiol* **285:** H493–H498.

Petzelbauer P, Pober JS, Keh A, Braverman IM. 1994. Inducibility and expression of microvascular endothelial adhesion molecules in lesional, perilesional, and uninvolved skin of psoriatic patients. *J Invest Dermatol* **103:** 300–305.

Pober JS, Sessa WC. 2007. Evolving functions of endothelial cells in inflammation. *Nat Rev Immunol* **7:** 803–815.

Pober JS, Tellides G. 2012. Participation of blood vessel cells in human adaptive immune responses. *Trends Immunol* **33:** 49–57.

Pober JS, Min W, Bradley JR. 2009. Mechanisms of endothelial dysfunction, injury, and death. *Annu Rev Pathol* **4:** 71–95.

Rondaij MG, Bierings R, Kragt A, van Mourik JA, Voorberg J. 2006. Dynamics and plasticity of Weibel-Palade bodies in endothelial cells. *Arterioscler Thromb Vasc Biol* **26:** 1002–1007.

Rous P, Smith F. 1931. The gradient of vascular permeability: III. The gradient along the capillaries and venules of frog skin. *J Exp Med* **53:** 219–242.

Sainson RC, Johnston DA, Chu HC, Holderfield MT, Nakatsu MN, Crampton SP, Davis J, Conn E, Hughes CC. 2008. TNF primes endothelial cells for angiogenic sprouting by inducing a tip cell phenotype. *Blood* **111:** 4997–5007.

Simionescu M, Simionescu N, Palade GE. 1975. Segmental differentiations of cell junctions in the vascular endothelium. The microvasculature. *J Cell Biol* **67:** 863–885.

Suarez Y, Wang C, Manes TD, Pober JS. 2010. Cutting edge: TNF-induced microRNAs regulate TNF-induced expression of E-selectin and intercellular adhesion molecule-1 on human endothelial cells: Feedback control of inflammation. *J Immunol* **184:** 21–25.

Sun HX, Zeng DY, Li RT, Pang RP, Yang H, Hu YL, Zhang Q, Jiang Y, Huang LY, Tang YB, et al. 2012. Essential role of microRNA-155 in regulating endothelium-dependent vasorelaxation by targeting endothelial nitric oxide synthase. *Hypertension* **60:** 1407–1414.

Zavoico GB, Ewenstein BM, Schafer AI, Pober JS. 1989. IL-1 and related cytokines enhance thrombin-stimulated PGI2 production in cultured endothelial cells without affecting thrombin-stimulated von Willebrand factor secretion or platelet-activating factor biosynthesis. *J Immunol* **142:** 3993–3999.

Zhu W, London NR, Gibson CC, Davis CT, Tong Z, Sorensen LK, Shi DS, Guo J, Smith MC, Grossmann AH, et al. 2012. Interleukin receptor activates a MYD88–ARNO–ARF6 cascade to disrupt vascular stability. *Nature* **492:** 252–255.

Sinusoidal Immunity: Macrophages at the Lymphohematopoietic Interface

Siamon Gordon[1], **Annette Plüddemann**[2], **and Subhankar Mukhopadhyay**[1,3]

[1]Sir William Dunn School of Pathology, University of Oxford, Oxford OX1 3RE, United Kingdom
[2]Department of Primary Care Health Sciences, University of Oxford, Oxford OX2 6GG, United Kingdom
[3]Wellcome Trust Sanger Institute, Hinxton, Cambridge CB10 1SA, United Kingdom

Correspondence: siamon.gordon@path.ox.ac.uk

Macrophages are widely distributed throughout the body, performing vital homeostatic and defense functions after local and systemic perturbation within tissues. In concert with closely related dendritic cells and other myeloid and lymphoid cells, which mediate the innate and adaptive immune response, macrophages determine the outcome of the inflammatory and repair processes that accompany sterile and infectious injury and microbial invasion. This article will describe and compare the role of specialized macrophage populations at two critical interfaces between the resident host lymphohematopoietic system and circulating blood and lymph, the carriers of cells, humoral components, microorganisms, and their products. Sinusoidal macrophages in the marginal zone of the spleen and subcapsular sinus and medulla of secondary lymph nodes contribute to the innate and adaptive responses of the host in health and disease. Although historically recognized as major constituents of the reticuloendothelial system, it has only recently become apparent that these specialized macrophages in close proximity to B and T lymphocytes play an indispensable role in recognition and responses to exogenous and endogenous ligands, thus shaping the nature and quality of immunity and inflammation. We review current understanding of these macrophages and identify gaps in our knowledge for further investigation.

The origin, distribution, and functions of cells of the mononuclear phagocyte system have been redefined by a resurgence of studies in the past decade, mostly in the mouse and to a limited extent in human (Davies et al. 2013). Progress has been made regarding the production and migration of monocytes and macrophages during embryonic and adult life, giving rise to resident cell populations in hematopoietic and nonlymphohematopoietic organs, such as liver, lung, gut, skin, and the central nervous system (CNS) (Schulz et al. 2012; Hashimoto et al. 2013; Yona et al. 2013). Blood-derived monocytes contribute new migrants, that give rise to tissue monocytes, macrophages, and dendritic cells (DCs), which all contribute to the inflammatory and immune response to infection and tissue injury (Cheong et al. 2010; Reizis 2012). A great deal has been learned regarding the versatility and plasticity of macrophages in response to microbial and self-constituents, their extensive endocytic and secretory capacity, and the ability to discriminate between host and foreign ligands (Gordon 2012). Much of our

Cite this article as *Cold Spring Harb Perspect Biol* doi: 10.1101/cshperspect.a016378

current understanding derives from in vitro cell culture, supplemented by in vivo genetic and infectious experimental models, mainly in the mouse.

Early studies of particle clearance, including bacterial capsular polysaccharides (Humphrey and Grennan 1981; Zamze et al. 2002) and other tracers from blood, revealed the presence of sinus-lining phagocytes in liver (Kupffer cells) (Jenne and Kubes 2013), in spleen, and in bone marrow (Moghimi et al. 2012). The discovery of the circulation of lymphocytes (Gowans 1991) and numerous experimental immunization protocols initiated extensive research into lymphoid structure, lymphocyte migration, subsets, and functions. The development of monoclonal antibodies and of fluorescent imaging techniques combined with traditional histological techniques helped to define the distribution of resident macrophages in mouse tissues and drew attention to specialized anatomical regions in spleen and lymph nodes (LNs), in which blood and lymph directly reach resident antigen capturing, processing, and presenting cells (APCs) (Figs. 1 and 2). Antigen marker studies in the mouse revealed strikingly different phenotypes of macrophage populations in the splenic MZ and LN SCS, compared with other resident tissue macrophages (Figs. 3 and 4). Although in situ studies provided useful glimpses of function, the intrinsic difficulty of isolating these resident macrophages from tissue has limited their characterization. Furthermore, the marginal zone, although well defined in rodents, is not identifiable as a distinct structure in human spleen, depriving it of apparent interest. Although the two specialized structures share properties and perform analogous functions in LNs of human and mouse, their relationship has not been established.

In this review, we compare the properties of metallophilic and marginal zone macrophages in mouse spleen, and the SCS and medullary macrophages in mouse LNs, with those of other tissue macrophages; their properties and close proximity to resident lymphoid and other myeloid cells distinguish them from mucosal, hepatic, and other sinus-lining cells. Their similarity and specialization provide insights into the regulation of innate and adaptive immunity,

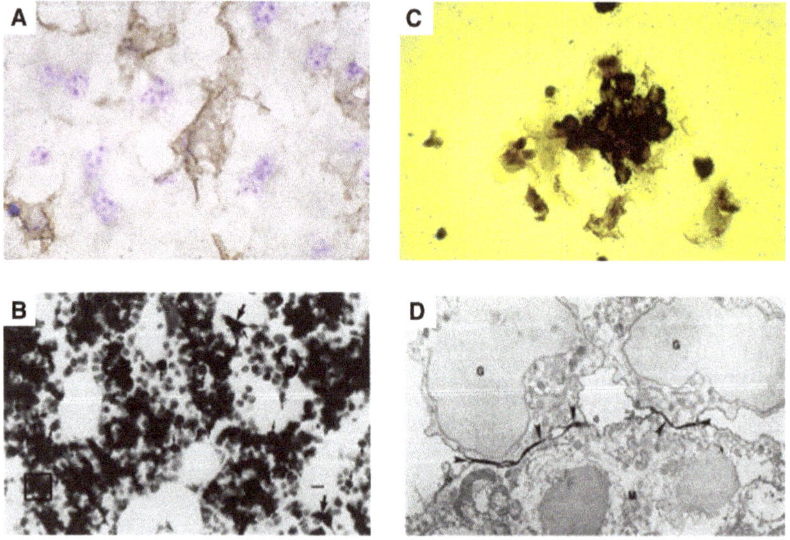

Figure 1. Sinusoidal resident tissue macrophages in the mouse. (*A*) F4/80$^+$ Kupffer cells, distinct from F4/80- sinusoidal endothelium and adjacent hepatocytes. Bone marrow stromal macrophages in situ (*B*), and in vitro (*C*). These F4/80$^+$ macrophages associate with clusters of hematopoietic cells during their differentiation and express CD169 at sites of contact with myeloid cells (*D*). (From Hume et al. 1983 and Crocker et al. 1990; reprinted, with permission.)

 Cite this article as *Cold Spring Harb Perspect Biol* doi: 10.1101/cshperspect.a016378

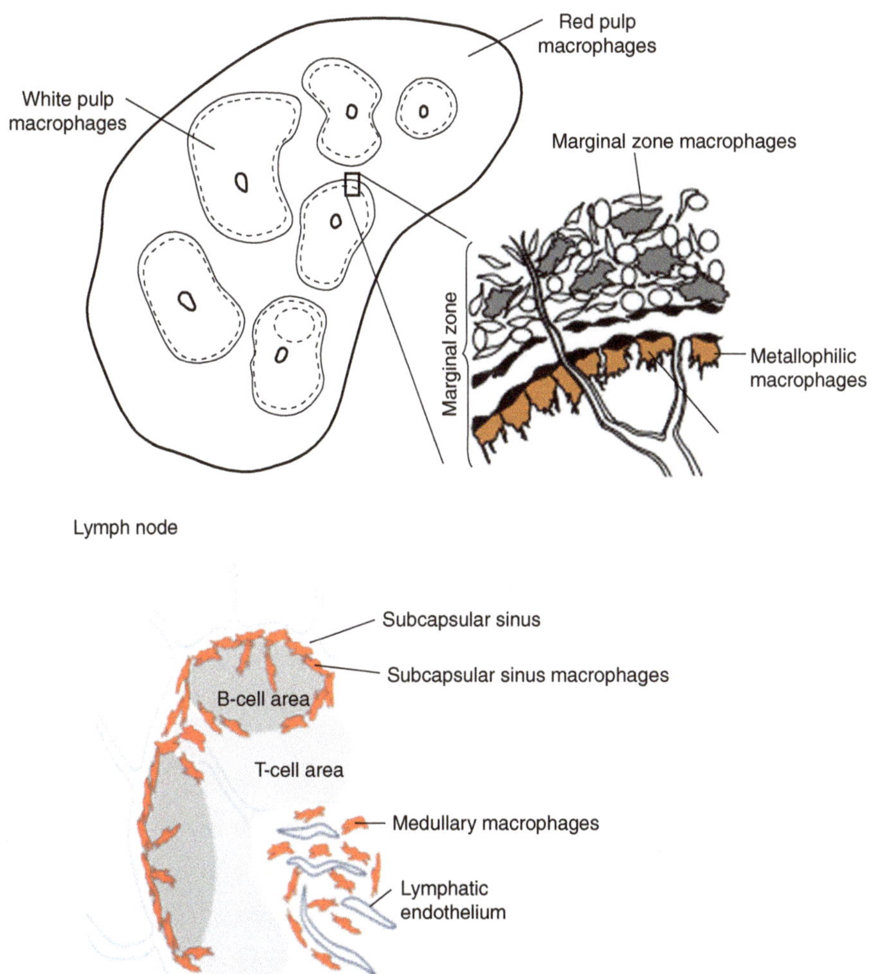

Figure 2. Schematic representation of mouse spleen and lymph node. Illustrating the lymphohematopoietic interface with specialized macrophages in close proximity to B- and T- lymphocyte populations. (From Gordon 2012 and Martinez-Pomares and Gordon 2012; adapted, with permission.)

but also indicate gaps in our knowledge to guide further study and manipulation. For further information, readers are referred to several excellent general reviews (Kraal 1992; den Haan and Kraal 2012; Gray and Cyster 2012), as well as selected references.

ANATOMY

Spleen Marginal Zone

The spleen is a complex organ consisting of hematopoietic and lymphoid compartments, the red and white pulp, respectively, separated in the normal adult mouse by the marginal zone (MZ), as illustrated in Figure 2. Arterial branches of the splenic artery are sheathed by T lymphocytes and B-cell follicles in the white pulp, and end in venous sinuses directly in the red pulp and MZ.

Antigen markers have aided the identification of distinct macrophage populations in the inner and outer MZs; characteristic sinus-lining $CD169^+$ F4/80-metallophilic (MMM) cells with an affinity for silver stains in the inner region (Fig. 4), and more phagocytic,

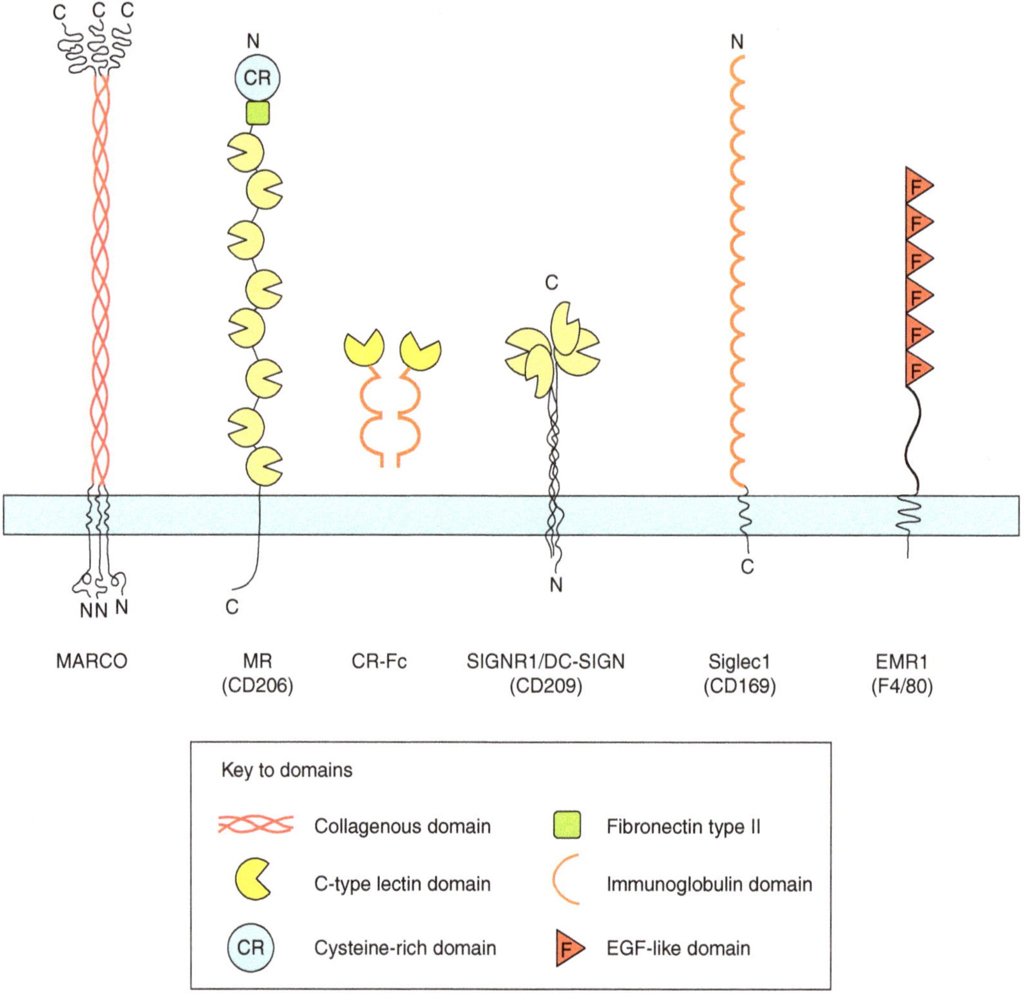

Figure 3. Schematic diagram of plasma membrane glycoproteins used to characterize sinusoidal and related molecules in mouse lympho-hematopoietic tissues. CR-Fc is a soluble chimaeric protein derived from the cysteine-rich domain of the mannose receptor and human Fc.

MARCO$^+$SIGNR1$^+$ macrophages in the outer zone (MZM). These macrophage populations can be distinguished from monocytes, mature macrophages, and DC in the red and white pulp, some of which also line sinusoids (Table 1). Other cells in the MZ include sinus-lining Madcam$^+$ vascular endothelial cells, DCs, and B1 lymphocytes, as well as stromal fibroblasts (Kraal et al. 1995; Nolte et al. 2004; Mueller and Germain 2009; You et al. 2011). Finally, there is evidence for heterogeneity among metallophils in mouse spleen, defined by their ability to bind Fc-chimaeric probes containing the cysteine-rich domain of the mannose receptor, CD206 (Martinez-Pomares et al. 1996), as discussed further below.

The MZ varies during development (Morris et al. 1991; Morris et al. 1992), is influenced by inflammatory/immunologic stimuli (Benedict et al. 2006), and shows striking species differences (Steiniger et al. 1997; Milicevic et al. 2011). Ontogeny studies have shown that the MZ of mouse spleen develops postnatally over several weeks; depletion of macrophages

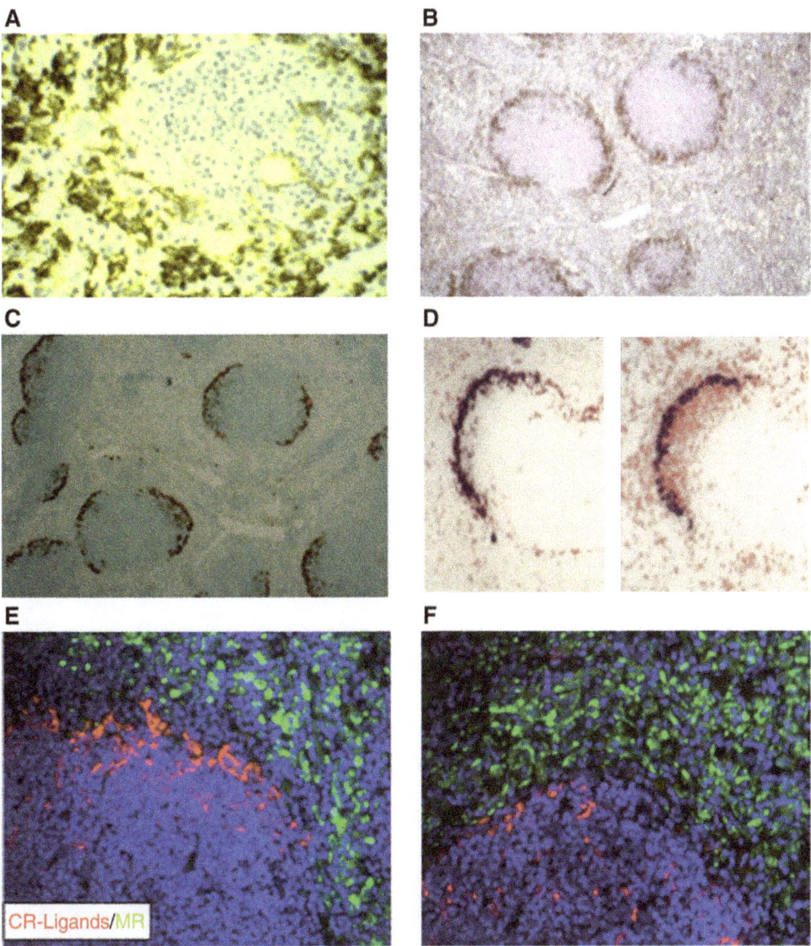

Figure 4. Immunocytochemistry of mouse spleen. (*A*) F4/80^{+} red pulp macrophages. Note the absence of F4/80 expression in marginal zone. (*B–F*) CD169^{+} marginal metallophilic macrophages. (*C–F*) Subpopulations of CD169^{+} bind CR-Fc and associate with IgD^{+} B lymphocytes (*D*, panel on *right*). *E* and *F* illustrate induced migration of CR-Fc^{+} macrophages into white pulp after immune activation. (From Crocker and Gordon 1989 and Taylor et al. 2004; reprinted, with permission, from The Rockefeller University ©1989 and the authors.)

by clodronate-loaded liposomes (Wijffels et al. 1994) is followed by relatively slow recovery of MZ MARCO^{+} macrophages, compared with that of other phagocytes (Kraal et al. 2000). The macrophage growth factor CSF-1 (M-CSF) is required for the appearance of CD169^{+} metallophils in wild-type mice, as shown by their absence in CSF-1 deficient osteopetrotic mutant mice (Fig. 5) (Witmer-Pack et al. 1993). The absence of F4/80 expression by MMM is striking (Figs. 4 and 6).

The CD169^{+} MMM line B-cell follicles at the border of the marginal sinus; their presence depends on lymphotoxin $\alpha 1\beta 2$ (LT$\alpha 1\beta 2$) produced by follicular B cells (Zindl et al. 2009), and LT may also be necessary for the appearance of MARCO^{+} macrophages in the outer MZ (Fig. 6).

The MZ is a prominent multilayered structure in the rat (Steiniger et al. 2006). The human spleen lacks the characteristic organization found in rodents (Steiniger et al. 1997), but

Table 1. Selected markers of murine sinusoidal macrophages

Spleen	Marginal zone metallophilic macrophages	CD169^{+} (sialoadhesin/MOMA-1) CR-L^{+}[a] F4/80^{-}
	Outer marginal zone macrophages	MARCO^{+} SIGN-R1^{+} F4/80^{-}
	Red pulp macrophages	F4/80^{+} MR^{+} CD169 dim
Lymph node	Subcapsular sinus macrophages	CD169^{+} CR-L^{+} F4/80^{-}
	Medullary macrophages	CD169$^{+/-}$ F4/80^{+} SIGN-R1^{+} MARCO^{+} MR^{+}
Bone marrow	Stromal macrophages	F4/80^{+} CD169^{+}
Liver	Kupffer cells	F4/80^{+} CD169$^{-/+}$ MR^{+} (also endothelium) SR-A^{+} (also endothelium)

[a]Ligand for cysteine-rich domain of MR.

CD169^{+} macrophages, which are present in interfollicular regions, may represent an homologous population, as discussed below.

Lymph Node Subcapsular Sinus

The structure of mouse LNs is illustrated in Figure 2. Several afferent lymphatics enter a single LN and pass into a narrow subcapsular sinus (SCS), which overlies the lymphocyte-rich cortex. Conduits (Roozendaal et al. 2008; Moussion and Sixt 2013) deliver lymph to the medullary region, which contains numerous irregularly shaped, interconnected sinuses, and medullary cords from which it is collected into an efferent lymphatic vessel. Larger LN in the human also contain trabecular sinuses, which radiate from SCS through parenchyma, providing an additional path for lymphatic flow.

Subcapsular sinus macrophages (SSMs) line the floor of the SCS, lying over lymphoid follicles. SSMs are quick to capture small amounts of lymph-borne substances (Figs. 7 and 8), but are poorly phagocytic compared with medullary sinus macrophages (MSMs), which are highly phagocytic, evident through their abundant phagolysosomes. SSMs display limited motility, unlike DCs, and have long stable tails extending across the lymphatic lining layer, as well as head and neck regions. MSMs adhere to the sinus walls and traversing reticular fibers in their lumen, which also contains lymphocytes. Cortical sinuses are packed with lymphocytes; as lymph flows toward the medullary sinus and cords, it contains macrophages, fewer lymphocytes, DCs, and variable numbers of plasma cells, depending on prior immune stimulation. The cords often contain a blood vessel. Neutrophils are present after inflammation.

Lymphoid follicles, germinal centers, and follicular dendritic cells (FDCs) are well described elsewhere (MacLennan 2005; Kastenmuller et al. 2012). The paracortex and interfollicular regions contain a range of immune cells: DCs, natural killer (NK), natural killer T (NKT), innate lymphoid cells (ILCs), and γ/δ lymphocytes, as well as scattered B and T (CD4 and innate-like CD8) cells, some of which are prepositioned to interact with SSM and antigens, as discussed below. Finally, the LN stroma contains heterogeneous mesenchymal cells (Castagnaro et al. 2013) and blood vessels, including High endothelial venules (HEV), for homing of recirculating lymphocytes.

LNs can vary substantially, depending on their location and drainage, for instance, in the mesentery or at peripheral sites, as well as after antigen stimulation. The origin and turnover of different LN macrophage populations has not been fully characterized. Resident SSMs develop postnatally, and is in part dependent on M-CSF (Witmer-Pack et al. 1993) and lymphotoxin. The expansion of macrophages after immunization involves recruitment and local proliferation; cells can emigrate or die by apoptosis. Experimental depletion by clodronate liposomes has revealed differential repopulation from bone marrow precursors and

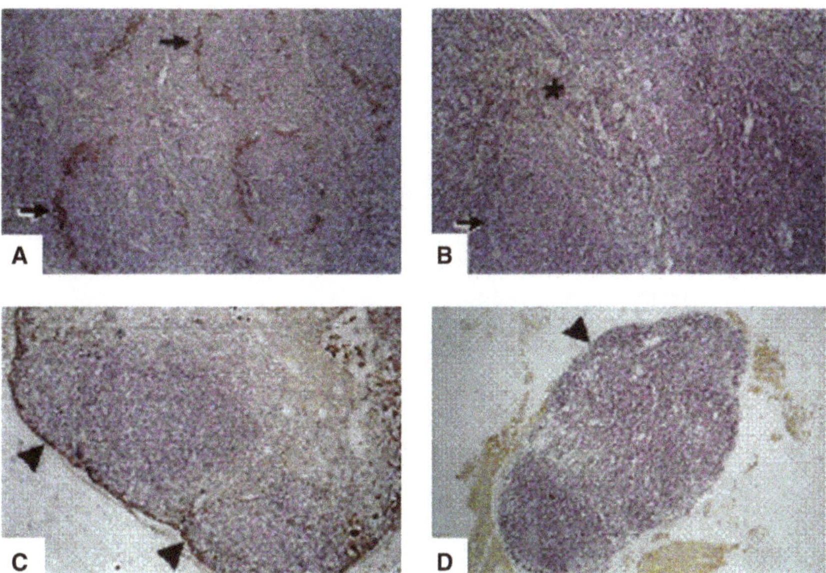

Figure 5. Spleen and lymph node in wild type and CSF-1-deficient op/op mice. CD169 expression by marginal metallophilic macrophages (*A,B*), and subcapsular macrophages (*C,D*) depends on CSF-1. (From Witmer-Pack et al. 1993; reprinted, with permission, from the authors.)

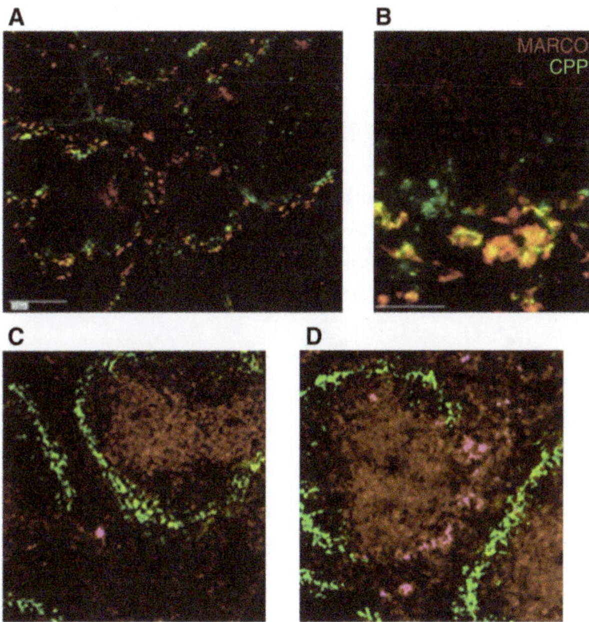

Figure 6. Functions of marginal zone macrophages. (*A,B*) Splenic MARCO$^+$ marginal zone macrophages clear calciprotein particles from the circulation. (From Herrmann et al. 2012; reprinted, with permission.) (*C,D*) F4/80$^+$ APC (magenta) localize to marginal zone of spleen after induction of tolerance to antigen delivered into the anterior chamber of the eye; CD169$^+$ marginal metallophilic macrophages in green. (From Lin et al. 2005; reprinted, with permission, from the authors.)

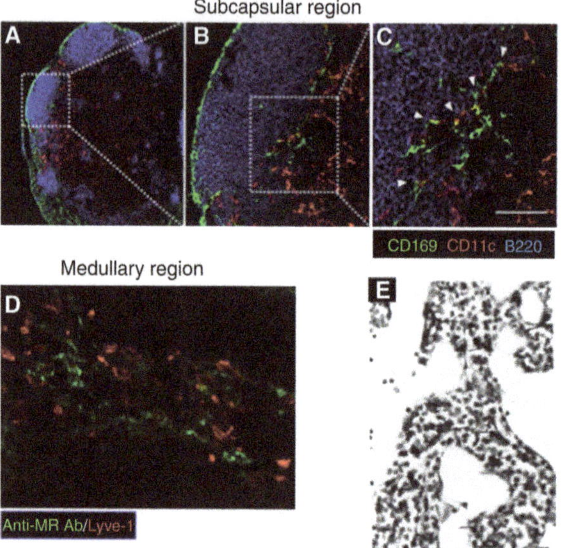

Figure 7. Sinusoidal macrophages in lymph node. (*A–C*) $CD169^+$ subcapsular sinus macrophages in proximity to $CD11c^+$ DC and $B220^+$ B lymphocytes. Scale bar, 50 μm. (*D*) Adjacent cells in the medullary sinus express MR and Lyve-1. (*E*) $F4/80^+$ macrophages in the medulla. (*A–D*, From Asano et al. 2011; reprinted, with permission, from the authors; *E*, from Hume et al. 1983; reprinted, with permission, from the authors.)

monocytes (Weisser et al. 2012). Unknown constituents of lymph may be required for LN maintenance, as shown by experimental lymphatic occlusion.

PHENOTYPIC CHARACTERIZATION

Studies have focused mainly on resident, rather than monocyte-derived recruited cells, in both spleen and LN. In this section, we describe some of the characteristic antigen markers that have been used (Fig. 3) (Taylor et al. 2005) and compare the lymphohematopoietic macrophages in mouse MZ and SCS with macrophages in splenic red pulp, bone marrow stroma, and liver Kupffer cells (Table 1).

F4/80 (Gordon et al. 2011), the widely expressed mouse macrophage plasma membrane marker, is strikingly absent on MMM, MZM, and SSM, but present on LN medullary macrophages. It is strongly exhibited by mouse red pulp macrophages, bone marrow stromal mac-

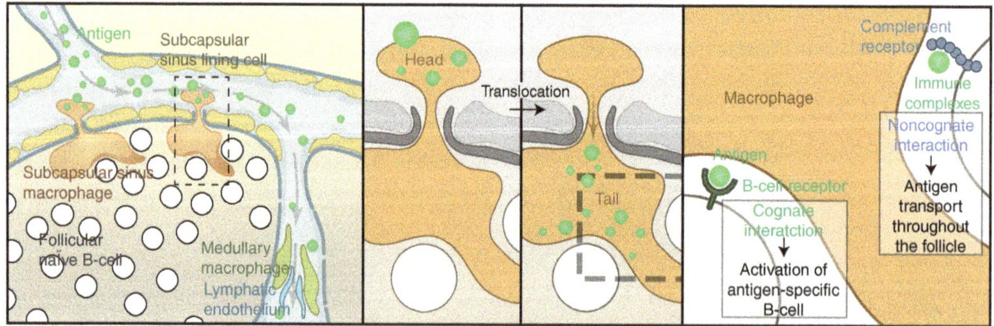

Figure 8. Schematic representation of subcapsular sinus macrophage function: antigen capture, delivery, and activation of B lymphocytes. For details, see text. (From Martinez-Pomares and Gordon 2007; reprinted, with permission, from the authors.)

rophages, Kupffer cells, as well as mature macrophages in most tissues, including microglia, Langerhans cells in epidermis, and lamina propria macrophages of the intestine. It is weakly present on DC and monocytes and dim or absent on macrophages in T-cell-rich areas, such as the white pulp of spleen, interstitial regions of LN, and in Peyer's patches. The F4/80 antigen (EMR1) is also present on eosinophils in mouse; in humans, the homolog is restricted to eosinophils. The molecule belongs to a small leukocyte family of EGF-TM7 seven transmembrane GPCR-like adhesion receptors. Studies on the anterior chamber associated immune deviation (ACAID) model of ocular peripheral tolerance revealed that F4/80 is essential for this process; F4/80^{+} cells from the anterior chamber of the eye, possibly DC, migrate to the marginal sinus of spleen (Fig. 6), where they interact with CD4 and CD8 lymphocytes, and NKT cells, to induce systemic tolerance (Lin et al. 2005; van den Berg and Kraal 2005). Our studies (M Stacey, HH Lin, and S Gordon, unpubl.) have shown an uncharacterized ligand for F4/80 in the MZ, consistent with the local absence of F4/80 antigen expression. Related myeloid molecules, EMR2 (human) and CD97 (mouse and human), have proteoglycan and complement regulatory membrane glycoprotein ligands.

CD169 (Klaas and Crocker 2012; O'Neill et al. 2013), although not unique to MMM and SSM, it is a characteristic, strongly expressed lectin-like marker for these specialized macrophages (Crocker and Gordon 1986, 1989). In spleen, it is dim on MZM and weakly present in red pulp macrophages, and variably present on medullary macrophages in LN. Originally termed sialoadhesin because of its ability to bind sialoconjugates on host and microbial ligands, CD169 is part of a leukocyte family of Siglecs, because it contains 17 Ig superfamily domains. In bone marrow, it is present on stromal macrophages at the center of hematopoietic cell clusters and plays a role in retention of immature myeloid cells (Chow et al. 2011). Expression on Kupffer cells is variable.

In our experience, it is not primarily an endocytic receptor, but well expressed on the plasma membrane of extensive macrophage cellular processes. Mouse serum/plasma contains an inducer of its expression by peritoneal macrophages in culture; in the CNS, only macrophages outside the blood – brain barrier are CD169^{+} (Perry et al. 1992). Our studies (P Tree and S Gordon, unpubl.) implicated β interferon (IFN) as a candidate inducer of its expression. It is plausible that exposure to plasma and possibly lymph contributes to its high expression in MMM and SSM, respectively; proximity to B-cell follicles may indicate a role in capture of potential tolerogenic/immunogenic antigens.

The best-characterized ligands of CD169 contain terminal α-2,3 sialylated residues (e.g., heavily *O*-glycosylated molecules, such as CD43 and Mucin-1). Sialic acid-independent interactions have been reported with the Cysteine-rich domain of the mannose receptor, described below, and the galactose-type lectin, MGL.

Apart from its value as a macrophage-restricted marker, the CD169 knockout mouse shows only minor phenotypic changes, such as less severe autoimmunity. CD169 presents a target for antigen-liposome delivered nanoparticles (Chen et al. 2012); transgenic targeting has provided an excellent reporter for imaging and migration studies in vivo, as well as conditional ablation of MZM via diphtheria toxin receptor (Saito et al. 2001).

MANNOSE RECEPTOR (MR, CD206)

This well-characterized lectin-like endocytic receptor (Martinez-Pomares and Gordon 2012) is widely present on mature macrophages in many tissues, including sinusoidal red pulp and Kupffer cells, but is not present on macrophages in the MZ or SCS. However, a ligand for its amino-terminal cysteine-rich domain (CR-L) is strongly expressed by subpopulations of MMM and SSM. CD206 is absent on resting human blood monocytes, but readily expressed after serum-induced maturation of macrophages in cell culture. The Th2 cytokines IL-4 and IL-13 are prototypic inducers of CD206, markers of an alternative form of macrophage activation in vitro and in vivo, in contrast with selective down-regulation by interferon-γ, the inducer of classical, antimicrobial activation. CD206 is present on DC, and is structurally related to DEC-205, whose ligands are uncharacterized, compared with the recognition of terminal mannose, fucose or GlcNac residues by the eight calcium-dependent carbohydrate-recognition domains (CRD) of the MR. CD206 is also present on lymphatic endothelium and sinusoidal endothelia, for example, in liver, where, together with macrophages, it is a potent clearance receptor for mannose-terminal lysosomal hydrolases, a nonredundant function in CD206 knockout mice. CD206 mediates receptor-dependent endocytosis and macropinocytosis of endogenous and exogenous ligands, and was the first macrophage receptor found to recycle extensively. Although it contributes to binding of several viruses (e.g., HIV-1 and Dengue), mycobacteria, yeasts (*Candida albicans*), and parasites, such as *Leishmania*, it has not been shown to mediate efficient phagocytosis directly, perhaps depending on cooperation with other receptors or uncoating of its ligands.

Structural analysis has identified the multiple CRD as responsible for the above endocytic functions. The fibronectin II domain binds collagen, consistent with an additional adherence function, but, remarkably, the conserved cysteine-rich domain has turned out to be a distinct lectin for clearance of endogenous sulfated galactosyl ligands found, for example, on lutropin, a pituitary hormone. The use of chimaeric probes consisting of the cysteine-rich domain and human Fc(CR-Fc), revealed selectivity for the MMM and SCSM, as well as FDC and tingible body macrophages in germinal centers (Martinez-Pomares et al. 1996; Taylor et al. 2004, 2005). The CR-Fc stains selectively those MMM in close proximity to B cells. This marker has been used to follow migration of MMM into white pulp follicles after immunization, and the reagent provides a tool for efficient and selective targeting of antigen to B cells (Berney et al. 1999).

DC-SIGN/SIGNR1(CD209)

SIGNR1 is one of a group of related mouse type II C-type lectins (CTLs), homologous to DC-SIGN of humans (Geijenbeek et al. 2002). SIGNR1 is highly expressed by macrophage subpopulations in the MZ of spleen, SCS, and medullary sinus of the mouse, as first shown with mab ERTR9. In addition, it is variably present on other macrophages and DC. First described as a "nonintegrin ICAM3 grabbing lectin," it binds mannose and fucose, for example, on capsular polysaccharides of *Streptococcus pneumoniae* and dextran, and interacts with C1Q to activate complement, thus promoting uptake of capsulated bacteria, viruses, such as HIV, and apoptotic cells. Fluorescent and blocking antibodies, and CD209 knockout mice have been used extensively to study SIGNR1 function in virus infection and imaging.

Other CTLs (Drummond and Brown 2013), less defined in lymphohematopoietic tissue expression, include Dectin-1, -2, and -3, MCL, MINCL, and galectins. Several of these have known pathogen ligands and use well-characterized signaling pathways, for instance, the β-glucan-dependent uptake of yeasts via Dectin-1 signals through its hemi-ITAM motif, syk, and Card 9.

MARCO

This class A scavenger receptor, a type II transmembrane collagenous molecule, is an excellent constitutive marker for MZM, the large phagocytic macrophages of the splenic outer MZ. Closely related to the macrophage SRAI/II, it is a distinct gene product with overlapping, but different binding preference for polyanionic, modified host, and microbial ligands. Regulation of MARCO expression also differs from that of SRAI/II, which is widely present on tissue macrophages and mainly regulated by CSF-1. In the unstimulated mouse, MARCO is strikingly restricted to the MZM and to a subpopulation of resident peritoneal macrophages. The MZM expression depends on the LTα1β2 receptor, whereas peritoneal macrophage expression depends on the microbial flora in the gut (S Mukhopadhyay and S Gordon, unpubl.). However, most tissue macrophages can be induced to express MARCO by TLR ligands (e.g., during bacterial sepsis). It can also be induced on DC by sterile and infectious stimuli.

MARCO is an adhesion, phagocytic, and endocytic receptor (Chen et al. 2006; Mukhopadhyay et al. 2006). MARCO readily binds and takes up bacteria, such as unopsonized *Neisseria meningitidis*, but also binds B cells in the MZ (Karlsson et al. 2003). Together with SRAI/II, it contributes to the rapid clearance of particles, such as calciprotein–fetuin complexes from the circulation (Herrmann et al. 2012).

CD68, the pan macrophage and DC endosomal ag, various plasma membrane receptors, FcR, CD11/CD18, CD11c, and other integrins, have all been reported on MZ and SCSM, but these markers are shared with other leukocytes in spleen and LN. LYVE-1 is on LN macrophages and lymphatic endothelium. Madcam is a valuable marker for the sinus endothelium in spleen.

In summary, the metallophils and SSM express strikingly similar markers; splenic MZ, and LN medullary macrophages also have several corresponding markers. Their properties set them apart from bone marrow stromal macrophages, Kupffer cells, and other sinus-lining macrophages in other tissues. The combination of a cluster of highly expressed macrophage markers, such as CD169, SIGNR1, MARCO, and CR-Fc binding, by a population of F4/80 negative cells in a specialized anatomic location exposed to blood and lymph, suggests that they represent a distinctive sublineage. Further studies are required to determine their relationship to monocyte/macrophages and DC, their production and turnover, as well as functions to be considered next.

FUNCTIONS OF MARGINAL ZONE AND SUBCAPSULAR SINUS MACROPHAGES

General Aspects

The macrophages in the MZ and SCS are strategically placed to interact with pathogens and host cells and their products through a range of recognition receptors (Table 2), at the interface between blood and lymph with the immune system. They provide an innate recognition and clearance barrier to infection and serve as sentinels to initiate an adaptive response, if required, by delivering microorganisms and antigens to DC and other innate and lymphoid cells, by mobilization of hematopoietic cells including monocytes, and initiation of inflammatory cytokines and plasma cascades (Martinez-Pomares and Gordon 2007; Barral et al. 2010; Gray and Cyster 2012; den Haan and Martinez-Pomares 2013). Local interactions with stromal and vascular elements provide distinct microenvironments, which influence macrophage gene expression and functional responses. Depletion studies with clodronate, for example, have shown enhanced sensitivity to infection and facilitation of spread (Moseman et al. 2012). Transgenic ablation of $CD169^{+}$ macrophages showed that they are required to cross-

Table 2. Receptors of marginal zone and subcapsular sinus macrophages

Receptor	Nature	Ligand	Function
Fc receptors[a]	Immunoglobulin superfamily	Immunoglobulin	Immune complex clearance
Complement receptor[a]	Integrin	Cleaved C3	Opsonic uptake
Toll-like receptors[a]	Leucine-rich repeats	Multiple	Sensing microbes, altered self
CD169	Immunoglobulin superfamily	Sialylated glycoconjugates	Adhesion, cell–cell/microbe interactions
SIGN-R1	C-type lectin	Terminal mannosyl	Clearance and cellular interactions
Mannose receptor	C-type lectin domains[a]	Mannose, fucose, GlcNac termini	Adhesion and clearance of glycoconjugates
	Cysteine-rich domain	Sulfated sugars	Targeting to MZ and LN
MARCO[a]	Transmembrane collagenous scavenger receptor	Selected polyanions	Adhesion, uptake of microbes, B-cell interactions

[a]Also expressed in other tissue macrophages.

present apoptotic cell-associated antigens during antitumour activity (Asano et al. 2011). We discuss present knowledge of cellular responses in situ in relation to the general properties of macrophages, and bring out the need for further studies on these particular macrophage populations.

ROLE OF MACROPHAGES IN THE MARGINAL ZONE

The enhanced susceptibility of immature and splenectomised hosts to capsulated bacterial infection is well known. The lectins of the MMM and MZM, CD169, SIGNR1, and possibly others, may account for the loss of clearance of capsulated bacteria, such as *S. pneumoniae*, after macrophage depletion by clodronate, as shown by mab inhibition (SIGNR1) and the effect of gene knockout on phenotype. Similarly, MARCO knockout diminishes clearance of sterile, noninfectious particulates (Fig. 6). *N. meningitidis* is an excellent nonopsonic ligand for MARCO, and, independently, for CD169. Studies with SRAI/II knockout (Plüddemann et al. 2009) have shown a phenotype for *Neisseria* bacteraemia even in the presence of complement, a likely enhancer of opsonic phagocytosis through CD11/CD18 receptors, after classical, alternative, and lectin-activation by bacterial capsular polysaccharides.

Lectins play a similar role in direct or complement-mediated uptake of viruses, such as HIV and serum hepatitis virus.

It should be noted that phagocytic recognition is not necessarily nonspecific, because these nonopsonic receptors have a very broad range of ligands, are promiscuous, and often collaborate with other low-affinity receptors. Specific antibodies will enhance clearance of microbes and immune complexes through various FcR (Phan et al. 2009).

TLR can also collaborate with other plasma membrane and intracellular endosomal recognition receptors; their expression on MMM and MZM has not been adequately defined. The same holds true for NLR, RLR, and other cytosolic receptors for nucleic acids. Inflammasome activation has not been defined in these macrophages, to our knowledge (for LN, see below).

Phagocytosis of microbes is more evident in MZM, compared with MMM, but little is known about the cellular pathways that regulate uptake of particles and surface receptor complexes, or the subsequent intracellular fusion, acidification, digestion, or antigen processing/presentation. The possible link of infection of MZ macrophages to autophagy has not been examined.

MMM express Tim-4 and Treml4 receptors for early- and late-stage apoptotic cells, in addition to uptake by tingible body macrophages. MFG-E8 and MER/Gas6 participate in phagocytic clearance of apoptotic B cells. MARCO on MZM has also been implicated in apoptotic cell uptake; deficient clearance by macrophages may contribute to autoimmunity. It is not clear whether uptake of apoptotic cells suppresses inflammation, as shown by IvIg administration, after which IL-33 up-regulates FcRγ2 expression (Anthony et al. 2011).

Both MZM and MMM show IRF7 recruitment to phagosomes after *Leishmania donovani* infection. IRF deficiency gives rise to a small increase in the intracellular parasite load. Production of type 1 interferon by MZM and MMM contributes substantially to innate resistance to viral infection (e.g., HSV); its role in bacterial infection, such as mycobacterial bacteraemia, may be deleterious to host resistance.

MZM and MZ B cells collaborate as a cellular network in T-independent IgM responses to capsular polysaccharide. SIGNR1 expression depends on the continuous presence of MZ B cells and is lost if the MZM are induced to migrate. SIGNR1 interacts with C1Q, initiating complement activation and binding by B-cell receptors and deposition on FDC. Attempts to distinguish the contribution of each macrophage subset to immune responses by conditional ablation of CD169^{+}MMM or of MZM^{+} SIGNR1 via diphtheria toxin receptor has not proved sufficiently selective. Note that MZM may also express CD11c, complicating the interpretation of CD11c-targeted ablation.

Apart from the requirement for CSF-1 and LT in regulating marker expression in MMM

 Cite this article as *Cold Spring Harb Perspect Biol* doi: 10.1101/cshperspect.a016378

and MZM, respectively, the cytokine-responsive pathways remain poorly defined, as are the transcription factors involved in selective gene and protein expression. The phenotype of CSF-1R deficiency is more severe than that for CSF-1; the possible role of IL-34 as a second ligand for CSF-1R has not been characterized in this context. It is likely that the proximity to B lymphocytes in the MZ, as in the SCS, provides the LT stimulus for induction via LT receptors, and continued expression of macrophage markers, whereas stromal fibroblasts are a likely source of CSF-1.

In addition to the induction of type 1 interferon in MZM and autocrine activation, the production of chemokines, cytokines, lysozyme, other enzymes, complement components, and low molecular weight metabolites (e.g., superoxide and NO) in response to microbial and apoptotic cell uptake, requires further study, with due consideration to the local in situ environment.

ANTIGEN TARGETING TO THE MARGINAL ZONE

Although confocal imaging methods have contributed greatly to research of dynamic antigen responses in LN, to be discussed below, this has lagged behind in the less accessible spleen. However, targeting via the blood has brought to light new aspects of antigen delivery to the immune resident macrophages of the MZ. Antigen targeted to MMM and MZM is transferred to $CD8^+$ DC, which depend on batf3 to induce strong CTL (den Haan and Kraal 2012); targeted antibody can also be cross presented to CD8T cells by DC. Therefore, macrophages of the MZ interact with both B cells and DC by uptake and transfer of ag.

We have exploited a novel mannose receptor-dependent pathway for clearance of glycoconjugates, to target antigen through the soluble cysteine-rich domain of the mannose receptor (Figs. 3 and 4) (Taylor et al. 2004). Human Fc-chimaeric proteins (CR-Fc) were used as antigen probes and for detection of delivery by immunocytochemistry. CR-Fc bound specifically to sulfated ligands expressed by a subpopulation of $CD169^+$ MMM associated with IgD^+ B lymphocytes. Although the normal function of this clearance pathway may promote noninflammatory disposal of host ligands bound by the MR through carbohydrate recognition domains, targeting by CR-Fc-activated B cells and induced specific IgG responses. Mutational analysis of the CR-Fc established the specificity of targeting. Soluble MR, shed by TACE-like metalloproteinases from the surface of APC (e.g., after phagocytic stimulation) (Gazi et al. 2011), could similarly interact with MMM subpopulations and modulate B-cell responses.

Antigen targeting to DC via the closely related DEC-205 has been used to enhance antibody responses; the endocytic compartment reached by CR-L targeting of macrophages and its possible role in antigen processing and presentation have not been determined. Because MMM are inefficient endocytic macrophages, it is possible that bound protein is not internalized and digested, but delivered intact to other APC, B cells, and DC. The expression and role of MHC class II and costimulatory molecules by MMM remain unclear.

ROLE OF MACROPHAGES IN SUBCAPSULAR SINUS AND MEDULLA OF LYMPH NODES

SSM on the floor of the SCS, over follicles, are quick to capture lymph-borne substances, including virus particles, through lectins, scavenger receptors, and CR3, but are poorly endocytic. In contrast, MSM take up large amounts of material into conspicuous phagolysosomes. The long tails of SSM are stable, unlike more motile DC, and are anchored by the extracellular matrix. In rodents, SSM readily pick up ferritin, colloidal carbon, liposomes, and opsonized antigen. Opsonization increases electron microscopic evidence of internalization and degradation. Depletion by clodronate is slower than for medullary macrophages, consistent with differences in endocytic activity. SSM retain phycoerythrin labeled immune complexes (IC), which move along their surface to follicles; two-photon imaging shows their capture and transfer to B cells. Viruses, ag-loaded 200-nm beads, and IC bind to SSM heads, which translocate without internalization, and present these particulates on their tails to follicular B cells randomly migrating through the dense net of tails. Cognate B cells are activated directly from SSM via BCR. Opsonized antigen also activates noncognate B cells via CR1 and CR2, delivered to FDC as IC aggregates. SSM express ICAM-1 and VCAM1, which may assist transfer.

The role of SSM and MSM in innate immunity has been studied extensively by several groups (Gonzalez et al. 2010a; Moseman et al. 2012). SSM may transfer antigen intact to B cells or via DC. Kastenmuller, Germain, and colleagues (Kastenmuller et al. 2012) have investigated the role of LN macrophages as innate sentinel cells. CD169$^+$ SSM and medullary macrophages (F4/80$^+$), trap a range of pathogens including VSV, MVA, *Pseudomonas*, and *S. typhi*. SSM express IL-18, generated from pro-IL-18 by caspase 1, and IL-1β, depending on extra/intracellular pathogen virulence. Type 1 IFN and IL-12 activate prepositioned NKT, NK, γ/δ T cells, plasmacytoid DC, and novel CD8$^+$ γδT innate cells to produce IFN-γ, which activates SSM and medullary macrophages to express iNOS. PMN recruitment depends on IL-1β production as a result of pyroptosis. This innate immune response limits systemic spread of pathogens to the spleen and can be destroyed by clodronate. DCs are present in a different location, for adaptive responses.

SSM and B cells are closely associated; B cells are a source of LTα1β2, also required for SSM maintenance, allowing viral replication, balanced by type 1 IFN production. A distinct pathway has been reported for influenza virus infection, by Carroll and colleagues (Gonzalez et al. 2010b), involving CD169$^+$SIGNR1$^+$ SSM, which depends on complement opsonization via a lectin activation pathway.

MEDULLARY SINUS MACROPHAGES (MSM)

After trafficking through conduits, lymph encounters the actively phagocytic MSM before exit through efferent lymphatics. MSM play an important role in determining survival and clearance of short-lived early plasma cells. MSM variably express CD169, CR3, SIGNR1, MARCO, MR, Lyve-1, as well as F4/80. Cyster and colleagues (Gray and Cyster 2012) distinguish macrophages in the sinuses (low acid phosphatase, high nonspecific esterase) from those in the cords, (opposite expression). In the mouse, medullary cord macrophages lack CD169, but express F4/80.

MSM adhere to sinus walls and reticular fibers in the lumen, which also contains lymphocytes, variable numbers of plasma cells, depending on immunization, and small numbers of DC.

Antigens accumulate in highly active large phagocytic MSM, with heterogeneous lysosomes, unlike the less phagocytic cord macrophages. MSM take up apoptotic plasma cells, but macrophages in cords also provide trophic products, such as APRIL, promoting plasma cell survival.

Medullary macrophages take up and destroy viruses, whereas SSM (and MMM) are permissive for a range of virus infections (e.g., VSV, vaccinia, CMV), as well as toxoplasma, especially if type 1 interferon levels are reduced experimentally.

CD8$^+$ T cells could be activated directly or via DC transfer, which is more effective. Presentation by SSM stimulates some CD4 T-cell proliferation, but is not good at generating CD8 effector cells. Toxoplasma infected SSM can present antigen in the context of MHCI for killing by CD8 T cells, but may lack sufficient costimulation. SSM are also able to present CD1d-restricted microspheres coated with α-galactosyl ceramide liposomes to NKT cells.

LNs contain additional CD68$^+$ macrophages, including interfollicular cells interacting with various innate and adaptive cells, especially after immune activation. Macrophages also extend more deeply along the B–T boundary, or into T-cell zones.

ROLE OF LIPIDS

Afferent lymph reaching mesenteric LN is rich in absorbed lipids and microbial products. The expression of macrophage surface and intracellular sensors is regulated by, and, in turn, influences their phenotype, as does their location along the path of lymphatic flow. The scavenger, Toll- and lectin-like receptors are responsive to lipid components in lymph. Knockout studies with Angptl4-deficient mice, for example, have revealed defective regulation of short chain fatty acids and dramatic changes in triglyceride metabolism. Mesenteric LN expand enormously, with extensive formation of multinucleated F4/80^{+} giant cells, the result of macrophage fusion. The role of steroid and other lipid ligands, vitamins A and D, and of transcription factors, such as PPARs in SCM, has not been explored.

Conclusion

Above all, the question remains of APC function of SSM and MSM, compared with DC and B cells, and activation of CD4 T-lymphocyte subsets, Th1, -2, -17, regulatory T cells, and NKT cells.

There is a striking similarity between LN SSM and MMM with regard to CD169 and other recognition molecule expression, their dependence on CSF-1, LT, and B lymphocytes, and proneness to viral infection at the lymphohematopoietic interface (Table 3); however, SSMs have been reported to capture immune complexes more readily than MMM. It is possible that MZM and MSM, both enhanced phagocytic populations, represent homologous populations in spleen and LN. These tissue-specific similarities and differences in macrophage function may arise by differentiation or local environmental influences, such as the extracellular stroma or vasculature.

Table 3. Shared properties of marginal zone, subcapsular sinus and medullary macrophages

Interface: blood/lymph
Proximity to innate and adaptive myeloid and lymphoid cells
Induction and regulation of B- and T-cell responses
Capture and disposal of apoptotic cells/role in autoimmunity/variably phagocytic
Postnatal development
Host defense: clearance of capsulated bacteria, polysaccharides, viruses, other bacteria
Clearance of sterile particles (outer MZ)
Antigen capture and delivery to APC
Inflammasome procaspase 1 store/activation (SSM, not reported for MZ)
Secretion of proinflammatory cytokines, IL-1β, type I interferon, IL-12, chemokines, NO (incomplete information)

SPECIES DIFFERENCES

The human and rodent spleen and LN display obvious species differences in marker expression; Steiniger and colleagues (1997) found CD169 expression in interfollicular regions of human spleen, which lack the characteristic MZ of rodents. The CR-Fc (mouse) probe also shows differences in expression of ligands in human spleen, as do the MR and probes for MR CRD ligands (Martinez-Pomares et al. 2005; Steiniger et al. 2006). Human red pulp contains macrophages in splenic cords, which express CD68 and CD163, but not MR or DC-SIGN. The marginal sinus is absent, but the perifollicular area contains sinuses and blood-filled spaces without an endothelial cell lining. MSMs express DC-SIGN, MARCO, and LYVE-1 on both macrophages and endothelium. SSM/MSM are not well defined, but express CD14 and CD68. Unlike mice, human SSM and MSM express DC-SIGN; CD169^{+}/DC-SIGN^{+} macrophages are perifollicular, surrounding capillaries.

We conclude that although there may well be homologous sinusoidal macrophage populations in human and mouse lymphohematopoietic organs, a great deal more has to be learned by the study of resting and activated human LN, as well as spleen, to establish bona fide equivalent functional populations.

CONCLUSIONS, QUESTIONS, AND SUGGESTIONS FOR FURTHER STUDY

In this article, we have focused on a group of resident sinusoidal macrophages in spleen and LN, reviewed what we know concerning their properties and functions, and assessed gaps in our knowledge. Their study over five decades owes much to pioneering studies by John Humphrey, Georg Kraal, Jason Cyster, Luisa Martinez-Pomares, and their collaborators, as well as other investigators interested in the initiation of innate and adaptive immunity to virus infection and immunogenic stimulation. We propose that these macrophages are at the center of a highly specialized and important immunological function, which is best studied in the context of their living environment. Unlike surface and mucosal epithelial barriers, they provide defense against blood- and lymph-borne pathogens with which they are in direct contact, but also monitor and clear endogenous host molecules and dying cells without necessarily activating immunity and inflammation. Of course, they share capture and disposal, as well as trophic and defense functions, with other resident tissue macrophages, especially in liver, bone marrow stroma, and splenic red pulp. Furthermore, they overlap with other myeloid cells, such as DC in antigen uptake, processing, and presentation, and in clearance and homeostasis with sinusoidal endothelial cells. Nevertheless, it is worthwhile to distinguish their nature and functions as important variations of the macrophage phenotype, and to approach their investigation in an appropriate way.

Importance

Pathogens, such as HIV, hepatitis virus, capsulated bacteria, such as *S. pneumoniae* and meningococci, as well as *Mycobacterium tuberculosis* and *Plasmodium falciparum* have potentially catastrophic effects on the host when disseminated through blood, as do needle stick injury, transfusions, and narcotic abuse. What makes the lymphohematopoietic interface special is the close proximity of the sinus lining and associated phagocytic macrophages to other cells of the innate and adaptive immune response, especially B lymphocytes and DCs. However, equally important is the limitation and suppression of inappropriate immunity and potential tissue injury to modified host components, a form of peripheral tolerance, which can give rise to autoimmunity. Persistence of sterile and infectious agents in macrophages that are compromised in antimicrobial resistance or degradation contribute to chronic infection and inflammation, which promotes further destruction of this specialized compartment. All vaccination strategies and the use of adjuvants depend on understanding of this critical tissue compartment. Finally, malignant cells can interact with macrophages at the lymphohematopoietic interface, limiting cancer spread to spleen, while favoring lymph node secondary invasion, whereas lymphoma metastasis, in contrast, involves both tissues.

Nature

We have focused our discussion on the macrophages of the splenic MZ, the metallophils, and their more phagocytic partner in the outer MZ, and on LN SSMs, and their phagocytic, albeit more separated, macrophages in the medulla, but in contact with the same lymphatic stream. In the mouse, there is good correspondence of markers between these populations, whereas in humans, there is limited knowledge of markers and anatomical differences but functional similarities. The MMM and SSM capture, but seem not to internalize, particles efficiently compared with MZM and MSM, respectively. Recognition by plasma membrane molecules may seem nonspecific ("flypaper model"), but we believe the multiplicity of nonopsonic and opsonic receptors (lectins, scavenger, and complement receptors, among others not yet defined) endow these macrophages with a remarkable capacity to recognize a broad range of soluble and particulate ligands. Single markers, such as CD169, SIGNR1, and MARCO are suggestive, but not sufficient to characterize such macrophages. The absence of F4/80 on both MMM and SSM is striking, reminiscent of its absence on osteoclasts. The origin of these populations needs further study; postnatal development of secondary lymphoid structures argues against yolk sac and foetal liver origin, as does reconstitution by bone marrow after clodronate depletion. The role of CSF-1 and perhaps IL-34, for MMM and SSM development and LT for MZM and SSM, needs to distinguish marker expression, from the presence or absence of macrophages and other components, such as

 Cite this article as *Cold Spring Harb Perspect Biol* doi: 10.1101/cshperspect.a016378

endothelium and B cells. The stromal microenvironment is also heterogeneous and still poorly defined. A further problem that needs more study is the life history of these distinct macrophage populations, their recruitment, survival, proliferation, emigration, and death. The fate of newly recruited monocytes after infection, for example, has not been fully determined.

The regulation of phagocytic uptake in macrophages is incompletely understood. Apart from the functional state and interactions of surface receptors, which contain or lack internalization and signaling motifs, the role of the cytoskeleton (Batista and Dustin 2013) and of cross talk among signaling pathways in these sinusoidal macrophages are complex and unknown. Interactions with TLR and other surface molecules are also undefined. In truth, we still do not fully understand the discrimination between "foreign" and "modified self" by any macrophages. The dynamics of surface receptor fluidity and interactions are relevant to transfer hypotheses; the delivery by MR CR-Fc provides a potential mechanism for targeting of circulating ligands. Shedding, bleb, and nanotube formation, or induced migration, are other possible mechanisms for interactions with DC and B cells, which are highly motile compared with resident MMM and SCM. Interactions with other innate and adaptive cells, endothelium, and stromal cells may use the same and additional surface receptors.

The intracellular recognition, induced secretory responses, and fate of internalized material is also poorly characterized in the phagocytic MZM and MSM; this is critical to the balance between immune activation and suppression (e.g., after uptake of apoptotic B cells and plasma cells), and the balance between humoral and cellular immunity. Pathogens evolve evasion mechanisms to avoid capture and ingestion, or to facilitate their uptake by those macrophages that are relatively deficient in antimicrobial resistance mechanisms, such as type I interferon, the respiratory burst and i-NOS. Finally, the entire pathway of signaling and gene expression (chromatin, transcription, microRNA, posttranscriptional and translational control) remains to be examined in these cells.

Investigation and Manipulation

An appreciation of anatomy is essential to the study of lymphohematopoietic immunobiology. Most of our knowledge derives from mab production and, in situ analysis in the mouse, traditionally by microscopy and immunocytochemistry, and, recently, by sophisticated two- and three-dimensional multiphoton, confocal analysis, pioneered by Ron Germain and his colleagues (Gerner et al. 2012). In vivo functional studies date from early clearance studies, more recently reinforced by genetic manipulation of mice. The development of a clodronate liposome depletion method by Van Rooijen and colleagues has been widely used, in spite of limitations; conditional ablation based on highly expressed receptors, such as CD 169 and SIGNR1, also may lack specificity in their ability to discriminate between MMM and MZM, for example, and efficacy. Mab and soluble receptors (e.g., for targeting LT and its receptor) have helped to avoid the problem of regional differences in LN development in transgenic animals. As noted, heterogeneity of LN arising from differing drainage of lymph from gut and peripheral sites, such as skin, should also be borne in mind.

Of course, and not always fully appreciated, is the difficulty of isolating a minor population of highly stellate MMM or SCM, in terms of yield, damage, and artifacts. Nevertheless, the striking expression of marker antigens in situ makes it possible to relate their structure to function. A present limitation is the need to use multiple markers on the same tissue sections, but there is ongoing progress in this area. In humans, more markers are needed, apart from material for investigation, perhaps more easily solved for LN than spleen, especially where normal controls are required.

Nevertheless, depending on the availability of specific antibodies, in situ single cell analysis of many cellular properties mentioned above, can be significantly expanded. Laser capture microscopy provides limited numbers of cells for microarray and proteomic analysis, but can be used to identify selectively expressed transcription factors and other molecules for further validation in mouse and man.

A new resource, relatively easy to undertake, would be to screen mutants generated by ENU mutagenesis or by programs of genome wide knockout. Hypomorphs, generated by mutagenesis, should be more useful in retaining local cellular interactions. The known immunocytochemical markers, such as CD169, SIGNR1, and MARCO could be used as first screen; follow up clearance and immunological assays should detect functional differences.

Induced pluripotent cell technology (IPS) would ultimately pose a demanding test to recapitulate the special phenotype of MMM and SCM in vitro. A fascinating new tool to begin to understand the species differences and the role of stroma and hematopoietic contributions, could take advantage of humanized mice (Manz 2007), where technology is progressing for myeloid as well as lymphoid reconstitution.

CONCLUSION

We propose the term sinusoidal immunity to emphasize the special functions of selected sinusoidal macrophages at the lymphohematopoietic interface in spleen and LNs. Their distinctive phenotype sets them apart from other sinusoidal macrophages in bone marrow, liver, and splenic red pulp, which share some of their functions. In a sense, it refines the now abandoned historic concept of the reticuloendothelial system, which emphasized the particle clearance function of these and other sinus-lining macrophages, irrespective of their close relationship to resident immune cells. The possible immunologic functions of sinusoidal true endothelial cells, which express similar clearance receptors remain to be established. We believe that the adoption of a different terminology will stimulate further studies of their unusual properties, similar to recognition of mucosal immunity as a distinct subject for investigation.

ACKNOWLEDGMENTS

We thank members of our group and many collaborators for helpful discussions. Work in the laboratory was supported by grants from the Medical Research Council, UK, the Wellcome Trust, and Arthritis Research, UK.

REFERENCES

Anthony RM, Kobayashi T, Wermeling F, Ravetch JV. 2011. Intravenous gammaglobulin suppresses inflammation through a novel T_H2 pathway. *Nature* **475:** 110–113.

Asano K, Nabeyama A, Miyake Y, Qiu CH, Kurita A, Tomura M, Kanagawa O, Fujii S, Tanaka M. 2011. CD169-positive macrophages dominate antitumor immunity by cross-presenting dead cell-associated antigens. *Immunity* **34:** 85–95.

Barral P, Polzella P, Bruckbauer A, van Rooijen N, Besra GS, Cerundolo V, Batista FD. 2010. $CD169^+$ macrophages present lipid antigens to mediate early activation of *i*NKT cells in lymph nodes. *Nat Immunol* **11:** 303–312.

Batista FD, Dustin ML. 2013. Cell:cell interactions in the immune system. *Immunol Rev* **251:** 7–12.

Benedict CA, De Trez C, Schneider K, Ha S, Patterson G, Ware CF. 2006. Specific remodeling of splenic architecture by cytomegalovirus. *PLoS Pathog* **2:** e16.

Berney C, Herren S, Power CA, Gordon S, Martinez-Pomares L, Kosco-Vilbois MH. 1999. A member of the dendritic cell family that enters B cell follicles and stimulates primary antibody responses identified by a mannose receptor fusion protein. *J Exp Med* **190:** 851–860.

Castagnaro L, Lenti E, Maruzzelli S, Spinardi L, Migliori E, Farinello D, Sitia G, Harrelson Z, Evans SM, Guidotti LG, et al. 2013. Nkx2-5^+islet1^+ mesenchymal precursors generate distinct spleen stromal cell subsets and participate in restoring stromal network integrity. *Immunity* **38:** 782–791.

Chen Y, Pikkarainen T, Elomaa O, Soininen R, Kodama T, Kraal G, Tryggvason K. 2005. Defective microarchitecture of the spleen marginal zone and impaired response to a thymus-independent type 2 antigen in mice lacking scavenger receptors MARCO and SR-A. *J Immunol* **175:** 8173–8180.

Chen WC, Kawasaki N, Nycholat CM, Han S, Pilotte J, Crocker PR, Paulson JC. 2012. Antigen delivery to macrophages using liposomal nanoparticles targeting sialoadhesin/CD169. *PLoS ONE* **7:** e39039.

Cheong C, Matos I, Choi JH, Dandamudi DB, Shrestha E, Longhi MP, Jeffrey KL, Anthony RM, Kluger C, Nchinda G, et al. 2010. Microbial stimulation fully differentiates monocytes to DC-SIGN/$CD209^+$ dendritic cells for immune T cell areas. *Cell* **143:** 416–429.

Chow A, Lucas D, Hidalgo A, Mendez-Ferrer S, Hashimoto D, Scheiermann C, Battista M, Leboeuf M, Prophete C, van Rooijen N, et al. 2011. Bone marrow $CD169^+$ macrophages promote the retention of hematopoietic stem and progenitor cells in the mesenchymal stem cell niche. *J Exp Med* **208:** 261–271.

Crocker PR, Gordon S. 1986. Properties and distribution of a lectin-like hemagglutinin differentially expressed by murine stromal tissue macrophages. *J Exp Med* **164:** 1862–1875.

Crocker PR, Gordon S. 1989. Mouse macrophage hemagglutinin (sheep erythrocyte receptor) with specificity for sialylated glycoconjugates characterized by a monoclonal antibody. *J Exp Med* **169:** 1333–1346.

Crocker PR, Werb Z, Gordon S, Bainton DF. 1990. Ultrastructural localization of a macrophage-restricted sialic acid binding hemagglutinin, SER, in macrophage-hematopoietic cell clusters. *Blood* **76:** 1131–1138.

Davies LC, Jenkins SJ, Allen JE, Taylor PR. 2013. Tissue-resident macrophages. *Nat Immunol* **14:** 986–995.

den Haan JM, Kraal G. 2012. Innate immune functions of macrophage subpopulations in the spleen. *J Innate Immun* **4:** 437–445.

den Haan JM, Martinez-Pomares L. 2013. Macrophage heterogeneity in lymphoid tissues. *Semin Immunopathol* **35:** 541–552.

Drummond RA, Brown GD. 2013. Signalling C-type lectins in antimicrobial immunity. *PLoS Pathog* **9:** e1003417.

Gazi U, Rosas M, Singh S, Heinsbroek S, Haq I, Johnson S, Brown GD, Williams DL, Taylor PR, Martinez-Pomares L. 2011. Fungal recognition enhances mannose receptor shedding through dectin-1 engagement. *J Biol Chem* **286:** 7822–7829.

Geijtenbeek TB, Groot PC, Nolte MA, van Vliet SJ, Gangaram-Panday ST, van Duijnhoven GC, Kraal G, van Oosterhout AJ, van Kooyk Y. 2002. Marginal zone macrophages express a murine homologue of DC-SIGN that captures blood-borne antigens in vivo. *Blood* **100:** 2908–2916.

Gerner MY, Kastenmuller W, Ifrim I, Kabat J, Germain RN. 2012. Histo-cytometry: A method for highly multiplex quantitative tissue imaging analysis applied to dendritic cell subset microanatomy in lymph nodes. *Immunity* **37:** 364–376.

Gonzalez SF, Kuligowski MP, Pitcher LA, Roozendaal R, Carroll MC. 2010a. The role of innate immunity in B cell acquisition of antigen within LNs. *Adv Immunol* **106:** 1–19.

Gonzalez SF, Lukacs-Kornek V, Kuligowski MP, Pitcher LA, Degn SE, Kim YA, Cloninger MJ, Martinez-Pomares L, Gordon S, Turley SJ, et al. 2010b. Capture of influenza by medullary dendritic cells via SIGN-R1 is essential for humoral immunity in draining lymph nodes. *Nat Immunol* **11:** 427–434.

Gordon S. 2012. Macrophages and phagocytosis. In *Fundamental immunology* (ed. Paul WE). Lippincott, Williams and Wilkins, Philadelphia.

Gordon S, Hamann J, Lin HH, Stacey M. 2011. F4/80 and the related adhesion-GPCRs. *Eur J Immunol* **41:** 2472–2476.

Gowans JL. 1991. First Medawar prize lecture. The recirculating small lymphocyte. *Transplant Proc* **23:** 7–8.

Gray EE, Cyster JG. 2012. Lymph node macrophages. *J Innate Immun* **4:** 424–436.

Hashimoto D, Chow A, Noizat C, Teo P, Beasley MB, Leboeuf M, Becker CD, See P, Price J, Lucas D, et al. 2013. Tissue-resident macrophages self-maintain locally throughout adult life with minimal contribution from circulating monocytes. *Immunity* **38:** 792–804.

Herrmann M, Schafer C, Heiss A, Graber S, Kinkeldey A, Buscher A, Schmitt MM, Bornemann J, Nimmerjahn F, Helming L, et al. 2012. Clearance of fetuin-A-containing calciprotein particles is mediated by scavenger receptor-A. *Circ Res* **111:** 575–584.

Hume DA, Robinson AP, MacPherson GG, Gordon S. 1983. The mononuclear phagocyte system of the mouse defined by immunohistochemical localization of antigen F4/80. Relationship between macrophages, Langerhans cells, reticular cells, and dendritic cells in lymphoid and hematopoietic organs. *J Exp Med* **158:** 1522–1536.

Humphrey JH, Grennan D. 1981. Different macrophage populations distinguished by means of fluorescent polysaccharides. Recognition and properties of marginal-zone macrophages. *Eur J Immunol* **11:** 221–228.

Jenne CN, Kubes P. 2013. Immune surveillance by the liver. *Nat Immunol* **14:** 996–1006.

Karlsson MC, Guinamard R, Bolland S, Sankala M, Steinman RM, Ravetch JV. 2003. Macrophages control the retention and trafficking of B lymphocytes in the splenic marginal zone. *J Exp Med* **198:** 333–340.

Kastenmuller W, Torabi-Parizi P, Subramanian N, Lammermann T, Germain RN. 2012. A spatially-organized multicellular innate immune response in lymph nodes limits systemic pathogen spread. *Cell* **150:** 1235–1248.

Klaas M, Crocker PR. 2012. Sialoadhesin in recognition of self and non-self. *Semin Immunopathol* **34:** 353–364.

Kraal G. 1992. Cells in the marginal zone of the spleen. *Int Rev Cytol* **132:** 31–74.

Kraal G, Schornagel K, Streeter PR, Holzmann B, Butcher EC. 1995. Expression of the mucosal vascular addressin, MAdCAM-1, on sinus-lining cells in the spleen. *Am J Pathol* **147:** 763–771.

Kraal G, van der Laan LJ, Elomaa O, Tryggvason K. 2000. The macrophage receptor MARCO. *Microbes Infect* **2:** 313–316.

Lin HH, Faunce DE, Stacey M, Terajewicz A, Nakamura T, Zhang-Hoover J, Kerley M, Mucenski ML, Gordon S, Stein-Streilein J. 2005. The macrophage F4/80 receptor is required for the induction of antigen-specific efferent regulatory T cells in peripheral tolerance. *J Exp Med* **201:** 1615–1625.

MacLennan IC. 2005. Germinal centers still hold secrets. *Immunity* **22:** 656–657.

Manz MG. 2007. Human-hemato-lymphoid-system mice: Opportunities and challenges. *Immunity* **26:** 537–541.

Martinez-Pomares L, Gordon S. 2007. Antigen presentation the macrophage way. *Cell* **131:** 641–643.

Martinez-Pomares L, Gordon S. 2012. $CD169^+$ macrophages at the crossroads of antigen presentation. *Trends Immunol* **33:** 66–70.

Martinez-Pomares L, Kosco-Vilbois M, Darley E, Tree P, Herren S, Bonnefoy JY, Gordon S. 1996. Fc chimeric protein containing the cysteine-rich domain of the murine mannose receptor binds to macrophages from splenic marginal zone and lymph node subcapsular sinus and to germinal centers. *J Exp Med* **184:** 1927–1937.

Martinez-Pomares L, Hanitsch LG, Stillion R, Keshav S, Gordon S. 2005. Expression of mannose receptor and ligands for its cysteine-rich domain in venous sinuses of human spleen. *Lab Invest* **85:** 1238–1249.

Milicevic NM, Klaperski K, Nohroudi K, Milicevic Z, Bieber K, Baraniec B, Blessenohl M, Kalies K, Ware CF, Westermann J. 2011. TNF receptor-1 is required for the formation of splenic compartments during adult, but not embryonic life. *J Immunol* **186:** 1486–1494.

Moghimi SM, Parhamifar L, Ahmadvand D, Wibroe PP, Andresen TL, Farhangrazi ZS, Hunter AC. 2012. Particulate systems for targeting of macrophages: Basic and therapeutic concepts. *J Innate Immun* **4:** 509–528.

Morris L, Graham CF, Gordon S. 1991. Macrophages in haemopoietic and other tissues of the developing mouse detected by the monoclonal antibody F4/80. *Development* **112:** 517–526.

Morris L, Crocker PR, Hill M, Gordon S. 1992. Developmental regulation of sialoadhesin (sheep erythrocyte receptor), a macrophage-cell interaction molecule expressed in lymphohemopoietic tissues. *Dev Immunol* **2:** 7–17.

Moseman EA, Iannacone M, Bosurgi L, Tonti E, Chevrier N, Tumanov A, Fu YX, Hacohen N, von Andrian UH. 2012. B cell maintenance of subcapsular sinus macrophages protects against a fatal viral infection independent of adaptive immunity. *Immunity* **36:** 415–426.

Moussion C, Sixt M. 2013. A conduit to amplify innate immunity. *Immunity* **38:** 853–854.

Mueller SN, Germain RN. 2009. Stromal cell contributions to the homeostasis and functionality of the immune system. *Nat Rev Immunol* **9:** 618–629.

Mukhopadhyay S, Chen Y, Sankala M, Peiser L, Pikkarainen T, Kraal G, Tryggvason K, Gordon S. 2006. MARCO, an innate activation marker of macrophages, is a class A scavenger receptor for *Neisseria meningitidis*. *Eur J Immunol* **36:** 940–949.

Nolte MA, Arens R, Kraus M, van Oers MH, Kraal G, van Lier RA, Mebius RE. 2004. B cells are crucial for both development and maintenance of the splenic marginal zone. *J Immunol* **172:** 3620–3627.

O'Neill AS, van den Berg TK, Mullen GE. 2013. Sialoadhesin—A macrophage-restricted marker of immunoregulation and inflammation. *Immunology* **138:** 198–207.

Perry VH, Crocker PR, Gordon S. 1992. The blood-brain barrier regulates the expression of a macrophage sialic acid-binding receptor on microglia. *J Cell Sci* **101:** 201–207.

Phan TG, Green JA, Gray EE, Xu Y, Cyster JG. 2009. Immune complex relay by subcapsular sinus macrophages and noncognate B cells drives antibody affinity maturation. *Nat Immunol* **10:** 786–793.

Plüddemann A, Hoe JC, Makepeace K, Moxon ER, Gordon S. 2009. The macrophage scavenger receptor a is host-protective in experimental meningococcal septicaemia. *PLoS Pathog* **5:** e1000297.

Reizis B. 2012. Classical dendritic cells as a unique immune cell lineage. *J Exp Med* **209:** 1053–1056.

Roozendaal R, Mebius RE, Kraal G. 2008. The conduit system of the lymph node. *Int Immunol* **20:** 1483–1487.

Saito M, Iwawaki T, Taya C, Yonekawa H, Noda M, Inui Y, Mekada E, Kimata Y, Tsuru A, Kohno K. 2001. Diphtheria toxin receptor-mediated conditional and targeted cell ablation in transgenic mice. *Nat Biotechnol* **19:** 746–750.

Schulz C, Gomez Perdiguero E, Chorro L, Szabo-Rogers H, Cagnard N, Kierdorf K, Prinz M, Wu B, Jacobsen SE, Pollard JW, et al. 2012. A lineage of myeloid cells independent of Myb and hematopoietic stem cells. *Science* **336:** 86–90.

Steiniger B, Barth P, Herbst B, Hartnell A, Crocker PR. 1997. The species-specific structure of microanatomical compartments in the human spleen: Strongly sialoadhesin-positive macrophages occur in the perifollicular zone, but not in the marginal zone. *Immunology* **92:** 307–316.

Steiniger B, Timphus EM, Barth PJ. 2006. The splenic marginal zone in humans and rodents: An enigmatic compartment and its inhabitants. *Histochem Cell Biol* **126:** 641–648.

Taylor PR, Zamze S, Stillion RJ, Wong SY, Gordon S, Martinez-Pomares L. 2004. Development of a specific system for targeting protein to metallophilic macrophages. *Proc Natl Acad Sci* **101:** 1963–1968.

Taylor PR, Martinez-Pomares L, Stacey M, Lin HH, Brown GD, Gordon S. 2005. Macrophage receptors and immune recognition. *Annu Rev Immunol* **23:** 901–944.

van den Berg TK, Kraal G. 2005. A function for the macrophage F4/80 molecule in tolerance induction. *Trends Immunol* **26:** 506–509.

Weisser SB, van Rooijen N, Sly LM. 2012. Depletion and reconstitution of macrophages in mice. *J Vis Exp* **66:** 4105.

Wijffels JF, de Rover Z, Beelen RH, Kraal G, van Rooijen N. 1994. Macrophage subpopulations in the mouse spleen renewed by local proliferation. *Immunobiology* **191:** 52–64.

Witmer-Pack MD, Hughes DA, Schuler G, Lawson L, McWilliam A, Inaba K, Steinman RM, Gordon S. 1993. Identification of macrophages and dendritic cells in the osteopetrotic (op/op) mouse. *J Cell Sci* **104:** 1021–1029.

Yona S, Kim KW, Wolf Y, Mildner A, Varol D, Breker M, Strauss-Ayali D, Viukov S, Guilliams M, Misharin A, et al. 2013. Fate mapping reveals origins and dynamics of monocytes and tissue macrophages under homeostasis. *Immunity* **38:** 79–91.

You Y, Myers RC, Freeberg L, Foote J, Kearney JF, Justement LB, Carter RH. 2011. Marginal zone B cells regulate antigen capture by marginal zone macrophages. *J Immunol* **186:** 2172–2181.

Zamze S, Martinez-Pomares L, Jones H, Taylor PR, Stillion RJ, Gordon S, Wong SY. 2002. Recognition of bacterial capsular polysaccharides and lipopolysaccharides by the macrophage mannose receptor. *J Biol Chem* **277:** 41613–41623.

Zindl CL, Kim TH, Zeng M, Archambault AS, Grayson MH, Choi K, Schreiber RD, Chaplin DD. 2009. The lymphotoxin $LT\alpha_1\beta_2$ controls postnatal and adult spleen marginal sinus vascular structure and function. *Immunity* **30:** 408–420.

Approaching the Next Revolution? Evolutionary Integration of Neural and Immune Pathogen Sensing and Response

Kevin J. Tracey

Feinstein Institute for Medical Research, Manhasset, New York 11030

Correspondence: kjtracey@nshs.edu

Mammalian immunity evolved by the process of natural selection that produced differential survival and reproduction advantages through combinations of hereditary traits underlying the response to pathogens. Primitive animals sense the presence of microbial pathogens through recognition of pathogen-derived molecules in their rudimentary immune and nervous systems. No molecular biological mechanism assigns primacy of pathogen sensing mechanisms to immune cells over neurons. Rather, in animals as diverse as *Caenorhabditis elegans* to mammals, neural reflexes are activated by the presence of pathogens and transduce neural mechanisms that control the development of immunity. A coming revolution in immunological thinking will require immunologists to incorporate neural circuits into understanding pathogen signal transduction, and the molecular mechanisms of learning, that culminate in immunity.

On considering memory, one finds an ironic perspective, tainted by shadows of two major scientific fields that, historically at least, did not collaborate. Immunological memory, mediated by lymphocytes, and neurological memory, mediated by neurons, evolved over millions of years in response to environmental changes. Closer inspection within both fields reveals key features of common origin between neural and immune information collection and retrieval. Major evolutionary advantages arose at points of intersection of these systems, manifested as beneficial physiological responses to environmental stimuli during infection, injury, and metabolic stress. In 1989, Charles Janeway proposed that the first revolution in immunological thinking led to the domination of the humoral theory of immunity, and that an approaching second revolution would integrate innate and adaptive immunity by understanding the role of pathogen-associated molecular pattern receptors (Janeway 1989). Today, I believe that the groundwork has been laid for a third revolution in immunological thinking that will integrate the role of neurological feedback circuits into innate and adaptive immunity, including the role of molecular mechanisms through which neurons sense microbes and regulate the output of hematopoietic-derived immune cells. I also suggest that costimulation of neural reflex circuits by microbial products plays a major role in the immune response to

Cite this article as *Cold Spring Harb Perspect Biol* doi: 10.1101/cshperspect.a016360

infection, and to the subsequent development of immunity (Fig. 1). If correct, this idea could revolutionize our thinking about immunity and take us beyond innate and adaptive immunity to an integrated view of neurological and immunological recognition and learning.

EVOLUTIONARILY ANCIENT SENSING MECHANISMS IN NEUROSCIENCE AND IMMUNOLOGY

It begins with sensing a change in the environment. Consider the evolutionarily primitive animal, *Caenorhabditis elegans*, which harbors rudimentary immune and nervous systems. Moving and feeding within a bacterial lawn, it uses neural and immune sensory mechanisms to detect the presence of pathogenic microbes and toxins (Pradel et al. 2007). On encountering specific molecules, sensory neurons transduce the chemical environmental information into action potentials, which initiate neural responses that modify the animal's behavior and motion. The presence of pathogens also activates its innate immune cells, which incite defensive reactions, including enhanced MAP kinase and XBP-1 signaling, accumulation of unfolded proteins in endoplasmic reticulum, and other cellular and metabolic responses (Styer et al. 2008; Aballay 2013). Sensory neural signals initiate primitive response circuits linked to corresponding motor neural signals that determine behavioral and physiological responses. Thus, the worm's neural and innate immune sensory systems independently detect the presence of the pathogens in the environment.

Primitive innate immune-like cells in *C. elegans*, which predate the evolution of competent Toll-like receptor and NF-κB signal transduction mechanisms, are also activated by the presence of pathogens or injury (Aballay 2013). Gene expression patterns in the innate immune system, and animal behavior mediated by the nervous system, are altered in a pathogen species-dependent manner. The result of these responses is that the animal can defend itself against certain pathogen encounters, although not all. The innate neural and immunological

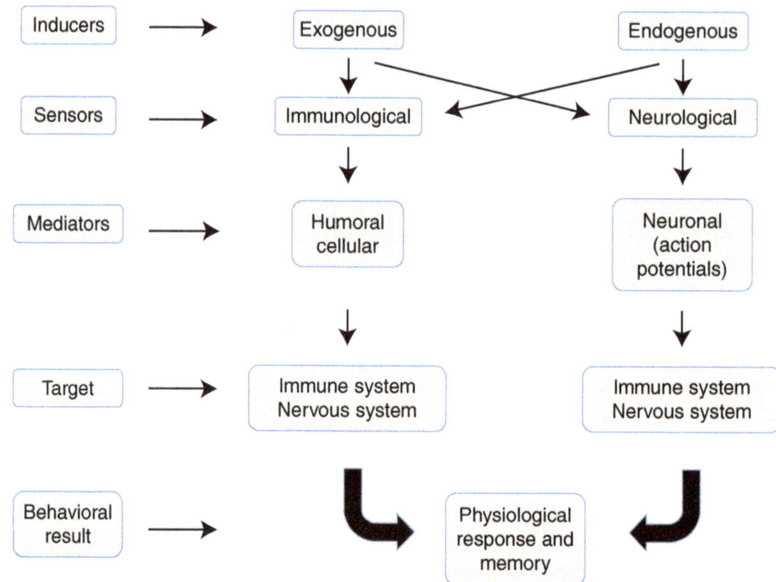

Figure 1. Pathogens or products of cellular inflammation and injury costimulate immune cells and neurons during the earliest stages of infection. Neural input activates reflex circuits, which modulate the nervous system and the immune system. Hematopoietic-derived cellular responses produce humoral and cellular signals that influence immune cells and neurons. These signal transduction pathways culminate in mediating the behavior of the animal and in initiating learning.

Cite this article as *Cold Spring Harb Perspect Biol* doi: 10.1101/cshperspect.a016360

sensing events are both operative at the earliest stage of the encounter, but the speed of neural transmission offers a key advantage to coordinating the animal's specific responses.

There are 302 neurons in *C. elegans*, and the interconnections and response characteristics of these have been elucidated. A characteristic feature of the worm's response to infection is that one family of neural circuits, activated when sensory neurons detect the presence of pathogenic pseudomonas, transmits neurotransmitter-dependent signals to mediate pathogen avoidance behaviors, and to suppress the unfolded protein response in innate immune cells (Sun et al. 2011). In light of today's advanced knowledge, and our detailed molecular mechanistic understanding about inhibitory reflex control of innate immune responses in mammals, perhaps the perspective becomes clearer. The regulatory intersection between the nervous and innate immune systems in *C. elegans*, which has been retained in today's mammals, originated from evolutionary pressure to optimize physiological homeostasis, and survival.

Beyond either field, there is little need to account for separateness in functional mechanisms that operate in a parallel time frame, and also intersect to control the magnitude of host defense and response. But immunology and neuroscience have only recently embraced this theory, so there is value in reviewing the concept from the perspectives originating within each. From an historic perspective in immunology, the principal focus has been to understand the development and maintenance of immunity, and to provide a complete molecular mechanistic basis for how lymphocytes acquire and retain the information necessary to remember the nature of infectious threats. This has been highly successful for establishing the principles to explain the diversity and repertoire of long-term humoral and cellular immune memory that results from antigen processing, presentation, and clonal development of B and T cells. A detailed molecular understanding for the basis of innate immunity was also produced. When likened to reflex or instinctual behavior, pathogen recognition by innate immune cells is encoded in the genome, enables a relatively rapid and consistent response to pathogen-associated and damaged-associated molecular patterns, and is highly conserved. These mechanisms of immunity have been exceedingly well described within a knowledge sphere circumscribed by cells of hematopoietic origin, operating in response to pathogens, absent control from other systems. It is axiomatic within immunological thinking that the earliest stages of sensing pathogens occurs through activation of innate immune pattern-recognition receptor signals, and that later sensing occurs through adaptive immune responses.

In an historic time line parallel with these unfolding events, the principle focus of neuroscience has been to understand the functions of the brain and nervous system. This established neurons as the principal cells controlling information, learning, and memory; the chemical basis of neurotransmission; and the principles of synaptic neural communication underlying neuronal signaling. In the mid-20th century, Sherrington famously recognized that the simple reflex, comprised of a sensory arc, interneurons, and a motor arc, is the fundamental unit for information processing on which the entire nervous system evolved (Sherrington 1906). From *C. elegans* and to modern mammals, sensory neurons respond to chemical, mechanical, or temperature changes in the environment. This information is transduced via ion channels and gradient potentials that generate action potentials that are relayed to other neurons. During pathogen threat, or in response to other environmental changes, information processing in the nervous system begins with sensory input transmitted in the afferent arc of a reflex circuit.

Sensory neural signals in mammals are routed through nuclei residing within the central nervous system, primarily in the brain stem. These nuclei produce the output that regulates organ function by transmission through the motor arc to complete the reflex circuit. The simple reflex unit is the basic building block that evolutionary pressure used to integrate and assemble the mammalian nervous system as the regulatory and coordinating system of host responses to infection, injury, and other threats. Nuclei that provide the relay stations for regulatory reflexes are housed in the evolutionarily ancient regions of the mammalian brain stem. These brain stem nuclei control the homeostatic set points of all basic physiological functions associated with health, and the response to environmental change.

Individual neural reflex circuits are stimulated in response to specific sensory inputs arising within the body organs. The reflexively controlled release of neurotransmitters from the motor arc controls the output of each organ. Ultimately, the physiological, or involuntary, behavior of the host is the result of the sum total action of all reflexes connecting its organs to its brain stem. These involuntary reflexes function with limited or no input from the cerebral cortex. They operate largely beneath conscious control. Primitive animals, like *C. elegans*, which (presumably) lack the cerebral faculties of consciousness and self-awareness, also show reflex behavior in response to changes in pressure, temperature, the chemical composition of the environment, and the presence of pathogens.

In considering the role of regulatory feedback on the molecular mechanisms underlying immunity, it bears special note here that there is a modern tendency to center discussions of mammalian neural information processing on the cerebral cortex. Termed "cortical conceit," one may view the sophisticated human cerebral encephalon as an undoubted source of inestimable pride, a grand evolutionary accomplishment (Dubos 1998). But the advanced cerebral cortex found in mammals is a relatively new addition to biology, one that is largely dispensable for understanding basic mechanisms of physiological and immunological homeostasis. Surgically or genetically decorticated mammals can well maintain physiological and immunological set points within a healthy range, as long as the brain stem nuclei and corresponding reflex circuits are functional. Basic reflex circuits, having developed in an evolutionary time, maintain physiological responses during infection and injury without requiring input from the cortex to operate competently. Signals descending from the higher brain can influence the functional output of the reflex nuclei, but these psychological implications need not be considered here to understand the basic physiological principles governing homeostasis. So, in defining rules that govern how the nervous system modulates the immune system to establish immunological homeostasis, we can dispense with the need to consider individual acts of conscious will, or choices made. The system operates in the background of conscious existence, driven by sensory inputs that elicit reflex responses.

These reflex circuits originating in the evolutionary precursors of the modern mammalian brain stem nuclei conferred the ability to modulate cardiac and pulmonary function in response to infection, and countless other protective and beneficial organ responses. Animals expressing such competent beneficial reflexes gained a preservation and reproductive advantage over breeding populations lacking it. In time, the accumulation of individual simple reflexes endowed evolving species with optimal flexibility necessary to use effective defensive responses and maintain homeostasis in response to a range of changes in the internal and external milieu. The identity and function of these regulatory reflexes in mammals was revealed in classic physiological studies of the anatomically accessible body systems. Consider the cardiovascular system, which is beautifully presented in the thorax when the sternum is divided, to behold the beating heart.

With each cardiac contraction it is possible to observe, measure, and record numerous experimental end points reflecting the intricate physiological working of the reflexively controlled heart. Cutting the vagus nerve to the heart produces an increase in heart rate, indicating that heart rate is tonically suppressed by descending neural signals from the brain stem. This descending inhibitory signal is transmitted in the motor arc of a reflex initiated when an increase in heart rate is sensed by afferent baroreceptor neurons. Action potentials arrive within brain stem nuclei, and from there, neural signals exit the brain stem to be transmitted back to the heart to decrease heart rate relative to a predetermined set point. Other reflexes, activated when heart rate decelerates below its set point, culminate in signaling via other motor arcs that increase heart rate. From observations of this sort, physiologists and neuroscientists built detailed models and developed rules of feedback control based on individual units of reflex action to explain how cardiovascular homeostasis is maintained during a range of changes in the internal and external milieu. Similar rules have been developed and applied to other organ systems. The regulatory neural systems that reflexively control physiological homeostasis operate in real time, are highly sensitive to changes in external and internal environment, and ultimately enable organ systems to function within a narrow range of metabolic, thermal, nutrient, and behavioral parameters that is compatible with cellular and animal health.

The vagus nerve is a principal nerve in mammals that transmits afferent, sensory information from the organs to the brain stem, and it is a mechanism for informing the nervous system about the status of the internal milieu. From the visceral organs, it travels a meandering course across the abdomen, thorax, and neck, entering the brain stem, where its sensory nerve endings synapse on, and relay information to, neurons in the nucleus tractus solatarius. Interneurons arising there relay signals to the dorsal motor nucleus of the vagus nerve, representing a major cholinergic outflow tract that returns to the peripheral organs. The motor arcs stimulated by vagus sensory afferent neurons return to the organs via efferent neurons that travel along two major routes: the sympathetic chain, and the vagus nerve itself. At the nerve terminus, motor neurons from these two major outflow paths release acetylcholine, norepinephrine, vasoactive peptide, and other neuropeptides that bind to cognate receptors in the target organ's epithelial and parenchymal cells, to modulate cellular metabolism, function, and secretory capacity. Studies of the cardiopulmonary, gastrointestinal, renal, hepatic, and endocrine systems established the anatomic and neurophysiological principles underlying reflex control of physiological homeostasis. Until quite recently, however, the existence of regulatory reflexes exerting control over the immune system had evaded description. Immunological thinking had not viewed immunity as a reflexively regulated response.

REFLEX REGULATION OF INNATE IMMUNITY

This began to change following the discovery that cytokines produced in response to infection and injury are necessary and sufficient mediators in disease pathogenesis. This provided experimental tools and strategies that enabled the illumination of neural circuits that regulate innate immunity. Termed "the cytokine theory of disease," cytokines produced in response to infection and injury mediate cellular, metabolic, and pathological effects underlying the major physiological and clinical manifestations of illness (Tracey 2007). Restated, disease occurs when there is a loss of immunological homeostasis producing an unbalanced or excessive cytokine response. In such cases, inhibiting or modulating the cytokine response to restore immunological homeostasis can prevent or reverse disease, and it should be stressed that the cytokine theory does not state that all cytokines are "bad." By analogy, the germ theory does not state that all microbes are pathogenic. Rather, the cytokine theory indicates that in some instances of cytokine imbalance or excess, treatment of diseases can be specifically accomplished with targeted therapeutics.

The cytokine theory of disease was validated in mammals, including humans, within a few years after early direct evidence became available that tumor necrosis factor (TNF) is a necessary and sufficient mediator of acute septic shock during infection (Tracey et al. 1986, 1987). The first principles for administering anti-TNF monoclonal antibodies as therapy were established in baboons infected with *Escherichia coli* (Tracey et al. 1987). Treated animals survived despite the presence of replicating bacteria in their bloodstream, proving that the disease (septic shock) required the activity of TNF. Disease was attributable to the activity of the cytokine, regardless of the status of the pathogen (Tracey et al. 1987). Today it is common to witness this principle being applied to patients receiving monoclonal anti-TNF for the treatment of rheumatoid arthritis, inflammatory bowel disease, and other conditions. From a practical point of view, the earliest underlying or inciting etiological agent can be dissociated from the disease. The physician standing at the patient's bedside is witness to signs and symptoms of inflammation, organ damage, and toxicity mediated by the direct action of cytokines. In many, if not most, diseases, excessive cytokine activity can be considered as a failure of immunological homeostasis.

It was from this perspective, in the 1990s, that my colleagues and I began to directly address the possibility that reflex neural circuits had a critical role in regulating cytokine release and maintaining immunological homeostasis. We focused on the TNF response in animal models of infection and injury. The choice of TNF as the target of these studies was important, because it represents an early-stage response, rapidly induced in response to pathogenic stimulation. The measurable TNF response falls within a time window peaking 90 min after exposure to endotoxin, quite compatible with making observations that could be interpreted in the context of experimental manipulation of neural circuits. To determine whether neural signals suppress TNF release from tissue macrophages, we applied stimulating electrodes to the vagus nerve where it traversed the neck on its descending course to the reticuloendothelial system (Borovikova et al. 2000). Electrical stimulation of the vagus nerve significantly inhibited TNF release in red pulp and marginal zone macrophages in spleen, suggesting that descending signals in the motor arc of an "inflammatory reflex" participate in regulating the innate immune response and maintaining immunological, or "cytokine homeostasis" (Tracey 2002; Rosas-Ballina et al. 2008).

When the vagus nerve was divided before exposing animals to endotoxin, we observed that TNF production by splenic macrophages was enhanced (Borovikova et al. 2000). By analogy to the tonic inhibitory influence of vagus nerve signaling, which inhibits heart rate, this fact indicated that constitutively active or tonic vagus nerve signals slow or inhibit the innate immune response to endotoxin. By following the track of the vagus nerve from the brain stem, down the neck, across the thorax, and into the abdomen, and either dividing the nerve in some cases, or electrically stimulating it along its course in others, we discovered that the point of convergence of the vagus nerve signals to regulate serum TNF during endotoxemia is in the spleen (Huston et al. 2006; Rosas-Ballina et al. 2008). Electrical signals applied to the cervical vagus nerve descend to the celiac ganglion, which is the origin of motor neurons that innervate the spleen. Vagus nerve and splenic nerve action potentials inhibit the release of TNF in red pulp and marginal zone macrophages, which together account for >90% of the peak serum TNF levels following the acute onset of endotoxemia.

Three implications from these findings are immediately obvious. First, the TNF producing innate immune response to endotoxin is regulated by neural signals. The constitutive inhibitory activity of neural circuits establishes a set point for the innate immune response to endotoxin. Second, the immune system is anatomically and functionally innervated. Although extensive prior anatomic evidence indicated that all of the major organs of the reticuloendothelial system receive neural connections from the brain stem, the prior anatomic findings had not provided a scientific framework for experimental study of the functional operation of these neurons. Approaching this as a functional, neurophysiological property of neuronal activity to regulate the innate immune response to endotoxin opened a door to delineating molecular mechanisms. Third, it is overwhelmingly probable that the basic principles of reflex action activated by sensory input also apply to other aspects of innate immunity and adaptive immunity. The evolutionary emergence of lymphocytes, and adaptive immunity, occurred roughly in parallel with the emergence of more sophisticated neural networks arising from evolutionarily expanding neural circuits and connectivity. It is nearly impossible to imagine an evolving situation in which the environmental changes caused by infection and injury did not intersect with signals that simultaneously influence both the immune and nervous systems.

As broadly defined, immunity is the state of acquiring (learning) and retaining (remembering) systems to confer protection against infection. On accepting the theory that neural signals, which regulate immune responses, are costimulated during the activation of innate immunity, and provide regulatory signals in the earliest stages of acquiring immune memory, then it can be strongly argued that a complete understanding of immunity requires incorporating neural mechanisms. To determine the mechanism for reflex inhibition of innate immunity, we experimentally followed the path of the neural signals arising in the vagus nerve and terminating in the spleen (Rosas-Ballina et al. 2011). Within the spleen, we observed splenic nerve endings terminating in synapse-like structures adjacent to lymphocytes. In transgenic mice, which coexpressed green fluorescent protein in cells that express choline acetyltransferase, the rate-limiting enzyme that catalyzes the biosynthesis of acetylcholine, we observed that T lymphocytes capable of synthesizing acetylcholine resided in close proximity to splenic nerve endings. Electrical stimulation of the vagus nerve stimulated the adrenergic splenic nerve to release norepinephrine, which in turn stimulated Chat-expressing T cells to release acetylcholine. Lymphocyte-derived acetylcholine is the neurally activated molecular mediator that binds to α7 nicotinic acetylcholine receptors expressed by red pulp and marginal zone macrophages (Wang et al. 2003). Ligand receptor interaction inhibits the release of TNF through a molecular mechanism that has been attributed to formation of a heteroprotein complex comprised of α7 binding to JAK-STAT that suppresses activation of the nuclear translocation of nuclear factor κB and to stabilizing mitochrondrial membrane permeability (de Jonge et al. 2005; Lu et al. 2014). The functional importance of these T cells in this neural circuit is striking, because T-cell-deficient nude mice lack a working inflammatory reflex. Electrical stimulation of the vagus nerve in nude mice fails to inhibit TNF. This phenotype is reversed by passive transfer of Chat-expressing T cells into naïve recipient nude mice, which restores the functional inflammatory reflex in these animals (Rosas-Ballina et al. 2011). Thus, the inflammatory reflex is defined by vagus nerve action potentials that activate the release of an inhibitory neurotransmitter, acetylcholine, from specialized lymphocytes to suppress the magnitude of the TNF response in spleen. A prototypical neural circuit prevents overexpression of TNF to establish healthy immunological homeostasis.

Our initial observations on the neural regulation of TNF release enabled us to formulate the fundamental principles for reflex control of immunity. Discrete neurons terminating in specific molecular mechanisms modulate the immune response to infection (Tracey 2009; Andersson and Tracey 2012a). Inhibitory neural control of innate immunity by acetylcholine is not restricted to controlling TNF, because acetylcholine interaction with α7 nicotinic acetylcholine receptors in macrophages significantly inhibits activation of the inflammasome, the protein complex required for the release of leaderless cytokines, including IL-1, HMGB1, and IL-18 (Borovikova et al. 2000; Lu et al. 2014). Expression of α7 nAChR is essential to the integrity of the inflammatory reflex, because α7 nAChR knockout mice have an impaired inflammatory reflex. Vagus nerve stimulation fails to suppress inflammasome activity in these animals; and, moreover, activation of innate immunity in the α7 nAChR knockout mice produces significantly higher levels of TNF and the leaderless cytokines (Wang et al. 2003; Lu et al. 2014).

SENSORY STIMULATION OF NEUROLOGICAL REFLEXES IN IMMUNITY

As is the case for all reflexes, the motor arc of the inflammatory reflex is activated in response to signals arising in the sensory arc. Beyond the evidence cited above that the interaction of *C. elegans* with pathogens begins by sensing signals transmitted in afferent neurons, a series of pivotal studies have provided insight into the origin and function of the sensory arc of the inflammatory reflex in mammals. Originally, Linda Watkins and her colleagues (1995) made the seminal discovery that the fever response induced by the intra-abdominal administration of IL-1 requires an intact and functional vagus nerve, because when they surgically divided the vagus nerve, animals maintained normothermia despite exposure to intra-abdominal IL-1. Thus, the presence of IL-1 in the abdomen, or more specifically in the hepatic circulation, is sensed by the vagus nerve, which propagates action potentials ascending to synapses in the nucleus tractus solatarius in the brain stem (Fairchild et al. 2011). Sensory vagus nerve action potentials mediated by IL-1 have been experimentally shown using recording electrodes applied to the vagus nerve at various points along its course from the abdomen to the brain stem (Niijima 1996). Compound action potentials arising in response to intrahepatic IL-1 have also been recorded in the splenic nerve, which is a pure motor nerve lacking sensory fibers (Niijima et al. 1991). This gives direct evidence that sensory vagus nerve signals activate compound action potential signals descending to the spleen via the motor arc of the inflammatory reflex.

Other inflammatory mediators have been implicated in stimulating sensory action potentials in the vagus nerve, including endotoxin, TNF, prostaglandins, substance P, and other endogenous and exogenous products of infection and injury (Chiu et al. 2012). The sensory and motor arcs of the inflammatory reflex within the vagus nerve are positioned to anatomically and functionally respond to and modulate the host response to infection and threat in the visceral organs of the body compartment. The importance of this inhibitory neural circuit to maintaining immunological homeostasis is indicated by the findings that animals deficient in this reflex are rendered increasingly sensitive to excessive TNF release. This raises the next question: Can reflexes stimulate, rather than inhibit, innate immune responses to maintain immunological homeostasis?

As far back as 1874, it had been proposed that inflammation is mediated by action potentials transmitted in peripheral neurons (Chiu et al. 2012). This stemmed from the observations that electrical stimulation of dorsal root neurons produced vasodilation in the innervated skin. Today we understand that the inflammatory effect of these motor neurons is attributable to neurotransmitters released into the tissue including neuropeptides, catecholamines, and substance P. These molecules interact with cognate receptors expressed on endothelial cells, smooth muscle cells, lymphocytes, monocytes, and macrophages that in turn mediate vasodilation, increased capillary permeability, neutrophil recruitment, and swelling. Signaling through these stimulating neural pathways produces an enhancement of inflammation attributable to the biological mechanisms of specific proinflammatory neurotransmitters activated by that specific neural circuit. As these motor circuits are stimulatory, defined by their activity to increase the output of innate immune responses, there is evidence supporting a functional approach to studying the regulation of innate immunity by combined inhibitory and excitatory neural signaling mechanisms.

A basic principle in neuroscience is that the interaction of inhibitory and excitatory neural circuits mediates the output of a neural network, and determines the final state of organ function. Major forms of information processing in the nervous system are routinely defined by the identity of the excitatory or inhibitory neurotransmitter released in response to specific action potentials. Accordingly, depending on the neurotransmitters released, neural circuits can either inhibit inflammation, as in the case of cholinergic signals, or enhance inflammation, as in the cases of proinflammatory neuropeptides, adrenergic signals, and substance P. The functional output depends on the neurotransmitter receptors, rather than the neurotransmitters per se, as in the case of adrenergic signaling by norepinephrine secreting neurons, which can enhance or inhibit inflammation, depending on whether the principle receptors engaged are α or β adrenergic GPRs. Indeed, most innervated tissues receive input from circuits that release immune inhibitory and immune excitatory signals, so the net effect of neural control on the immune response will depend on the sum total interaction of molecular mechanisms resulting from ligand receptor signal transduction in a cell-specific and neurotransmitter-specific path.

To this point, I have argued that the output of neural reflex circuits regulates the innate immune response to pathogens, and that these circuits can be activated by IL-1 and other endogenous products of inflammation and tissue damage. This is in keeping with pervasive immunological thinking that the innate immune system is the primary sensor of the presence of pathogens. Next, consider that the regulatory intersection between the nervous and immune systems can also originate with pathogen sensing. It is clear that innate immune system receptors respond to pathogen-associated molecular patterns during early phases of infection, and that clonal expansion of lymphocytes expressing specific pathogen sensing receptors confers long-term protection against subsequent infections. But how early is "early," and what role might the nervous system play if it detects the presence of microbes or tissue damage first, before innate immune responses? Can neurons respond to bacteria directly, or must there be an intermediate sensing step of microbial recognition by the innate immune system?

Consider that peripheral sensory neurons form a dense meshlike network that envelops and covers all tissues exposed to the external environment, including the epithelial lining of the skin and soft tissues, as well as the pulmonary, genitourinary, and gastrointestinal surfaces. Wherever microbes might gain entry, a sensory neural net is in place to respond. Specialized "nociceptor" neurons can be activated to transmit afferent action potentials by exposure to specific molecules. In *C. elegans*, with its limited repertoire of neurons, glycolipids containing ascarylose (3,6-dideoxy-L-arabino-hexose), termed "ascarosides," stimulate specific nocioreceptor neurons that lead to discrete behavioral responses (Ludewig and Schroeder 2013). For example, exposure of worms to ascr#3 mediates either attraction or repulsion of hermaphrodites and males, whereas chemical modification of ascr#3 by addition of a tryptophan adduct (producing "Icas#3") produces entirely different aggregation and attraction behavior. In this example, discrete chemical modification of a well-characterized molecular entity provides a differential sensory input into reflex circuits that produce visibly perceptible behavioral responses.

Stimulation of nocioreceptors produces graded potentials that propagate short distances, and can lead to stimulation of action potentials in a frequency that reflects the intensity of the original stimulus. These action potentials travel longer distances along axons from the peripheral tissue into the central nervous system, where the incoming information can be relayed to the brain stem, and can activate an interneuron to stimulate a motor arc of a simple reflex that returns to the periphery. Another feature of some nocioceptive neurons is that signals propagating toward the central nervous system can be diverted at branch points to travel back toward the periphery in a phenomenon termed an "axon reflex" (Yaprak 2008). By this mechanism, TNF, IL-1, prostaglandins, substance P, and other molecular products of inflammation stimulate nocioceptive neurons to stream signals to the brain stem and back into the regional tissues. Functional expression of receptors for cytokines (e.g., IL-1R, TNFR), pathogen-associated molecules (e.g., TLR3, TLR4, TLR7, TLR9), and damage-associated molecules (e.g., RAGE, P2X3) have all been implicated in mediating neuronal hyperpolarization and signaling (Chiu et al. 2012).

From the neuroscience perspective, there is nothing particularly curious about the nervous system sensing molecular input from the milieu interior and exterior. Changes in the molecular composition of the environment directly mediate the generation of action potentials, and initiate the neural platform for responding to threat, which stimulates learning. If one approaches this from within a narrow immunology perspective, there will be significant bias in favor of theories that cells within the innate immune system, not the nervous system, are the primary sensing apparatus to detect pathogens. From an evolutionary perspective, however, there is no reason to ascribe innate sensing primacy to the immune system. Indeed, the facts argue to the contrary. Atomic structural mechanistic studies of molecular signaling through neurons in *C. elegans*, the evolutionary precursors to mammalian neuronal reflexes, react differentially to subtle changes in molecular structure. It can be strongly argued that neurons, as well as innate immune cells, evolved to "sense" molecules derived from pathogens in a manner that initiates highly specific and coordinated responses that benefit survival, and that evolution would preserve these advantages.

The recent discovery by Clifford Woolf and colleagues that mammalian nocioceptors sense pathogens independently from the innate immune system indicates that some adjustment will have to be made to prior immune-centric sensing theories (Chiu et al. 2013). Inoculation of bacteria into mice produced pain and inflammatory responses, even when animals were rendered genetically deficient in TLR and MyD88-dependent signal transduction mechanisms. *N*-formylated peptides derived from Gram-negative and Gram-positive bacteria are sensed by G-protein-coupled formyl peptide receptors (FPRs) on neurons. Administration of *E. coli*–derived peptide (fMLF), and *Staphylococcus aureus*–derived peptide (fMIFL) to mice-activated calcium flux in nocioceptive neurons that mediate pain and mechanical hypersensitivity. Exposure of the *S. aureus*–derived pore-forming toxin, α-hemolysin, induced a concentration-dependent calcium flux in sensory neurons, which express A disintegrin and metalloprotease 10 (ADAM10), implicated in membrane pore assembly. The assembly of molecular pores in response to α-hemolysin is sufficient to mediate ion flux and generate action potentials in sensory neurons. Mice rendered deficient in nocioceptive neurons (Nav1.8-Cre/diphtheria toxin A mice) and exposed to *S. aureus*–developed enhanced inflammatory responses, including increased neutrophil and monocyte infiltration, increased lymph node swelling, and increased production of TNF. Thus, the interaction of pathogen-associated molecules on neurons directly activated sensory information, transmitted as action potentials, to mediate a locally inhibitory neural circuit that suppresses inflammation (Chiu et al. 2013). This direct mechanistic evidence establishes that pathogen-derived molecules can be sensed by neurons, that specific molecular products activate specific neuronal mechanisms that culminate in the generation of action potentials, that localized neural reflexes contribute to regional regulation of innate immunity, and that all of this occurs at the earliest, immediate stages of host responses that are associated with learning and, eventually, memory.

CONCLUSIONS

Viewing the onset of immunity as commencing with neural sensing independent from, and co-stimulatory with, immune cell sensing has profound implications to immunological thinking. For students interested in the functional activity of hematopoietic cells, it will remain an option, of course, to continue investigation into the interaction between specific microbes and pathogen-associated molecules with specific cognate receptors expressed in cells from the myeloid lineage. This can be furthered in exquisite molecular detail by pursuing signal transduction mechanisms that converge on gene expression patterns. Students interested in understanding how immunological memory develops in mammals during infection, and those interested in understanding how the innervated host operates as a physiologically regulated animal within constraints of a stable internal milieu, will take a different approach. The revolutionary new approach to immunological thinking will embrace the mechanisms of inhibitory and excitatory neural feedback loops that function from the onset of infection, and are coordinated by brain stem reflexes and regional axonal reflexes, to modulate the output of the innate immune cells that independently sense the presence of microbial agents.

Revolutions can be messy business, and predicting the outcome even messier. But sometimes, as perhaps illustrated by Janeway's prescience in defining the second revolution in immunological thinking, the outcome can be predetermined by facts. At present, nearly 25 years after Janeway's prediction, innate immunity has been elevated to heightened preeminence in immunological thinking. There is an intense and appropriate focus on understanding the role of innate immunity in the ac-

Cite this article as *Cold Spring Harb Perspect Biol* doi: 10.1101/cshperspect.a016360

quisition of adaptive immunity. Looking to the coming revolution, there is clear evidence today that innate immunity is regulated by neural reflexes activated by pathogens and host-derived mediators of infection, inflammation, and injury; and there is strongly suggestive evidence that neural signals modulate adaptive immunity as well (Wong et al. 2011; Arima et al. 2012; Mina-Osorio et al. 2012). From my vantage point, this has all the makings of a revolution in immunological thinking that is under way owing to a convergence, not a clashing, of two major fields. A complete understanding of immunity requires fully understanding the neural response to infection and injury (Andersson and Tracey 2012b). Tools provided by molecular biology, neurophysiology, and cell biology render this a particularly exciting time to embrace this intersection of neuroscience and immunology, and to consider that advances from worms to mammals hold promise for advances in the clinic.

REFERENCES

Aballay A. 2013. Role of the nervous system in the control of proteostasis during innate immune activation: Insights from *C. elegans*. *PLoS Pathog* 9: e1003433.

Andersson U, Tracey KJ. 2012a. Neural reflexes in inflammation and immunity. *J Exp Med* **209:** 1057–1068.

Andersson U, Tracey KJ. 2012b. Reflex principles of immunological homeostasis. *Annu Rev Immunol* **30:** 313–335.

Arima Y, Harada M, Kamimura D, Park J-H, Kawano F, Yull FE, Kawamoto T, Iwakura Y, Betz U, Márquez G, et al. 2012. Regional neural activation defines a gateway for autoreactive T cells to cross the blood–brain barrier. *Cell* **148:** 447–457.

Borovikova LV, Ivanova S, Zhang M, Yang H, Botchkina GI, Watkins LR, Wang H, Abumrad N, Eaton JW, Tracey KJ. 2000. Vagus nerve stimulation attenuates the systemic inflammatory response to endotoxin. *Nature* **405:** 458–462.

Chiu IM, Von Hehn CA, Woolf CJ. 2012. Neurogenic inflammation and the peripheral nervous system in host defense and immunopathology. *Nat Neurosci* **15:** 1063–1067.

Chiu IM, Heesters B, Ghasemlou N, Von Hehn C, Zhao F, Tran J, Wainger B, Strominger A, Muralidharan S, Horswill AR, et al. 2013. Bacteria activate sensory neurons that modulate pain and inflammation. *Nature* **501:** 52–57.

De Jonge WJ, van der Zanden EP, The FO, Bijlsma MF, van Westerloo DJ, Bennink RJ, Berthoud H-R, Uematsu S, Akira S, van den Wijngaard RM, et al. 2005. Stimulation of the vagus nerve attenuates macrophage activation by activating the Jak2-STAT3 signaling pathway. *Nat Immunol* **6:** 844–851.

Dubos R. 1998. *So human an animal: How we are shaped by surroundings and events*. Transaction, Piscataway, NJ.

Fairchild KD, Srinivasan V, Moorman JR, Gaykema RP, Goehler LE. 2011. Pathogen-induced heart rate changes associated with cholinergic nervous system activation. *Am J Physiol Regul Integr Comp Physiol* **300:** R330–R339.

Huston JM, Ochani M, Rosas-Ballina M, Liao H, Ochani K, Pavlov VA, Gallowitsch-Puerta M, Ashok M, Czura CJ, Foxwell B, et al. 2006. Splenectomy inactivates the cholinergic antiinflammatory pathway during lethal endotoxemia and polymicrobial sepsis. *J Exp Med* **203:** 1623–1628.

Janeway C. 1989. Approaching the asymptote? Evolution and revolution in immunology. *Cold Spring Harb Symp Quant Biol* **54:** 1–13.

Lu B, Kwan K, Levine YA, Olofsson PS, Yang H, Li J, Joshi S, Wang H, Andersson U, Chavan SS, Tracey KJ. 2014. $\alpha7$ nicotinic acetylcholine receptor signaling inhibits inflammasome activation by preventing mitochondrial DNA release. *Mol Med* doi: 10.2119/molmed.2013.00117.

Ludewig AH and Schroeder FC. 2013. Ascaroside signaling in *C. elegans*. In *WormBook* (ed. The *C. elegans* Research Community), doi/10.1895/wormbook.1.155.1.

Mina-Osorio P, Rosas-Ballina M, Valdes-Ferrer SI, Al-Abed Y, Tracey KJ, Diamond B. 2012. Neural signaling in the spleen controls B-cell responses to blood-borne antigen. *Mol Med* **18:** 618–627.

Niijima A. 1996. The afferent discharges from sensors for interleukin 1β in the hepatoportal system in the anesthetized rat. *J Auton Nerv Syst* **61:** 287–291.

Niijima A, Hori T, Aou S, Oomura Y. 1991. The effects of interleukin-1β on the activity of adrenal, splenic and renal sympathetic nerves in the rat. *J Auton Nerv Syst* **36:** 183–192.

Pradel E, Zhang Y, Pujol N, Matsuyama T, Bargmann CI, Ewbank JJ. 2007. Detection and avoidance of a natural product from the pathogenic bacterium *Serratia marcescens* by *Caenorhabditis elegans*. *Proc Natl Acad Sci* **104:** 2295–300.

Rosas-Ballina M, Ochani M, Parrish WR, Ochani K, Harris YT, Huston JM, Chavan S, Tracey KJ. 2008. Splenic nerve is required for cholinergic antiinflammatory pathway control of TNF in endotoxemia. *Proc Natl Acad Sci* **105:** 11008–11013.

Rosas-Ballina M, Olofsson PS, Ochani M, Valdés-Ferrer SI, Levine Y, Reardon C, Tusche MW, Pavlov VA, Andersson U, Chavan S, et al. 2011. Acetylcholine-synthesizing T cells relay neural signals in a vagus nerve circuit. *Science* **334:** 98–101.

Sherrington SCS. 1906. *The integrative action of the nervous system*. Yale University Press, New Haven, CT.

Styer KL, Singh V, Macosko E, Steele SE, Bargmann CI, Aballay A. 2008. Innate immunity in *Caenorhabditis elegans* is regulated by neurons expressing NPR-1/GPCR. *Science* **322:** 460–464.

Sun J, Singh V, Kajino-Sakamoto R, Aballay A. 2011. Neuronal GPCR controls innate immunity by regulating noncanonical unfolded protein response genes. *Science* **332:** 729–732.

Tracey KJ. 2002. The inflammatory reflex. *Nature* **420:** 853–859.

Tracey KJ. 2007. Physiology and immunology of the cholinergic antiinflammatory pathway. *J Clin Invest* **117:** 289–296.

Tracey KJ. 2009. Reflex control of immunity. *Nat Rev Immunol* **9:** 418–428.

Tracey KJ, Beutler B, Lowry SF, Merryweather J, Wolpe S, Milsark IW, Hariri RJ, Fahey TJ, Zentella A, Albert JD. 1986. Shock and tissue injury induced by recombinant human cachectin. *Science* **234:** 470–474.

Tracey KJ, Fong Y, Hesse DG, Manogue KR, Lee AT, Kuo GC, Lowry SF, Cerami A. 1987. Anti-cachectin/TNF monoclonal antibodies prevent septic shock during lethal bacteraemia. *Nature* **330:** 662–664.

Wang HH, Yu M, Ochani M, Amella CA, Tanovic M, Susarla S, Li JH, Yang H, Ulloa L, Al-Abed Y, et al. 2003. Nicotinic acetylcholine receptor α7 subunit is an essential regulator of inflammation. *Nature* **421:** 384–388.

Wang H, Liao H, Ochani M, Justiniani M, Lin X, Hong L, Al-abed Y, Wang H, Metz C, Miller EJ, et al. 2004. Cholinergic agonists inhibit HMGB1 release and improve survival in experimental sepsis. *Nat Med* **10:** 1216–1221.

Watkins L, Goehler L, Relton J. 1995. Blockade of interleukin-1 induced hyperthermia by subdiaphragmatic vagotomy: Evidence for vagal mediation of immune-brain communication. *Neuroscience* **183:** 27–31.

Wong CHY, Jenne CN, Lee W-Y, Léger C, Kubes P. 2011. Functional innervation of hepatic iNKT cells is immunosuppressive following stroke. *Science* **334:** 101–105.

Yaprak M. 2008. The axon reflex. *Neuroanatomy* **7:** 17–19.

Index

A

AAM. *See* Alternatively activated macrophage
AD. *See* Alzheimer's disease
ADAM10, 224
Ahr. *See* Aryl hydrocarbon receptor
AIM2
    activation, 50, 53
    autoimmune disease, 57
    DNA sensing, 150
    lung, 52
Allergy
    components in host defense and allergy
        alternatively activated macrophage, 174
        basophil, 172
        eosinophil, 174–175
        group 2 innate lymphoid cell, 173–174
        immunoglobulin E, 171–172
        mast cell, 172
        prospects for study, 177–178
        T cells, 172–173
    group 2 innate lymphoid cell studies
        allergic airway inflammation, 159–160
        atopic dermatitis, 162
    inflammasome activation, 57
    metabolic functions of allergic module components, 176–177
    type 2 immunity and allergy paradox, 170–171
    wound healing role of allergic module components, 177
ALS. *See* Amyotrophic lateral sclerosis
Alternatively activated macrophage (AAM). *See* Macrophage
ALX/FPR2, 131–132
Alzheimer's disease (AD), inflammasome activation, 55
Amyotrophic lateral sclerosis (ALS), inflammasome activation, 55
Angiogenesis, chronic inflammation, 191
AP-1, inflammatory response, 37–40, 88
Apoptosis, DNases
    caspase-activated DNase, 141–143
    DNase II, 144–145
APRIL, 208
Areg, 157
Arid5A, 90–92
Arnt, 89
Arteriole. *See* Microvascular system
Aryl hydrocarbon receptor (Ahr), 89–90
ASC, 45, 49–50
Aspirin-triggered lipotoxins (ATL), inflammation resolution, 125–126
Atherosclerosis, inflammasome activation, 54–55
ATL. *See* Aspirin-triggered lipotoxins
Atopic dermatitis. *See* Allergy
Autophagy, Toll-like receptor linking, 26–27

B

BAI1, 144
Basophil
    allergy and immune response role, 172
    chemokine system in homeostasis, 107
B cell, innate B cells and sentinel pathogen permissive cells, 71–72
Blood vessels. *See* Microvascular system
BLT1, 131
BSF-2, 85
BTLA
    defects and disease, 76–78
    dendritic cell regulation, 76
    γδ T cell regulation, 74–76
    innate immunity role, 68, 72–73
    natural killer cell antiviral activation, 73–74

C

CAD. *See* Caspase-activated DNase
Capillary. *See* Microvascular system
CAPS. *See* Cryoprin-associated periodic syndromes
Caspase-activated DNase (CAD)
    apoptosis, 141–143
    defects
        autoinflammation activation, 145–146
        cancer and autoimmune disease, 143–144
    inhibitor complex, 142
    pyrenocyte DNA digestion, 145
Cathepsins, Toll-like receptor cleavage, 23
CCL21, 71
CCR7, dendritic cell function, 36
CD14, Toll-like receptor signaling, 6, 22–23
CD25, 176
CD99, 188
CD160
    defects and disease, 76–78
    innate immunity role, 72–73
    natural killer cell antiviral activation, 73–74
CD169, macrophage expression at lymphohematopoietic interface, 203
cGAS. *See* Cyclic GMP-AMP synthase

Chemokines. *See also specific chemokines and receptors*
acute inflammation role
induction by resident immune cells, 109–110
overview, 108–109
recruitment of cells
dendritic cell, 113
eosinophil, 113
innate lymphoid cell, 114
monocyte, 112–113
neutrophil, 110–112
evolution, 103, 105
innate immune cell homeostasis role
basophil, 107
dendritic cell, 107–108
developing cells, 105
eosinophil, 107
innate lymphoid cell, 108
mast cell, 107
monocyte, 105–107
neutrophil, 105
leukocyte recruitment, 186–187
prospects for study, 115
receptor expression and function, 103–104
T-cell priming
CD8 cell, 115
Tfh cell, 114–115
Th1 cell, 114
Th2 cell, 114–115
types and receptors, 101–103
ChemR23, 131
Chronic rhinosinusitis (CRS), group 2 innate lymphoid cell studies, 161–162
cis-regulatory elements, identification in inflammatory gene expression, 40
CLRs. *See* C-type lectin receptors
CpG island, nucleosome positioning, 39–40
C-reactive protein (CRP), 86, 93
CRP. *See* C-reactive protein
CRS. *See* Chronic rhinosinusitis
Cryoprin-associated periodic syndromes (CAPS), inflammasome activation, 54
CS36, Toll-like receptor signaling, 6
C-type lectin receptors (CLRs), 20–21
CXCL13, 69, 71
Cyclic GMP-AMP synthase (cGAS), intracellular DNA sensing, 11–12, 27–29, 150

D

DAI, intracellular DNA sensing, 11, 150
DC. *See* Dendritic cell
DC-SIGN (SIGNR1), macrophage expression at lymphohematopoietic interface, 204–206
DDX41, intracellular DNA sensing, 11
Death receptors, 142
Dectins, 21
Dendritic cell (DC)
antigen targeting to marginal zone, 207
aryl hydrocarbon receptor, 90
chemokine system
homeostasis, 107–108
recruitment in acute inflammation, 113
follicular, 200
inflammatory gene expression program, 36–37
TNFSF regulation, 76
DHA. *See* Docosahexaenoic acid
DHX9, intracellular DNA sensing, 11
DHX36, intracellular DNA sensing, 11
Diabetes, inflammasome activation in type 2 disease, 55
DLAD, 146
DNA-PK, intracellular DNA sensing, 11
DNases
caspase-activated DNase
apoptosis, 141–143
defects in cancer and autoimmune disease, 143–144
inhibitor complex, 142
cytosolic DNA sensing, 150
DNase II
apoptotic cell DNA digestion, 144–145
defects in autoinflammation activation, 145–146
pyrenocyte DNA digestion, 145
DNase II–like acid DNase, 146
prospects for study, 150–151
signal transduction, 148–150
Trex1, 146–148
Docosahexaenoic acid (DHA), inflammation resolution, 124

E

Eicosapentaenoic acid (EPA), inflammation resolution, 124
Endothelial cell
capillary leak, 191
homeostatic function, 185
microvascular system, 184
postcapillary venules, plasma protein extravasation, and leukocyte recruitment, 186–189
thrombosis, 190
Enhancers, inflammatory gene expression, 40–42
Eosinophil
allergy and immune response role, 174–175
chemokine system
homeostasis, 107
recruitment in acute inflammation, 113
metabolic homeostasis role, 176
Eosinophilic pleural effusion (EPE), group 2 innate lymphoid cell studies, 161
EPA. *See* Eicosapentaenoic acid
EPE. *See* Eosinophilic pleural effusion

F

F4/80, macrophage expression at lymphohematopoietic interface, 202–203
FACS. *See* Familial cold autoinflamatory syndrome

Familial cold autoinflamatory syndrome (FACS), inflammasome activation, 54
FoxA, inflammatory response, 38
FOXP3, 173
FPR1, 110, 112

G

γδ T cell, BTLA regulation, 74–76
GATA3, 157, 173
Gout, inflammasome activation, 54
gp96, 22
gp130, 92
GPR32, 131–132

H

HAT. *See* Histone acetyltransferase
Hepatocyte-stimulating factor (HSF), 85
Herpes virus entry mediator (HVEM)
  defects and disease, 76–78
  innate immunity role, 68, 72–75
HGF. *See* Hybridoma growth factor
Histone acetyltransferase (HAT), inflammatory gene expression regulation, 40–42
HMGB1, 87, 150, 221
HSF. *See* Hepatocyte-stimulating factor
HVEM. *See* Herpes virus entry mediator
Hybridoma growth factor (HGF), 85

I

ICAM-1. *See* Intercellular adhesion molecule-1
IFI16, 50
IFIX, 50
IgE. *See* Immunoglobulin E
IL-1. *See* Interleukin-1
IL-6. *See* Interleukin-6
ILC. *See* Innate lymphoid cell
Immunoglobulin E (IgE), allergy and immune response role, 170–172
Inflammasome
  activation
    AIM2, 50
    allergens and particulates, 57
    Alzheimer's disease, 55
    amyotrophic lateral sclerosis, 55
    atherosclerosis, 54–55
    autoimmune disease, 56–57
    cryoprin-associated periodic syndromes, 54
    diabetes, 56
    gout, 54
    metabolic syndrome, 56
    NLRC4, 49–50
    NLRP1, 47
    NLRP3, 47–49
    NLRP6, 49
    NLRP12, 49
    RIG-I, 50
  CARD domains, 46–47
  commensal microbiota response, 53
  functional overview, 43–45
  host defense against infection
    intestine, 51–52
    lung, 52–53
    overview, 50–51
  prospects for study, 57–58
Inflammation resolution
  anti-inflammation versus proresolution processes, 124
  docosahexaenoic acid, 124
  eicosapentaenoic acid, 124
  G-protein-coupled receptors, 131–132
  infection relationship, 130–131
  microRNA roles, 132
  specialized proresolving mediators
    animal disease model studies, 132–134
    human disease studies, 125, 127
    identification, 122–124
    maresins, 128–130
    metabololipidomics profiling, 130
    overview, 121–122
    structure elucidation, 124–126, 128
Influenza, group 2 innate lymphoid cell studies, 159
Innate lymphoid cell (ILC)
  chemokine system
    homeostasis, 108
    recruitment in acute inflammation, 114
  group 2 cells
    adipose tissue function, 163–164
    allergy and immune response role, 173–174
    atopic dermatitis studies, 162
    effector functions, 157–158
    functional overview, 157
    gut function, 158–159
    identification, 157
    liver function, 162–163
    prospects for study, 164–165
    regulation, 157–158
    respiratory tract function
      allergic airway inflammation, 159–160
      chronic rhinosinusitis, 161–162
      eosinophilic pleural effusion, 161
      influenza, 159
      pulmonary fibrosis, 161
  lymphotoxin-αβ and TNFR-dependent immunity, 69–71
  markers, 155–156
  subsets, 156–157
Intercellular adhesion molecule-1 (ICAM-1), leukocyte recruitment, 187–189
Interferon regulatory factors (IRFs)
  inflammatory response, 37–39
  IRF7, 206
Interferon-α, innate B cells and sentinel pathogen permissive cells, 71–72

Interferon-β
  innate B cells and sentinel pathogen permissive cells, 71–72
  macrophage expression, 141
Interleukin-1 (IL-1)
  atherosclerosis studies, 55
  reflex regulation of innate immunity, 222–223
Interleukin-6 (IL-6)
  functional overview, 85–87
  prospects for study, 95–96
  receptor
    expression and function, 89–90
    signaling, 92–93
  synthesis regulation, 87–89
  therapeutic targeting, 93–94
  transcript stabilization and regulation, 90–92
IPEX, 173
IRAKs, Toll-like receptor signaling, 6–7
IRFs. *See* Interferon regulatory factors

J

JAK, interleukin-6 signaling, 92–93

K

KLF-4, 189

L

LBP. *See* Lipopolysaccharide-binding protein
LC3, 26–27, 29
Leukotriene B_4 (LTB_4), neutrophil production, 110–112
LFA-1, 187
LGP2, intracellular virus recognition, 8–10
LIGHT (TNSF-14)
  defects and disease, 76–78
  innate immunity role, 67–68, 72–73
Lipopolysaccharide-binding protein (LBP), 23
Lipoxin A_4 (LXA_4)
  group 2 innate lymphoid cell studies, 158, 160
  inflammation resolution, 123, 125–126, 131, 134
LTB_4. *See* Leukotriene B_4
LTi. *See* Lymphoid tissue inducer
LXA_4. *See* Lipoxin A_4
Lymph node subcapsular sinus (SCS), macrophages
  functions, 205–209
  lipid regulation, 208–209
  overview, 195–197, 201
  phenotypic characterization
    DC-SIGN, 204
    F4/80, 202–203CD169, 203
    mannose receptor, 203–204
    MARCO, 204–205
  prospects for study, 209–212
  species differences, 209
Lymphoid tissue inducer (LTi), 108
Lymphotoxin-α, 68, 70
Lymphotoxin-αβ, 69–71
Lymphotoxin-β receptor, 68, 70–71, 76

M

Macrophage
  alternatively activated macrophage, allergy and immune response role, 174
  apoptotic cell engulfment, 144
  inflammation resolution, 128–130
  lymphohematopoietic interface. *See* Lymph node subcapsular sinus; Medullary sinus macrophage; Spleen marginal zone
Mannose-binding lectin (MBL), 24
Mannose receptor, macrophage expression at lymphohematopoietic interface, 203–204
MARCO, macrophage expression at lymphohematopoietic interface, 204–205
Maresins, inflammation resolution, 125, 128–130
Marginal zone. *See* Spleen marginal zone
Mast cell
  allergy and immune response role, 172
  chemokine system in homeostasis, 107
MAVS, 28
MBL. *See* Mannose-binding lectin
MD2, Toll-like receptor signaling, 22–23
MDA5
  intracellular virus recognition, 8–10
  RNA binding, 27–28
MDP. *See* Muramyl dipeptide
Medullary sinus macrophage (MSM), functions, 207–208, 211
Metabolic syndrome, inflammasome activation, 56
MicroRNA (miRNA)
  inflammation resolution, 132
  interleukin-6 expression regulation, 89
  Toll-like receptor signaling modulation, 8
Microvascular system
  angiogenesis in chronic inflammation, 191
  arterioles and control of blood flow, 185–186
  capillary leak, 190–191
  flow dysregulation, 190
  homeostatic function, 185
  organization, 183–185
  postcapillary venules, plasma protein extravasation, and leukocyte recruitment, 186–190
  thrombosis, 190
MIF, 112
miRNA. *See* MicroRNA
MNDA, 50
Monocyte, chemokine system
  homeostasis, 105–107
  recruitment in acute inflammation, 112–113
Mre11, 150
MSM. *See* Medullary sinus macrophage
Muckle–Wells syndrome, inflammasome activation, 54
Muramyl dipeptide (MDP), 47
MyD88, Toll-like receptor signaling, 6, 24–25, 149
MZ. *See* Spleen marginal zone

N

nAChR. *See* Nicotinic acetylcholine receptor
Neutrophil, chemokine system
  homeostasis, 105
  recruitment in acute inflammation, 110–112
NF-κB. *See* Nuclear factor-κB
Nicotinic acetylcholine receptor (nAChR), reflex regulation of innate immunity, 221
NLRC4
  activation, 49–50
  intestine, 51
  lung, 52
NLRP1
  activation, 47
  autoimmune disease, 56
  lung, 52
NLRP3
  activation, 47–49, 53
  atherosclerosis studies, 55
  autoimmune disease, 56–57
  commensal microbiota response, 53
  cryoprin-associated periodic syndrome role, 54
  gout response, 54
  insulin resistance, 56
  intestine, 51
  lung, 52
  neurodegenerative disease, 55
  particulate activation, 57
  viral RNA activation, 10
NLRP6
  activation, 49
  commensal microbiota response, 53
NLRP12, activation, 49
Nrdp-1, Toll-like receptor signaling, 6
Nuclear factor-κB (NF-κB)
  interleukin-6 transcription regulation, 89
  nucleosome/transcription factor interplay in inflammatory response, 39
  Toll-like receptor signaling, 6
Nucleosome
  CpG islands in positioning, 39–40
  evolution, 216
  regulatory elements controlling inflammatory gene expression
    cis-regulatory elements, 40
    enhancers, 40–42
  transcription factor interplay in inflammatory response, 38–39

O

Oxidative stress, NLRP3 activation, 48

P

p53, 89
p300, 40
Pattern recognition receptors. *See specific receptors*
PC. *See* Pericyte
PECAM-1, 188
Pericyte (PC)
  chemokines, 189
  microvascular system, 184–185
PMN. *See* Polymorphonuclear neutrophil
Pol II. *See* RNA polymerase II
Polo-like kinases, Toll-like receptor signaling, 8
Polymorphonuclear neutrophil (PMN), inflammation induction and resolution, 122–123
PRAT4A, 22
Protectin, inflammation resolution, 125
Pu.1, inflammatory response, 40, 42
Pulmonary fibrosis, 161
Pyrenocyte, DNA digestion, 145

R

RA. *See* Rheumatoid arthritis
Rab5, 144
RAGE, 87
Regnase-1, 90–92
Resolvins, inflammation resolution, 125–134
Rheumatoid arthritis (RA), interleukin-6 role and targeting, 93–95
Rhinosinusitis. *See* Chronic rhinosinusitis
RIG-I
  activation, 50
  intracellular virus recognition, 8–10
  lung, 52
  RNA binding, 27–28
RNA polymerase II (Pol II), inflammatory gene expression regulation, 39–40
RNF135, RIG-I-like receptor signaling modulation, 10

S

S100, 87
SAA. *See* Serum amyloid A
SCS. *See* Lymph node subcapsular sinus
Sec5, 29
Serum amyloid A (SAA), 86
SHP-1, 75
SHP-2, 73, 75
SIGNR1. *See* DC-SIGN
SLE. *See* Systemic lupus erythematosus
SMC. *See* Smooth muscle cell
Smooth muscle cell (SMC), microvascular system, 183–185
SOCS, 93
SP1, 39, 88
Specialized proresolving mediators (SPM). *See* Inflammation resolution
Sphingosine-1-phosphate receptor, 108
Spleen marginal zone (MZ)
  anatomy, 197–200, 202
  antigen targeting, 207
  macrophages
    functions, 205–207
    overview, 195–197, 201

Spleen marginal zone (MZ) (*Continued*)
phenotypic characterization
CD169, 203
DC-SIGN, 204
F4/80, 202–203
mannose receptor, 203–204
MARCO, 204–205
prospects for study, 209–212
species differences, 209
SPM. *See* Specialized proresolving mediators
Stabilin2, 144
STATs
aryl hydrocarbon receptor interactions, 89–90
inflammatory response, 37–39
interleukin-6 signaling, 92–93
STING, intracellular DNA sensing, 11–13, 29, 149–150
Subcapsular sinus. *See* Lymph node subcapsular sinus
Systemic lupus erythematosus (SLE), 144

T

TAT, 89
TBK1, intracellular DNA sensing, 11, 13, 29
T cell. *See also* γδ T cell
allergy and immune response role, 172–173
chemokines in priming
CD8 cell, 115
Tfh cell, 114–115
Th1 cell, 114
Th2 cell, 114–115
Thymic stromal lymphopoietin (TSLP), 157–160
Tim4, 144, 206
TIRAP, Toll-like receptor signaling, 25–26
TLRs. *See* Toll-like receptors
TNF. *See* Tumor necrosis factor
TNFR1, 67–68
TNFR2, 67–68
TNSF-14. *See* LIGHT
Tocilizumab, 93–94
Toll-like receptors (TLRs)
cleavage, 23
evolution, 216
expression, structure, and ligands
cell surface receptors, 3–4, 22–23
endosomal receptors, 4–6
intracellular DNA sensing, 10–11
overview, 1–3, 20–22
phagocytosis and autophagy linkage, 26–27
signaling
overview, 6–8
sorting adaptor proteins, 25–26
subcellular localization dependence, 24–25
subcellular distribution, 22–24
TRAF, Toll-like receptor signaling, 6–7
TRAIL, 142
TRAM, Toll-like receptor signaling, 25–26
Trem14, 206
Trex1, 146–148
TRIF, Toll-like receptor signaling, 6, 24, 149
TRIM25, RIG-I-like receptor signaling modulation, 10
TSLP. *See* Thymic stromal lymphopoietin
Tumor necrosis factor (TNF)
angiogenesis in chronic inflammation, 191
reflex regulation of innate immunity, 219–221

U

UL144, 74
UNC93b1, 22–23
Unilateral ureteric obstruction (UUO), 134
UUO. *See* Unilateral ureteric obstruction

V

Vagus nerve, reflex regulation of innate immunity, 219–223
Vascular cell adhesion molecule-1 (VCAM-1), leukocyte recruitment, 187–189
Vascular endothelial growth factor (VEGF), 87, 191
Vasculature. *See* Microvascular system
VCAM-1. *See* Vascular cell adhesion molecule-1
VEGF. *See* Vascular endothelial growth factor
VLA-4, 187–188

W

WHIM syndrome, 105
Wound healing, role of allergic module components, 177

X

Xkr8, 144

Z

ZIP14, 86